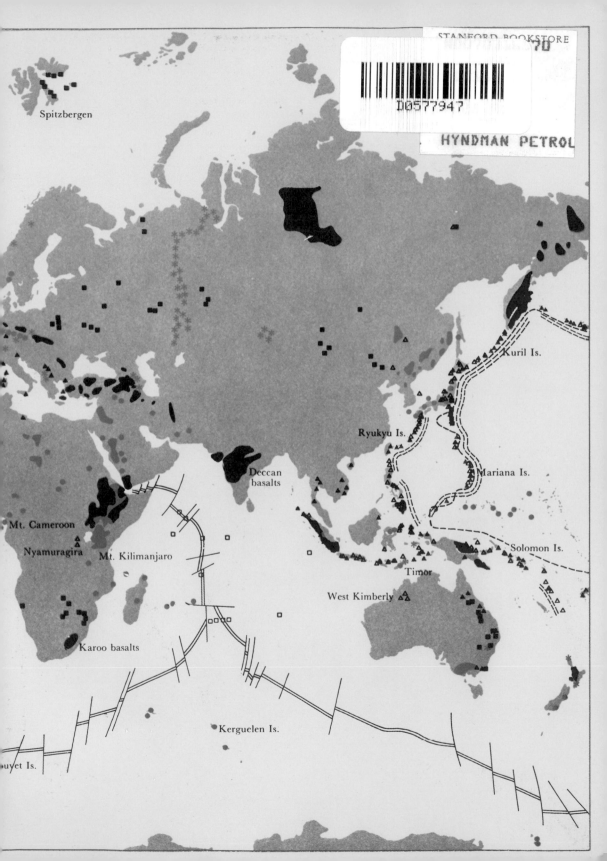

Spitzbergen

Kuril Is.

Ryukyu Is.

Mariana Is.

Deccan
basalts

Mt. Cameroon

Nyamuragira

Mt. Kilimanjaro

Solomon Is.

Timor

West Kimberly

Karoo basalts

Kerguelen Is.

...uvet Is.

PETROLOGY
OF IGNEOUS
AND METAMORPHIC
ROCKS

PETROLOGY OF IGNEOUS AND METAMORPHIC ROCKS

DONALD W. HYNDMAN
Professor of Geology
University of Montana

McGRAW-HILL BOOK COMPANY
New York   San Francisco   St. Louis   Düsseldorf   Johannesburg
Kuala Lumpur   London   Mexico   Montreal   New Delhi
Panama   Rio do Janeiro   Singapore   Sydney   Toronto

This book was set in Baskerville by Black Dot, Inc. The editors were Jack L. Farnsworth and Eva Marie Strock, the designer was Janet Bollow, and the production supervisor was Michael A. Ungersma. The drawings were done by David A. Strassman.

The printer and binder was Kingsport Press, Inc.

PETROLOGY OF IGNEOUS AND METAMORPHIC ROCKS

Library of Congress Cataloging in Publication Data

Hyndman, Donald W.
  Petrology of igneous and metamorphic rocks.

  (International series in the earth and planetary sciences)
  Bibliography: p.
  1. Rocks, Igneous. 2. Rocks, Metamorphic.
I. Title. II. Series.
QE461.H98   552'.1   79-37097
ISBN 0-07-031657-0
    15 MAMM 98765432

CONTENTS

PREFACE

"Petrology of Igneous and Metamorphic Rocks" is intended as a text for the lecture part of an introductory course in petrology and as a review of recent literature and concepts for those wishing to update their background. Description and identification of hand specimens or thin sections is, for the most part, left to the lab. Such a course presupposes knowledge of the general principles of physical geology and of mineralogy, along with beginning college chemistry.

In conformity with common current practice in colleges and universities in North America, only igneous and metamorphic rocks have been included. Sedimentary rocks are left for a separate course in stratigraphy or sedimentary petrology. This text is concerned with the description, problems, and processes of igneous and metamorphic rocks. It emphasizes the practical aspects of the subject and their relationships to the theoretical in an attempt to give the student a working understanding of petrology and a basis for critical judgment. It emphasizes those features that can be used in the field, in hand specimen, and in thin section, in conformity with the belief that many petrological problems are best solved by combining the field study of the rocks themselves with some of the more theoretical aspects of experimental and thermodynamic geochemistry. Geological knowledge and principles are related to the knowledge and principles of chemistry, geophysics, and other peripheral fields. Although these fields are not emphasized, they are drawn on where their contributions seem appropriate for the subject. In many instances, physical chemical or thermodynamic rationales are used in support of an argument, without explaining in detail why these should be correct. Addition of such detail would be outside the scope of this text. A few footnotes relating concepts to thermodynamics are included as a guide for those familiar with thermodynamic principles. Readers who desire additional information should refer to standard works such as Krauskopf's excellent recent text, "Introduction to Geochemistry"; Kern and Weisbrod's "Thermodynamics for Geologists"; or Fyfe, Turner, and Verhoogen's "Metamorphic Reactions and Metamorphic Facies." A short section on the most important petrological aspects of mineralogy is included as a review and for amplification of topics that may not have been covered in a course in mineralogy.

It will be clear to workers in the field that, as the name implies, the book is no more than an introduction to the extensive field of petrology. Numerous references are included, both for the support of statements and for additional information to permit the reader to delve deeper into the subject. Photographs have been used in the hope that they clarify some of the description, for few students at this stage will have much familiarity with igneous and metamorphic rocks and features in the field.

Terminology has been kept to a minimum, the emphasis being placed on a brief description of the materials, and understanding of the processes, and a feeling for the problems in petrology. Of course a certain amount of nomenclature is necessary for communication, both at this level and for future more advanced reading. Topics for more advanced consideration are included in later parts of the book in the expectation that students will be able to handle them as their knowledge and petrologic sophistication progresses.

One of the most intriguing directions in which modern geology has been advancing in the past ten years is in the study of the upper part of the earth's mantle and its bearing on structures, compositions, and processes in the earth's crust. As more information comes to light, the considerations can be expected to have a revolutionary effect on modern petrology. For this reason, the first few pages are devoted to a brief outline of a presently acceptable picture of the earth's crust and upper mantle on the continental scale and the petrological processes operating in them.

The writer of a review of this sort owes a debt of thanks to numerous persons who have helped in the development of the facts, concepts, and ideas presented. Authors of professional articles and books have been acknowledged in the text but others too numerous to mention have aided in less formal ways. The final product reflects in part the views of and discussions with former teachers and colleagues, along with students who have suffered through the earlier stages of its development. Noteworthy are K. C. McTaggert, F. J. Turner, W. S. Fyfe, D. Alt, and an unjustifiably tolerant typist and family.

DONALD W. HYNDMAN

CHAPTER 1
ENVIRONMENT
AND
MATERIALS

This book on the petrology of igneous and metamorphic rocks is concerned with the description and development of (1) the complex crystalline rocks that make up the vast ancient and until recently nearly unknown continental shield areas; (2) the large to small masses of different kinds of relatively homogeneous igneous rocks that have invaded sedimentary and metamorphic rocks elsewhere; (3) the equally variable crystalline to glassy volcanic rocks that have solidified near the earth's surface; and (4) the widespread metamorphic rocks that exhibit all transitions from sedimentary and volcanic rocks to the complex crystalline rocks of the continental shields. The study of the origin of this variety of nonsedimentary rocks and of the processes that formed them is an intriguing one. It differs from much of the study of sedimentary rocks in that, except for the later stages of formation of volcanic rocks, the processes of development are inaccessible to direct observation. The methods of attack on the problem revolve around careful observation and description, geological reasoning, and analogy with experimentally determined occurrences.

A study of petrology presupposes a general knowledge of physical geology, especially mineralogy. It draws on these and related fields within and peripheral to geology. Although the formation of igneous, metamorphic, and sedimentary rocks is closely related, one with the others, it is convenient to separate the groups for purposes of discussion. Before doing this, however, a sketch will be developed of an overall geological framework into which to fit the discussion.

Studies of the earth's gravitational field and earthquakes indicate that the earth is divisible into three major shells: (1) a dense nickel-iron core; (2) a less-dense peridotite mantle, a lower crust that is high in SiO_2, MgO, and FeO (hence the name Sima); and (3) a still less-dense "granitic" crust that is very high in SiO_2 and Al_2O_3 (hence the name Sial) and high in K_2O, Na_2O, and CaO. The composition of the crust, about which most is known and which, because of its proximity, is most important, is about like an alkali-rich andesite (see Table 1-2). More than 99% of the crust is composed of the oxides of 10 common metals, which are listed in Table 1-2. These few elements combine to form the common minerals making up the earth's crust and mantle. Most of the remaining 83 elements are present in minor amounts, substituting for these major elements in the common rock-forming minerals or making up minor minerals of their own. Most of the latter involve special or peculiar circumstances for their formation. The preponderance of oxygen, the only abundant element with a negative charge, and its large size favor its role as the major unit around and between which most of the other elements must fit. This subject will be developed further in connection with certain aspects of the common minerals that make up most rocks.

DISTRIBUTION AND DEVELOPMENT OF SEDIMENTARY ROCKS
OF THE CONTINENTAL MARGINS

The bulk of sediments that are forming today and probably also those that formed in the past are found where great thicknesses accumulate on continental shelves. Since from these sedimentary rocks many of the igneous and metamorphic rocks develop and since through and into these sedimentary rocks many of the remainder of the igneous rocks are injected and emplaced, it will be advantageous to review some of the general and modern aspects of sediment distribution. The kind, thicknesses, location, and other characteristics of the sedimentary rocks affect the development and final characteristics of the igneous and metamorphic rocks developed from and within them. A cross section across an idealized continental shelf–island arc–oceanic trench, in an early orogenic stage of development, will suffice for this purpose (see Fig. 1-1). This sketch also fits, in a general way, the bulk of sedimentary rocks in the geological record, especially those in the North American cordillera and the Appalachian "geosynclines"[1] where igneous activity and metamorphism also play a major role.

[1] The term "geosyncline" is set off by quotation marks throughout this book because the classical concept of a "geosyncline" as an elongate depositional trough that receives a thick deposit of sediment has been shown to be incorrect. In the past decade, it has become clear that these thick deposits formed along continental margins, for example, as continental shelf and slope deposits, as discussed below. The terms are retained because of their usefulness in referring to the descriptive entities and distinction between separate parts.

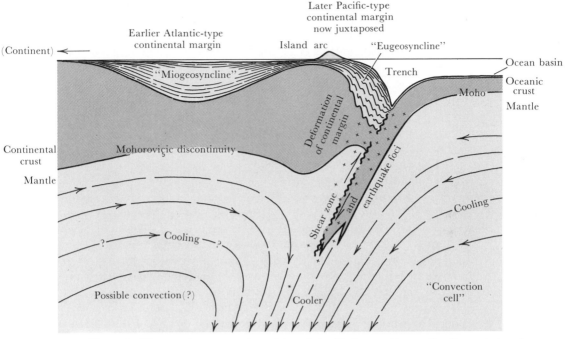

Fig. 1-1 *Diagramatic representation of the relationship between "geosynclines," oceanic trenches, earthquake foci, the Mohorovičic discontinuity, and inferred "convection cells" in the earth's mantle. It is not clear whether "convection" also occurs under the continental crust.*

EVOLUTION OF THE SEDIMENTARY PILE

Most geologists acknowledge that these very thick prisms of sediments, the "geosynclines," exist. Most also agree that these sedimentary piles occur at the margins of continents or between continents (e.g., Aubouin, 1965) and in pairs—the "miogeosyncline" and the "eugeosyncline." Most, but certainly not all, agree that such "geosynclines" exist today as on the continental shelf and continental slope of the Atlantic Coast of North America (Dietz, 1963b). It must be emphasized that not all continental margins or "geosynclinal" pairs have all of the characteristics shown in Fig. 1-1, especially considering that the characteristics seem to change significantly as the sedimentary accumulations evolve. Although most "geosynclines" go through grossly similar stages of evolution, not all have volcanic island arcs or oceanic trenches as shown here for this stage; and rock types and their thicknesses vary to a major degree. The most important general characteristics appear to be as shown here. One of the intriguing questions confronting petrologists today is why these sedimentary piles exist and why they tend to be spatially associated with major igneous activity, deformation, and regional metamorphism. Some answers are now coming to light. Studies of earthquake waves and gravity data lead to the notion that mountains have "roots" and to the concept of isostacy (floating equilibrium). The lighter rocks of the earth's crust float on the heavier or more dense rocks of the earth's mantle. The thicker parts of the crust, the continents and especially the major mountain ranges, tend to sink deeper into the underlying mantle, forming the mountain roots, as, for example, under the Sierra Nevada Range and the Rocky Mountains (Eaton, 1963; Pakiser, 1963; Pakiser and Robinson, 1966; Healy and Warren, 1969; Prodehl, 1970).

This floating equilibrium seems to be maintained in a general way except for a few situations. Most important for purposes of this discussion is the pronounced negative-gravity anomaly under most deep-ocean trenches, first noted by Hess (1938a) and Vening Meinesz (1948, 1954; see also, for example, Woolard and Strange, 1962; Isacks et al., 1969). This anomaly suggests that the deep-ocean trench seems to be somehow held down or even pulled down by some powerful force in the mantle. In the past few years, several other lines of evidence have been brought to bear on the problem, e.g., distribution of oceanic heat-flow data (von Herzen and Uyeda, 1963; Lee and Uyeda, 1965; Lee, Uyeda, and Taylor, 1966; Langseth et al., 1966; von Herzen and Lee, 1969), patterns of magnetic anomalies on the ocean floor (Vine and Mathews, 1963; Raff and Mason, 1961; Wilson, 1965; Vine and Wilson, 1965; Vacquier, 1969), correlation of these anomalies with radiometric ages of magnetic reversals in the recent geological record (Cox and Doell, 1964; Vine, 1966), increase in maximum age and thickness of sediment and increase in age of islands away from the crests of midoceanic ridges and rises (Hess, 1962; Wilson, 1963a; Ewing and Ewing, 1967), and sense of first motion calculated for earthquakes (Sykes, 1966, 1967; Isacks et al., 1968). This evidence has led most geologists to believe in the presence of huge, slowly moving intermittent convection currents in the earth's mantle and in "ocean-floor spreading." Accompanying this belief is a strong revitalization of the old concept of continental drift, which then forms a natural accompaniment.

Convection cells will, of course, develop in almost any liquid that is heated at the bottom and is cooling at the top. Convection currents have also been inferred in the formation of mafic layered intrusions, from mineral distribution, and from "flow" structures (Wager and Deer, 1939; Brothers, 1964). The earth's mantle is not liquid, as shown by the behavior of earthquake waves; but it will deform plastically like a very viscous liquid when subject to continued stress over a long period of time, as shown by isostatic rebound of parts of the crust.

How convection currents in the earth's mantle may be inferred as producing the oceanic trenches is shown diagrammatically in Fig. 1-1.

Higher temperatures in the lower part of the mantle would cause a slow rise in hotter mantle material. On reaching the cooler crust, this material would spread sideways, cool, and then slowly fall, pulling with it the overlying crust, as first inferred by Holmes (1931) and later by Griggs (1939). A reasonable calculated rate of movement is only 1 to 5 cm (1/2 to 2 in.) per year. This slow rate, however, amounts to about 6 to 30 vertical miles in the geologically short time of 1 million years.

The size, distribution (see Fig. 1-2), and depth of the convection cells and their bearing on igneous and metamorphic processes in the earth's crust are persistent questions. Only in the past few years has the most convincing evidence accumulated, as noted above. As inferred by Menard (1958), Wilson (1963b), Hess (1962, 1965), and Cann (1968), it now appears that convection currents rise and spread apart along zones of high-heat flow such as the mid-Atlantic ridge, the East Pacific rise, and some major rift valleys. Carrying the ocean bottom (and continents) along on their surface, the currents descend the oceanic trenches such as the Japan Trench. It is now well established that earthquake foci are concentrated along surfaces ("Benioff zones") that begin in the trenches and dip back

Fig. 1-2 *Global tectonic map showing axes of actively spreading midocean ridges (thick lines), fracture zones (single lines), active trenches (double-dashed lines), and spreading directions parallel to fracture zones (shaded arrows). Numbers near ridges indicate spreading rates in centimeters per year. Slightly modified from Le Pichon (1968) with minor additions from Hiertzler (1968).*

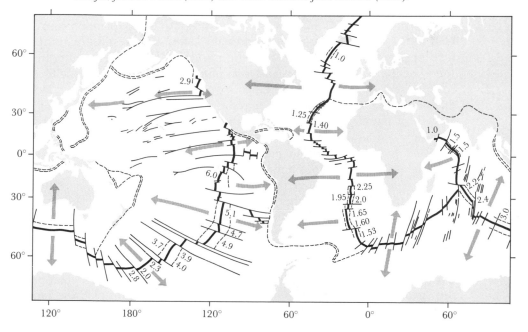

under the continents at about 45° (Benioff, 1954; Sykes, 1966; Isacks et al., 1968). Dips range from about 30 to 80°, the lower dips corresponding to high rates of convergence between lithospheric plates at the corresponding Benioff zone (Luyendyk, 1970). Since these foci extend to depths of about 400 to 700 km, it may be inferred that the convection cells extend to about this depth (Hess, 1965). The cells may also be of global size, a single cell having a downward limb approximating the form of the "stitching" of the cover on a tennis ball, corresponding fairly well with the pattern of trenches on the earth (Hess, 1965). The rate of movement of the top of the convecting cell, figured from the width and relative age of magnetic anomalies and magnetic reversals preserved in the ocean floor, is about 1 to 6 cm/year (e.g., Hess, 1966a; LePichon, 1968; Morgan, 1968; Heirtzler, 1969), the same order of magnitude as that calculated for the rate of vertical rise. The continents have been likened to rigid rafts of slag or froth on pots of convecting molten iron or hot water (Wilson, 1966). They are carried along on the top of convection cells, the leading edge of the continent being crumpled against the oceanic crust moving in the opposite direction at the oceanic trench (see Figs. 1-1 and 1-2). Unlike the oceanic crust, the floating continental crust is too buoyant to be dragged down the oceanic trench (Hess, 1962).

If, as seems likely, this convection current continues for millions of years, some interesting consequences may develop. The "geosynclinal" pile of sediments is gradually thickened by sinking and continued sedimentation. Radioactivity, mainly from radioactive K, U, and Th that are concentrated in the sediments, has been shown to be the primary source of heat in the earth's crust (Hyndman et al., 1968). As a result, the temperature begins to rise in the tens of thousands of feet of sediments in the "geosyncline." The sediments and associated volcanic rocks become heated, compressed, and metamorphosed to a greater and greater degree. Details of the process are less clear. It may be that the higher temperatures and growth of parallel micaceous minerals result in a decrease in strength and a greater tendency to flow and to yield to compressional stresses that are imposed by the convection currents. Another possibility is that partial melting and upward movement of granitic magma create disturbances and zones of weakness related to deformation. Alternatively, the reduction in strength, related to either of the above, may reduce the permissible downward drag by the convection current and permit isostatic rebound, folding, and faulting of the "geosynclinal" pile. Ultimately, the movement of hot mantle material to the upper part of the convection cell may remove the driving force behind the convection, resulting in cessation of movement (Griggs, 1939). This would permit isostatic rebound and deformation of the "geosyncline." Perhaps most likely, some combination of these factors plays a role.

In the past few years, the concepts of sea-floor spreading and continental drift have been further developed with the study of the geometric patterns and motions of separate semirigid "plates" (say 100 km thick) that make up the oceanic and continental crust of the earth (McKenzie and Parker, 1967; see Fig. 1-2). Although there is some disagreement on whether the "convection" is driven by rise of hot material under the ridges or sinking of cold plates in the trenches (Isacks et al., 1969; McKenzie, 1969; Ave'Lallemant and Carter, 1970), the plates can conveniently be considered as riding on slowly moving convection cells deeper in the mantle. As the plates slide past and against one another near the surface, they give rise to most of the major tectonic and petrologic features now expressed at and near the earth's surface.

Oceanic crustal plates are now thought to consist of a thin layer of young sediments partially covering oceanic basalt that probably grades down to gabbro (e.g., Gass, 1967; Davies, 1968; Davies and Milson, 1969; Thayer, 1969; Varne et al., 1969). Below the Moho, this gives way to peridotite in the upper mantle. These oceanic plates are considered to be continuously formed under the midoceanic ridges (or rises), perhaps by basaltic-dike injection, and to be continuously consumed under the oceanic trenches. There the oceanic crust and upper mantle slide under the adjacent continental margin along what are now known as subduction zones (White et al., 1970).

Continental margins may be separated into two types distinguished on the basis of the behavior of the crustal plates in these areas (Dewey, 1969; Gilluly, 1969; Dickinson, 1971). Margins of the "Atlantic type," such as along the Atlantic Coast of North America, are aseismic (no earthquakes) and nonvolcanic; and they show thick accumulations of sediment. Those of the "Pacific type," such as along the Pacific Coast of South America and probably in the past along the Pacific Coast of North America, overlie active Benioff seismic zones; these include adjacent oceanic trenches, active deformation of sediments and continental rocks, volcanic island arcs or chains of coastal volcanoes, and thin accumulations of sediment (for example, Fig. 1-1). Regional metamorphism and most kinds of igneous activity are related to activity along the Pacific type of margins. Oceanic volcanism is primarily associated with extension and upwelling at the midoceanic ridges. Continental margins of the Pacific type may revert to the Atlantic type with dying out of underthrusting of the oceanic plate under the continent, cessation of seismic activity, filling and uplift of the trench sediments, and welding of the continental to the oceanic plate. Completion of the cycle by initiation of Pacific type of activity may occur with the beginnings of underthrusting, development of a shear separation between oceanic and continental plates, and development of an oceanic trench (Dewey, 1969; Dewey and Bird, 1970).

DEVELOPMENT OF METAMORPHIC ROCKS

We shall turn now to the effect of these inferred occurrences on the sedimentary and volcanic rocks in the "geosynclines." The increase in temperature and the added pressure resulting from depression of these rocks to greater depths take the rocks far from the physical conditions under which they were formed. Being out of equilibrium with one another, the minerals react to form new minerals more nearly in equilibrium with the conditions of their new environment. These changes in the solid state are part of the processes of regional metamorphism. Such metamorphic rocks show the clear imprint of increased temperatures and widespread intense deformation occurring in the late stages of development of the "geosyncline." They become hardened and coarser grained, and they typically develop a strong parallelism of newly formed minerals.

Such widespread regional metamorphism, nevertheless, has no clearly defined source of heat. The metamorphism merely increases gradually in one direction—in a general way with depth, though intensity of metamorphism increases either along individual beds or oblique to them. At first sight, this seems reasonable enough, but it does not explain the common association of regional metamorphic rocks with streaks and masses of apparently igneous, granitic rock. Nor does it explain the high thermal gradients in regional metamorphic rocks spatially associated with some granitic batholiths. It may be that intrusion of hot igneous magma forms at least a partial additional heat source for those rocks above the crustal levels of partial melting. The answers to these problems will be considered more fully below after discussion of additional information on which to draw.

FORMATION OF IGNEOUS MAGMAS

As the metamorphic rocks are heated to higher and higher temperatures, they presumably reach a state at which some of them begin to melt. However, the whole rock would melt only under a wider range of temperature. Since the silicate chemical composition corresponding to melting of common sedimentary rocks at minimum temperatures corresponds to granite, streaks of molten granite form in the high-grade metamorphic rocks. Presumably almost all granite that occurs in discrete bodies[1] originates in this way—by extreme metamorphism resulting in partial melting of rocks in the earth's crust. An alternative origin, thought in the past to be the main source of granitic magma, will be discussed in a later section.

These streaks of granite magma are squeezed out as a result of their lower density and greater mobility and with the aid of the folding and de-

[1]This does not necessarily include the large areas of gneissic granite that are prominent in the Precambrian "basement" complexes and shield areas.

formation accompanying metamorphism. Migrating upward to lower pressures higher in the crust, they tend to come together to form larger bodies such as batholiths and stocks. Granitic rocks, then, form some of the most abundant and widespread rocks in the earth's crust. Why these streaks of granitic magma accumulate to form such large bodies, why they tend to rise to a common level seldom reaching the earth's surface, and many other related questions must await more detailed consideration of the properties of magmas and their behavior under crustal influences. These are considered below.

The other very abundant and widespread igneous rock, basalt, presumably forms in a very similar fashion by extreme metamorphism and partial melting of the peridotite or similar rocks below the crust in the earth's mantle. Evidence pointing to a subcrustal origin for basalt at 60 to 100 km below the surface (Eaton and Murata, 1960) includes data on depth of earthquake foci, gravity and density measurements, and occurrence of inclusions of peridotite and eclogite associated with some volcanic eruptions. Why basaltic magmas tend to rise all the way to the earth's surface, whereas granitic magmas seldom reach the surface, must also await more detailed discussion of the properties and behavior of different magmas. It may be noted at this point that granitic magmas have long been associated with development in a compressional environment, whereas basaltic magmas have been associated with development in a tensional environment.[1]

We are now in a position to consider the most general aspects of the evolution of the "geosyncline," especially the "eugeosyncline" and adjacent parts of the "miogeosyncline," in terms of igneous and metamorphic activity (see Table 1-1). It should be noted that not every detail of this evolutionary scheme appears in every "geosyncline" or in every part of any individual one. It does, however, follow quite well the development of many, including those in North America. Even within an individual "geosyncline," there is some indication that the stage of activity is earlier in some parts than others, that is, migrating with time from the core of the belt in the "eugeosyncline" toward the craton (e.g., Aubouin, 1965). It should also be noted that there is considerable overlap between some kinds of activity. For example, regional metamorphism may begin at depth well before the end of "Flysch"-type sedimentation. Similarly, volcanism of intermediate composition may overlap with eruption of flood basalts at another locality and with emplacement of granitic plutons at depth.

---

[1]As will be seen below, an *increase* in water pressure lowers the melting (or crystallization) temperature, whereas a *decrease* in load pressure (dry) lowers the melting temperature. Water-saturated magmas (e.g., most granite) therefore would be expected to form by melting in a compressional environment, whereas dry magmas (c.g., basalt) would be expected to form in a tensional environment. Compression and tension may therefore contribute to the formation of wet and dry magmas from rocks that are otherwise near their melting point.

Table 1-1 *Igneous and metamorphic evolution of the "geosyncline"**

Phase	Igneous Activity	Other Activity
"Geosynclinal" (pre-orogenic) ophiolitic suite ("greenstones")	Extrusion of spilitic and basaltic flows and tuffs; intrusion of serpentinite, peridotite, and gabbro.	"Flysch"-type sedimentation: turbidite graywacke, shale, chert, and ± limestone. Or pelagic limestone and chert. (More mature sandstone, shale, and limestone in the "miogeosyncline.")
Orogenic (folding and thrusting)	Anatectic development of granitic melts; upward rise and emplacement of granitic plutons. Volcanism of intermediate composition.	Regional metamorphism and folding beginning in the core of the belt and migrating outward toward the craton. Contact metamorphism.
Epeirogenic (uplift)	Flood basalts; intermediate volcanism, including basaltic to trachyandesitic and some rhyolitic eruptions. Emplacement of some granitic plutons.	"Molasse"-type sedimentation: marine to non-marine siltstone, conglomerate, and coal. With uplift and block faulting.
Postorogenic	Flood basalts. Emplacement of small alkaline plutons.	

*The igneous and metamorphic activity is concentrated in the "eugeosyncline" and adjacent parts of the "miogeosyncline." Different kinds of structural, igneous, metamorphic, or sedimentary activity are localized in different places at different times. Some of these types of activity may not develop in some "geosynclines" or parts thereof.
SOURCE: Modified from Stille (1940); Tyrell (1955); Turner and Verhoogen (1960); Dietz (1963); Aubouin (1965); and Hermes (1968).

This brief discussion of the origin of some of the most prominent kinds of igneous and metamorphic rocks has been deliberately somewhat dogmatic in order to build a coherent framework into which we can merge a more detailed discussion of rocks and their formation. It represents an oversimplified version of the views of most petrologists on the origin of these rocks. It is not intended to represent the views of all geologists on the subject or the only origin for some of these rocks. Some rocks, such as granite, may form in more than one way, though the view presented probably accounts for the formation of a larger proportion than any other. Divergent views on the origin of many rocks, evidence bearing on the subject, and descriptions of important rock types are now considered.

THE THREE MAIN CLASSES OF ROCKS

Distinction in practice among igneous, metamorphic, and sedimentary rocks, although simple in most instances, may be very difficult and subjective for other instances. Igneous rocks may be considered generally to be those that crystallized from a hot rock melt or magma. They include volcanic tuffs that are deposited through air or water and that may be cold when deposited. The same materials when reworked and redeposited become sedimentary rocks. Metamorphic rocks are those that remain in the solid state while being changed by heat and/or pressure, with or without overall chemical change. Sedimentary rocks are those that settled or precipitated from air or water. These long-used major categories are necessarily in part genetic, but they are universally accepted by essentially all geologists.

Igneous rocks most commonly crystallize in a stress-free environment, which is reflected in their massive texture. A few, however, crystallize during flow, giving a parallelism of constituent minerals. The grains are interlocking except in fragmental types and range in size from very coarse in granite pegmatites to glassy in obsidian. Mineralogically, igneous rocks are composed of silicate minerals, with the single notable exception of carbonatites, which contain an appreciable percentage of calcite. The most common and abundant of these minerals are those on Bowen's Reaction Series,[1] including the feldspars, quartz, biotite, hornblende, augite, and olivine. Characteristic metamorphic minerals such as sillimanite or staurolite are absent. In many instances, igneous rocks show evidence of having moved and intruded other rocks during their formation.

Most *metamorphic rocks* crystallize under stress, resulting in a characteristic foliation or parallelism of the constituent grains, especially micas. The contact metamorphic rocks, however, more commonly form without deformation, resulting in a massive texture. Their proximity to a heat source, such as an adjacent igneous intrusive, and their characteristic spotted appearance resulting from the growth of new minerals aid in their recognition. Sedimentary layering, pebbles, or other structures from pre-existing rocks may be preserved. As in igneous rocks, the grains are interlocking and range in size from coarse to very fine and, in some mylonites (intensely sheared rocks), glassy. Mineralogically, most metamorphic rocks are composed of silicate minerals and/or carbonate. By contrast with the igneous rocks, feldspars may be, but are not normally, dominant. Common minerals include quartz, feldspars, micas, chlorite, amphiboles, diopside, epidote, and calcite. Less-abundant but characteristic metamorphic minerals include sillimanite, staurolite, and most garnets.

[1]Bowen's Reaction Series gives an approximate order of formation of the common igneous minerals in a magma, from olivine at the highest temperatures to quartz at low temperatures (see Fig. 3-6, p. 76).

Most *sedimentary rocks* show evidence of their origin in the presence of somewhat rounded or abraided grains cemented together and in deposition in layers. The grains are not interlocking except in fine-grained chemical precipitates, which, except for chert, tend to be composed of non-silicate materials. Quartz and carbonates are the most obvious minerals except for the often abundant clay minerals, which are not stable at the elevated temperatures of formation of igneous and metamorphic rocks. Because they may be derived from these rocks, sedimentary rocks may contain, in addition, small amounts of most minerals found in igneous and metamorphic rocks.

Chemical distinctions may also be made among these major classes of rocks; but there is considerable overlap, especially between the sedimentary and metamorphic rocks. Igneous rocks may also overlap in composition, but they rarely contain more than about 78% SiO_2, 27% Al_2O_3, 13% CaO, or less than about 34% SiO_2. Average chemical compositions of common igneous rock types are listed in Table 1-2.

THE GROWTH OF MINERALS

The whys and wherefores of the growth of the individual minerals within rocks is a fascinating subject. Why one mineral should grow instead of another has plagued and intrigued geologists for hundreds of years. With the support of experimental and theoretical chemistry, some answers are now becoming apparent.

Table 1-2 *Average chemical compositions of common igneous rock types** *

	Alkali Granite	Grano-diorite	Quartz Diorite (Tonalite)	Andesite	Basalt (Tholeiitic)	Basalt (Alkali Olivine)	Peridotite	Nepheline Syenite
SiO_2	73.86	66.88	66.15	54.20	50.83	45.78	43.54	55.38
TiO_2	0.20	0.57	0.62	1.31	2.03	2.63	0.81	0.66
Al_2O_3	13.75	15.66	15.56	17.17	14.07	14.64	3.99	21.30
Fe_2O_3	0.78	1.33	1.36	3.48	2.88	3.16	2.51	2.42
FeO	1.13	2.59	3.42	5.49	9.06	8.73	9.84	2.00
MnO	0.05	0.07	0.08	0.15	0.18	0.20	0.21	0.19
MgO	0.26	1.57	1.94	4.36	6.34	9.39	34.02	0.57
CaO	0.72	3.56	4.65	7.92	10.42	10.74	3.46	1.98
Na_2O	3.51	3.84	3.90	3.67	2.23	2.63	0.028[+]	8.84
K_2O	5.13	3.07	1.42	1.11	0.82	0.95	0.005[+]	5.34
P_2O_5	0.14	0.21	0.21	0.28	0.23	0.39	0.05	0.19
H_2O	0.47	0.65	0.69	0.86	0.91	0.76	0.76	0.96

*Note the gradational variation in composition from granite to basalt and peridotite, especially the decrease in SiO_2, K_2O, and to some extent Na_2O, and the increase in FeO, MgO, MnO, CaO, TiO_2, and P_2O_5.
[+]From Hamilton and Mountjoy (1965).
SOURCE: Nockolds (1954).

Particular mineral structures grow as a result of their being more stable in the prevailing environment than all other minerals. The most important factors in igneous and metamorphic environments are the total chemical composition in a small volume, the temperature, and the fluid pressures. Other factors, although playing their roles, are thought to be of relatively minor importance.

Because of the great abundance of Si and O in the earth's crust, most of the common rock-forming minerals are silicates of the elements next in abundance. Most important of these, in order of abundance, are Al, Fe, Ca, Mg, Na, K, and Ti. As a first approximation, the earth's crust can be considered to be a thick layer of closely packed oxygen atoms, with the metal ions in the interstices. If we calculate the density of packing of oxygen, using the theoretical closest packing of oxygen as a reference, common silicate minerals range from about 90% of this maximum in kyanite to about 56% of this maximum in nepheline. The very abundant feldspars have about 80% of the maximum oxygen density (Lacy, 1965). The specific kind of regular arrangement of atoms depends on the relative size of oxygen and cation. Given the SiO_4^{4-} tetrahedron as the stable basic building unit, the tetrahedra and metal ions arrange themselves in only a few stable configurations. These arrangements depend mainly on the kind (size and charge) of cations present and their abundance. Consider some of the common arrangements and a highly simplified explanation of why they develop. It must be emphasized, of course, that the situations discussed below are extremes. Natural situations are almost always intermediate between the situations presented. As a result, most natural rocks consist of two or three or more major minerals instead of only one.

If no cations are present in addition to the silica tetrahedra, these cluster together so that each corner oxygen is shared with another tetrahedron. This forms a three-dimensional network of silica tetrahedra—SiO_2. Depending on the most stable arrangement for the prevailing temperature and pressure, the different structures—quartz, tridymite, or cristobalite—will tend to form.

If K and Al are also present, a three-dimensional framework, with the small Al^{3+} taking the place of the small Si^{4+} in one out of four tetrahedra, forms in addition. The positive-charge deficit is balanced by K^+ entering some of the large holes in the "rings" of tetrahedra. The resulting mineral is orthoclase, $KAlSi_3O_8$. Na^+ takes the same role as K^+ in the formation of albite. Similarly, substitution of $2(Al^{3+})$ for $2(Si^{4+})$ may be accommodated by a single divalent cation Ca^{++} to form anorthite, $CaAl_2Si_2O_8$.

If water is present, some of the strong bonds between silica tetrahedra may be weakened, one out of six oxygens being replaced by $(OH)^-$. The silica tetrahedra may then take the form of sheets. Generally, a part of the Si^{4+} is replaced by Al^{3+}, the charge deficit being taken care of by

K^+, which also serves to bind the layers together. The resulting mineral is muscovite, $KAl_2(AlSi_3O_{10})(OH)_2$. If the environment also contains appreciable Mg, three Mg^{++} will do about the same job as two Al^{3+}, since three equivalent spaces are available. The result is biotite, $K(Mg, Fe^{++})AlSi_3O_{10}(OH)_2$.

If some water and appreciable Mg, Fe, and Ca are present, commonly with some Al in a low-alkali environment, only about half of the tetrahedral corners join directly with other tetrahedra. Long chains or double chains are formed. These are linked together by Mg and/or the other smaller divalent cations to form the pyroxenes [for example, $(Mg, Fe, Ca)SiO_3$] or amphiboles [for example, $Ca_2(Mg, Fe)_5Si_8O_{22}(OH)_2$]. If much Na is also available, a soda pyroxene or soda amphibole may develop.

If the environment is very high in Mg and Fe and low in the larger cations, Ca and the alkalis, and low in water, the silica tetrahedra may not be able to link directly with one another at all. The resulting Mg silicate is olivine, $(Mg, Fe)_2SiO_4$. If much Al and/or Ca is also available, a garnet may be formed, $(Ca, Mg, Fe)_3(Al, Fe^{+3})_2(SiO_4)_3$.

Superimposed on this dominant chemical control of mineralogy is a temperature effect and, to a still lesser extent, a pressure effect. Two examples will suffice for purposes of illustration. At higher temperatures, the double-chain amphibole structure (e.g., actinolite) can accommodate some Al^{+3} in the place of Si^{+4} in the silica tetrahedra (e.g., hornblende). Probably the greater thermal vibration of the ions at higher temperature permits accommodation of some of the larger Al that is almost always present in the environment. At higher temperatures, Fe tends to be lost from Mg silicates. Stated in another way, Fe^{++} tends to lower the upper temperature limits of stability of such minerals. Olivine, pyroxenes, and amphiboles are well-known examples.

One means of attack on the origin of igneous and metamorphic rocks is to study some aspect of their formation that can actually be seen in the process. Probably the closest we can come is to watch the solidification of a molten rock after it pours out of a volcanic vent. If a basalt magma is examined shortly after it leaves the vent, we may find it to be extremely hot, fluid, and lacking in crystals. Several days or weeks later, we may find that the still-molten parts are somewhat cooler, are less fluid or more viscous, and contain some tiny crystal grains. After months or years, the still-molten, cooler, and pasty remainder, perhaps deep under the solidified crust of a lava lake, may contain many small crystal grains. Clearly, these small crystals were not in the original magma. They formed during solidification of the lava.

As a molten rock slowly cooled, some of the components must have gradually reached saturation, as in any cooling solution, and then supersaturation. Tiny nuclei of the supersaturated mineral appeared and began to grow in the melt. They grew in size by addition of material to the outer surface of the mineral. As noted above, the composition of the growing mineral commonly depends on the composition of the environment from which it is derived. Since removal of the mineral from the magma by crystallization must cause a change in the composition of the magma, the mineral composition may also be affected by the change. Obviously, each crystallizing mineral affects the magma in a different manner and if more than one is crystallizing, each indirectly affects the other. Needless to say, the situation is exceedingly complex in this process, which is most nearly accessible to direct observation.

In order to clearly understand what is going on during the crystallization of a magma, it would be a big advantage to be able to separate the behavior of each of the constituent minerals. To do this, something must be known about the physical and chemical properties of each of the common rock-forming minerals and about their behavior during crystallization of melts of simple composition under known and controlled conditions.

COMMON ROCK-FORMING MINERALS

Feldspars

Feldspars are the commonest and most important minerals in igneous and metamorphic rocks. They are potassium-, sodium-, and calcium-aluminum silicates with partial to complete solid solution between the end members. Complete solid solution exists at almost all temperatures in the plagioclase series from albite ($NaAlSi_3O_8$) to anorthite ($CaAl_2Si_2O_8$). Complete solid solution exists at only elevated temperatures in the alkali feldspars from albite ($NaAlSi_3O_8$) to potassium feldspar ($KAlSi_3O_8$). Partial solid solution exists at lower temperatures. In other words, under these conditions, the albite crystal structure can accommodate only a small amount of K^+, and the potassium feldspar structure can accommodate only a small amount of Na^+. What happens between these high and low temperatures, in fact, throughout the crystallization of the feldspars, is important in understanding their behavior during crystallization.

Synthetic studies on silicate melts and solid-state reactions in silicate systems have given us a great deal of insight into the behavior of minerals during crystallization and recrystallization. At first sight, this looks like the solution to most petrological problems. The specific physical conditions such as temperature and pressure cannot normally be applied directly to natural rocks, however, because of several complicating factors. Natural minerals commonly contain small amounts of other elements that are not present in the simple synthetic systems, and the natural rocks normally contain other minerals, such as quartz, that affect the crystallization conditions, e.g., through a reaction or eutectic relation. These studies do, however, give an excellent idea of the trends and processes during crystallization and an approximation of the temperatures and pressures involved for some rocks. In fact, so similar are the artificial products to the natural minerals and mineral associations that we must conclude that minor amounts of additional elements do not significantly affect the final product or the process by which it is formed.

As a practical example, consider the crystallization behavior of the plagioclase solid solution series as deduced from an important series of experiments (see Fig. 1-3).

Slow crystallization of plagioclase from a melt of a given composition (X) first gives crystals more calcic in composition (C_1) at temperature (T_1). This removal of calcium causes the melt to become more sodic.

With falling temperature, the melt (L_1), which becomes gradually more sodic with falling temperature, continuously reacts with the

Fig. 1-3   *The plagioclase series. Based on Bowen (1913, p. 583) and Yoder et al. (1957).*

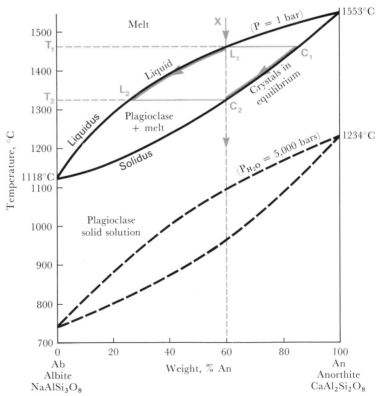
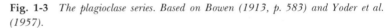

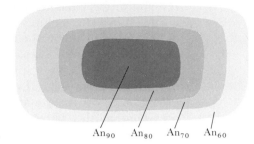

Fig. 1-4 *Zoned crystal seen under the microscope.* An_{90} An_{80} An_{70} An_{60}

more calcic crystals making them more sodic and tending to maintain equilibrium at any given temperature.

With continued falling temperature, reaction continues until the crystals have all been made over into the composition (C_2) of the original melt at temperature (T_2) when the liquid (L_2) is all used up.

With rapid cooling, the melt may have only enough time to react with the rim of the crystal so that the core of the crystal remains more calcic and the crystal becomes progressively zoned to more and more sodic rims (Fig. 1-4). This effectively removes the calcic cores from further reaction, permitting the final melt (rims or final crystals) to become still more sodic. With reference to Fig. 1-3, for example, if equilibrium was maintained down to 1400°C (crystals about An_{75}), then more rapid cooling isolated these crystal cores from further reaction, the remaining melt (about An_{37}) would crystallize as if it were the original melt. Rims on the crystals would therefore be An_{37} or more sodic. At higher water pressures (P_{H_2O}), crystallization begins and is completed at lower temperatures. That is, the temperatures for liquidus and solidus are lowered with increased water pressure.

Given such a temperature-composition diagram, which was worked out by some of the experimental petrologists at the Geophysical Laboratory in Washington, D.C., it might be well to consider the procedure involved in preparing such a diagram. Although some of the practical difficulties, especially at high pressures, were formidable until recently, the procedure is simple in principle.

A mixture (for example, of oxides) is weighed out in the proportions to give the desired starting composition. The finely ground mixture is heated to a temperature above that necessary for complete melting. This temperature, generally measured by a thermocouple, lies of course in the melt field above the liquidus. To make sure that the melt is completely liquid, the melting capsule is suddenly chilled by plunging it into cold water. Since it takes time for crystals to grow, it is intended that none will form during the rapid cooling and a check under the microscope will reveal if any crystals were present in the melt. The same mixture is then heated above the melting point, then cooled to a temperature suspected of being slightly below the temperature for complete melting. After holding this temperature for some time (hours, days, or even weeks) to permit growth of crystals, the capsule is again suddenly chilled and checked for the growth of crystals. If crystals are present, that temperature is interpreted as below the liquidus. The composition of the liquid and crystals coexisting at this known temperature can be determined by chemical analysis or from optical properties. This gives one point on the liquidus and one on the solidus. Repetition of the procedure for different temperatures of crystallization and for different compositions of initial melts permits one to draw the liquidus and solidus curves.

The manipulations are the same for experimental studies under high water pressure. Only the equipment differs. Most commonly, the starting material is enclosed in a hydrothermal "bomb"—a high-strength steel cylinder permanently sealed at one end and with a screw-in seal at the other end. This bomb is fitted with a thin tubing through which to apply steam under pressure and a thermocouple by which to measure the temperature. The heat is applied externally by a heating coil surrounding the bomb. For further details see Tuttle (1948, 1949). Numerous modifications and improvements have been made in experimental equipment and methods in the past 20 years (e.g., in application and determination of pressure and temperature in the experiment, in determination of resulting phases, and in recognition of possible metastable growth), but for present purposes this simplified description will be sufficient. Further discussion of experimental problems will be deferred to the section on metamorphism.

Fig. 1-5 *The system orthoclase-albite. Based on Bowen and Tuttle (1950, p. 497) and Yoder, Stewart, and Smith (1957), modified from Turner and Verhoogen (1960, p. 109), Morse (1970).*

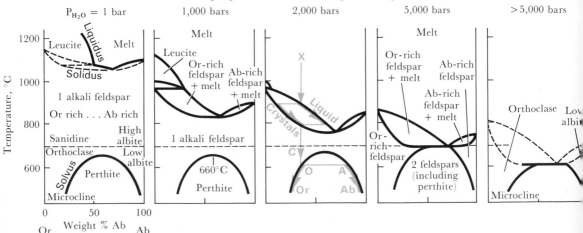

Crystallization of the alkali feldspars differs from that of the plagio-clases, largely in the presence of a liquidus minimum and of a pronounced solvus. Crystallization behavior on either side of the liquidus minimum is just as in the plagioclase series. For example:

Near-equilibrium (slow) crystallization of a melt of a given composi-tion (X) as shown in Fig. 1-5 results in uniform alkali feldspar crystals of the same composition (C).

Continued slow cooling of these crystals may result, at temperatures below the solvus, in exsolution (separation of different mineral phases from the earlier solid crystals) of albite (A) from orthoclase (O).

With continued cooling, orthoclase and albite components are able to hold less and less of the other end member, becoming purer and purer end members. The result is perthite—streaks of exsolved al-bite in orthoclase.

Just as complete solid solution exists at certain temperatures within the plagioclase and alkali feldspar series, partial solid solution exists be-tween the two series—that is, some orthoclase component exists in solid solution in the plagioclase series and some anorthite component exists in solid solution in the alkali feldspar series (see Fig. 1-6). Note the fol-lowing:

Ab and An form a complete solid solution series (are completely mis-cible) at essentially all temperatures—neglecting the peristerite exsolution field, which is shown on Fig. 1-6.

Ab and Or show complete solid solution only at elevated temperatures —say, above 660°C as shown on Fig. 1-5.

An and Or show very limited solid solution at all temperatures.

A rise in temperature tends to increase the solid solution between all components.

A rise in pressure (as shown in Fig. 1-5) tends to reduce the solubility slightly because the solvus tends to rise (or expand) with increasing pressure by about 14°C/1,000 bars (Yoder et al., 1957). Since the solvus is bounded on both sides by solids, Le Châtelier's rule leads one to expect very little pressure effect. This is why the pressure effect here is so much smaller than that for the liquidus-solidus curves—between solid and melt.

The crystallization behavior of the feldspars and its dependence on

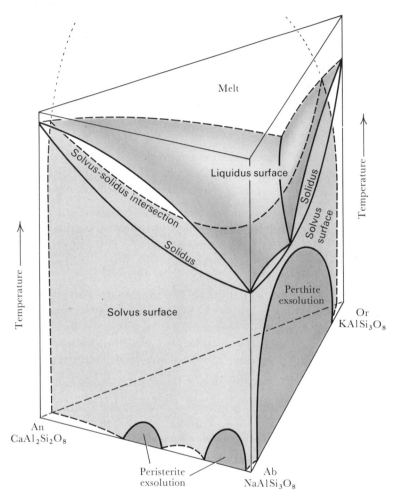

1-6 *Diagramatic feldspar system at about 2,000 bars* P_{H2O}. *The composition of all feldspars lies on or above the surface of the solvus and below the surface of the solidus. The left-front side corresponds to Fig. 1-3; the right-front side corresponds to Fig. 1-5.*

pressure are shown in Fig. 1-7. Note that with increasing water pressure (as also shown in Fig. 1-5) the liquidus or surface of initial crystallization is strongly depressed. Feldspar melts falling in the two-feldspar field will crystallize two feldspars from the melt directly (a plagioclase and a potassium feldspar). Reasons for this are readily apparent from a consideration of Fig. 1-5 under high-pressure conditions in which the solidus intersects the alkali feldspar solvus. Conversely, feldspar melts falling in the one-feldspar field will crystallize only one feldspar (either plagioclase or potassium feldspar depending on composition of the melt) if crystallization proceeds slowly under near-equilibrium conditions.

Consider, for example, a rock such as a syenite, containing one potassium feldspar and no plagioclase. The magma from which it crystallized

Fig. 1-7 *Crystallization of the feldspars at 1 bar and 5,000 bars water pressure. The solidus-solvus intersection for 1 bar in the orthoclase corner has been extended for high SiO_2, so that the field of leucite is eliminated. Temperature contours are drawn on the liquidus. After Franco and Schairer (1951, p. 264, copyright 1951 by the University of Chicago); Yoder, Stewart, and Smith (1957, p. 211); Turner and Verhoogen (1960). The liquidus at 1,000 bars has been determined by James and Hamilton (1969).*

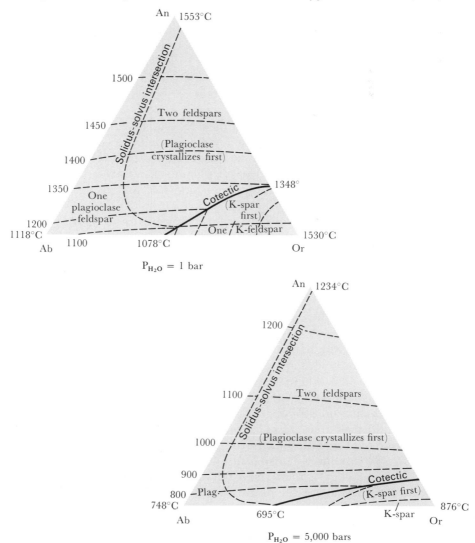

must have had a composition in the rather restricted Ca-poor one − K-feldspar field. Slow cooling below the solvus could and commonly does result in exsolution of excess sodium in the form of the typically observed streaks of perthitic albite in the potassium feldspar. On the other hand, a rock such as a porphyritic granite, containing phenocrysts of orthoclase with orthoclase and plagioclase in the groundmass, would be interpreted as having crystallized from a melt in the K-feldspar-first field.

Figure 1-7 at 1 bar has been simplified for purposes of illustration by elimination of the field of leucite. This elimination commonly occurs in nature, not only by increased pressure but also by addition of excess silica, as in common granites and rhyolites. As shown in Fig. 1-8, the field of leucite is reduced to elimination at about 30% excess SiO_2. At water pressures somewhat higher than 1 bar, less SiO_2 is required for elimination of leucite. These two factors, along with the high potassium content required of the melt, account for the rarity of leucite except in high-potassium, low-silica volcanic rocks.

Cooling of silica-rich melt X (Fig. 1-8) results first in the crystallization of tridymite at about 1280°C. As the melt cools, continued removal of silica (tridymite) leaves the melt enriched in $KAlSi_3O_8$. At 1200°C, the system consists of about 1 part tridymite crystals and 6 parts melt (of composition about 55% silica component and 45% K-feldspar component).

Note that the coexisting phases are found by drawing a tie line between the liquidus and solidus at the desired temperature, just as in the plagioclase feldspar system (Fig. 1-3). The amounts of these coexisting phases can most easily be obtained by noting on the tie line how close the total composition (i.e., the original composition in a near-equilibrium situation) is to either end of the tie line, in this case the liquidus or the solidus.

As the temperature continues to fall, tridymite continues to crystallize, the melt moving away from silica until at about 1000°C the eutectic is reached. There the system of total composition X consists of about 1 part tridymite crystals and 2 parts melt (of composition about 42% silica component and 58% K-feldspar component). At the eutectic, both K-feldspar and tridymite crystallize until the melt is completely used up. Note that below the eutectic, the tie line across the total composition X (e.g., at 980°C) joins K-feldspar at one end and silica at the other end. The final mineralogy is then about 62% tridymite and 38% K-feldspar (same total composition as the original melt X).

Cooling of silica-poor melt Y (Fig. 1-8) presents somewhat different problems because of the presence of a peritectic. The first crystals to form are leucite at about 1470°C. As the temperature falls, leucite continues

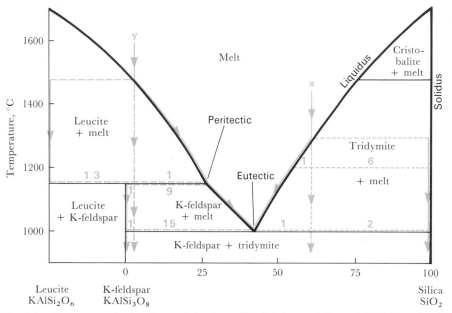

Fig. 1-8 *Binary system leucite-silica at 1 bar* P_{H_2O}*. After Schairer and Bowen (1947). Note that the melt-cristobalite boundary may be calculated approximately from the van't Hoff equation:*

$$\log K_{SiO_2} = \frac{\Delta H}{2.303R}\left(\frac{1}{T_{SiO_2}} - \frac{1}{T}\right)$$

where the equilibrium constant K_{SiO_2} *equals the activity of* SiO_2 *in the melt. If at very high concentrations, the activity of* SiO_2 *is approximately equal to the concentration of* SiO_2 *in the melt.* ΔH *is the heat of fusion of* SiO_2*; R is a constant (1.99 cal/mole);* T_{SiO_2} *is the melting temperature of* SiO_2*; T is the melting temperature of the mixture. Or the liquidus may be located approximately using* $\mu_{melt} = \mu_{SiO_2} +$ *RT* $\ln X_{K\text{-spar}}$*, where* μ = *chemical potential and X = fraction of component in melt.*

to crystallize, the composition of the melt moving away from leucite and becoming enriched in silica. At about 1160°C, just above the horizontal phase boundary at the peritectic, the system consists of about 1 part (42%) leucite crystals and 1.3 parts (58%) melt (of composition about 26% silica component and 74% K-feldspar component[1]). At about 1140°C, just below the horizontal phase boundary at the peritectic, the system consists of about 1 part melt (of composition about 27% silica component and 73% K-feldspar component) and 9 parts K-feldspar. Again the temperature tie line joins the two coexisting phases. Because of the peritectic reaction, the silica-deficient leucite reacts with the silica-rich melt of the peritectic composition to form K-feldspar. Also, since the percentage of melt drops from about 58 to 10%, K-feldspar must crystallize from the

[1]Note that if somehow the leucite crystals were separated from the melt at this stage (e.g., by floating, since leucite has a lower density than the melt, or by lack of complete reaction with the melt), the remaining melt could crystallize free silica.

melt while the reaction is going on. As the temperature continues to fall below the peritectic, K-feldspar continues to crystallize, the melt becoming still richer in silica until at about 1000°C the eutectic is reached. There the system of total composition Y consists of about 1 part melt (of composition 42% silica component and 58% K-feldspar component) and 15 parts K-feldspar. At the eutectic, both K-feldspar and tridymite crystallize until the melt is completely used up. Note that below the eutectic, the tie line across total composition Y (e.g., at 980°C) gives 3% tridymite and 97% K-feldspar. Although the melt initially crystallized leucite, none remains in the product. By the same token, melting of K-feldspar does not lead to a melt of K-feldspar composition (i.e., it melts incongruently) but to leucite and a more silica-rich melt.

A knowledge, therefore, of the compositions of the two feldspars in igneous or metamorphic rocks should give some idea of the temperature of formation of the rocks containing them. The temperature of crystallization cannot be read directly from the feldspar diagrams, even if the confining water pressure is known, however, because the presence of quartz or excess silica in the melt, for example, 25% excess silica, may lower the initial temperature of crystallization by as much as 250°C. A much closer estimate of crystallization temperature of granitic rocks may be obtained

Fig. 1-9 *Composition of plagioclase in common igneous and metamorphic rocks. (Occurrence and type of potassium feldspar shown on the diagonal.)*

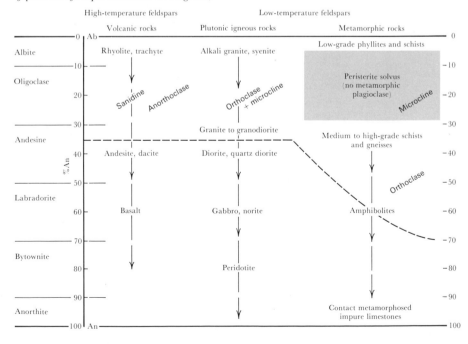

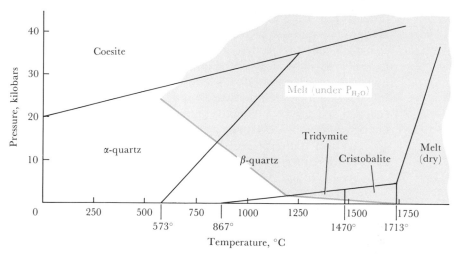

Fig. 1-10 *The silica system. After Boyd and England (1960, p. 752). Note that the slope of the phase boundary may be simply calculated using the Clausius-Clapeyron equation:* $dP/dT = \Delta S/\Delta V = H/T\Delta V$. *Thus the slope of the boundary equals the change in entropy (ΔS) divided by the change in volume (ΔV) or the heat of reaction (ΔH) divided by the temperature times change in volume. Note also that a high-temperature polymorph, such as tridymite or cristobalite, has a higher entropy (more disordered and therefore higher symmetry—hexagonal and cubic, respectively) than a lower-temperature polymorph such as β-quartz (more ordered and lower symmetry—less symmetrical hexagonal).*

from a consideration of ternary diagrams in which silica is a component, as will be seen below. Other complicating factors involve the presence of a few percent of FeO and MgO and lesser amounts of other elements, for which very few experimental data are available. Preliminary studies on the effect of these components (FeO, MgO) on the partial melting of shales (Von Platen, 1965b), however, suggest that they have but a minor effect on that process. From this it might be surmised that in small amounts they would also have but a minor effect on crystallization behavior. The evidence is insufficient, however, to give more than a notion in this direction. Even if this turns out to be valid, there is a question about at what percentage these components become important in controlling the temperature of crystallization or in otherwise affecting the behavior of crystallization.

The occurrence of the feldspars, as with most other minerals, is a major aid in recognition of different varieties. The most common occurrences of several varieties are shown in Fig. 1-9.

Quartz and Other Silica Minerals: SiO₂

Next to the feldspars, quartz is the most common and widespread mineral in nature. As may be seen from the phase diagram (Fig. 1-10), quartz is the stable form of silica under almost all geological conditions. Although β-quartz crystallizes during the formation of volcanic and some plutonic igneous rocks, it always inverts to α-quartz on cooling. The transformation is the very rapid displacive type that involves only a slight change in crystal structure and properties, so β-quartz does not persist at room temperatures. Tridymite and cristobalite, crystallizing at high temperatures and low pressures in volcanic rocks, normally do persist on cooling and do occur in rocks at room temperatures. The transformation is the very sluggish reconstructive type involving breakage of tetrahedral bonds and a major change in crystal structure.

α- and β-quartz are distinguished with difficulty except by crystal form and nature of twinning (Frondel, 1945). The presence of crystals with prism faces indicates α-quartz. Bipyramidal crystals without prism faces strongly suggest β-quartz. The quartz in volcanic rocks is always the high-temperature variety (β), indicating temperatures of crystallization of the mineral above 573°C. Tridymite and cristobalite suggest low pressures of crystallization but are not indicative of temperature. They can form metastabily outside their stability fields at much lower temperatures due to rapid crystallization (e.g., in the presence of volatiles) or due to incorporation of certain impurities (for example, Na and Al). Quartz forms only within its stability field.

Coesite and stishovite are rare high-pressure forms resulting only from extreme shock, such as meteoritic impact. The occurrence of the other silica polymorphs is outlined in Table 1-3.

Table 1-3 *Appearance and occurrence of common silica polymorphs*

Polymorph	Appearance	Occurrence
α-quartz	Crystals with prism faces with horizontal striations	Silica-rich to intermediate igneous rocks, hydrothermal veins, sandstones, common metamorphic rocks; as chalcedony, chert, agate, etc.; not with feldspathoids (e.g., nepheline) or common (Mg) olivine.
β-quartz	Crystals without prism faces	
Tridymite	Small, platy, hexagonal crystals ($\leq$ 1mm across)	Silica-rich to intermediate volcanic rocks, especially in cavities and fractures.
Cristobalite	Small octahedral crystals ($\leq$ 1 mm) or small rounded aggregates	

Higher Temperature (arrow indicating increasing downward)

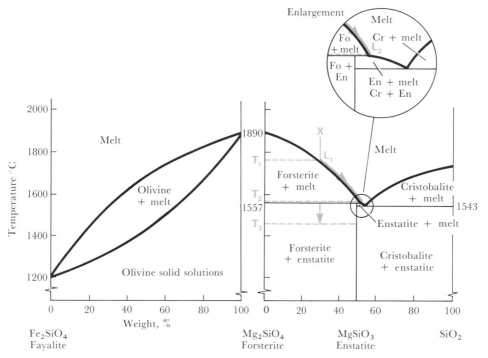

Fig. 1-11   *Temperature-composition diagram for olivine. After Bowen and Schairer (1935, p. 163); Bowen and Anderson (1914, p. 488). Pressure 1 bar.*

Olivine: (Mg, Fe)$_2$SiO$_4$

The olivine phase diagram is much like that for the plagioclase series. As with plagioclase, the first crystals to form from the melt are richer in the high-temperature end member, in this case forsterite (Mg). With slow crystallization, later crystals are enriched in fayalite (Fe). The effect of pressure is to lower the crystallization temperatures.

Mg-rich olivine (about Fo^*_{60-100}) is by far the most common. This is the common olivine in mafic-rich igneous rocks such as dunite, peridotite, gabbro, and basalt, and in metamorphosed impure dolomitic limestones. It is never found in equilibrium with quartz. As shown in Fig. 1-11, additional silica would develop enstatite (or another pyroxene) in place of olivine.

Fe-rich olivine (about Fo_{0-60}) occurs in small amounts in some iron-rich igneous rocks, some carbonatites, and metamorphosed iron-rich, siliceous sediments. Not uncommonly, it is found with quartz.

The antipathy of forsterite (Mg-rich) olivine for quartz results from the existence of the mineral enstatite, compositionally intermediate be-

*Fo = forsterite molecule, Mg$_2$SiO$_4$.

tween them (Fig. 1-11), just as orthoclase is compositionally intermediate between leucite and quartz. Like orthoclase, enstatite is an incongruent melting mineral, melting to a liquid of a different composition and at a much lower temperature than either forsterite or quartz. The binary system forsterite-silica has been studied by Bowen and Anderson (1914) and Grieg (1927).

Crystallization of olivine-rich melts containing excess silica may, with slow cooling, lead to rims of enstatite on forsterite. This is an example of Bowen's Reaction Principle (see Fig. 3-6, p. 76), as shown in the following example:

A melt of composition X (Fig. 1-11) cools to a temperature T_1 when pure forsterite (Mg_2SiO_4) begins to crystallize.

As forsterite separates with further cooling, the melt becomes richer in SiO_2 until the system reaches temperature T_2. At T_2, the system consists of pure forsterite in equilibrium with a melt of composition L_2.

Continued cooling results in reaction of forsterite with the silica-rich melt to form enstatite ($MgSiO_3$). Temperature T_2 is maintained until all the melt is used up.

At temperatures below T_2 (for example, T_3), the system consists of forsterite and enstatite, the latter sometimes taking the form of reaction rims on the former. Crystallization behavior in this system is closely comparable to that in the leucite−K-feldspar−silica system (Fig. 1-8).

Olivine alters and weathers easily to a variety of hydrous mixtures. Dark green to yellow "serpentine" rocks not uncommonly consist largely of olivine, which is not seen on broken surfaces because the fractures are localized in the serpentine. Reddish brown to orange colors on the weathered surface of, or sometimes scattered through, many peridotites and basalts result from coatings of the iron-rich mineraloid "iddingsite" on the olivine.

Pyroxenes: Orthopyroxenes $(Mg, Fe^{++})_2Si_2O_6$
Clinopyroxenes $Ca(\underline{Mg, Fe^{++}})\underline{Si_2O_6}$ less prominent substitutions include
$\updownarrow$ $NaFe^{3+}$ for $(Ca, Mg, Fe)_2$
Mg, Fe   Al, Al   NaAl

The pyroxenes in common iron-poor metamorphic rocks (deep seated) are diopside, if white to green in color, and enstatite to hypersthene, if white to brown in color. By contrast, the pyroxenes in high-temperature volcanic rocks (e.g., basalts) are likely to be clinopyroxenes, especially augite. At somewhat lower temperatures, such as those during crystallization of more

silica-rich volcanic rocks (e.g., pyroxene andesites) or more deep-seated
mafic igneous rocks (e.g., diabase or norite), hypersthene or pigeonite may
crystallize with the augite. Inspection of Fig. 1-12 gives an indication of
why this should be so. Most common mafic rocks contain enough Al, and

Fig. 1-12 *Phase relations in the pyroxenes. Modified from Deer, Howie, and Zussman (1963, 2, p. 5).
Subcalcic augites (between augite and pigeonite) in volcanic rocks may be unstable quench crystals so
that exsolution field may remain in this area, even under volcanic conditions (H. Greenwood, written
communication, June, 1969).*

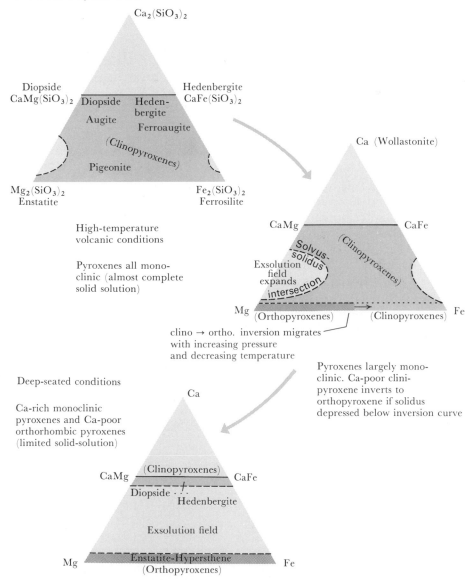

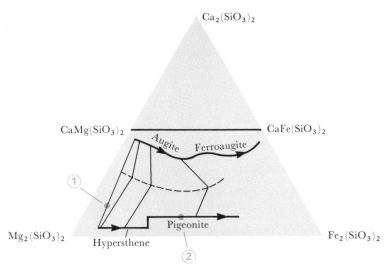

Fig. 1-13 *Normal course of crystallization in pyroxenes. After Hess (1941); Boyd and Schairer (1964, p. 302, after Muir [1951] and Brown [1957]).*

the structure of the common clinopyroxenes is flexible enough, to accommodate a variety of elements so that augite can form. Under deep-seated, higher-pressure conditions, the solidus is depressed to intersect the solvus, thereby restricting the solid-solution field (between solidus and solvus) to high- and low-calcium varieties. Slow cooling below the solvus may result in exsolution of augite from hypersthene or of hypersthene from augite much as in the feldspars (Poldervaart and Hess, 1951; Brown, 1957; Boyd and Brown, 1969). As shown in Fig. 1-13, augite and hypersthene crystallize together (in equilibrium), both becoming more iron rich as the temperature falls. As the hypersthene becomes more Fe rich and reaches the inversion curve, it becomes increasingly able to accommodate Ca in the crystal lattice, thereby inverting to the higher-Ca monoclinic form, pigeonite. With continued falling temperature, augite crystallizes in equilibrium with pigeonite. Note that with an initial Fe-poor composition (1), hypersthene will crystallize first, whereas with an initial Fe-rich composition (2), crystallization will begin at a lower temperature with pigeonite.

CHAPTER 2
PROPERTIES
AND PROCESSES
IN IGNEOUS
ROCKS

CLASSIFICATION OF IGNEOUS ROCKS

Igneous rocks are subdivided and given descriptive names for purposes of description, communication, and comparison. These subdivisions or classifications are made in different ways for convenience. Different classifications are based on mineral content, occurrence (plutonic, hypabyssal, volcanic), chemical composition, and color index. For many purposes a classification based on mineral content is most convenient and tells most about the rock.

Other classifications used to subdivide igneous rocks on different bases give less commonly used, more generalized names, or no names at all. A gross subdivision of igneous rocks is made on the basis of *general occurrence:*

Volcanic—ideally those igneous rocks formed at or very near the earth's surface. In practice this generally corresponds to very fine grained to glassy rocks.

Hypabyssal—intrusive igneous rocks formed near the earth's surface. Characteristically they are porphyritic.

Plutonic—intrusive igneous rocks of deep-seated origin. Commonly medium to coarse grained.

Extrusive—those igneous rocks, fragmental or otherwise, that are erupted onto the earth's surface (extruded).

Intrusive—those igneous rocks that are emplaced below the earth's surface (intruded).

Mineralogical Classifications

Mineralogical classifications are normally based on one or more important variables:

Percent and type of the feldspars

Presence or absence of quartz, feldspathoids, or olivine

Percent and type of dark minerals

Grain size and texture

Some of these variables are not independent of one another, that is, they show a sympathetic or antipathetic relationship. For example:

An increase in anorthite content of plagioclase is usually correlated with an increase in ferromagnesian minerals such as hornblende and a decrease in quartz and alkali feldspars.

The presence of quartz is almost always correlated with an absence of feldspathoidal minerals and olivine.

Minerals *essential* to naming the rock fall into two classes: (1) those that form the *specific* name of the rock (such as quartz, orthoclase, and plagioclase in a granite) and (2) those that are important enough to indicate the variety of that rock (such as biotite and lesser hornblende in a hornblende-biotite granite). These *varietal minerals* are used as modifiers of the specific name, the least abundant being listed first. Other less-abundant minerals formed by primary crystallization are called *accessory minerals* (e.g., sphene, magnetite). Minerals formed by later alteration are called secondary (e.g., chlorite altered from biotite).

Several mineralogical classifications are in everyday use; most of them have much in common, using similar criteria and having similar limits for the standard rock names. Many commonly used variations depend on the anorthite content of the plagioclase as a variable—a practice largely contingent on use of the polarizing microscope. For our purposes, a hand-specimen classification suitable for use in the field and closely conforming to the microscope-based variation is desirable.

The multiplicity of somewhat different mineralogical classifications as used by different authors in different countries has given rise to international attempts at standardization. The International Union of Geological Sciences Commission of Petrology has established a "working group on rock nomenclature," the final report of which has recently been published (Streckeisen, 1967). A large number of geologists in many countries have participated in the discussions, and the final conclusions as embodied in this paper are to be reviewed by the International Geological Congress (Montreal) in 1972. In this light we will follow, in this book, the recent modification of Streckeisen's igneous rock classification, which is presently under discussion by the IUGS Commission. Presumably, this classification, or some minor modification of it, will be the standard in the future. The discussion below is based directly on Streckeisen (1967; written communication, November, 1970).

The following main criteria have been used in setting up the classification:

1. Igneous rocks should be named according to their actual ("modal") mineral composition (measured in volume percent). "Only when the exact determination of the mineral composition presents difficulties, a chemical analysis will help to calculate the corresponding mode" (Streckeisen, 1967, p. 150).
2. "Igneous rocks" for this classification includes all "igneous-looking rocks" regardless of their origin. They may have formed by magmatic, metamorphic, or metasomatic crystallization.
3. The centers of natural distribution of the various rock groups should fall within the corresponding classification fields, not on their borders.
4. To cause minimal confusion or change, historical tradition should be followed as much as possible.
5. It should be simple.
6. The following minerals and mineral groups are considered most important and will be used as the corners of the classification double triangle:

   Q = silica minerals (quartz, tridymite, cristobalite)
   A = alkali feldspars (orthoclase, microcline, sanidine, perthite, anorthoclase, albite An_{0-5})
   P = plagioclase (An_{5-100}), scapolite
   F = feldspathoids (leucite, pseudoleucite, nepheline, sodalite, nosean, hauyne, analcime, cancrinite, etc.)

This classification considers only rocks with a mafic mineral content (M, color index) of less than 90%.

M = mafic minerals (micas, amphiboles, pyroxenes, olivines, ore minerals, accessories, epidote, garnets, melilites, monticellite, primary carbonates, etc.)

Rocks in which M = 75 to 90 should be designated *mafitic* or should be given compound names (e.g., gabbro-peridotite or hornblendite-diorite). Rocks in which M > 90 are classified according to their predominant mafic minerals (see below). Rocks in which M < 90 are represented on

Fig. 2-1a *Classification of plutonic rocks in the double triangle* Q–A–P–F, *according to their actual mineral composition. After Streckeisen (1967; written communication, November, 1970).*[1]

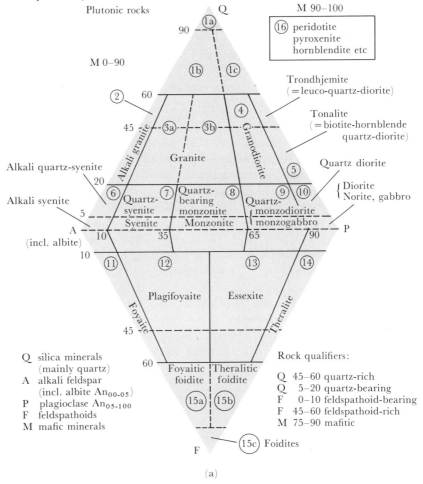

(a)

[1]See *Geotimes*, Oct. 1973, p. 26, for final report of the IUGS.

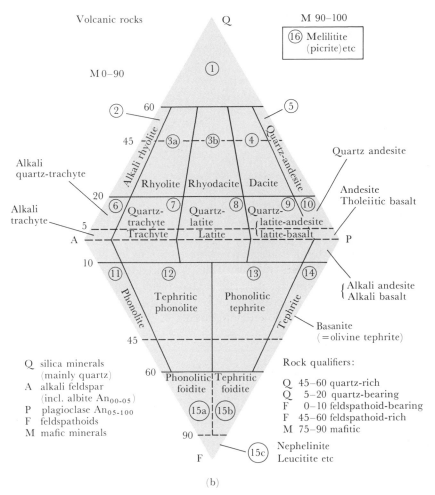

Fig. 2-1b *Classification of volcanic rocks in the double triangle* Q–A–P–F, *according to their actual or calculated mineral composition. After Streckeisen (1967, p. 161; written communication, November, 1970).*

the double triangle Q–A–P–F (see Figs. 2-1a and b). The relative amounts of quartz, feldspars, and feldspathoids should be calculated from the light-colored constituents only (that is, $Q + A + P = 100$ or $A + P + F = 100$).

Although the diagram is largely self-explanatory, some points should be clarified, concerning the limits chosen. The lower limit for granite at $Q = 20\%$ has been chosen because the large majority of granitic rocks fall within the limits $Q = 20$ to 45 (Figs. 2-2a and b). Most of the older classifications placed lower limits at $Q = 5$, 10, or 15; but, as proposed by Chayes

(1957), $Q = 20$ better satisfies the natural relationships (see Fig. 2-2a and b).

Granitic rocks in which $Q = 5$ to 20 should be designated as "quartz-bearing" syenites, monzonites, or diorites. The widely used name quartz monzonite is to be abandoned (Streckeisen, 1967), since it is used in quite different ways by different workers, in some cases with little overlap of meaning. As such, the term "granite" includes much of what has been used by many North American and Western European authors as quartz monzonite. Originally, however, the term was used for monzonites with a small quartz content (field 8). Since it is still used this way by Soviet geologists and in still other ways by other geologists, the name is now somewhat ambiguous.

The alkali granites of field 2 are commonly hypersolvus (one-feldspar, crystallized above the solvus) granites containing a perthitic alkali feldspar. With soda amphibole or pyroxene, they are called soda granites. Trondhjemites of field 5 are leucocratic (color index less than 15) quartz diorites. They are quite common. Field 10 includes not only diorite and gabbro but also anorthosite—the leucocratic ($< 10\%$ mafic minerals), plagioclase rocks. The distinction between diorite and gabbro is based on several criteria in the following approximate order of reliability:

Criterion	Gabbro	Diorite
Plagioclase mineral composition (determined microscopically)	$An_{>50}$	$An_{<50}$
Associated rocks	Pyroxenites and anorthosites	Granodiorites and quartz diorites
Mafic minerals	Clinopyroxene or orthopyroxene,* $\pm$ olivine, $\pm$ hornblende	Hornblende or biotite, $\pm$ augite
Plagioclase color	Gray to greenish gray	White or nearly so

*Norites are gabbros in which predominant pyroxene is orthopyroxene instead of clinopyroxene.

A single criterion is not always sufficient to distinguish gabbro from diorite, and preferably two or more criteria should be used.

The *feldspathoidal rocks* of fields 6, 7, 8, 9, and 10 are named according to the feldspathoid contained; for example, nepheline syenite, sodalite syenite, nepheline monzonite. The feldspathoid-bearing gabbroic rocks, for which numerous names have been proposed, may be grouped under alkali gabbro. The rare plutonic rocks of field 15 are very high in feldspathoids (hence foidolites) and include mafic-rich rocks such as ijolite (nepheline-pyroxene rocks with 30 to 60% mafic minerals).

The *ultramafitites* (ultramafic rocks in which M > 90) are subdivided as follows (field 16):

90–100% olivine	Dunite
30–90% olivine	Peridotite (olivine + pyroxenes):
	Lherzolite (olivine + orthopyroxene and clinopyroxene in subequal amounts)
	Harzburgite (olivine + orthopyroxene)
	Kimberlite (olivine + pyroxene, melilite, biotite)
0–30% olivine	Pyroxenite (mostly pyroxenes)
	Hornblendite (mostly hornblende—not to be confused with amphibolite, which is a metamorphic rock)

It must be recognized that the rock names of this classification (or, for that matter, any other descriptive classification) are artificial and exist only for descriptive convenience. Although some kinds of rocks predominate and the names are intended to encompass high natural concentrations of similar rocks (see Fig. 2-2), all gradations exist between the different kinds.

Volcanic rocks are classified in exactly the same way. The parameters on the double triangle (Q–A–P–F) are the same; but, since the rocks are very fine grained to glassy and presumably originate in different ways, the rock names are different (Fig. 2-1b). Where the mineralogy of a rock is inexact or impossible to determine directly, a chemical analysis may be recalculated to a probable mode (mineral percentage). This is not to mean recalculation to a standard set such as of normative[1] minerals, but recalculation to a set of minerals that would have appeared in the rock had it crystallized to a coarser grain size. Such minerals are chosen by consideration of those that appear in coarser-grained equivalents or in associated rocks of the same association (see, for example, Jung and Brousse, 1959).

In the absence of a chemical analysis, a fine-grained volcanic rock must be named from those minerals that are recognizable. If these amount to only phenocrysts, it is advisable to prefix the resultant rock with "pheno-," for example, phenoandesite. Inasmuch as the indeterminent ground-

---

[1]Normative minerals are a set of artificial "minerals" obtained by recalculation (using a prescribed set of rules) from a chemical analysis of a rock (see p. 42).

mass could contain an abundance of quartz and potassium feldspar, the rock could actually be a dacite.

Figure 2-1b, like the corresponding plutonic rock classification, is largely self-explanatory. By contrast with the plutonic rocks, field 3 is divided into two parts—rhyodacite (quartz latite) and rhyolite—which together are mineralogically equivalent to granite. Alkali rhyolites (field 2) may be called soda rhyolites if they contain soda amphibole or soda pyroxene. The distinction between andesite and basalt, as with diorite and gabbro, has been based on different criteria by different workers. The following criteria are useful, in approximate order of reliability (note the change from the plutonic rocks):

Fig. 2-2 *Estimated mineralogical distribution of igneous rocks. After Streckeisen (1967, p. 236). Q = silica minerals (mainly quartz); A = alkali feldspar (including albite An_{0-5}); P = plagioclase An_{5-100}; F = feldspathoids; M = mafic minerals. (a) Plutonic rocks. (b) Volcanic rocks.*

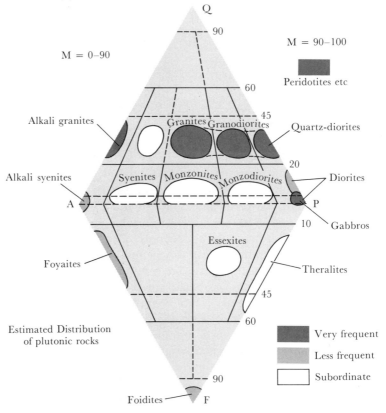

(a)

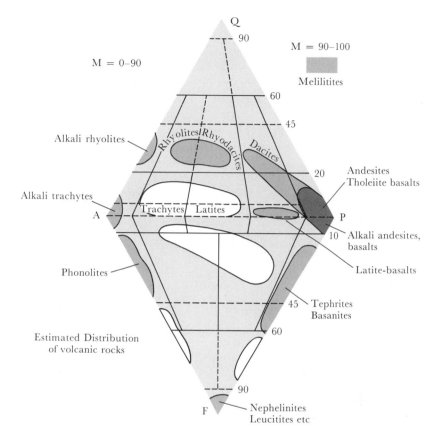

(b)

Criterion	Basalt	Andesite
Color index (% dark minerals may be reliably determined in thin section; corresponds to "black" basalts and intermediate-colored andesites in hand specimen)	>40	<40
Plagioclase mineral composition (determined microscopically, but crystals commonly zoned; phenocryst composition may be determined but groundmass composition commonly different and indeterminate)	$An_{>50}$	$An_{<50}$
Plagioclase composition in the norm (by chemical analysis)	$An_{>50}$	$An_{<50}$
Silica percentage of the rock (by chemical analysis)	<52	>52
Mafic minerals	Pyroxene (augite or hypersthene) or olivine	Hornblende or hypersthene, ± augite

Tholeiitic basalts are commonly low in (or lacking) olivine. They or their late differentiates, such as rhyolite, may contain a small amount of free quartz. Alkali basalts (and andesites), as the name implies, are higher in alkalis that appear as a small amount of nepheline or other feldspathoid, alkali feldspar, titan-augite, or possibly barkevikitic amphibole. They usually contain olivine but never contain primary quartz, even in the late differentiates such as trachyte or phonolite.

Although field 15c may appear self-explanatory, note that these rocks are named according to their feldspathoidal constituents only, even though pyroxene (generally titan-augite) commonly predominates. The approximate "normal" mafic mineral contents for each of the common igneous rock types is plotted by Streckeisen (1967, pp. 194–195).

Whenever a general term for dark, fine-grained volcanic rocks seems appropriate, "basaltic rocks" may be used. A general term for similarly abundant coarser-grained plutonic igneous rocks would be "granitic rocks." Less commonly used, varietal, or older rock names not used in the present classification of igneous rocks may be found in more detailed petrographic works such as Johannsen (1939) and Williams, Turner, and Gilbert (1958).

The following rocks do not fit naturally into the above classification:

Keratophyre—a sodic trachyte in which albite or oligoclase is the main constituent but which may also contain chlorite, epidote, and calcite. It is characterized by a highly altered appearance and is generally associated with spilites. Quartz-bearing varieties are called quartz keratophyre.

Spilite—a sodic basalt in which albite or oligoclase is the main feldspar but which also contains chlorite, epidote, or calcite, and sometimes phenocrysts of unaltered augite. It is characterized by a highly altered appearance—a green to gray, fine-grained fuzzy-looking rock, not dark gray or black like most fresh basalts.

Lamprophyre—a dark-colored dike rock containing euhedral mafic phenocrysts. Similar dark grains also occur in the groundmass, and light-colored minerals are confined to the groundmass. The mafic phenocrysts include biotite, amphibole, pyroxene, or olivine.

Serpentinite—a rock composed largely of serpentine. Olivine and pyroxene may occur as relict minerals.

Prominent varietal minerals that are not necessary to the name (e.g., biotite in a granite) are commonly used to indicate varieties of the rock name (e.g., biotite granite). Where the rock has a very distinctive texture, such as extremely coarse grained as in pegmatite or fragmental as in tuff, the mineralogical name is used to modify the textural name as in granite pegmatite or rhyolite tuff.

Textures also used as rock names include the following (the textures are described in more detail below):

Pegmatite—extremely coarse grained; granite to granodiorite compositions most common, but other compositions known

Aplite—fine, even, sugary grained, commonly lacking mafic (dark) minerals and associated with pegmatite; granitic mineralogy

Obsidian—volcanic glass; white (rare) to gray or black and maroon in color

Porphyry—coarse grains distinctly larger than those of remainder of the rock make up more than 50% of the rock

Pumice—frothy glass; generally with a very low specific gravity— may even float on water

Scoria—numerous vesicles (gas holes) in volcanic rock

Tuff—finely fragmental volcanic rock

Breccia—coarsely fragmental volcanic rock

The so-called *color index* of a rock is the same as its percentage of dark (ferromagnesian) minerals. This gives rise to several descriptive terms, most commonly used in the field for a generalized designation but also used to record specific percentages:

	% of Dark Minerals*	Color Index (C. I.)
Leucocratic†	<30	0–30 (a light-colored rock)
Mesotype	30–60	30–60 (an intermediate rock)
Melanocratic‡	>60	60–100 (a dark-colored rock)

*Dark minerals ("mafics") include mica, amphiboles, pyroxenes, and olivine; light minerals ("felsics") include quartz, feldspars, and feldspathoids.
†"Leuco-" is commonly used in rock names as a prefix to indicate abnormally light colored.
‡"Melano-" is used as a prefix to indicate abnormally dark colored.

Chemical Classifications

The chemical designation in most common use is that of *silica percentage*. Names, especially "acid" and "basic," applied to ranges of silica percentage, although still in common use, have fallen into some disfavor because of the implied correlation with solution chemistry (pH). Although originally intended, this correlation is erroneous; and such terms should no longer be used. Preferable terms are indicated:

% SiO_2	Old Designation	Preferred Designation	Example Rock
>66	Acid	Felsic	Granite, rhyolite
52–66	Intermediate	Intermediate	Diorite, andesite
45–52	Basic	Mafic	Gabbro, basalt
<45	Ultrabasic	Ultramafic	Dunite, peridotite

Comparison of these silica percentages with those of chemical analyses of igneous rocks gives some indication of the kind of rock involved (see, for example, Table 1-2, p. 12).

Other chemical designations now less commonly used include:

Norm Classification (CIPW Classification)

This classification is more widely used for volcanic rocks (Cross et al., 1902), in which the chemical analysis is recalculated to a standard set of "normative" minerals whose presence and percentage may be compared from one rock to another (for details, see Johannsen, 1939, p. 89, or Barth, 1962, p. 65). The only major advantages that arise from the calculation of such theoretical minerals are in the comparison of extremely fine-grained or glassy volcanic rocks with one another and with coarser-grained rocks in which the actual minerals are identifiable.

Niggli Values (More Widely Used in Europe)

As in the Norm classification, the chemical analysis is recalculated in a standard way. But instead of converting the analysis into a set of standard minerals, the oxides are grouped according to chemical similarity: Al_2O_3 + Cr_2O_3 + rare earths, FeO + MnO + MgO, CaO + BaO + SrO, Na_2O + K_2O + Li_2O. The groups of oxides are calculated according to cation percentage, the sum for the rock being recalculated to 100. Normally, these chemical numbers are plotted individually against the silica number for visual comparison with associated rocks (for details, see Barth, 1962, p. 63).

Alumina Saturation

The excess or deficiency of Al_2O_3 with respect to Na_2O + K_2O + CaO as compared with that in the feldspars is reflected in the kind of ferromagnesian minerals present. Where $Al_2O_3 > Na_2O + K_2O + CaO$ (peraluminous), the excess alumina appears in muscovite, biotite, corundum, topaz, or garnet. Where $Al_2O_3 > Na_2O + K_2O$ but $< Na_2O + K_2O + CaO$ (metaluminous), hornblende, epidote, melilite, or biotite + pyroxene appear. Where $Al_2O_3 \approx Na_2O + K_2O$ (subaluminous), alumina-poor minerals—olivine, orthopyroxene, or clinopyroxene—appear. Where $Al_2O_3 < Na_2O + K_2O$ (peralkaline), the excess alkali appears in soda pyroxenes and soda amphiboles (for details, see Shand, 1943, p. 189).

Alkali-lime Index

An excellent means of comparison of the alkalinity of different igneous rock series has been devised by Peacock (1931). He plots a graph of CaO vs. SiO_2 and one of $Na_2O + K_2O$ vs. SiO_2 for a related series of igneous rocks. Since CaO commonly decreases and $Na_2O + K_2O$ increases with respect to SiO_2, the two curves cross. Curves for more alkaline rocks intersect at lower silica contents, so that the silica contents are used to subdivide rock series as follows:

Type of Rock	% SiO_2	Illustrative Rock Series*
Calcic	>61	Basalt, andesite, rhyolite series
Calc-alkalic	56–61	Tholeiitic basalt; basalt, andesite, rhyolite series
Alkali-calcic	51–56	Alkaline olivine basalt, phonolite series
Alkalic	<51	Alkali syenite complexes

*Igneous rock associations are listed and explained on the following pages.

Geological Occurrence and Association

The geological occurrence and association of igneous rocks form one of the few classifications that does not suffer materially from the somewhat arbitrary nature of the foregoing laboratory-oriented classifications. The rocks are grouped according to their association in nature, in both space and time. Rocks that characteristically occur together and presumably evolve together, perhaps from one another, are described together. Such a classification is conducive to the understanding of the origins of and relationships between rocks because the relationships between spatially and in most cases genetically related rocks are emphasized.

A single rock association generally has chemical and mineralogical characteristics that are either nearly constant or vary regularly from one member of the series to the next. For example, the gabbroic layered intrusions generally range from peridotite at the base, up through gabbro and anorthosite, to a granophyre or quartz ferrogabbro toward the top; there is a great diversity in chemical and mineralogical composition, but all are part of the same body and exhibit a striking gradation in characteristics from bottom to top. Somewhat differently, the alkaline olivine basalts develop extensive flows with nearly fixed chemical and mineralogical characteristics and usually occur with small amounts of trachyte or phonolite that develop in the late stages.

The plutonic and volcanic associations outlined in Table 2-1 are modified from Tyrrell (1929) and Turner and Verhoogen (1960). They have been rearranged here according to the time of their development in an orogenic cycle. As noted in Chap. 1, the phase of "geosynclinal" sedimentation is generally followed by deformation of the "geosynclinal" pile

Table 2-1 *Igneous rock associations*

Suites	Volcanic Associations	Plutonic Associations
Preorogenic and early orogenic (ophiolites)	Spilite, keratophyre sequences: Spilite, keratophyre (with graywacke, arkosic sandstone, shale, ± bedded chert)	Peridotite-serpentine bodies: Peridotite, serpentinite, ± gabbro, ± troctolite (olivine-plagioclase rock)
Synorogenic	Basalt, andesite, rhyolite flows and tuffs: Basalt, pyroxene andesite, andesite, rhyodacite, dacite and rhyolite welded tuffs	Granitic batholiths: Granodiorite, quartz diorite, granite, ± gabbro, granitic pegmatite and aplite Hornblende granite, ± leucogranite, granitic pegmatite and aplite Augite syenite, quartz soda-syenite, ± granite, granitic pegmatite and aplite Anorthosite plutons: Anorthosite, hypersthene gabbro, pyroxene syenite, hypersthene granite (charnockite)
Postorogenic	Alkaline basalt flows and dikes: Alkaline olivine basalt, nepheline or melilite basalt, ± trachyte, ± phonolite Tholeiitic basalt flows: Tholeiitic basalt, ± andesite, ± rhyolite, ± trachyte Quartz diabase sills and dikes: Quartz diabase, ± granophyre, ± olivine diabase Lamprophyre dikes: Lamprophyres with granitic intrusions Potassium-rich basalts: Leucite basalt, potash trachybasalt, trachyandesite, trachyte	Granite stocks, ring dikes: Granite, ± quartz syenite, ± syenite, granitic pegmatite and aplite Alkali syenite plutons, ring complexes: Nepheline syenite, syenite, granite, ± gabbro, ± ijolite (nepheline-pyroxene rock) Carbonatite, mafic alkaline rock plutons, and ring complexes: Carbonatite ijolite, alkali gabbro, alkali syenite Gabbroic layered intrusions: Peridotite, olivine gabbro, gabbro, ferrogabbro, anorthosite, granophyre

(orogeny) and building of a mountain range. Accordingly, the tectonic aspects of "geosynclinal" evolution may be divided into three phases: (1) preorogenic and early orogenic, (2) orogenic, and (3) epeirogenic (uplift or postorogenic).

Textures of Igneous Rocks

The textures of igneous rocks, that is, the degree of crystallization, grain size, shape, and mutual relationships of the grains, provide direct evidence as to the origin of the rock and always form a part of the rock description.

The *crystallinity* of the rock, or degree of crystallization, is the amount of glass versus crystals and is governed largely by the rate of cooling, composition, and viscosity of the magma. Tiny crystals in the glass are called microlites. All volcanic glasses tend in time, because of their inherent instability, to spontaneously crystallize in the solid state, or to *devitrify* (see Fig. 2-3). The result is generally a mass of very fine radiating or unoriented crystals. Why some magmas solidify as glass, rather than as crystals, depends on several factors, some of which also control grain size, as noted below. Glassy rocks are found only where they have solidified at or near

Fig. 2-3 *Obsidian with layers of numerous white spherulites (radiating structures composed of fibers of feldspar and silica minerals—probably produced by devitrification of the volcanic glass). A few larger lithophysae appear in upper left (concentric structures). Nickel (white circle) indicates scale. Obsidian Cliff, Yellowstone Park.*

the earth's surface or at the margins of near-surface dikes where they would be expected to cool much more rapidly than rocks solidifying at depth or within the dikes. Extremely rapid cooling would thus be expected to be a significant factor in the formation of volcanic glass. Confirmation of this expectation derives from experimental crystallization of rock melts, as we have already seen. Slow cooling below the initial temperature of crystallization permits crystal growth, but rapid chilling of the melt produces only glass. Similarly, the viscosity of rhyolitic, which is higher than that of basaltic magmas, inhibits movement of ions to crystallization sites, thereby inhibiting the formation of crystals.

The *granularity,* or grain size, of a rock is governed by the rate of cooling, composition and viscosity of the magma, number of crystal nuclei, and movement of the magma. Rocks in which the grains are too small to be individually visible are called *aphanitic.* Coarser-grained *phaneritic* rocks may be further subdivided on the basis of grain size:

Fine grained: <1 mm to aphanitic

Medium grained: 1 to 5 mm

Coarse grained: >5 mm

Coarse grains are favored by (1) slow cooling, to permit time for additional ions to become attached to the growing crystals; (2) low magma viscosity, to permit more rapid migration of ions toward the crystals; and slow (i.e., difficult) nucleation, or a small number of crystal nuclei, to permit the few crystals to grow larger before adjacent grains begin to interfere with one another.

The fine grains of a basalt, for example, are formed by the rapid nucleation and rapid crystallization produced by rapid cooling at the surface, counteracted by the low viscosity of the magma. The glassy texture of many rhyolites is produced by more rapid cooling aided by fragmentation of the more viscous siliceous magma by release of the higher gas content on extrusion. The development of a glassy texture is aided by the higher degree of polymerization (linkage of bonds in the magma before crystallization) of more siliceous magmas, which would contribute to the ease and therefore the rate of crystallization. Slower ion migration due to the higher viscosity of more siliceous magmas may be a counteracting effect.

Turning to a comparison of grain sizes of igneous rock emplaced at the same depth, intrusion of magmas of basaltic composition results in generally coarse-grained rocks (gabbros), whereas intrusion of those of rhyolitic composition results in somewhat finer-grained (generally medium-grained) rocks. With a similar environment of crystallization, the most obvious difference between the two magmas is the higher viscosity of the more siliceous granitic magma, which would inhibit the movement of ions and therefore also the growth of crystals. The relative rates of nucleation are uncertain.[1] It may be, as suggested above, that the greater degree of polymerization (more linkages between silica tetrahedra to form chains, sheets, and frameworks) with the more siliceous granitic magma represents a major step toward nucleation. Resultant more rapid nucleation would produce more, and therefore smaller, grains.

The extremely coarse grains of granitic pegmatites are apparently dependent on the very high volatile content developed in the late stages of crystallization of the magma. The water and other volatiles appear to have two significant effects that lead to coarser grains. They inhibit nucleation by weakening the bonds between silica tetrahedra, and they increase the rate of growth by drastically lowering the viscosity and permitting more rapid movement of ions. Thus few nuclei are formed and these grow to few large crystals.

It might be well at this point to note that the great bulk of all igneous rocks correspond to two contrasting types. The felsic igneous rocks occur mainly as the medium- to coarse-grained granitic rocks that make up the deep-seated great batholiths. The mafic igneous rocks occur mainly as the very fine-grained basalts crystallized at the surface. There are good reasons why felsic magmas seldom reach the earth's surface but mafic magmas almost always do so.

We might expect, at the outset, either that the physical properties of basalt magma are more conducive to its upward movement or that the change in conditions between subsurface and surface is more likely to arrest the upward movement of granitic magma. As noted above, the viscosity of basaltic magma is less than that of a granitic magma. Other things being equal, this would permit basalt to move upward more easily. Perhaps a more important effect is related to the origin of the two types of magma. It was also noted above that basalt magma probably originates in the earth's mantle by partial melting under nearly dry conditions (see curve 3, Fig. 2-4). Granite, on the other hand, probably originates in the crust by partial melting of wet "geosynclinal" sediments and volcanic rocks (see curve 1, Fig. 2-4).

Consider now the fate of granite melt (g) and basalt melt (b) slightly above their respective melting points, as each rises rapidly to lower pressures toward the earth's surface. The granite melt soon crosses its melting curve, resulting in crystallization well below the surface. The basalt melt moves even farther from its melting curve, thereby remaining molten and

[1]Nucleation is fastest with a greater degree of cooling below the crystallization temperature and with the smallest change in entropy S (Fyfe, Turner, and Verhoogen, 1958, pp. 72–73).

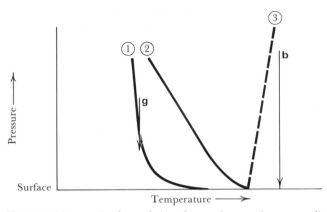

Fig. 2-4 *Diagramatic phase relations for granite assuming wet conditions (water saturation) and for basalt assuming wet (2) and dry (3)
conditions.*

permitting it to rise all the way to the surface. Slight cooling due to expansion with decreasing pressure[1] does not affect the argument.

Crystal shapes refer to the degree of development of crystal faces and
relative dimensions of the crystals. Crystals having well-developed crystal
faces are referred to as *euhedral* (e.g., in lamprophyres). Those with partly
developed crystal faces are *subhedral* (as in most igneous rocks). Crystals
lacking crystal faces are *anhedral* (e.g., in aplite). Equidimensional minerals
are referred to as *equant;* elongate or needle-shaped ones as *prismatic;*
platy ones as *tabular.*

Some minerals characteristically show euhedral grains, others anhedral, and still others euhedral in some rocks and subhedral or anhedral
in other rocks. Two factors seem to be important. The first is the order
of crystallization. An early formed magmatic mineral surrounded only
by magma generally has well-developed faces, as seen in phenocrysts
in aphanitic or glassy lavas. Late-formed minerals can grow only in the
spaces left between the earlier minerals. Their boundaries are dictated by
the boundaries of the space rather than by the geometry of the crystal
lattice. The property of some minerals that permits them to characteristically show euhedral grains regardless of the order of crystallization, perhaps more important in metamorphic rocks, has been called the *force of
crystallization.* It depends on the surface energy of the mineral relative to
those minerals that surround it, just as the surface tension of a droplet
of liquid governs whether or not that droplet will rest as a nearly spherical
bubble or will spread out on the surface. Some metamorphic minerals such
as garnet and sillimanite tend to form euhedral grains, whereas others
such as quartz and orthoclase are almost always anhedral.

[1] Adiabatic cooling: expansion without loss of heat.

The mutual relationships of grains give rise to the textural names, some of which are used as rock names or parts of names as noted above. Rocks that have about equal-sized essential minerals (small accessory minerals are ignored) are called *equigranular*, whereas those that have large crystals (phenocrysts) surrounded by smaller crystals (groundmass) are called *porphyritic*. If the phenocrysts are surrounded by glass, the rock has a *vitrophyric* texture and is called vitrophyre. In some rocks, especially granitic rocks of granite or granodiorite composition, some doubt may arise as to whether the large crystals are phenocrysts (crystallized from the magma) or porphyroblasts (crystallized in the solid state by replacement of the surrounding minerals). In this instance, the term *megacrysts* ("large crystals") may be used (see Fig. 2-5). Where phenocrysts in a porphyry bunch together to form aggregates, the term *glomeroporphyritic* is used.

Porphyritic textures, in general, result from two stages of cooling of the magma—slow cooling well below the surface to form larger crystals (phenocrysts), followed by more rapid cooling on intrusion to shallower levels or extrusion (groundmass).

Some megacrysts (especially potassium feldspars in certain granitic rocks) contain numerous inclusions of small crystals like those in the groundmass, resulting in a *poikilitic* texture. In some instances, these crystals seem to form by late magmatic growth of the megacryst and re-placement of the surrounding grains (e.g., Hyndman, 1968b, pp. 57–60, 71–72). In others, the inclusions may have become incorporated in a phenocryst growing in the magma (Hibbard, 1965; Bateman and Wahr-haftig, 1966).

Preferred orientation of tabular or prismatic mineral grains may re-sult from flow in the magma during crystallization or from later deforma-tion. Such preferred orientation (e.g., of biotite) that results in a planar structure in the rock is called *foliation*. Preferred orientation (e.g., of horn-blende) that results in a linear structure is called *lineation*. A foliation re-sulting from a parallelism of numerous tabular potassium feldspars (e.g., in a syenite) is referred to as a *trachytoid* texture. Any rock that lacks pre-ferred orientation (most igneous rocks) is said to be massive. A special type of preferred orientation that occurs in some pegmatites and related rocks results in a geometrical or cuneiform intergrowth (coarser than 1 mm) between quartz and alkali feldspar. This *graphic* texture consists of parallel triangular or polyhedral prisms of quartz in the alkali feldspar. Its origin is uncertain, but arguments have been proposed in favor of origin by replacement and by crystallization at a eutectic (see, e.g., Barker, 1967, 1970).

Distinctive textures are also formed by the presence of cavities in igneous rocks. Gas holes (vesicles) in volcanic rocks are most commonly (not invariably) nearly spherical like bubbles in water, unless flattened by

(a)

(b)

Fig. 2-5 *Euhedral megacrysts of potassium feldspar in biotite granite of Mount Powell batholith, Flint Creek Range, western Montana. (a) Fresh surface: Bowman Lakes cirques. (b) Megacrysts weathering out: Racetrack Creek.*

flow. The resulting texture is called *vesicular*. If the vesicles are so numerous as to dominate the rock, the texture is called *scoriaceous*. If the vesicles in a rock are filled by later minerals (e.g., zeolite, chalcedony), they are called *amygdules,* and the texture is *amygdaloidal.* Gas holes in plutonic igneous rocks are almost always angular, being bounded by the angular faces of crystals (e.g., quartz and feldspars) in the rock. Textures so characterized are *miarolitic,* as in some shallow granitic stocks.

Volcanic igneous rocks are often characterized by fragmental textures that fall under the general designation of *pyroclastics.* Pyroclastic materials result from gas-driven explosive volcanic activity. There is a complete range of fragment sizes from volcanic dust (<0.0625 mm), through ash (0.0625 to 4 mm), lapilli (0.4 to 3.2 cm), and bombs (3.2 to 25.6 cm), to blocks (>25.6 cm or 10 in.). The finer three categories form the rock *tuff.* The larger two categories form the rock *agglomerate* (see Fig. 2-6). Some geologists use this latter term to designate pyroclastic materials deposited in and near a volcanic vent. Such a genetic definition, however, is unreliable and is difficult to apply because, more often than not, the location of the vent remains uncertain. The term "vent agglomerate" has been used to avoid any ambiguity of meaning. Parsons (1969) has reviewed the characteristics and recognition of the various types of volcanic breccias.

Fig. 2-6 *Heterolithologic agglomerate. Lacks sorting and bedding. Hammer handle is 15 in. long. West side of Highway 89, 34 miles south of Livingston, Montana.*

Fig. 2-7 *Roof of Mount Powell batholith. Contact between quartz monzonite of batholith and overlying intruded Paleozoic sedimentary rocks is shown by black line. Bedding in sedimentary rocks dips to left (north) and away from observer. Pegmatite-aplite dike swarm is visible below low point in ridge—presumably derived from batholith. Distance to ridge crest in center of photo is about 1 1/2 miles. Exposed in cirque on northeast side of Mount Powell, Flint Creek Range, Montana.*

OCCURRENCE

Igneous bodies take several different forms, depending on many factors, such as the composition and viscosity of the magma, its volume and rate of intrusion, and the structure and composition of the intruded rocks. Granitic magmas are generally emplaced as large masses (plutons) that crystallize below the surface. Some of these masses reach immense surface dimensions of hundreds of miles in length along "geosynclinal" belts and tens of miles in width, such as in the Sierra Nevada Range of California and in the Coast Mountains of British Columbia. Such large bodies that have a surface area of more than 40 sq miles are called *batholiths* (see Fig. 2-7). Those with surface area less than 40 miles are called *stocks*. Extension of these plutons to depth is problematic, for few areas have enough vertical exposure to permit a reliable extrapolation of the contacts to depth. Classically, batholiths were considered to be "bottomless," apparently because they supposedly widened to depth. As much as 40 to 50 years ago, however, some evidence began to come to light to indicate that batholiths do bottom at relatively shallow depths (J. P. Iddings, 1914; H. Cloos, 1923). They may be, in fact, much like gigantic sills, as suggested by recent seismic and gravity data (Thompson and Talwani, 1964; Hamilton and Myers, 1967; Biehler and Bonini, 1969). Stocks may also be sill-like or lenslike (Hyndman, 1968b). More extensive discussion of granitic plutons is reserved for the granitic batholith section of Chap. 4.

Laccoliths are igneous plutons formed when magma that has risen through rocks in the crust until reaching a more resistant layer or a weaker zone between layers spreads laterally, bulging up the overlying strata to form a mushroom-shaped mass (Fig. 2-8). Well-known laccoliths are those of the Henry Mountains, Utah (Gilbert, 1880) and the Shonkin Sag laccolith of central Montana (Barksdale, 1937; Hurlbut, 1939). Most laccoliths are only a few miles across.

Gabbroic plutons are less common but do occur as fairly small to large masses. *Lopoliths* are very large, commonly plate-shaped masses (Fig. 2-9) as much as tens of thousands of feet thick and more than 100 miles across. They generally form sedimentlike layers in which different minerals are concentrated. Many lines of evidence suggest that crystal settling played an important part during crystallization of the lopolith. Similar gabbroic magmas in other instances form smaller cone-shaped (Fig. 2-9) or large dikelike intrusions, otherwise similar to lopoliths. Well-known mafic layered intrusions are the Stillwater complex of Montana (Howland et al., 1936; Hess, 1960a) and the Skaergaard intrusion of Greenland (Wager and Deer, 1939).

Smaller tabular igneous bodies that measure an inch to thousands of feet across are generally aphanitic to medium grained, if of simatic composition, and fine grained to very coarse grained (pegmatitic), if of sialic composition. If the magma is emplaced along a fracture across (discordant to) the bedding or foliation of the rock, a *dike* is formed. If it follows along (concordant to) the bedding or foliation, a *sill* is formed. Dikes commonly, but not invariably, occur in parallel sets as in dike swarms, where the magma has injected a fracture zone.

Other dikes are emplaced along cone-shaped fractures associated with the intrusion of small, near-surface igneous plugs. When filled with magma, downward-tapering cone-shaped fractures formed by upward forces form *cone sheets*. Upward-tapering domal fractures formed by release of pressure or by settling of the magma form *ring dikes*. Anderson (1942) discusses the formation of these features, which he first explained some 30 years earlier. The best-known examples in North America are associated with the White Mountain Magma Series of New Hampshire (Chapman and Chapman, 1940; Billings, 1943).

The fractures or joints, into which these dikes are intruded, are in

Fig. 2-8 *Diagramatic form of a laccolith.*

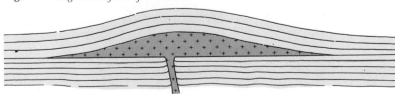

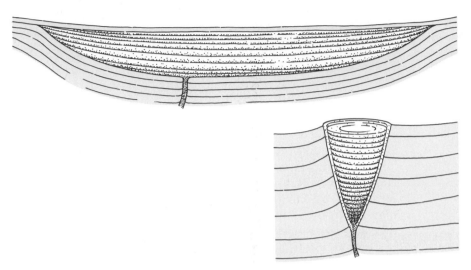

Fig. 2-9   *Diagramatic plate- and cone-shaped forms of mafic layered intrusions.*

both cases tensional features caused by movement of the magma. Most other joints are also tensional features, some formed by movement of magma, others by cooling of magma, and still others by release of lithostatic pressure by erosion.

Rapid cooling of basaltic flows or shallow sills normally results in formation of *columnar joints* (see Figs. 2-10a, b, and c)—distinctive polygonal columns oriented with their long axis perpendicular to the cooling surfaces. These most commonly five- or six-sided columns are thought to be tensional joints formed by shrinkage stresses relieved at nearly regular intervals along the most prominent cooling surface, extending inward as

Fig. 2-10a   *Well-developed columnar jointing in Columbia River basalt. Most columns are six-sided. Road cut is 10 to 15 ft high. On Highway 2 about 1 mile west of Spokane, Washington.*

Fig. 2-10b *Columnar jointing in 50-ft-thick Upper Cenozoic basalt flow. Flow is underlain by volcanic conglomerates of Cathedral Cliffs formation. East wall of Yellowstone gorge between Tower Junction and Tower Falls, Yellowstone Park.*

Fig. 2-10c *Base of same flow on west side of gorge. Note baking of underlying conglomerate and alteration of base of flow by heated groundwater. Hammer handle in lower center is 15 in. long.*

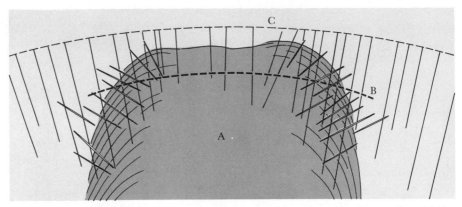

Fig. 2-11   *Ideal superposition of flow structure types and fracture systems in an igneous pluton. "Oldest structure is a dome of flow layers (A); next younger are marginal fissures, with or without upthrusts, frequently accompanied by dikes (double lines). The position of an arch of flow lines is indicated by the line of heavy dashes (B); it may transgress somewhat the main contact of the massif. Cross joints may be arranged perpendicular to this arch or may be referable to a still larger (imaginary) arch (C), which indicates harmonious arching of a considerable portion of the wall rocks." After Balk (1937, p. 103).*

the body cools. The size of the columns may be related to the rate of cooling —more rapid cooling giving rise to more rapid release of stresses built up and to formation of narrower columns. Thick flows may show two tiers of columns, those in the lower tier being more regular in shape, five- or six-sided, and larger in diameter than those in the upper tier, which tend to be orthogonal, more curved, and smaller (Spry, 1962). Recent studies of columnar-joint formation in Hawaiian lava lakes (Peck and Minakami, 1968) show that the joints begin forming within minutes after appearance of fresh lava when the crust is only 1/4 in. thick and still glowing red (900°C). Cracks propagate to depth at temperatures up to about 1000°C, essentially at or slightly above the temperature of the solidus.

Joints also form in bodies of plutonic igneous rock (Balk, 1937), such as granitic stocks, where they normally intersect in two or more sets. Here they are less easily interpreted, especially if in a very large body such as a batholith. In some instances, they seem related to upward pressures by or withdrawal of subjacent, not completely crystallized, magma. In others, they may be caused by tectonic stresses operating long after crystallization of the body. Joint sets, especially gently dipping ones, nearly parallel to a major topographic surface, may even be caused by release of lithostatic pressure by erosion of overlying rocks. A somewhat idealized representation of tectonically produced joints in an igneous pluton is shown in Fig. 2-11.

Other types of occurrence of igneous rocks are related to volcanism or surface eruption of magmas. Most basalts are erupted from long fissures as great floods of lava that spread out in all directions, filling areas topo-graphically lower than the sites of eruption. These basalt *flows* commonly give rise to nearly flat basalt "plateaus" (e.g., the Columbia River Plateau covering much of eastern Washington, Oregon, and western Idaho) that cover many thousands of square miles. Other flows occur on the gently sloping flanks of huge basaltic *shield volcanoes* (such as Mauna Loa and Kilauea in Hawaii) or on the more steeply sloping flanks of *composite* or *stratovolcanoes* of basalt and andesite (such as Mount Rainier in Washington and Mount Shasta in California). These latter composite volcanic cones also contain large quantities of pyroclastic material—tuffs and breccias—interbedded with the flows. Magma that erupts from these steeper-sided volcanoes most commonly comes out of a more confined vent or breach in a crater at or near the top of the volcano, though alignment of cones in some areas such as Nicaragua (McBirney and Williams, 1965) suggests a fissure control over the site of the eruptions.

    Pillow lavas are accumulations of pillow-shaped pods of lava, probably formed when basaltic lava flows into the ocean (e.g., Moore, 1970) or other body of water or into wet sediments (see Figs. 2-12a and b). Droplike fingers

Fig. 2-12a *Pillow lavas in the Franciscan formation, California. Pillow-lava tubes flowed downward toward observer. Man on right gives scale. In quarry 9 miles north of Ukiah, northern California.*

Fig. 2-12b   *Cross section of small pillow in spilitic basalt, same locality. Note chilled rim and wedges of fissile muddy matrix above and to right of pillow.*

of magma ooze out over the crests of their predecessors, tending to sink into the spaces between the underlying pillows. The finer-grained, smoothly undulose, crack-free surfaces of the pillows suggest chilling of the basalt against cold water or wet mud to form a tough, elastic skin that maintains the droplike form (Snyder and Fraser, 1963; Yagi, 1969). Concentric compositional variations in each pod imply some contamination of the magma by the intruded material. The downward-pointing rounded protrusions permit identification of the original tops of the flows, even where later overturned. The matrix between the pillows consists of an unsorted greenish breccia of volcanic debris and mud (Fig. 2-12b). The number and size of vesicles in pillow lavas may give an indication of the depth of water at the site of eruption (Moore, 1965; Jones, 1969). Vesicles range from 1 to 2 mm in diameter at water depths of 0.15 to 0.4 km, to smaller diameters at shallower or deeper depths, to about 0.5 mm at 1 km, and disappearance below about 5 km.

Features produced in lavas depend on many factors, including chemical composition, gas content, rate of cooling, temperature, slope of the ground, and characteristics of the vent. If the gas content of a lava is lost explosively, the result is a fragmental deposit (Fig. 2-13), but if it is lost more slowly, it may bubble out like bubbles in a carbonated drink. As the lava cools, becomes more viscous, and finally freezes, these gas bubbles are trapped in the form of *vesicles,* often producing a scoria near the upper or lower surface where they are trapped by chilling from the outer surface (Fig. 2-14).

Layering and lamination may be produced by flow in a congealing

Fig. 2-13 *Crater in top of Pacaya Volcano, Guatemala. Crater (about 25 m across) was formed by explosive eruptions such as this, November 26, 1967. Note breach in far wall of crater, through which lava flows intermittently. The scoria is largely olivine basalt.*

Fig. 2-14 *Base of olivine basalt flow. Note chilled, finer-grained base and stretched vesicles (elongate parallel base) concentrated above this chilled base. Hammer handle, below base, is 15 in. long. In quarry on Highway 89, 25 miles south of Livingston, Montana.*

lava, the laminae being outlined by stretched and flattened vesicles. Tiny newly formed crystals or microlites are also aligned parallel to the direction of flow. Smoothly flowing fluid lava, most viscous at the smooth cooler surface, is commonly dragged into striking "ropy" forms, called *pahoehoe* lava in Hawaii. Further movement of the rigid solidified scoriaceous crust of the lava or of more viscous lava breaks the crust into slabs or blocks called *aa* lava (Macdonald, 1953). The internal parts of fluid (low-viscosity) lava flows are commonly crystalline though fine grained. The outer parts are usually in part glass. Flows of pahoehoe are generally about 10 ft thick and of aa about 25 ft; the more viscous blocky flows range up to more than 100 ft thick (Thornton, 1964). Many shield-volcano flows thicken in the direction of flow because of increase in viscosity (Stearns and Macdonald, 1946). Those of plateau flood basalts may do the opposite, possibly because of flow ponding (White, 1960; Lefebvre, 1970).

Further information on structures on and within lava flows and their use in determining the direction of flow may be found in Macdonald (1953), Wentworth and Macdonald (1953), Waters (1960), and Lefebvre (1970).

Fragmental deposits or pyroclastics are produced when the gas content of a magma is lost explosively, as in most felsic magmas. They consist of both glassy or crystal-bearing disintegrated magma and fragments of the volcanic vent and shallow part of the conduit. The finest products are volcanic ash, which, if glassy, commonly consists of *shards*, the fine three- or four-pointed splinters bounded by concave surfaces and formed by comminution of pumice during eruption. This fine ash is often carried far downwind from the volcano, the coarser fragments being deposited in and around the vent. The resulting tuff is generally bedded and at least somewhat sorted, often showing graded bedding, except in the immediate (say, a few miles) vicinity of a volcanic vent (Ross and Smith, 1960). *Pumice* is formed by frothing of the gas-rich magma shortly before or at the time of ejection from the vent. Aggregation of ash to form small spherical pisoliths that fall like hail is considered sure evidence of a subaerial eruption. Crystal-rich pyroclastic materials are commonly called *crystal tuffs*.

Less vesicular cinderlike masses of *scoria*, usually a few inches or less in size, are much darker, heavier, and more irregular than pumice. They are formed from more fluid, mafic magmas poorer in volatiles. Near the vent, the still-plastic scoria may be flattened and welded together where they fall, often producing spatter cones (Rittmann, 1962, p. 77).

Ejection of masses of fluid lava sometimes results in rounded, twisted, or spindle-shaped volcanic *bombs* with streaky ridged surfaces formed by spinning through the air (see Fig. 2-15). More viscous lavas tend to form bombs with deeply cracked surfaces like that on a loaf of French bread; hence the name breadcrust bombs. Fragments of solidified lavas or other rocks may also be ejected in the form of millimeter- to centimeter-size *lapilli* or larger *blocks*.

Fig. 2-15 *Volcanic bomb 12 in. long in thick-bedded agglomerate. Note twisted shape, longitudinal grooves in sides, and vesicular nature. West side of Highway 89, 34 miles south of Livingston, Montana.*

Fig. 2-16 *Consolidated lahar—volcanic mudflow. Note bleached rim of many of the boulders (e.g., center of upper left quadrant). Hammer handle is 12 in. long. North rim of Lake Atitlán caldera, Guatemala.*

Fig. 2-17 *Dacite flow breccia. Monolithologic breccia formed by fragmentation of solid-ified crust of flow and cementation by the still-molten lava. The larger dark seams were formed by later brecciation and cementation by later lava. Knife is 3 in. (7½ cm) long. West side of Highway 89, 33 miles south of Livingston, Montana.*

Deposits from volcanic *mudflows* (see Fig. 2-16) tend to be very poorly sorted debris with a wide range of grain sizes up to the coarsest of blocks even far from the source area. They are generally confined to topographically low areas. Compare also flow breccia (Fig. 2-17). Criteria for recognition of different types of volcanic breccias are reviewed by Parsons (1969).

Deposits from ash flows (sometimes called glowing clouds or glowing avalanches) also tend to lack sorting and bedding (Ross and Smith, 1960). Whether the rapid flow of the particles is due to the buoyant effect of gasses given off by each particle or entrapment and expansion of air is not clear (e.g., Perret, 1935; McTaggart, 1960, 1962; Gibson, 1970). These *ash-flow tuffs* or *ignimbrites*, as they are sometimes called, commonly show a gradation from a loose, relatively unconsolidated tuff at the top of the sheet (Fig. 2-18) to a thoroughly welded, massive, jointed, lavalike rock toward the base (Gilbert, 1938; Gibson, 1970), and often a thinner unwelded tuff at the base (see Fig. 2-19a and b). These rhyolitic deposits generally cover moderate to large areas where they level off the underlying topography. As seen under the microscope, they are characterized by well-preserved shards and pumice fragments at the top, grading to extremely flattened

Fig. 2-18 *Unconsolidated ash-flow tuff. Note lack of sorting, layering, or welding. Hammer handle is 12 in. long. Quarry face 13 km south of Pologuá, Guatemala.*

(a)   (b)

Fig. 2-19   *(a) Lower welded portion of ash-flow tuff. Note good jointing and numerous flattened pumice streaks increasing downward in strongly welded rocks in the upper half of photo. Note gradation to poor jointing and few pumice streaks in more poorly welded basal part of ash flow (bottom half of photo). Road cut on Pan-American Highway, 5 km south of Huehuetenango, Guatemala. (b) Closeup of lower welded part of same flow showing flattened lenses of darker pumice. Hammer head is 6½ in. (16 cm) long.*

and streaked-out fragments often squashed around rigid crystals near the base. The welding of these rhyolitic deposits is caused by the high temperature of the tuff on deposition, in conjunction with the weight of the overlying material. They are characterized by absence of bedding (within individual flows) and presence of columnar jointing and densely welded textures in the lower part of sections a few hundred feet thick (e.g., Fenner, 1923; Boyd, 1961). The transition between the soft unwelded upper part and the hard, jointed, welded lower part is commonly gradational but only over a very narrow interval, giving the appearance of a contact between two separate layers (Ross and Smith, 1960). Colors of ash-flow tuffs are variable, but shades of light gray to brown are most common.

Alteration of volcanic rocks, new or old, sometimes produces areas in which the rocks are intensely altered to a dark olive green to pale yellow, brick red, or white claylike material. These zones tend to have poorly defined boundaries, grading from relatively little altered greenish "propylitized" rocks to intensely altered, light-colored zones consisting of kaolinite, opal, alunite, and hydrates of ferric iron. Even after the most intense alteration, the original textures commonly survive. Such an effect is commonly attributed to the "solfataric" action of sulphuric-acid-bearing hot springs in the dormant or dying stages of volcanic activity. Areas of alteration up to several square miles in extent, sometimes concentrated along fault zones, are apparently supplied by surface waters heated by magmatic steam.

CHAPTER 3
CHEMICAL
BEHAVIOR
OF IGNEOUS
ROCKS

PHASE EQUILIBRIA AND COMPOSITION

To this point, consideration has been limited to the crystallization behavior of a few simple two-component systems, as in the alkali and plagioclase feldspars. Rocks, however, are commonly composed of three or more
important components. Data for several ternary systems have been worked
out experimentally, mainly at 1 atm (1 bar) pressure but in some cases at
high water pressures. As examples, we shall consider a few ternary systems
that pertain most closely to the two most important groups of igneous
rocks—granitic rocks and basaltic rocks.

Granitic Systems: Quartz and Alkali Feldspars

Crystallization in the ternary system silica (SiO_2)–albite $(NaAlSi_3O_8)$–
potassium feldspar $(KAlSi_3O_8)$ (see Fig. 3-1) and its extensions to the silica-
deficient minerals nepheline $(NaAlSiO_4)$ and kalsilite $(KAlSiO_4)$ is most
easily visualized by first considering the binary systems (for example, SiO_2–
$NaAlSi_3O_8$) and then modifying the binary system behavior for the ternary
system. In these diagrams, the binary temperature-composition sides of
the ternary system have been rotated outward so that the binary behavior

can be followed more easily. For example, consider crystallization of a melt of composition M:*

> On cooling, the system remains the same (all melt) until the liquidus is reached (compare M′ on the binary diagram).

> On reaching the liquidus, the first crystals to form are silica.

> As silica (e.g., tridymite) is removed from the melt, the remaining melt moves compositionally away from SiO_2 down the liquidus until the cotectic trough is reached at M_1 (compare M_1 at the binary eutectic). At this point, the alkali feldspar component of the liquid has the composition of M_1 as seen on the alkali feldspar binary system.

*The example chosen here begins crystallization within the silica field for ease of presentation. As will become apparent below, however, most granitic rocks appear to begin crystallization with the feldspars. The principles of handling the diagram are the same.

Fig. 3-1   *The ternary system* $NaAlSiO_4$–$KAlSiO_4$–SiO_2, *at 1 bar. Modified from Schairer (1950, p. 514). Copyright 1950 by the University of Chicago.*

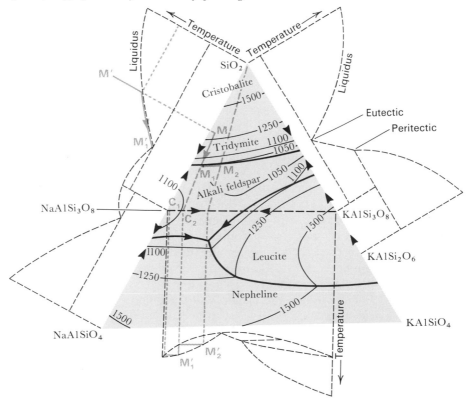

With continued cooling, silica and alkali feldspar of composition C_1, nearly pure albite (compare first crystals—on solidus—to form from liquid of composition M_1' —on liquidus), crystallize simultaneously.

As crystallization proceeds, the melt and the alkali feldspar crystallizing from it both become richer in potassium.

Finally, the melt is all used up as it reaches M_2; the approximate feldspar composition of the final melt being M_2' and the alkali feldspar crystals having the composition C_2, assuming crystallization under near-equilibrium conditions.

Just as in the binary systems, water-vapor pressure affects the temperatures of crystallization. Under water pressure (P_{H_2O}) of several thousand bars, corresponding to crystallization under plutonic conditions, crystallization occurs at lower temperatures. The field of leucite is eliminated and the silica-alkali feldspar cotectic shifts somewhat away from silica, the low point being nearer albite in composition (Tuttle and Bowen, 1958; Luth, Jahns, and Tuttle, 1964). The phases formed are the lower-temperature polymorphs such as quartz in place of tridymite and orthoclase in place of sanidine. Otherwise, relationships are much as described above for 1 bar. The effect of addition of calcium (or anorthite) to the system can be inferred from the An–Ab–Or system described in the section on the feldspars. Most obvious is the increased temperatures of crystallization. For further details on the calcium-bearing system, see Winkler (1965, 1967) and James and Hamilton (1969).

So far, discussion has involved only crystallization under equilibrium conditions—crystallization slow enough to permit complete reaction between crystals and remaining melt at all stages. Normally, this means crystallization of the magma under deep-seated, higher-pressure conditions, under the insulation effect of overlying rock. Somewhat different behavior occurs under surface or near surface conditions, where cooling is more rapid, and insufficient time is available for complete reaction between crystals and surrounding melt. In effect, this results in removal of early formed crystals (especially the cores of crystals) from the melt-crystal system because insufficient time is available for equilibrium with the melt. The melt then behaves as if the system did not contain these crystals.

To consider the example of the melt of composition M that was followed above, crystallization proceeds as described until alkali feldspar ("albite" of composition C_1) begins to come out. As reaction between albite and melt is inhibited, less potassium goes into the crystallized "albite," more being retained in the melt. As a result, the melt can proceed farther down the cotectic trough than M_2, and the final crystals (or crystal rims) to form are richer in potassium than C_2. The situation to be expected in the final rock is one of zoned "albite" crystals with cores poorer and rims richer in potassium. Crystallization in this quartz-albite-orthoclase system always trends toward the low-temperature "basin" (see Fig. 3-1), with subequal amounts of the three minerals—that is, toward "granite" in composition.

At this point it may be well to clarify the effect of pressure on the crystallization behavior of magmas. If the effect is merely load pressure—the weight of the overlying rocks—Le Châtelier's rule from general chemistry (if a stress is placed on a system in equilibrium, the system will tend to move in the direction that will help to relieve that stress) tells us to expect that an increase in pressure should favor a decrease in volume, that is, it should favor the higher density form. Magmas, like most melts (water is an exception), are less dense than the equivalent solid rock, so they contract on crystallization. An increase in rock pressure at a given temperature tends to cause an increase in crystallization temperature (or tends to aid crystallization), as shown in Fig. 3-2.

We know from experimental work under high water pressures (P_{H_2O}), however, that if enough water is present to transmit the pressure through water-vapor pressure, an increase in pressure causes a pronounced *decrease* in crystallization temperature (or inhibits crystallization) as shown in Fig. 3-3.

That water pressure is the more important variable, especially for felsic (e.g., granitic) magmas, seems apparent from several points of view, including:

> Plausible means of formation of granitic (and to a lesser extent other) magmas suggest that appreciable water should be present. As discussed in a later section, most granitic rocks probably form by partial melting of water-saturated sedimentary rocks.

Crystallization temperatures (e.g., of granite) under high pressures,

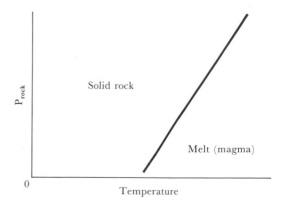

Fig. 3-2 *The effect of rock pressure on the crystallization temperature of a magma. The slope of the boundary may be calculated using the Clapeyron equation: $dP/dT = \Delta S/\Delta V$. Note that the entropy change (ΔS) is positive in going from solid (more ordered) to melt (more disordered). Also the volume change (ΔV) is positive in going from solid (more dense) to melt (less dense). Also $(\partial \Delta G/\partial P)_T = \Delta V$ and $(\partial \Delta G/\partial T)_P = \Delta S$ so that ΔG is negative (reaction will proceed) when on increasing pressure P at constant temperature the volume change is negative and when on increasing temperature T at constant pressure the entropy change is positive.*

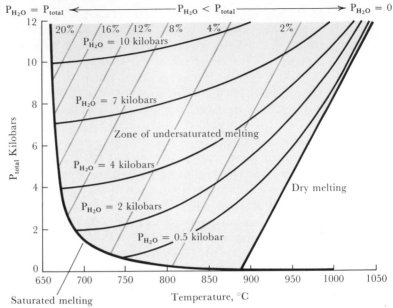

Fig. 3-3 *Schematic illustration of the probable minimum percentage water solubility in granitic liquids formed during water-undersaturated melting processes together with suggested water-pressure isobars. After Brown (1970, p. 356).*

from consideration of the temperatures indicated by metamorphic minerals under conditions of partial melting, suggest lower temperatures commensurate with those determined experimentally under P_{H_2O}.

The exact mechanism of lowering of the temperature of crystallization by water pressure is complex and is not well understood, but in a general way addition of water is like any other minor component in a mixture in which the constituents do not combine chemically. The minor component lowers the crystallization temperature of the mixture. The lowering of temperature is much like a water-mineral eutectic situation with the eutectic point shifted so close to the water side as to be indistinguishable from the melting point for pure water (Turner and Verhoogen, 1960, pp. 412–416).

Basaltic Systems: Olivine, Pyroxene, Ca-rich Plagioclase

Crystallization in basalts can be visualized perhaps best by consideration of the ternary system forsterite (Mg-olivine)–diopside–anorthite (Fig. 3-4), with modification by the presence of sodium in anorthite to form intermediate plagioclase and by silica with forsterite to form enstatite

($MgSi_2O_6$). As with the quartz-alkali feldspar ternary diagram, crystallization behavior in this system can be most easily visualized by considering the
binary sides (for example, Mg_2SiO_4–SiO_2) and then modifying the binary
system behavior for the ternary system (Fig. 3-4).

As with any ternary system, the first mineral to crystallize, and therefore the one forming phenocrysts if any, will be determined by the compositional field in which the total melt composition lies. If the first mineral
to crystallize has a fixed composition (e.g., forsterite), the remaining
melt moves in a straight line away from that mineral. If the first mineral
to crystallize is part of a solid-solution series (e.g., plagioclase), the remaining melt moves away from the solid-solution composition of the first crystals (refer to the binary system), its path gradually curving in response to
the changing solid-solution composition of the crystallizing mineral. The

Fig. 3-4 *The ternary system $Mg_2Si_2O_6$-$CaAl_2Si_2O_8$ and the quaternary apices SiO_2 and $NaAlSi_3O_8$,
at 1 bar. After Bowen (1914), Osborn and Tait (1952), Kushiro and Schairer (1963). S is diagramatic
representation of composition of a single basalt magma erupted during the 1959–1960 summit eruption
of Kilauea Volcano, Hawaii [from data in Richter and Murata (1966); Richter and Moore (1966)].
Separation of the several minerals and temperature intervals are about as predicted from the artificial
system but actual temperatures are about 150 to 200°C lower than predicted (see text). Several binary
sides (temperature-composition) to these ternary diagrams have been rotated down to aid visualization of
the liquidus surface. See also Fig. 2-8.*

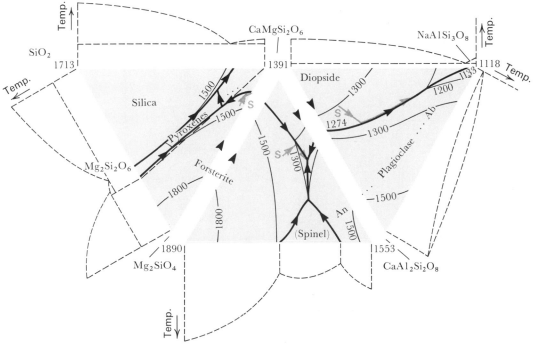

path always moves away from the composition of the grains crystallizing from the melt. For further details and examples of crystallization behavior, see the excellent discussions in Turner and Verhoogen (1960, chaps. 5 and 6) and in Krauskopf (1967, chaps. 13 and 14).

To add a fourth component, that is, if the albite (or SiO$_2$) corner is rotated up to form the apex of the albite-forsterite-diopside-anorthite tetrahedron, each mineral (e.g., plagioclase) controls the initial crystallization of a volume within the tetrahedron rather than an area on the ternary diagram (Fig. 3-5). The cotectic lines (e.g., the curved line from the diopside-anorthite join to near the albite corner) become surfaces within the tetrahedron. The melt, then, moves away from the composition of the mineral(s) crystallizing and into the volume of the tetrahedron. It continues along this path (curved if the mineral is part of a solid-solution series)

Fig. 3-5 *Schematic diagram to illustrate phase equilibrium relations in the system diopside-forsterite-albite-anorthite (after Yoder and Tilley, 1962). From Yagi (1967, p. 384). S is diagramatic representation of 1959–1960 basalt of Kilauea (as in Fig. 3-4). Note that S lies within field of forsterite, behind ruled Di + Fo surface. Path followed by magma is shown. Olivine (Fo) continues to crystallize from S until melt reaches Di + Fo surface. Then diopside begins crystallization along with the olivine, whereupon the melt moves down the Di + Fo surface until the Di + Fo + Pl line is reached. Then plagioclase begins to crystallize along with diopside and olivine, until all the melt is used up.*

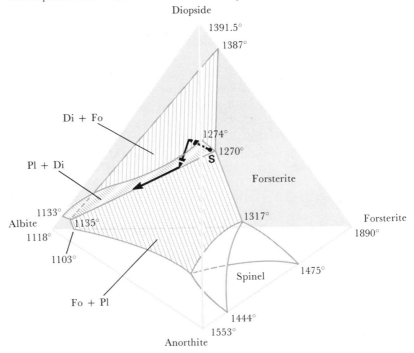

until it intersects a cotectic surface, where the first mineral is joined by the mineral that controls the volume on the other side of the surface. From there, the melt composition moves as a curved line on the surface, away from the two minerals that are crystallizing together.

Comparison with the natural crystallization of olivine basalt during the 1959–1960 summit eruption of Kilauea Volcano in Hawaii may be made from the data of Richter and Moore (1966) and Richter and Murata (1966). See Figs. 3-4 and 3-5. They find an eruption temperature of 1192°C, followed by separation of the first augite ("diopside" on the ternary diagram) at 1120°C, the first plagioclase (An_{64}) at 1090°C, and complete solidification at 1070°C. Although the order of crystallization and the temperature intervals between appearances of successive major minerals are essentially as would be predicted, the absolute values of the real temperatures are about 150 to 200°C lower than would be predicted. This is presumably due to several factors not taken into account in using the simple ternary diagrams of Fig. 3-4. Using any single ternary diagram (for example, $CaMgSi_2O_6$–Mg_2SiO_4–$CaAl_2Si_2O_8$) neglects other significant components, such as SiO_2 or $NaAlSi_3O_8$ (or others not here represented, such as FeO, H_2O), that characteristically lower the crystallization temperatures. The same difficulties are involved in prediction of natural temperatures from other phase diagrams. This does not thwart their use, however, in setting upper limits on temperatures and predicting the order and behavior of crystallization in many rocks.

Characteristics of a Silicate Magma

X-ray studies of silicate glasses (frozen melts) show that silicon-oxygen networks somewhat like those in silicate minerals are also present in the magma before crystallization. This is true of most if not all liquids just above the melting point (e.g., Bernal, 1960, 1964; Whittaker, 1967). Variable linkages of silica tetrahedra in the melt account for their variable viscosity. Consideration of the degree of linking of silica tetrahedra in the essential minerals making up a granite (quartz, orthoclase, plagioclase), all framework silicates with every corner of each tetrahedron linked to the corner of another tetrahedron, explains the high *viscosity* of granite magmas. By contrast, the minerals in a basalt (olivine, pyroxene, plagioclase), partly unlinked tetrahedra and single chains of tetrahedra and partly framework silicates, explain the lesser viscosity of basalt magmas. Viscosities decrease with rise in temperature above the melting point. Early experimental work was described by Bowen (1934). The viscosities measured on erupting tholeiitic basalt in Hawaii range from about 3×10^3 poises near the vent to about 3×10^4 poises (10 times as viscous) away from the vent (Macdonald, 1963). Viscosities of about 0.5×10^3 poises have been measured at liquidus temperatures of 1200°C (no crystals) by Shaw et al.

(1968). Viscosities measured on rhyolite are orders of magnitude higher —about 10^8 to 10^{11} poises at the melting point (Friedman et al., 1963; Shaw, 1963, 1965). For comparison, at 20°C (room temperature), water has a viscosity of 0.01 poise, and glycerin 15, or 0.015×10^3, poises (like thick syrup)—about 1/200 as viscous as erupting basalt. Only when glycerin is −20°C does it have a viscosity of 1.34×10^3 poises, approaching the viscosity of basalt. The viscosity of silicate melts thus depends primarily on temperature and composition.[1]

In general, SiO_2 and Al_2O_3, which make up the tetrahedra, increase the number of linkages (the *polymerization*) and therefore the viscosity. An increase in most of the other common metal ions decreases the linkages. In order of effectiveness (listing most effective in decreasing the viscosity first), these are: H_2O, Fe, Mg, Ca, Sr, Ba, Li, Na, K, Rb. Water owes its great effectiveness to removal of oxygen from silica tetrahedra to form $(OH)^-$. Loss of such water and other volatiles from the release in pressure and sudden boiling in a magma during extrusion results in a sharp increase in viscosity and an increase in melting point (compare Figs. 3-2 and 3-3). The viscosity also increases with decreasing temperature, as ions tend to move more slowly and the linkages become more abundant.

The *temperatures of crystallization* of magmas range from about 700°C for rhyolitic melts to 900 to 1200°C for most basaltic melts (e.g., Macdonald, 1963). They crystallize, not at a single temperature as with ice, but over a range in temperature because of the presence of several components and of members of solid-solution series. Simplified versions of their behavior in consideration of binary and ternary silicate systems have been discussed above.

Because of their probable origin, largely by partial melting of other rocks, granitic magmas presumably have little or no superheat—that is, their temperature is no higher than that for complete melting. Should influences act to raise this temperature, more solids would melt, mixing with any magma already present. In fact, most natural glasses contain at least a few crystals, indicating no superheat. Movement of a water-saturated magma toward the surface (to lower pressures) would enhance crystallization even more (as seen from Fig. 3-3). If, as in the possible case of basalt, the magma was nearly dry (Fig. 3-2), movement toward the surface would inhibit crystallization and would permit the development of superheat. Since even basalt magmas generally contain some crystals on eruption (O'Hara, 1965), evidence suggests lack of significant superheat there also.

The best estimates of *water content* of erupting tholeiitic basalt magma in Hawaii range from 1 to 2.5 weight percent in the early gas-rich phase of an eruption to 0.2 to 0.7% in the later phases, the magma at depth probably

---

[1]Bottinga and Weill (1970) have recently shown that the logarithm of viscosity of anhydrous silicate melts at constant temperature is a linear function of compostion within composition ranges defined by Si + Al/O.

averaging 0.5% water (Macdonald, 1963). The solubility, or maximum water content, of water in igneous melts has been determined by experiment: for Columbia River (Picture Gorge) basalt, 3.1 weight percent at 1,000 bars and 8.5% at 5,000 bars; for Cascade Range pyroxene andesite, 4.5 weight percent at 1,000 bars and 9.8% at 5,000 bars (Hamilton et al., 1964); for pegmatitic granite, 4 weight percent at 1,000 bars and 11.5% at 5,000 bars (Burnham and Jahns, 1962). See also Fig. 3-3.

The *pressure*, to a first approximation, neglecting complicating factors such as tectonic overpressures, is a function of the depth and is equal to the weight of the overlying rock. This lithostatic pressure amounts to about 260 bars/km of depth or 5 kbars at a depth of 19 km (for a rock density of about 2.7). The possibility of additional pressures resulting from tectonic forces, especially at shallow depths (a few kilometers) where the rocks are more rigid, has been considered by many workers (e.g., Rutland, 1965, pp. 119-139). Although overpressures of 2 or 3 kbars (Birch, 1955, p. 116) may be theoretically possible, there seems to be little concrete evidence for their natural existence at deeper levels, for petrologically important lengths of time.

The *time required for crystallization* of a large mass of magma such as a granitic batholith has been estimated as of the order of 1 million years. Such calculations have been based on the rate of heat conduction (Larsen, 1945) and on studies of metamorphic facies surrounding the intrusion (Bederke, 1947a). Jaeger (1957) presents calculations that suggest about 10 million years for the time of crystallization of a large batholith about 5 miles thick. Jaeger's calculations, like Larsen's, are based on the rate of heat conduction away from the intrusion, but in addition Jaeger takes into account the heat of crystallization of the magma. The time of crystallization of smaller intrusions is much shorter, as may be seen from the excellent discussion in Turner (1968, pp. 17-22). For example, a diabase sill 700 m thick intruded under 350 m of cover into dry rocks may be expected to solidify in 9,000 years, whereas a sill 200 m thick may be expected to solidify in only 700 years. Perhaps related is the time of emplacement of a granitic batholith. Potassium argon determinations on the composite Boulder batholith suggest 7 million or 8 million years (Knopf, 1964) or 10 million years (Tilling and Klepper, 1968; Tilling et al., 1968) for its emplacement.

CRYSTALLIZATION BEHAVIOR AND TRENDS

Crystallization of silicate magmas is not yet well understood. However, studies during the past 50 or 60 years of synthetic silicate systems by N. L. Bowen, his coworkers, and others, have given considerable insight into their crystallization behavior and trends. *Bowen's Reaction Series* (1922) provides us with a descriptive approximation for the order and behavior of crystallization of the common minerals from silicate melts (Fig. 3-6). Although expanded slightly here, the reaction series is essentially as presented in 1922 by Bowen. Reaction series C, from calcic plagioclase to sodic plagioclases, is a continuous solid-solution series called the continuous reaction series. Reaction series D, from olivine to biotite, is not a solid-solution series (each mineral has a different crystal structure) and is called the discontinuous reaction series.

In a very general way, minerals higher on the reaction series crystallize before minerals lower on the series. More correctly, each crystallized mineral on a reaction series tends to react with the magma to form a mineral lower on that series. However, the process ends when the magma has completely crystallized, so that the complete reaction series is almost never realized in a single rock.

Still, the description of order of crystallization from the reaction series is too simple. Actually, the *order of crystallization* of minerals from the magma is the order in which the minerals reach saturation. As the temperature falls, the first mineral to become supersaturated begins to crystallize from the melt. Its continued removal from the melt increases the concentration of other constituents in the melt. When another constituent reaches supersaturation, it too begins to crystallize, and the original mineral may continue to crystallize. These are just the relationships found in the crystallization of synthetic melts (for example, $SiO_2-Ab-Or$) above. Available experimental data for common rocks are as yet limited to three or four component systems such as the feldspars, quartz, and pyroxenes. Even then, the constituent minerals are largely of idealized "end-member"

Fig. 3-6 *Reaction series modified from Bowen (1922), Barth (1962, p. 110).*

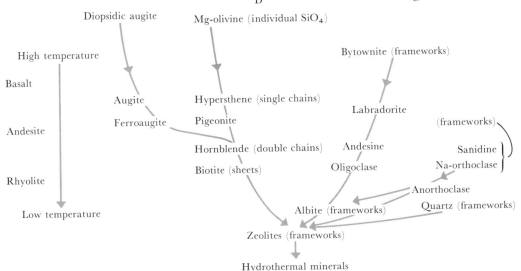

compositions. The more complex micas and hornblendes are almost completely unknowns, and their effect on the crystallization behavior of granitic magmas is likewise unknown. Recent experimental work on natural rocks by von Platen (1965a) suggests, however, that in moderate amounts, biotite and hornblende do not have a major effect on the crystallization of granitic magmas.

Bowen's Empirical Reaction Series (based on combined experimental and petrographic work) gives the reaction trend and generalized order of crystallization. The series do not really represent eutectic crystallization; but the discontinuous series does superficially resemble a prolonged cotectic trough, and a member of one series does lower the crystallization temperature of a member of the other series.

Differentiation

Consider some of the implications of Bowen's Reaction Series. Given a basaltic magma, minerals that crystallize early may include diopsidic augite and Mg olivine. Slow cooling under near-equilibrium conditions may give rise to a rock containing olivine, hypersthene, augite, and plagioclase. If, however, the early formed diopsidic augite and olivine are separated from the melt (e.g., settling out under the influence of gravity or squeezing off of the remaining melt by orogenic movements), the remaining melt will be enriched in the more "felsic" constituents lower on the reaction series. If this separation of earlier-formed "mafic" constituents (including calcic plagioclase) recurs or continues, the melt is gradually changed toward a rhyolitic composition. This gradual evolution of the melt and separation of more felsic constituents is magmatic differentiation, or fractional crystallization.

Differentiation may be separated into different types, for purposes of discussion: (1) fractional crystallization, (2) liquid immiscibility, and (3) movement of volatiles.

Fractional Crystallization

One of the most important means of differentiation is through *gravity settling* (Bowen, 1915, 1919). Early crystallization of minerals of specific gravity higher than that of the remaining magma, given time, commonly results in settling of these grains toward the bottom of the magma chamber. In a few rare cases, early formed low-specific-gravity minerals (e.g., leucite) may float upward (compare discussion of Fig. 1-8). *Filter-press action* (Bowen, 1919; Emmons, 1940) may also be important. During the later stages of crystallization, the residual liquid may be squeezed out of the crystal mush, much as water is squeezed out of wet sand by a footstep. The squeezed-out liquid may float to the top of the magma chamber to appear as an upper layer or as late dikes. Through this deformation or some other means, the grains in the crystal mush develop a more efficient close packing. *Flowage differentiation* (Fahrig, 1962; Bhattacharji and Smith, 1964) may result in concentration of solid grains away from the walls in a moving fluid. The early formed minerals are subjected to shear in addition to the longitudinal movement, resulting in a component of lateral movement into the zone of more rapid flow. This in turn may result in concentration of early formed minerals in the core of a vertical conduit. *Zoning* (Bowen, 1919) of early formed crystals through reaction with the remaining melt may result in gradual removal of the early crystal cores from further reaction (especially common in plagioclase). The residual fluid is thus further removed from the composition of the original magma. The final differentiate crystallizes in the interstices of the earlier grains unless separated through one of the above three means.

Liquid Immiscibility[1]

Just as oil is immiscible with water, it has been argued that different liquid fractions of a silicate magma may separate with cooling. In fact, sulfide melts may separate from a silicate magma at an early stage (Vogt, 1916–1918; Wager, Vincent, and Smales, 1957). Although there is little experimental evidence to suggest that most silicate magmas may separate in this way so as to contribute to differentiation, a significant amount of field petrographic and chemical evidence may be interpreted in this way (Holgate, 1954). Possible exceptions may exist in criteria interpreted as resulting from mixing of basic and acid magmas (Walker and Skelhorn, 1966). Under particular conditions in a few magmas (olivine-rich basalt, mafic alkaline), immiscibility has been proposed on textural and chemical evidence (Drever, 1960; Philpotts, 1968a, 1970b; Philpotts and Hodgson, 1968; Moore and Calk, 1971). Separation of carbonate liquids from mafic alkaline magmas has been demonstrated experimentally (Koster van Groos and Wyllie, 1963, 1966). Convincing examples of immiscibility of high-silica ("potassic granite") in high-iron silicate liquid ("pyroxenite") in lunar rocks have been described by Roedder and Weiblen (1970a, b). The common occurrence of voluminous flood basalt with minor rhyolite or intrusive gabbro with minor granophyre and virtually no intermediate rocks also suggests magmatic immiscibility (D. Alt, verbal communication, 1968). Lack of mixing of artificial melts suggests the same (see Yoder, 1971).

Movement of Volatiles

Although not as well documented and probably not as important quantitatively as fractional crystallization, movement of volatiles may also contribute to differentiation. *Gaseous transfer* (e.g., Fenner, 1926; Bowen, 1928) or the upward movement of gas bubbles through the magma may result in

[1]For example, Daly (1918); Grout (1918b); see also Bowen (1919, 1928, pp. 7–19) for comments.

concentration of volatiles in the gas phase and their concomitant transfer upward. Presumably this process is important only where vesiculation
can occur—that is, under shallow conditions and with a significant gas
content in the magma. The rise of such gas bubbles might in the later stages
of crystallization push the residual interstitial fluids upward through the
pores between the earlier-formed crystals (Shand, 1933). *Volatile streaming*
(e.g., Kennedy, 1955; Hamilton, 1965a), or the streaming of water, alkalis,
and other volatiles to areas of lower pressure, may perhaps occur without
the development of a gas phase. The volatiles may float upward due to
gravity or laterally into fractures.

As the crystals form in the melt, their lattices act as sorters for the
cations still in the melt. If the cations in the melt are similar to the cations
controlling the geometry of a growing crystal lattice, they will also enter the
lattice to a certain extent. The extent of this substitution depends mainly
on the *relative size and charge* of the substituting ion. Readily accessible lists
of ionic radii and charges are in Ahrens (1952) and Krauskopf (1967,
pp. 643–645). Recently refined ionic radii are in Shannon and Prewitt
(1969) and Whittaker and Muntus (1970). A set of empirical rules were
devised by V. M. Goldschmidt (1937) to explain this behavior. These have
since been modified slightly (e.g., Ringwood, 1955), but his basic ideas
remain unchanged. For mutual replacement:

> The ionic radii must differ by less than about 15% (for example,
> radius for $Mg^{++} = 0.66$, for $Fe^{++} = 0.74$).

> The ionic charges should be the same or nearly so (for example,
> Mg^{++}—Fe^{++}).

> Other things being equal, the smaller radius (ion fits in space more
> easily), the higher charge (ion pulled into space with greater force),
> or the smaller electronegativity is preferred. For example, Mg^{++}
> smaller than Fe^{++} so that Mg is concentrated in early pyroxenes and
> Fe in late ones.

In general, these rules for substitution of minor elements in crystal
lattices work very well. Many of the transition metals (for example, Ti^{3+},
Cr^{3+}, Mn^{3+}, Fe^{++}, Ni^{++}, Cu^{++}), however, exhibit behavior inconsistent with
that predicted by Goldschmidt's rules. An explanation (e.g., Curtis, 1964;
Burns and Fyfe, 1964, 1966; Burns, 1970a, b) has recently been formulated, which very nicely explains the anomalies in terms of "crystal field
theory." The theory is too involved to discuss in detail here; but, in essence,
many transition metal ions exhibit large preference differences (or "site
preference energy") for octahedral coordination positions because of the
peculiarities of their electrostatic fields. For divalent ions, the order of
preference is Ni > Cu > Co > Fe > Mn. For trivalent ions, the order of

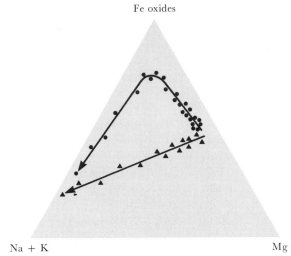

Fe oxides

Na + K   Mg

▲ ▲ ▲ ▲   ● ● ● ●

Basalt, andesite, rhyolite association   Tholeiitic flood basalts, diabase sills, mafic
                                                layered intrusions
(orogenic; assimilation)   (post-orogenic; no assimilation)

High P_{O_2} (inferred as due to high water   Low P_{O_2} (inferred as due to low water content)
content)
$\therefore$ iron $\rightarrow$ Fe^{3+} $\therefore$ iron $\rightarrow$ Fe^{++}

goes into magnetite   goes into silicates to minor extent during early
          $(Fe^{++}O \cdot Fe_2^{3+}O_3)$ formed minerals, for example:
                                                          $(Mg,Fe)_2^{+}SiO_4, (Mg,Fe^{++})SiO_3$

even crystallization of magnetite during   since Mg^{++} preferred over Fe^{++} in early
differentiation, $\therefore$ even removal of iron and   formed minerals, Fe^{++} is concentrated in the
no iron enrichment   liquid fraction to come out with the alkali-rich
                                                fractions later (in the limiting case with all the
                                                Fe as Fe^{++}, the trend would be directly away
                                                from Mg on the Mg-Fe-Alk diagram)

Fig. 3-7 *Effect of oxidation state of iron on the crystallization trend of basaltic magmas. With intermediate P_{O_2} (or water content), the trend lines should be intermediate between these two limiting cases
(plotted values idealized).*

preference is Cr > V > Mn > Ti > Fe. Goldschmidt's third point above
has been criticized (Shaw, 1953; Whittaker, 1967) for lacking any valid
theoretical foundation, but if recognized for its empirical base, it generally
gives qualitatively correct results. Whittaker (1967) shows that the structure of the parent liquid exerts about as much effect on the composition
of a crystal as does the structure of the crystal.

    The trend of differentiation in the magma is also affected to some ex-
tent by the oxygen partial pressure in the magma (e.g., Kennedy, 1955;
Osborn, 1959, 1962; Roeder and Osborn, 1966) as shown in Fig. 3-7.

With a low P_{O_2}, iron is not oxidized and is removed early (e.g., as magnetite); but it remains longer in the melt as Fe^{++} to continue crystallization of iron-rich silicate to a late stage. Such a situation may arise when a magma is emplaced at shallow levels so that water is lost and P_{O_2} is low. During crystallization, then, the magma is enriched in iron to a late stage (see Fig. 3-7). With a high P_{O_2}, iron is oxidized, thereby being removed as magnetite continuously during crystallization of the magma (Fig. 3-7). The variation in the chemistry and mineralogy of basaltic magmas, resulting from these considerations, has been discussed by Kuno (1968) and Coats (1968).

Late-magmatic to Postmagmatic Processes

As a result of these controls, the ions that do not easily substitute for the ions in common rock-forming minerals are concentrated in the *residual melt* along with the silica, alkalis, and water. These residual melts or gases, high in water and other volatiles, crystallize in irregular patches or dikes as very coarse-grained pegmatite. Presumably, the volatiles inhibit nucleation of grains (resulting in fewer grains) and/or aid the rate of diffusion and growth (resulting in larger grains). If the volatile phase is absent (either by never being present or by being lost) during crystallization of this residual melt, it crystallizes as finer-grained aplite (see Figs. 3-8a, b, and c).

In reality, the situation is not this simple. At some stage and under some conditions this water-rich gas phase separates from the rock melt. Try to imagine a granitic magma beginning to crystallize deep below the surface under conditions of high temperature and pressure. The magma

Fig. 3-8a *Unzoned pegmatite dike cutting contorted biotite-plagioclase-quartz gneiss. Hammer handle is 12 in. long. South Kootenai Lake, Bitterroot Range, western Montana.*

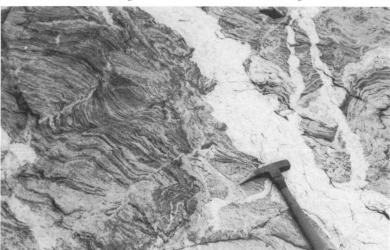

Fig. 3-8b *Zoned pegmatite-aplite dike cutting granodiorite. Dark gray quartz, white plagio-clase in coarse-grained core. Same minerals in fine-grained aplite margins. Highway 3, about 4 miles east of Osoyoos, British Columbia.*

Fig. 3-8c *Nonsymmetrical textural variations in pegmatite-aplite dike cutting fine-grained hornblende diorite. Dark gray quartz, white feldspars. One mile west of Bohn Lake, Flint Creek Range, Montana.*

consists of only one phase, a liquid. The volatiles are all dissolved in the melt, held in by the high pressure. As the magma continues to cool and crystallize, the volatiles become more and more concentrated—the prevailing external pressure becoming potentially less able to hold them in. At the same time, of course, the decrease in temperature causes a decrease in vapor pressure. If, as seems to be common with magmas that produce pegmatites, the increase in volatile concentration prevails, the liquid will at some stage "boil" when the vapor pressure exceeds the pressure tending to hold it in. This phenomenon of boiling due to an increase in volatile concentration and vapor pressure during cooling is called the *second boiling point*. The water-rich fluid that boils off crystallizes in fractures as the pegmatite dikes seen around many granitic bodies. Clearly the temperature at which this boiling occurs can be no higher than the liquidus (temperature of first appearance of crystals) for the appropriate magma composition. Whether or not the material boiled off is gas depends on whether the temperature is above the critical temperature of the water-rich fluid (at the appropriate pressure), which in turn depends on the amount and composition of dissolved materials. If so, the properties of a gas and liquid become indistinguishable and the general term "fluid" becomes more appropriate. The concept of a second boiling point is realistic only if the initial fluid is below its critical temperature. However, available information is insufficient to show what critical temperature pertains in real granitic magmas (Krauskopf, 1967).

Experimental work by Burnham and Jahns (1962) on a typical pegmatitic granite shows that the maximum amount of water dissolved in such a granite (saturated with water) increases from 0% at 1 bar (atmospheric pressure) to about 4% at 1 kbar (1,000 bars), 6.5% at 2 kbars, and 11% at 5 kbars. Therefore, if a granitic magma saturated in water vapor under a pressure of 5 kbars (about 19-km depth) rises toward the surface until the pressure is only 2 kbars (about 8-km depth), the water content must drop from 11 to 6.5%.

We are now in a better position to consider the difference in conditions necessary for the crystallization of pegmatite and aplite. A series of experiments by Jahns and Burnham (1957, 1969), intended to duplicate the natural conditions of pegmatite formation, shows that a fine-grained, sugary aggregate of quartz and feldspar (aplite) separates above the second boiling point where a water-rich phase separates. This water-rich phase provides space for large crystals (pegmatite) to grow from the magma. The high water content also inhibits nucleation, thereby developing only a few crystals, and lowers the viscosity, thereby promoting more rapid movement of ions and growth of crystals. Pegmatite and aplite could thus form in the same dike—a very common situation in nature. Variation in the conditions of temperature, water pressure, and composition could easily account for the natural variations encountered. A solid pegmatite dike could, for example, form by the development of a fracture in the cooling granite, below the second boiling point. Subsequent fracturing causing a sudden drop in pressure could result in loss of volatiles and renewed crystallization of fine-grained aplite. It is also possible that aplite could crystallize from the water-poor granite magma at the same time as pegmatite crystallizes from the water-rich fluid (Jahns and Burnham, 1969).

During the latest stages and after crystallization of the magma is virtually complete, including the pegmatites and aplites, hot corrosive watery solutions may remain. These account for the so-called *deuteric* (late magmatic) alteration of many igneous rocks. Changes effected by the magma's own late-stage solutions include the alteration of biotite to chlorite and less commonly amphiboles and pyroxenes to chlorite or actinolite, the alteration of the rims of or veins in some plagioclase or orthoclase grains to albite (e.g., Ramberg, 1962; Peng, 1970), the alteration of feldspars to analcite or to sericite. The microscopic "wormy" intergrowths of quartz in oligoclase feldspar called *myrmekite* are probably also a late-stage replacement phenomenon. It has long been recognized that myrmekitic plagioclase is characteristically found adjacent to potassium feldspar, suggesting some causal relationship (e.g., Sederholm, 1916).

Assimilation

Bowen's Reaction Series also gives information, in a general way, on the behavior of the magma when it is contaminated by material from the intruded rocks. The contamination or assimilation commonly involves both mechanical disintegration and chemical reaction between the magma and country rocks, both being affected (Fig. 3-9a and b; see also Greenwood and Mc-Taggart, 1957). The magma composition trends toward the country-rock composition and vice versa. More mafic-rich border zones in a granitic pluton, especially if accompanied by inclusions (or xenoliths, i.e., fragments of the surrounding country rocks) of higher-Ca or -Mg country rocks, most commonly result from assimilation (see Fig. 3-10). Only if the country rocks fall lower on the reaction series than the minerals crystallizing from the magma can the country rocks be melted. Under such conditions, the country rock melts at a lower temperature than the existing temperature of the magma. The magma cannot melt minerals higher in the reaction series than those with which it is in equilibrium (i.e., those that it is crystallizing). The magma will, however, react with the rocks to convert them into minerals more like those "with which the magma is saturated" (Bowen, 1928). Simultaneously, of course, the overall composition of the magma will trend toward that of the rock being assimilated (see, for example, Woodard,

Fig. 3-9 *Intrusion breccia. Stoping of biotite quartzofeldspathic schist by biotite-hornblende quartz diorite of the Coast Range batholith, British Columbia. Caulfield Cove, Vancouver, British Columbia. Hammer handle is 13½ in. long. (a) Note that schist blocks below hammer have pulled away from main mass and that quartz diorite has filled the resulting spaces (dilational dikes). (b) Note schist blocks to left of hammer beginning to pull away. Angular to rounded inclusions of schist swim in larger quartz diorite mass to right of hammer.*

1957; Leake and Skirrow, 1960). As a result, it should not be surprising to find the same minerals in a xenolith of country rock as in the magmatic rock (e.g., Reesor, 1958). Only the proportions should be different if equilibrium is approached. In fact, this is precisely the case for the common xenoliths found in granitic plutons everywhere.

An alternative hypothesis to explain the rather common more mafic-rich border zones of granitic plutons has been proposed by Vance (1961). He suggests that crystallization beginning from the borders, against

Fig. 3-10 *Mafic-rich inclusions in biotite-hornblende granodiorite of the Coast Range batholith more mafic-rich than that in Fig. 3-9a and b, 20 miles away. More mafic constituents probably developed by assimilation of inclusions such as these. Roadcut 1 mile south of Britannia Beach, British Columbia.*

the cooler country rocks, seals in the released volatiles that are forced inward and downward with further crystallization. The alkalis and silica, enriched with the volatiles during crystallization, also migrate downward, becoming enriched in the magma at depth.

VARIATION DIAGRAMS

The question often arises, in studying a closely associated group of igneous rocks, whether or not these rocks were derived from one another or from a common parent magma, and if so, how. The answer may sometimes be obtained through field or microscopic studies; but, more often than not, uncertainties still exist. Another approach is to compare chemical analyses, from the associated rocks, in the form of variation diagrams. If related, the individual rocks should show a smooth variation from one to the next, presumably in order of their time of development or degree of development from the parent material. Usually the variation diagrams take the form of an independent variable (or relatively so) such as time or order of emplacement or percentage of silica (the most abundant oxide) plotted against the dependent variables such as the percentages of individual chemical oxides (Fig. 3-11).

Commonly used variation diagrams include plots of the following:

Independent Variable	Dependent Variable	Example Reference
Position in a layered series	% individual chemical oxides	Wager and Deer, 1939
Order of emplacement	% individual chemical oxides	
% magma solidified	% trace elements	Wager and Mitchell, 1951
% SiO_2	% other oxides	Harker, 1909; Williams, 1942; Huber and Rinehart, 1967; Wright, 1970 (also called a Harker diagram)
% $\frac{1}{3}$ SiO_2 + K_2O−FeO−MgO−CaO	% other oxides	Larsen, 1938, 1940
($\frac{1}{3}$ Si + K)−(Ca + Mg)	% other ions	Nockolds and Allen, 1953
% Al/Si	% MgO (for basaltic suites)	Murata, 1960
"Si" number	Niggli number	Simonen, 1960
Color index	% oxides	

Fig. 3-11   *Examples of variation diagrams.*

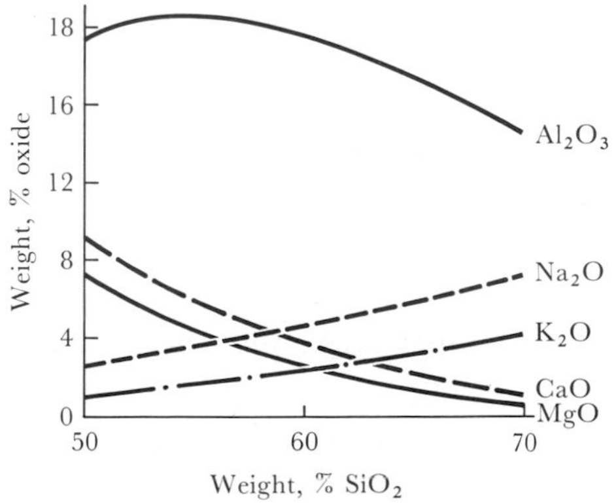

Variation diagram for Crater Lake, Ore.
(Williams, 1942, p. 154)

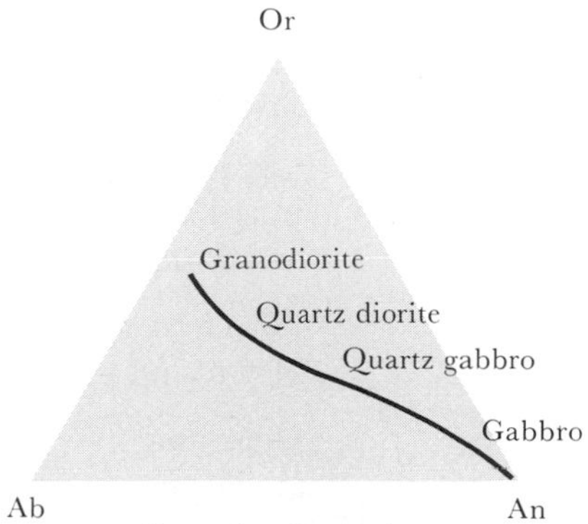

Normative Or-Ab-An
for granodiorite province
of SW Finland (Simonen, 1960)

Still other variation diagrams take the form of triangular plots:

Or (norm)	Ab (norm)	An (norm)	Larsen, 1938; Simonen, 1960
Quartz (norm)	Feldspar (norm)	Femic (norm)	Larsen, 1938
% Fe	% Na + K	% Mg	Nockolds and Allen, 1953
% K	% Na	% Ca	Nockolds and Allen, 1953

The choice of parameters for variation diagrams depends largely on the use to which they are to be put. Most of the more complex parameters are attempts to show trends of differentiation within the crystallizing magma, assuming that differentiation has controlled the variation. Some of the simpler parameters have been chosen mainly to illustrate trends of chemical variation and to suggest genetic affinities rather than some more rigorous, perhaps incorrect interpretation. Certain interpretations have been made from variation diagrams:

"Smooth" variation curve: trend of differentiation of a magma (curve not actually smooth; kinks where phase begins or ends crystallization).

Breaks or kinks in curve: points of separation of a new mineral phase.

Scatter of points at mafic end of curve: due to crystal accumulation. Composition of parent magma at this end.

Straight-line variation: assimilation of country rock (of a single type) or of another magma, in varying proportions.

These interpretations assume ideal conditions and should be approached with caution:

Enough analyses are required to make the curve statistically valid.

Curve for differentiation is valid for glassy or fine-grained rocks but porphyritic or coarse-grained rocks may be formed partly by crystal accumulation (Bowen, 1928).

A curved variation diagram may be produced by assimilation of rocks of varying composition.

Field and microscopic observations should point to differentiation or at least not point to a different process such as assimilation.

Recent attempts to make quantitiative the position within a differentiation sequence are the proposals of a "differentiation index" (normative quartz + orthoclase + albite + nepheline + kalsilite + leucite) by Thornton and Tuttle (1960) and, by Poldervaart and Parker (1964), a "crystallization index" (normative anorthite + Mg-diopside + forsterite + enstatite converted to forsterite + Mg-spinel calculated from normative corundum in ultramafic rocks).

The emphasis on differentiation and assimilation here is not meant to imply that these are the only means of production of magmas or of igneous rocks of varying composition. As will be seen in the following section on granitic rocks, the felsic rocks especially may be produced by other means, such as partial melting of rocks of varying composition or at varying temperatures, or by granitization.

Variation diagrams may also be used to compare trends of variation between separate series of igneous rocks to help decide whether or not the different series are genetically related. A final word of caution in the use and interpretation of "Harker diagrams" (SiO_2 versus other individual oxides) is in order. Chayes (1964b) has pointed out that the dominance of the variance of SiO_2 over the sum of the other variances requires a strong negative correlation between SiO_2 and some other oxides. This in itself negates the usefulness of such diagrams in the discrimination between different possible processes involved in variation between rocks of a suite.

STRONTIUM ISOTOPES AND THE ORIGIN OF IGNEOUS ROCKS

The ratio between strontium isotopes Sr^{87}/Sr^{86} has been shown to be low for most oceanic ("mantle-derived") igneous rocks and higher for most old continental rocks ("continent-derived") (Hurley et al., 1962; Faure and Hurley, 1963). This ratio can be useful in deciphering the origin of igneous rocks in the continents.

Because of their identical chemical characteristics and nearly identical weight, strontium isotopes are not fractionated to any significant extent by most natural processes such as metamorphism or igneous differentiation. Thus the isotopic ratio should not be changed by such processes.[1] Sr^{87} is produced, however, by radioactive decay of Rb^{87}, which, because of its chemical similarity to K (similar charge and size), is most abundant in K-rich minerals and rocks that are concentrated in the continents. Because

---

[1]A case involving selective migration of Sr^{87} in Precambrian basement rocks has recently been described (Naylor et al., 1970). The effects are most obvious in Rb-rich minor rocks. The possibility of selective depletion of residual basement rocks in Sr^{87} by partial melting during high-grade regional metamorphism has been suggested by Heier (1964b). He argues that since Rb (including Rb^{87}, which decays to form Sr^{87}) occurs in K minerals, Sr^{87} should also be concentrated in K minerals. The first melts during metamorphism, being of granitic composition, would tend to remove such K minerals and their Sr^{87}.

of this, the ratio Sr^{87}/Sr^{86} is higher in K-rich rocks, which have more Rb^{87} for decay, and in older rocks, which have had longer to decay.

The earth's mantle (predominantly peridotite) and oceanic basalts derived by partial melting of the mantle, being very low in K and associated elements, have a very low Sr^{87}/Sr^{86}, generally between 0.701 and 0.705 (see Fig. 3-12). This is close to the "primitive" value of 0.698 for meteorites, which are considered nonevolved. The old continental (shield) crust by contrast is largely granitic, containing a large amount of K and thus Rb, and it evolved between 1 billion and 3 billion years ago. Radioactive decay of Rb^{87} to Sr^{87} since that time has raised the Sr^{87}/Sr^{86} to values as high as 0.850 or more. Although the differences between these numbers are small, they are internally consistent, outside possible experimental errors, and are considered significant (see Table 3-1).

Strontium isotopic ratios are most useful in interpreting the nature of the parent rock from which magmas in the continent were derived and whether they have suffered any contamination. Young igneous rocks, such as most basalts, derived directly from the mantle should have a low Sr^{87}/Sr^{86} close to that of the mantle. Older rocks can be corrected according

Fig. 3-12 *Variation of* Sr^{87}/Sr^{86} *with time in common igneous rocks. Low* Rb/Sr *of basalts and mantle rocks gives a slight increase in* Sr^{87} *with time. Granitic magmas with much higher* Rb/Sr *(for example, 0.25) that separated from mantle during last 3 billion years (three representative separation times shown) increase in* Sr^{87} *much more rapidly with time.*

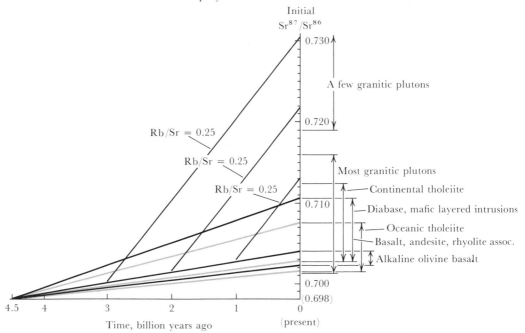

to their age and Rb/Sr for the amount of Sr^{87} produced since their formation. Igneous rocks such as continental basalts derived from the mantle but possibly subject to a minor amount of contamination (or possibly derived from a more Rb-rich part of the mantle) should have a slightly higher initial Sr^{87}/Sr^{86} (see Fig. 3-12).

Granitic rocks formed by some process within the old crystalline basement of the crust should have the high Sr^{87}/Sr^{86} of the basement. Thus granitic plutons having low Sr^{87}/Sr^{86} ratios of 0.705 to 0.708 have been considered to be derived from the mantle or "gabbroic" layer at the base of the crust (e.g., Fairbairn et al., 1964; Hurley et al., 1965). It is apparent, however, that granitic magmas could equally well be formed within the crust by partial melting of "eugeosynclinal" graywackes if these are fairly young and are derived primarily from volcanic and other low-Rb young rocks from island arcs, rather than from the old high-Rb crystalline basement. The feasibility of this has been demonstrated by Peterman et al. (1967) in a study of graywackes from Oregon and California. These 42 graywackes had Sr^{87}/Sr^{86} ratios ranging from 0.7035 to 0.7103. Thus major granitic batholiths such as the Sierra Nevada of California and the Coast Range of British Columbia have moderately low Sr-isotope ratios and could be derived from the "eugeosynclinal" crust or possibly (from isotope data alone) from the mantle but not from the old crystalline crust or from the "miogeosynclinal" pile mostly derived from it. Some old crystalline basement or "miogeosyncline" could be included in the partial melting, especially where the boundary between it and the "eugeosynclinal" deposits is blurred by regional metamorphism and accompanying deformation. The position of most major granitic batholiths within "eugeosynclinal" deposits bordering the "miogeosyncline" fits such an interpretation.

A possible alternative for the formation of the granitic magmas has been suggested by Doe et al. (1968). They suggest major partial melting of old lower-crust or upper-mantle source material such as granulites of about quartz diorite composition. Although such rocks presumably do exist in the lower crust (see, for example, p. 363 and variable differentiation of such a magma could give rise to the range of Sr^{87}/Sr^{86} values found, there are objections to such a model on petrologic grounds. As noted below (see p. 363), second-cycle regional metamorphism of dry granulite facies rocks probably involves a water pressure less than rock (or load) pressure. Under such conditions, the beginning of melting would require higher temperatures than rocks of equivalent composition under wet conditions such as in the adjacent "geosynclinal" deposits. Also granulite facies metamorphic rocks are deficient in the radioactive constituents apparently necessary to help raise the temperatures into the melting range.

Andesitic or alkali-rich "basaltic" volcanic rocks having intermediate Sr^{87}/Sr^{86} ratios (for example, 0.705 to 0.710) could be formed by crustal

Table 3-1 *Strontium isotope ratios for different rock associations.*

Rock Type	Sr^{87}/Sr^{86}	Number of Analyses	Reference
Tholeiitic basalt (oceanic)	0.7016–0.7076	46	Faure & Hurley, 1963;
Alkaline olivine basalt (oceanic)	0.7023–0.7040	16	Moorbath & Bell, 1965; Tatsumoto et al., 1965; Hamilton, 1965, 1968; McDougall & Compston, 1968; Powell et al., 1965; Powell & DeLong, 1966; Hedge, 1966; Heier et al., 1966; Gast, 1967; Armstrong, 1968; Hart et al., 1971
Phonolite-trachyte	0.7034–0.7094	14	Gast et al., 1964; Hamilton, 1965; Powell et al., 1965; Hedge, 1966
Tholeiitic basalt (continental)	0.7033–0.711	13	Hedge & Walthall, 1963; Faure & Hurley, 1963;
Alkaline olivine basalt (continental)	0.7041–0.7075	3	Hedge, 1966; Armstrong, 1968; Ewart & Stipp, 1968; Doe et al., 1969; Leeman, 1970; Damon, 1969; Hedge et al., 1970
Diabase	0.7028–0.7123	10	Faure & Hurley, 1963; Heier et al., 1965; Hedge & Walthall, 1963
Mafic layered intrusions	0.7033–0.7101	32	Hurley et al., 1962; Steuber & Murthy, 1966
Granophyre	0.7022–0.7327	12	Hamilton, 1963; Heier et al., 1965
Alpine peridotite (continental)	0.7068–0.7151, 0.7276	39	Steuber & Murthy, 1966
Peridotite (oceanic)	0.7034–0.7138, 0.7227	14	Bonatti et al., 1970
Kimberlite (mica peridotite)	0.705–0.729	14	Powell, 1966
Basalt, andesite, rhyolite assoc.: Basalts	0.7032–0.7044	23	Ewart & Stipp, 1968; Peterman et al., 1970; Hedge et al., 1970
Andesite, dacite	0.7029–0.7104	55	Faure & Hurley, 1963; Hedge, 1966; Doe et al., 1968; Ewart & Stipp, 1968; Pushkar, 1968; Peterman et al., 1970; Hedge et al., 1970
Rhyolite	0.7033–0.7068	28	Hedge, 1966; Ewart & Stipp, 1968; Pushkar, 1968

Table 3-1   *Strontium isotope ratios for different rock associations (Continued)*

Rock Type	Sr^{87}/Sr^{86}	Number of Analyses	Reference
Anorthosite	0.6997–0.7061	57	Hedge & Walthall, 1963; Heath & Fairbairn, 1968
Granitic plutons:		108	Hedge & Walthall, 1963;
Most:	0.7012–0.7162		Fairbairn et al., 1964b;
A few:	0.7191–0.7694		Hurley et al., 1965; Moorbath & Bell, 1965; Kistler et al., 1965; Moorbath et al., 1967; Pushkar, 1968; Doe et al., 1968; Menzer, 1970
Alkali gabbro	0.7031–0.7063	24	Fairbairn et al., 1963;
Alkali syenite plutons	0.7031–0.7156	17	Bell & Powell, 1970
Carbonatite	0.7016–0.7088	64	Faure & Hurley, 1963; Fairbairn et al., 1963; Hamilton & Deans, 1963; Powell, 1966; Bell & Powell, 1970
Fenite adjacent to carbonatite	0.7058–0.7527	8	Bell & Powell, 1970
K-rich basaltic rocks	0.7037–0.7082	81	Faure & Hurley, 1963; Bell & Powell, 1969
Graywacke and related sediments	0.7035–0.7089	48	Peterman et al., 1967; Ewart & Stripp, 1968
Shale, argillite ("eugeo-syclinal")	0.7077–0.7183	20	Peterman et al., 1967; Ewart & Stipp, 1968; Barker & Long, 1969
Calcareous ooze (oceanic)	0.7083–0.7088	2	Moorbath et al., 1967
Limestone	0.7004–0.7167	17	Hedge & Walthall, 1963; Hamilton & Deans, 1963; Powell, 1966
Shale ("miogeosyncli-nal" or platform)	0.7204	1	Faure & Hurley, 1963
Metamorphic rocks ("basement")	0.6947–0.8496, 1.176	75	Hurley et al., 1962; Hedge & Walthall, 1963; Zartman, 1965; Moorbath et al., 1967; Pushkar, 1968; Naylor et al., 1970

contamination of a mantle-derived basaltic magma or by partial melting of an old, low-K (low-Rb) crystalline basement in the lower crust or of a younger higher-K "geosynclinal" source that lacks much continental material (e.g. Peterman et al., 1969).

Granitic and volcanic rocks with high Sr^{87}/Sr^{86} ratios presumably form by some process such as anatexis, from old crystalline basement crust (Moorbath and Bell, 1965), or possibly from "miogeosynclinal" rocks derived largely from it.

Alkali gabbro, alkali syenite, and carbonatite occur in the same tectonic environment, commonly are associated with one another, and characteristically overlay old crystalline basement rocks. All have low to moderately low Sr^{87}/Sr^{86} ratios, indicating derivation from the mantle either directly or by differentiation from a magma derived from the mantle, rather than from the underlying Sr^{87}-rich crystalline basement.

It should be recognized that although Sr^{87}/Sr^{86} ratios appear to be a powerful tool in interpretation of the history of certain igneous rocks, more than one interpretation in some cases will fit the isotopic data. In all instances, the isotopic data should be compatible with available petrologic data.

CHAPTER 4
THE IGNEOUS ROCK ASSOCIATIONS: DESCRIPTION, OCCURRENCE, AND ORIGIN

The most prominent line of separation between different kinds of igneous rocks is between the continents and the ocean basins. Igneous rocks in the true ocean basins are almost entirely basalts and related rocks. Igneous rocks on the continents are dominated by great masses of granitic rock and are characterized by a more felsic nature, although all other igneous types are represented. A fairly well-defined line, called the "andesite line" (see inside front cover, right), separates these two major petrographic regions. This line surrounds, in a general way, the major ocean basins, most major islands, and island arcs lying on the continental side of the line.

The Family classification discussed earlier groups rocks that are associated in space and time and presumably are genetically related. Spatial association is not sufficient. Spatially associated rocks may be essentially unrelated, especially if they were emplaced at different times. Such seems to be the case, for example, with lamprophyres that commonly occur in the same areas as granitic plutons. The lamprophyres are almost always younger, however, and apparently lack any close genetic relationship. The distribution of the various families of igneous rocks will be described separately in the next section, along with their occurrence and origin. The most common kinds of igneous rocks probably evolved as "primary magmas," whatever their ultimate origin, rather than as derivatives of other magmas. For this reason, the most common plutonic igneous rock, "granite," and the most common volcanic igneous rock, basalt, will be described first in their respective groups.

The individual igneous rock associations have been plotted on maps of North America and the world. (See inside front and back covers.) The igneous rocks plotted on one map were not all emplaced at one time. However, within a single continental region such as a "geosynclinal" belt, the rocks of a single family or association were emplaced largely within a single major geotectonic cycle (see p. 10). To avoid confusion and to show relative spatial distribution, only the *more recent igneous activity* in a single continental region is plotted. In the case of the North American cordillera, Mesozoic and Cenozoic plutonic and predominantly Cenozoic to Recent volcanic igneous rocks are plotted; in the Appalachians, Paleozoic rocks are plotted. Anorthosites, gabbroic layered intrusions, peridotite bodies, and carbonatites and their associated alkalic rocks are plotted regardless of age. With the exception of these last four distinctive plutonic associations, rocks in the shield areas have not been included.

As noted in Table 2-1, the igneous rocks of a given area can often be grouped into those that were emplaced at about the same time, under somewhat similar conditions, that have similar or related mineralogical and chemical compositions, and that are presumably genetically related. This breakdown groups the igneous rocks according to rock association or kindred. It should be emphasized at the outset that some kinds of rocks occur in more than one association, suggesting that a given kind of rock may originate in more than one way. For example, rhyolite occurs in the basalt-andesite-rhyolite association and in the tholeiitic basalt association.

The different igneous rock associations will be considered in turn, beginning with the preorogenic and early orogenic associations, followed by the igneous rocks associated with the orogenic or deformational phase, and then those that typically follow the orogenic phase of the "geosyncline." For each association, a typical area, well described in the literature, will be reviewed so as to present a picture in some detail. The general features of the association as a whole—field, mineralogy, and chemistry—will then be assembled, followed by particular problems, the most prominent and feasible origins that have been suggested for the association; finally, a choice will be made of the origin or origins that best seem to fit presently available data.

PREOROGENIC AND EARLY OROGENIC SUITES ("OPHIOLITES")

SPILITE-KERATOPHYRE SEQUENCES

The Spilitic Association of the Olympic Peninsula, Northwestern Washington[1]

The most distinctive volcanic members developed within the sedimentary rocks in the "eugeosynclines" are the sodic basalts of the spilitic association.

[1]Park, 1946; Weaver, 1916.

The Eocene basalts interbedded with thousands of feet of argillites and graywackes of the Olympic Peninsula (see Fig. 8-3, p. 334) are an excellent example of this association. The volcanic rocks of this area consist dominantly of resistant lavas and agglomerates (mostly spilites), with lesser amounts of softer tuffs, agglomerates, and tuffaceous sediments. Pillow lavas (see igneous rock occurrences, p. 57) are abundantly developed, accumulating in places to tremendous piles thousands of feet thick. Fossils in the associated tuffs indicate a submarine, probably deep-water origin. Massive basalts, especially toward the top of the section, locally contain unaltered labradorite and may or may not be submarine flows. Forming a small but distinctive part of the association are reddish limestones, siliceous argillites, and red jasper, generally with small, irregular lenses and pockets of manganese minerals. Always in close association with the pillows, they are most abundant near the tops of flows. The pillows are commonly green in color, grading to dull reddish or purplish at the rims. The cores are almost uniformly vesicular or amygdaloidal.

The alteration so characteristic of the spilites is best seen under the microscope. The original plagioclase (labradorite) is largely changed to albite. The remainder of the rock has largely changed to chlorite, calcite, jasper, hematite, and zeolites. Most olivine and pyroxene is completely altered, but fresh grains of the former occur in some quickly chilled and glassy tuffs and agglomerates.

The codistribution of spilites, red siliceous and limey rocks, and manganese mineralization and the similarity of the association along and across strike is not chance (Park, 1946, p. 317). It suggests a common genetic process, "best explained by considering the associated products to have resulted from the alteration of the flows" or "introduced by solutions that accompanied the flows at the time of their formation." Park (1946) argues a largely submarine origin involving extrusion of lava on the sea floor, subdivision of the flow into pillows, rapid covering by new flows, and thermal agitation of the overlying water, resulting in deposition of limestone and volcanic debris. The limestone would be formed by release of CO_2 from bicarbonate in solution by agitation and a rise in temperature (Kania, 1929). Calcium was supplied in large quantities by alteration of the calcium-rich plagioclase in the lavas to albite. Retention of heat in the thick pile of lavas, insulated by a blanket of limestone, would keep the interiors of the pillows molten while seawater trapped in the interspaces or in the underlying muds circulated upward. Exchange of sodium in the seawater for the calcium in the plagioclase would accomplish the albitization.

$$2 \, Na^+ + CaAl_2Si_2O_8 + 4 \, SiO_2 \rightarrow Ca^{++} + 2 \, NaAlSi_3O_8$$

Silica would be derived by solution from the associated muds and decomposition of ferromagnesian minerals such as pyroxenes. Manganese and iron liberated by the same decomposition would migrate upward to form the manganese oxides and silicates and the red oxide coloration of the lime-stones and jasper. The vesicular nature of the pillows attests to the presence of the large quantities of gas expelled during these reactions. "The extrusion of lavas, the accumulation of the red rocks, and the alteration produced, are thought to result from one continuous process" (Park, 1946). Pardee (1927) came to a similar conclusion after studying the spilites of Vancouver Island a few miles to the north.

General Features of Spilitic Rocks[1]

Spilite-keratophyre sequences are a necessary member of the well-known *Steinmann trinity*—pillow basalts, serpentinites and bedded radiolarian cherts (Steinmann, 1927). These strongly sodic lavas are prominently developed in the oceanic "eugeosynclinal" environment with thick sections of gray-wacke.[2] Recently they also have been reported from the ocean floor (Cann and Vine, 1966). Although the spatially associated serpentinites are commonly thought to be emplaced somewhat later, the appearance of serpentinite detritus in the cherts and interbedded shales suggests contemporaneity, at least in part (Bailey and McCallien, 1953, 1960). The mafic members of the spilitic association, spilites, are sodic, albite-rich "basalts." The less-prominent felsic members, keratophyres, are highly sodic, albite-rich "andesites" and "dacites." They consist of submarine lavas and equivalent mafic intrusives and felsic tuffs. Spilites have also been dredged from the deep-ocean bottoms, where they appear to be clearly subordinate to the normal oceanic tholeiitic basalts (Mathews et al., 1965; Melson and Van Andel, 1966; Cann, 1969; Wiseman, 1937). Many pillow-lava accumulations now exposed along the continental margins may have been erupted on or near midoceanic ridges and later crumpled against or thrust up onto the continental margin in response to ocean-floor spreading (Moores and Vine, 1969; Vine and Moores, 1969; Bailey et al., 1970). The spilites often (but not always) form *pillow lavas* especially along the borders of flows, and most (but not all, e.g., Wells and Waters, 1935; Battey, 1956; Carlisle, 1963; Yagi, 1964; Lidiak, 1965) pillow lavas are spilitic, especially older examples. The nonspilitic pillow lavas are of normal basaltic type (Shand, 1949; Waters, 1955a; Davies, 1968; Varne et al., 1969). Some spilites alternate with nonspilitic flows (Gilluly, 1935), indicating that the upper spilitic flows at least were not developed by later metamorphism or hydrothermal alteration of normal basalts. Pillow lavas in the Franciscan formation of Cali-

---

[1]Ophiolites, a term long used in Europe and becoming more widely used in North America, include spilites, gabbros, and peridotite-serpentinite bodies, in many cases in that sequence (e.g., Vuagnat, 1952; Brunn, 1952; Gees, 1956; Dietrich, 1967; Thayer, 1967; Rocci and Juteau, 1968; Bailey et al., 1970).

[2]"Eugeosynclines" have been characterized, in contrast to "miogeosynclines," by the presence of ophiolites (e.g., Aubouin, 1965, pp. 151-159).

fornia, and elsewhere, in many cases have spilitic cores and tholeiitic basalt (normal nonspilitic) rims (Bailey et al., 1964; Vallance, 1960). The pillows are generally one to several feet across, the bottom of each bulbous pillow sagging into the spaces between the underlying pillows. Although less common, andesitic, rhyolitic, and other felsic pillow lavas are widely distributed (see Snyder and Fraser, 1963). Although metamorphosed and deformed spilites are common, many spilites are completely undeformed.

Mineralogically, the pillows generally consist of augite (or its alteration product actinolite, chlorite, epidote, calcite) and albite (typically An_{0-8}) or oligoclase of a low to intermediate structural state (low temperature). The larger pillows tend to have a coarser, diabasic core of albite, augite or chlorite, and sphene and a chilled, glassy to fine-grained variolitic (millimeter-sized spherulites of radiating albite) chlorite-rich rim surrounded by chloritic material. Minor quartz may be present, especially in cavities. The cores are often vesicular or amygdaloidal and show a radial columnar jointing.

Spilites are commonly dominant and may be associated with keratophyres (e.g., Gilluly, 1935; Lidiak, 1965), but in a given area either may be dominant. Mineralogically, *keratophyres* consist largely of albite with a felty or trachytic texture, lesser chlorite, and epidote. In some instances, quartz or potassium feldspar forms a major component. Phenocrysts of albite, hornblende, diopsidic augite, or even quartz may be present.

Chemically, spilites and keratophyres have the following characteris-

Table 4-1 *Average chemical composition of spilites (124 analyses)**

	Spilites		Oceanic Tholeiitic Basalt	Alkaline Olivine Basalt
	AVERAGE	RANGE		
SiO_2	48.8	32.08–63.58	49.5	47.1
TiO_2	1.3	0.05– 4.27	1.8	2.7
Al_2O_3	15.7	10.41–23.25	15.2	15.3
Fe_2O_3	3.8	0.05–13.31	2.4	4.3
FeO	6.6	0.35–16.52	8.0	8.3
MnO	0.15	0– 1.36	0.17	0.17
MgO	6.1	1.37–14.29	7.9	7.0
CaO	7.1	0.07–18.54	10.9	9.0
Na_2O	4.4	0.31– 6.95	2.7	3.4
K_2O	1.0	0– 4.55	0.26	1.2
P_2O_5	0.34	0– 1.22	0.21	0.41

*Analyses from those listed in Vallance (1960, 1965), Wiseman (1937), Hopgood (1962), Cann and Vine (1966), Yoder (1967), Cann (1969), and Hekinian (1971). Since individual samples are taken from different parts of pillows, the "range" probably depends in part on variation between core and rim of single pillows. Oceanic tholeiite and alkaline olivine basalt included for comparison (from Table 4-8, p. 171).

Table 4-2 *Average chemical composition of core and rim of spilitic basalt pillows (12 samples)**

	Cores of Pillows		Rims of Pillows	
	AVERAGE	RANGE	AVERAGE	RANGE
SiO_2	50.16	38.27–56.16	40.34	32.08–52.74
TiO_2	1.78	0.36– 2.88	2.49	0.27– 4.44, 15.78
Al_2O_3	15.15	10.62–18.81	15.95	12.80–20.29, 2.46
Fe_2O_3	3.22	1.73– 4.63, 13.31	5.28	1.10– 9.15, 49.86
FeO	6.59	4.20– 9.20, 1.73	10.34	1.58–16.52
MnO	0.16	0.09– 0.29	0.20	0.03– 0.31
MgO	5.08	1.75– 7.96	8.21	4.52–14.29, 1.86
CaO	7.16	2.17–14.06, 0.44	7.70	2.75–16.44, 0.17
Na_2O	4.93	2.90– 6.70	1.75	0.31– 2.97, 4.28, 4.58
K_2O	0.38	0– 1.10	0.71	0– 0.96, 1.76, 4.14
P_2O_5	0.32	0.02– 0.52, 1.22	0.16	0– 0.47, 1.47

Extreme values noted in range are all from one sample that is not included in average	Extreme values except for Na_2O, K_2O, and MgO are all from two samples that are not included in average

*Compiled from analyses in Vallance (1965, 1960), Vuagnat (1949), and Hopgood (1962).

tics compared with normal flood basalts and andesitic volcanics. Although their silica content is normal (about 50%), the spilites are commonly high in *Na₂O*, low in MgO, and low in CaO (see Table 4-1). Keratophyres show somewhat similar differences compared with more felsic volcanic rocks. With respect to individual pillows, the central part tends to be higher in Na_2O, SiO_2, and in many cases CO_3 and lower in Fe_2O_3, MnO, TiO_2, and H_2O than the rims, whereas the matrix is very high in FeO, MgO, and H_2O, and low in Na_2O, CaO, and silica (Vuagnat, 1949; Vallance, 1960, 1965; Hopgood, 1962). Compared with "normal" oceanic tholeiites, the cores of spilite pillows (Tables 4-1 and 4-2) are clearly "spilitic," being high in Na_2O and low in CaO and MgO. The thin rims, on the other hand, are lower than oceanic tholeiites in Na_2O, and SiO_2, and a little higher in FeO, Fe_2O_3, and MgO. In addition to these internal variations, spilites have a wide, apparently irregular chemical variation from place to place, presumably a reflection of their origin.

Suggested Origins of Spilitic Rocks

The mineralogical, chemical, and textural peculiarities of spilites and keratophyres seem to point to some mode of origin involving modification of more normal basalts, andesites, and related rocks by soda-rich solutions. Most suggested origins for spilites and keratophyres are variations on this. *Differentiation of a parent basalt* of either tholeiitic or alkaline olivine affinities could be followed by:

>*Assimilative reaction* with rocks deep in geosyncline (Turner and Verhoogen, 1960; Hopgood, 1962).

Concentration of Na^+ and water in magmatic or deuteric fluids (Eskola, 1925; King, 1937; Vuagnat, 1946; Niggli, 1952; Battey, 1956; Wilshire, 1959; Lidiak, 1965; Yoder, 1967) causing later alteration of the earlier crystallized basalt (i.e., autometamorphism).

Contamination by flowing into (or intruding) *wet sediments* or into *seawater*, entrapping the Na^+ and giving off Si to form chert, Mn and Fe to form Mn ores, and Ca and Mg to form limestone—rocks commonly associated with spilites (Clapp, 1927; Eskola et al., 1937; Park, 1946; Turner and Verhoogen, 1960; Hopgood, 1962; Yoder and Tilley, 1962; Lidiak, 1965; Donnelly, 1963, 1966; Yoder, 1967).

Redistribution of elements with adjustment to low temperature, hydrous conditions, in conjunction with any of the above processes (Daly, 1933; Vallance, 1960) or during diagenesis or low-grade metamorphism under the ocean bottom (Vallance, 1965; Cann and Vine, 1966; Melson and Van Andel, 1966; Cann, 1969; Hekinian, 1971).

Soda metasomatism of crystallized basalt by saline water squeezed out of buried sediments (Turner and Verhoogen, 1960).

Later low-grade terrestrial *regional metamorphism and metasomatism* of submarine volcanic rocks (Sundius, 1930; Waters, 1955a; Battey, 1955; Lidiak, 1965; Smith, 1968; Coombs et al., 1970). Some geologists would not call these rocks spilites.

Hydrothermal alteration of preexisting rocks (Knopf, 1912; Fairbairn, 1934; Schwartz, 1939). Again, some would not call these spilites.

One alternative, suggested for some spilites in which the albite and chlorite appear texturally to be magmatic (e.g., Benson, 1915; Scott, 1951; Battey, 1956), is that of a *parent spilite magma* that would give keratophyre by differentiation. Specialized conditions different from normal crystallization and obtained only irregularly from place to place in orogenic belts would have to be proposed. Even under such peculiar conditions, it seems unlikely that augite would crystallize along with albite and chlorite. The unlikelihood of a spilitic magma has been confirmed experimentally by Yoder and Tilley (1962) and Yoder (1967).

Any suggested origin for the spilite rocks must take into account the characteristic associated rocks, and chemical and textural variation within the spilites, including:

Associated chert, limestone, serpentinite, graywacke, and in some cases manganese mineralization.

Associated with and gradational to unrelated tholeiitic basalts.

Very common pillow form.

Mineralogy consisting of albite and chlorite with or without augite in the core, grading to albite with more abundant chlorite at the rim. These are characteristically low-temperature "metamorphic" minerals.

High Na_2O especially in the cores of the pillows, low MgO, CaO, as compared with normal (oceanic tholeiitic) basalts.

Developed in "eugeosynclinal" or submarine environment.

Irregular distribution from place to place.

One origin that seems to fit most of these characteristics involves contamination of a normal basalt by flowing into saline water in the ocean, near-surface marine sediments, or buried sediments. As suggested by Park and others, the heated and trapped water would percolate throughout the still-molten lava, producing lower-temperature alterations and exchanging materials. Na from the water would affect the pillows, removing Ca and some Mg to combine with carbonate in the seawater to form limestone. Silica dissolved especially from the muddy matrix would precipitate as chert. Mg and Fe from the pillows would appear in the chloritic matrix, the Mn forming oxides and silicates of its own. A diagramatic representation of these relationships is shown in Fig. 4-1. Why so few of the ocean-bottom basalts appear to be spilitic and why the rims of pillows are so low in Na_2O are difficulties with this interpretation.

Perhaps a better explanation is low-grade metamorphism of solid basalt at shallow depths under the ocean floor as suggested by Vallance (1965), Cann and Vine (1966), and Cann (1969). The reactions and associated secondary products would presumably be much like those described above and in Fig. 4-1. Heat for metamorphism could be provided from later overlying flows (Cann, 1969) or by later dike injection near the crest of a midoceanic ridge. Prolific dike injection in the vicinity of midoceanic ridges, where much basalt is erupted, now seems quite well established (Henson et al., 1949; Moores, 1969; Vine and Moores, 1969; Moores and Vine, 1969; Thayer, 1969) and could provide both the necessary heat for a long enough time and the irregular distribution of spilitization. An exposed example of such a midoceanic dike zone may exist in the form of the bulk of the Troodos ophiolite complex in Cyprus (Bishopp, 1952; Gass, 1967, 1968) and on Macquarie Island south of New Zealand (Varne et al., 1969). About 90% of this part of the Troodos complex consists of tens of thousands of nearly vertical basaltic dikes intruded into basaltic lava and overlying peridotite. The cross section is about 60 miles (100 km) long. Under this spilitization model, Na^+ and H_2O, which are needed to alter each pillow throughout, would be supplied by contamination of the still-hot or reheated basalt by seawater and/or soda-rich sediments under

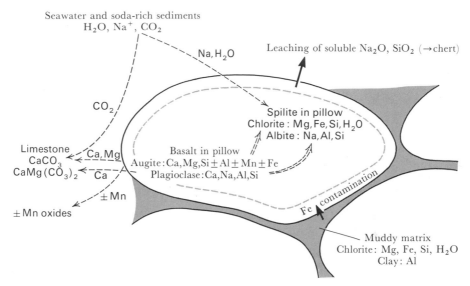

Fig. 4-1 *Diagramatic representation of the main reactions inferred as occurring during the formation of spilite from normal basalt. The muddy matrix between the pillows is shaded.*

the ocean bottom. Na^+ and SiO_2 could be leached out of the originally fine-grained or glassy rims of the pillows during cooling or possibly subsequent submarine weathering. Iron would be oxidized and concentrated in the pillow rims at the same time. The possibility of such submarine leaching has been confirmed by the experiments of Iiyama (1961). He showed that SiO_2 and Na_2O are preferentially leached from basalt into initially pure water at 200 to 400°C. Solubility was augmented by the presence of CO_2. Chlorite (as in the rims of spilite pillows) was found in the altered basalt at 200°C and amphibole at 400°C. Submarine weathering of tholeiitic basalt has been shown to decrease SiO_2 and to increase H_2O and oxidation of iron (Miyashiro et al., 1969b). Lack of decrease in Na_2O during weathering suggests that spilite pillow rims may lose Na_2O during cooling.

ALPINE PERIDOTITE-SERPENTINITE BODIES

The Alpine Peridotite-Serpentinite Association of the Central Dominican Republic[1]

The West Indies island arc, in which the Dominican Republic is situated, presumably represents a stage in the development of a "submarine volcanic geanticlinal pile" or a "eugeosyncline" along the lines suggested in Chap. 1 (see also Hess, 1960b; Mattson, 1960). The most prominent plutonic igneous feature in the core of the island arc is a mass of serpentinized peri-

[1] Bowin, 1966.

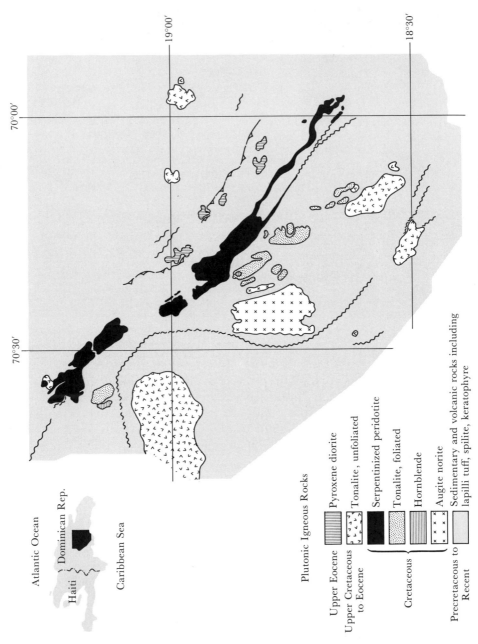

Fig. 4-2 *Serpentinite and associated igneous rocks in the central Dominican Republic (after Bowin, 1966).*

Plutonic Igneous Rocks

Upper Eocene — Pyroxene diorite

Upper Cretaceous to Eocene — Tonalite, unfoliated

Cretaceous:
- Serpentinized peridotite
- Tonalite, foliated
- Hornblende
- Augite norite

Precretaceous to Recent — Sedimentary and volcanic rocks including lapilli tuff, spilite, keratophyre

Atlantic Ocean

Dominican Rep.

Haiti

Caribbean Sea

70°30' 70°00' 19°00' 18°30'

104

dotite about 2 miles wide and 60 miles long (Fig. 4-2). It apparently lies along a fault that has a trend parallel that of the flanking regional metamorphic rocks. The steeply dipping body is black on fresh outcrops and commonly shows blue to green shear surfaces. The degree of serpentinization averages more than 50% and is complete in many places. Least-serpentinized samples of the peridotite contain major olivine, clinopyroxene, and/ or orthopyroxene, with minor fine-grained magnetite or chromite. The serpentinized peridotite weathers to a reddish-brown color.

Distributed irregularly throughout this intrusion are fine- to medium-grained inclusions, a few centimeters to 370 m across. They most commonly show rounded corners and are bordered by a reaction rim of black serpentine a few centimeters wide. The relict texture, consisting of "felted laths of plagioclase" (sodic andesine) with interstitial amphibole and chlorite, indicates that the parent rocks were hypabyssal dioritic intrusive rocks. The lack of a foliation or cataclastic (fractured or broken mineral grains) texture suggests thermal alteration of the inclusions without deformation.

Bowin (1966) notes that the evidence favoring a magmatic versus a cold tectonic emplacement is conflicting. That favoring a magmatic emplacement includes: (1) irregularities in the outline; (2) an irregular, bulbous northwest end; (3) lack of cataclastic texture in the least-serpentinized peridotite; (4) a nonsheared sample of porphyritic peridotite and of coarse-grained olivine clinopyroxenite; and (5) the apparent thermal metamorphism of dioritic inclusions. That favoring a cold tectonic emplacement includes: (1) different metamorphic rocks on opposite sides of the peridotite, (2) straight overall trend, (3) shearing in the serpentinized peridotite, (4) breaking up of diorite into blocks, and (5) apparent lack of contact metamorphism adjacent to most of the body. Bowin suggests magmatic intrusion at slightly greater depth, with later tectonic movement up to its present position.

Plutons of hornblendite, augite norite, and tonalite were intruded in the region of the peridotite-serpentinite (ultramafic) mass prior to emplacement of the latter. All were intruded as igneous bodies at moderate to shallow depths. Hornblendite is coarse grained to porphyritic and forms small stocks of less than 1 sq km in area. The augite norite is commonly a gray to black, coarse-grained rock, forming bodies up to a small batholith in size. Cataclastic textures are found near the borders of the batholith. Tonalite (quartz-rich quartz diorite) is a light gray to white, medium- to coarse-grained foliated rock, containing muscovite and biotite as the main accessory minerals. It forms several small to medium-sized stocks.

Intrusions of unfoliated tonalite, pyroxene diorite, and gabbro are thought to postdate the ultramafic mass. The unfoliated tonalite is a light to dark gray, medium- to coarse-grained hornblende-bearing rock, forming stock- to small-batholith-sized plutons. Pyroxene diorite and gabbro are medium-grained gray-green rocks intruded as small stocks and sills.

The peridotite-serpentinite mass was apparently emplaced after deposition and low-grade regional metamorphism of mafic volcanic rocks and quartz keratophyre and following submarine extrusion of basalt and deposition of tuff, graywacke, and red chert. This latter unmetamorphosed association, including the pillowed nature of many of the basalts, indicates development of the spilitic rock association considered in the previous section. Following emplacement of the peridotite-serpentinite body, volcanism recommenced with keratophyre, dacite, pyroxene andesite, tuff, and minor limestone and siltstone. Still later deposition involved especially limestone, clastic sedimentary rocks, and tuffs.

General Features of Alpine Peridotite-Serpentinite Bodies

Partly to wholly serpentinized peridotite occurs in several plutonic rock associations, generally with or in place of peridotite. These include peridotites in the basal parts of gabbroic layered intrusions, in zoned ultramafic "stocks," as minor associates of granitic complexes, and as large sheets and lenses in folded "geosynclinal" sediments of orogenic belts where they are called "alpine peridotites."

These alpine peridotites are the most important and widespread of the peridotite-serpentinite occurrences. They occur as concordant sheets to irregular bodies, a few feet to several miles thick and a few to several hundred miles long and commonly in en echelon swarms or in two linear series about 120 miles apart (Hess, 1955) along the axis of a fold belt. Characteristically, they occur with spilites and cherts (Steinmann, 1927), where they make up the so-called Steinmann trinity, and with graywackes, gabbro, and basalt as part of ophiolite complexes (e.g., Thayer, 1967, 1969; Moores, 1969; Moores and Vine, 1971; Dewey and Bird, 1971; Bailey et al., 1970). They tend to be associated with major faulting, but whether the serpentinite controlled the faulting or a preexisting fault guided the emplacement of the serpentinite is seldom certain.

Serpentinization of the peridotite is variable—in some instances evenly distributed, in others irregularly so, and in still others increasing toward the margins of the body (e.g., Benson, 1918; Challis, 1965b; Chidester, 1968). The serpentine appears as a green to black, highly schistose (Fig. 4-3) to massive material that commonly weathers to a distinctive orange to red-brown color. Common minerals are the serpentines (platy, Fe^{3+}-rich lizardite; fibrous chrysotile; platy, SiO_2-rich antigorite), brucite, magnetite, and to a lesser extent magnesite and talc (Coleman, 1971). Although commonly massive, it may show somewhat irregular to lenticular but sharply bounded "gneissic" layering. Foliation and lineation are common parallel to the layering but may cross it (Thayer, 1960). Contact meta-

Fig. 4-3 *Typical shearing (upper half of outcrop) and blocky jointing (lower half of outcrop) in serpentinite. On ridge east of highway, 9 miles north of Ukiah, northern California. Hammer with 1-ft-long handle on sheared part of outcrop.*

morphism adjacent to the peridotite is generally absent but may be slight to moderate. Forming thin selvages, several inches to a few feet thick, on many serpentinite bodies is a zone rich in hydrous calcium-aluminum silicates such as hydrogarnet, diopside, idocrase, prehnite, and epidote ("rodingite"), along with a chloritic zone perhaps an inch thick adjacent to the serpentine. These seem indicative of a relatively low temperature of final emplacement (e.g., Suzuki, 1954; Coleman, 1967b; Leonardos and Fyfe, 1967). Other selvages consist of a thin zone of chlorite rock ("black-wall") adjacent to the country rocks, with a talc rock ("steatite") and in turn a talc-carbonate rock adjacent to the serpentine (e.g., Hess, 1933; Cady et al., 1963; Thayer, 1966; Coleman, 1967b; Chidester, 1968). See Fig. 4-4.

Some peridotite bodies consist solely of peridotite or serpentinized peridotite, but others contain associated gabbro in notable proportions (see Thayer, 1960). Dikes of gabbroic or pyroxenite composition are quite common, cutting either the peridotite or the wall rock; and inclusions of peridotite in gabbroic phases are locally abundant. Normally, therefore, the associated gabbroic phases follow crystallization of the peridotite. Close interlayering and lack of chilling of gabbro in dikes that cut the peridotite indicate that in at least some instances, the gabbro and peridotite are comagmatic (Flett, 1946; Thayer and Brown, 1961; Thayer, 1967, 1969; Thayer and Himmelberg, 1968).

Peridotites and serpentinized peridotite have also been dredged from midoceanic ridges and oceanic trenches (Quon and Ehlers, 1963; Hess, 1964; Bowin et al., 1966; Melson et al., 1967; Bonatti, 1968; Aumento,

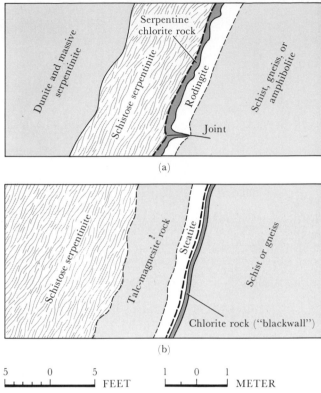

Fig. 4-4 *Diagramatic sketch maps illustrating the two principal types of zonal relations at the contacts of ultramafic complexes. (a) Zonation of rodingite (calc-silicate rock) and serpentine-chlorite rock; (b) zonation of chlorite rock ("blackwall") and steatite. Heavy broken line marks inferred position of original contact of the intrusive rock. After Chidester (1968, p. 346).*

1969; Engel and Fisher, 1969; Fisher and Engel, 1969; Miyashiro et al., 1969a; Thayer, 1969). Their characteristics in this oceanic and in the continental "alpine" environment are so similar as to imply a common origin (Thayer, 1967, 1969; Gass, 1967; Bonatti et al., 1970).

Mineralogically, the unserpentinized peridotite averages about 80% olivine and 20% pyroxene and contains no plagioclase (Thayer, 1960). Dunite is common in large masses. Gradations from peridotite to pyroxenite and gabbro may be found. Texturally, they are allotriomorphic (anhedral grains) to cataclastic or crushed and rarely poikilitic (e.g., Turner, 1942; Challis, 1965b). Anhedral grains of chromite a few millimeters in diameter are common in dunite. Streaks and pods or masses of chromite up to a few hundred feet thick may form ore bodies.

Serpentinization

Although the opposite has been proposed (e.g., Marmo, 1958), serpentinization is generally thought to postdate crystallization of the olivine and other peridotite minerals. The simplest reaction involving formation of serpentine from olivine requires the addition of water and silica:

$$3\ Mg_2SiO_4 + SiO_2 + H_2O \rightleftharpoons 2\ Mg_3Si_2O_5(OH)_4$$

$\qquad$ (olivine) $\qquad\qquad\qquad\qquad\qquad$ (serpentine)

Where serpentinization is localized at the margins of the peridotite body, the required water could be derived from the intruded wet rocks, the reaction occurring at about 430 to 470°C and 1 to 3 kbars (Bowen and Tuttle, 1949; Yoder, 1967; Scarfe and Wyllie, 1967). Talc and other minerals tend to form above this temperature. Where serpentinization is localized in the central portions of the body, it would seem more reasonable to appeal to some type of autometasomatism—alteration of the early formed crystals by a residual watery solution, in this case containing silica in addition (Hess, 1933; Francis, 1956).

Certain other reactions that involve no addition of silica or include the introduction of CO_2 require the removal of other components (e.g., $MgCO_3$) in solution. Such components would be expected to show up in large quantity in the surrounding country rock. However, such does not normally appear the case (e.g., Challis, 1965b). The above reaction involves a large increase in volume of about 70% over that of the original peridotite (Turner and Verhoogen, 1960). Reactions involving no increase in volume also require removal of components such as MgO and SiO_2 in solution. It would seem, however, that since these combine to form olivine, we would be appealing to solution of olivine to maintain constant volume. We are confronted with the dilemma of maintaining constant volume chemically and the apparent lack of field evidence in favor of expansion during serpentinization. The arguments in favor of minimal change in composition with major expansion (e.g., Milovanovic and Karamata, 1960; Coleman, 1963, 1966, 1967b; Green, 1964; Hess and Otalora, 1964; Hostetler et al., 1966; Page, 1967, 1968; Condie and Madison, 1969) versus those in favor of constant volume with major compositional change (e.g., Turner and Verhoogen, 1960; Thayer, 1966, 1967) are not conclusive. Coleman (1971) suggests that static conditions favor expansion, whereas shearing favors constant-volume serpentinization. Postemplacement serpentinization through metamorphism of an ultramafic body has been suggested by Miyashiro (1966). Episodes of diapiric rise of peridotite-serpentinite bodies may accompany further serpentinization and expansion with successive periods of orogeny (Milovanovic and Karamata, 1960; Coleman, 1967b). Serpentinization resulting from weathering or alteration by groundwater has been documented by Barnes et al. (1967) and Barnes and O'Neil (1969).

Dikes of pyroxenite and related rocks cutting the peridotite may result from similar addition of water along fractures at temperatures above the stability limits of serpentine (McTaggart, 1971).

The Problem of Emplacement Temperature

Pure magnesian olivine crystallizes at 1890°C. Admitting, as in nature, about 10% fayalite (Fe member) in the olivine formed, lowers the final crystallization temperature by about 150°C, as seen from the olivine phase diagram (Fig. 1-11). The presence of 5% more silica (to form 10% enstatite) would be expected to lower the final crystallization temperature to about 150°C below 1557°C, as seen from the right-hand side of Fig. 1-11, or about 1400°C. How much a few thousand bars water pressure, corresponding to the depth of emplacement of peridotite-serpentinite bodies, would lower the final crystallization temperature is a matter of conjecture (compare, for example, Figs. 1-3 and 1-5). A rough comparison with the effect on the crystallization temperature of plagioclase suggests that it may amount to a few hundred degrees Centigrade. If so, we might expect that a peridotite "magma" could remain mobile to temperatures down to perhaps 1200°C but not much lower even if the magma contained significant water. Experimental melting of a typical serpentine under 500 to 1,000 bars P_{H_2O} (Clark and Fyfe, 1961) showed the beginning of melting at 1300°C. With intrusion at greater depths under higher water pressures, partial melting should occur at still lower temperatures.

A few high-temperature peridotite intrusions have been described (e.g., Roots, 1954; MacKenzie, 1960; Chesterman, 1960; Green, 1964; Challis, 1965b; Melson et al., 1967; Kornprobst, 1969). MacKenzie (1960) estimates, for example, that a Venezuelan peridotite was emplaced at about 800 to 1000°C. High-temperature intrusions emplaced at the deepest levels contain Al-rich pyroxenes and have high-grade metamorphic aureoles. Intrusions emplaced at shallower levels contain Al-poor pyroxenes and have low-grade aureoles (Moores and MacGregor, 1968). Most bodies, however, seem to be emplaced at much lower temperatures, at least at the level at which we now see them.

Suggested Origins of Peridotite-Serpentinite Bodies[1]

Melting of peridotite in the mantle under a downfolded "geosyncline" (Hess, 1938b)

---

[1]Note that some of these suggestions are not incompatible, even for a single occurrence. Compare, for example, the last six possibilities.

Accumulation of olivine in the pipe or magma chamber of a basaltic volcano (Lauder, 1965; Challis, 1965b; Challis and Lauder, 1966; Osborn, 1969b; McTaggart, 1971)

Intrusion of a hot (800 to 1000°C) largely crystalline body, where strong contact metamorphism of the country rocks is indicated (MacKenzie, 1960)

Intrusion of an olivine crystal mush, lubricated by a small quantity of intergranular magmatic liquid or water absorbed from invaded sediments (Bowen, 1917, 1928; Bowen and Tuttle, 1949; Thayer, 1960; Turner and Verhoogen, 1960; Barth, 1962; Yoder, 1967; Thayer and Himmelberg, 1968)

Diapiric rise of hot mantle peridotite, the decrease in pressure resulting in partial melting and segregation of basaltic and other lower temperature magmas (Green and Ringwood, 1967b; Oxburgh and Turcotte, 1968a; Ringwood, 1969; Maxwell, 1969; Green, 1970)

Downward intrusion of ultramafic mass during tectonic activity because of greater density than surrounding crustal rocks (McTaggart, 1971)

Submarine eruption of ultramafic magma under deep-ocean conditions (Bailey and McCallien, 1953; Gass, 1958; Grindley, 1958; Clark and Fyfe, 1961)

Submarine landslides, mudflows, or turbidity currents from serpentinite protrusions on the sea floor or continents (Lockwood, 1971)

Extrusion and flow of "solid" cold serpentinite under surface conditions (Dickinson, 1966; Cowan and Mansfield, 1970)

Intrusion of "cold" solid or nearly solid bodies along preexisting faults (e.g., Benson, 1926; Turner, 1930; Turner and Verhoogen, 1960; Coleman, 1962; Ernst, 1965; Chidester, 1968)

Slivers of solid mantle caught up in the deep orogenic belts (Mattson, 1960; Hess, 1954, 1960b, 1966b; [?] Steuber and Murthy, 1966; Coleman, 1967b)

Segment of solid oceanic mantle thrust up onto or into continental plate (Dietz, 1963a; Gass, 1967, 1968; Davis, 1968; Davies, 1968; Davies and Milsom, 1969; Thayer, 1969; Hamilton, 1969; Bailey, Blake, and Jones, 1970; Medaris and Dott, 1970; Page, 1970; Coleman, 1971; Gilluly, 1971; Dewey and Bird, 1971)

Any origin proposed for peridotite-serpentinite bodies must account for the following characteristic features:

Most commonly found in "eugeosynclinal" environment (or island arc).

Associated spilites and cherts (Steinmann trinity).

Associated gabbroic and basaltic (or spilitic) phases in the form of sharply bounded layers, dikes, or inclusions quite common (ophio-lites).

Localization along fault or zone of displacement—typical but not invariable.

Mineralogy consisting of dominant olivine with or without pyroxene and chromite. Normally no plagioclase.

Wide range of amount and distribution of serpentine—usually part serpentinization but may be 0 to 100%. Serpentine localized at margins of body or in core or uneven.

Foliation and lineation may be well developed.

Strong deformational evidence typical but not invariable.

Contact metamorphism of country rocks typically absent or minor; strong adjacent to a few intrusions.

There seems to be no single hypothesis that best fits all the available information on origin and development of alpine peridotite-serpentinite bodies. Clearly, some bodies were emplaced at high temperatures. Most, however, were apparently emplaced at relatively low temperatures, but the lack of effect on the surrounding rocks could also be attributed to inflow of water from the country rock into the hot intrusion (e.g., Turner and Verhoogen, 1960). Almost certainly, the peridotite was not 100% liquid when emplaced, as the temperature required would be prohibitive. If only a small amount of interstitial fluid was present as a lubricant, permitting lower temperatures of intrusions as previously noted, fractured and crushed grains would be expected and are found.

A reasonable sequence of development of peridotite-serpentinite bodies, which could be terminated at any stage to give rise to any individual body, is as follows:

Formation of an olivine crystal mush through partial melting in the mantle and removal of basaltic constituents (e.g., spilitic suite) or through accumulation of olivine from a crystallizing basaltic magma.

Deformation and injection of this crystal mush along a zone of weakness such as a fault, or possibly eruption onto or somewhat below the sea floor.

Absorption of water from the country rocks if wet, with resulting marginal to irregular serpentinization, with or without some exchange and modification of the country rocks.

Intermittent movement due to tectonic activity, with gradual cooling; internal to irregular serpentinization due to residual magmatic fluids. Possibly descent because of greater density of peridotite or rise because of expansion and decreasing density during serpentinization.

Further movement of "cold" mass.

Perhaps a better explanation for the large sheetlike peridotite-serpentinite bodies that occur along major thrust faults in the "eugeosynclinal" orogenic belts is that they are slices of oceanic crust and uppermost mantle (e.g., Hess, 1954, 1960b, 1966b; [?] Barth, 1956; Mattson, 1960; Dietz, 1963a; Gass, 1967, 1968; Davies, 1968; Thayer, 1969; Bailey et al., 1970; Gilluly, 1971; Dewey and Bird, 1971). The oceanic crust probably consists essentially of a thin layer of sediments overlying oceanic basalts (primarily oceanic tholeiite[1] and spilite), which grade downward into gabbro and, in turn, a few kilometers below, into the peridotite or serpentinized peridotite—that is, an "ophiolite" sequence. The recently developed concept of ocean-floor spreading (see Chap. 1) suggests that the oceanic crust moves toward a continental margin and eventually down a subduction zone, most commonly near the continental margin. There this basalt-gabbro-peridotite sequence may become incorporated in the lowermost crust of the continental margin. The upper, lighter, basaltic layers may be largely scraped off to form parts of the "eugeosynclinal" assemblage along the continental margin. The lower layers, consisting primarily of peridotite, may become further serpentinized by addition of water during this deformation. Their density would drop from about 3.2 down to as low as 2.6, depending on the degree of serpentinization. At the same time, the accompanying expansion, loss of strength, and decrease in coefficient of friction would aid their rise into the core of the orogenic belt along major thrust faults. Associated gabbros and even basalts would thereby be related by common derivation from the oceanic crust as suggested by the authors listed above.

[1]See section on flood basalts below.

SYNOROGENIC SUITES

BASALT, ANDESITE, RHYOLITE ASSOCIATIONS

The Andesitic Volcanic Rocks of Mount Rainier, Washington[1]

Forming the high volcanic cones and earlier Tertiary volcanic rocks of the Cascade Range of Washington, Oregon, and northern California are members of the basalt, andesite, rhyolite association. Mount Rainier and its associated volcanic rocks provide a magnificent, well-exposed example (Fig. 4-5). The early to mid-Tertiary igneous activity can be divided into three major episodes: (1) eruption of an immense pile of submarine to subaerial fragmental, andesitic volcanic debris and flows; (2) eruption, following uplift and erosion, of a few thousand feet of rhyodacite ash flows; and (3) eruption of olivine basalt to basaltic andesite and minor rhyolite lava flows and mudflows.

Flows of the first unit are basaltic andesites, containing phenocrysts of plagioclase, augite, and hypersthene in a fine-grained to glassy groundmass. Massive unsorted mudflows and flow-top breccias are interstratified with the flows. Tuff breccia, consisting largely of unsorted, angular fragments up to a few inches in diameter in a volcanic graywacke matrix, forms the greatest part of the formation. Plugs of brecciated basaltic andesite apparently represent volcanic vents. All these rocks are dark gray to maroon in color.

Ash flows of the second unit are white to dark gray rocks containing abundant flattened and aligned pumice fragments and rounded and embayed phenocrysts of quartz. Subhedral to euhedral phenocrysts of plagioclase and a few of potassium feldspar also occur. Some flows of this unit were deposited as a "swirling hot avalanche of pumice and glass dust" that swept across the "landscape with catastrophic violence" (Fiske et al., 1963, p. 2). Fragments of bedrock saprolite (clay resulting from deep weathering), stream-rounded pebbles, and bits of wood were carried tens of feet up into the lower part of the lowermost ash flow.

Brown to black flows of the third unit contain conspicuous phenocrysts of plagioclase, olivine, augite, and hypersthene. They "covered much of this part of the Cascade Range," though individual flows were not extensive. Some formed small, elongate, shield volcanoes fed by dike swarms of diabase and basalt of similar composition. All these units are characterized by greenish-gray colors resulting from a later low-grade regional alteration (zeolite facies metamorphism).

Following folding, faulting, uplift, and erosion of these units, a

[1]Fiske, Hopson, and Waters, 1963; Coombs, 1936.

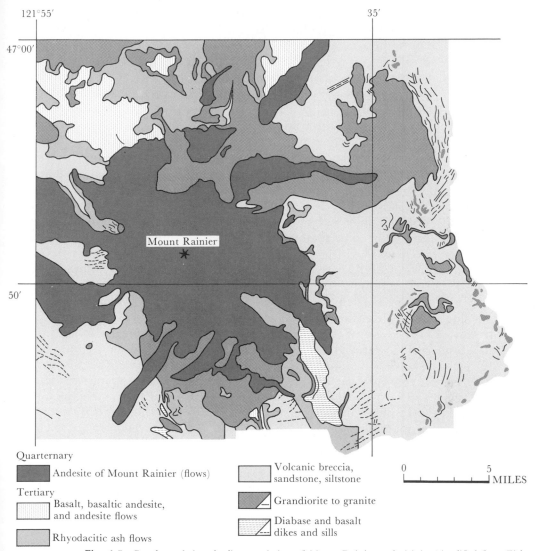

Quarternary

■ Andesite of Mount Rainier (flows)

Tertiary

Basalt, basaltic andesite,
and andesite flows

Rhyodacitic ash flows

Volcanic breccia,
sandstone, siltstone

Grandiorite to granite

Diabase and basalt
dikes and sills

0   5
|__|__|__|__|__| MILES

Fig. 4-5 *Basalt, andesite, rhyolite association of Mount Rainier and vicinity (simplified from Fiske, Hopson, and Waters, 1963).*

shallow mid-Tertiary granitic pluton and associated stocks, sills, and dikes were intruded. The small batholith consists mainly of medium-grained hornblende and biotite granodiorite, grading irregularly upward to somewhat finer-grained quartz monzonite. The pluton was so shallow that the near-surface parts of the magma contain vugs and swarms of large vesicles through bubbling-out of the contained gases. In places, the magma actually broke through to the surface "with explosive violence, shattering the partly congealed granitic rocks intruded earlier, erupting pyroclastic material on the surface, and solidifying within the vent as inclusion-filled welded tuff, vitrophyre, and rhyodacite" (Fiske et al., 1963, p. 40). Such loss of water in the volatiles would cause sudden crystallization of the remaining magma as suggested above by consideration of Fig. 2-4. At least one of the vesiculating plugs that broke through to the surface gave rise to a several-hundred-foot-thick flow of rhyodacite welded tuff. The latter is a glassy to aphanitic rock consisting, as seen under the microscope, "mainly of glass shards and tiny fragments of pumice, so thoroughly compacted and welded that before devitrification the rock must have resembled obsidian or perlite. The shards are draped around abundant rock fragments and broken minerals" (Fiske et al., 1963, p. 55).

Mount Rainier volcano grew during Pleistocene time on the extensively eroded granitic pluton. It developed as a huge composite volcano, now reaching 14,410 ft above sea level. The cone is built primarily from lava flows of uniform pyroxene andesite composition. Thick flows consist of porphyritic light-gray lava with chilled, black glassy bases and sides. The basal parts of thick flows show columnar jointing normal to the cooling surface. Interflow slaggy reddened breccias formed by steam explosions at the base of the lava as it flowed over mud, melting snow, or ice. Some such breccias grade gradually upward into massive flows. Thin flows tend to be vesicular and partly oxidized to shades of red, brown, and gray. "Where thin lava streams slid downhill mixing with large amounts of slushy snow and melt water, the entire flow was disrupted by steam explosions, and the resulting mixture of mud and blocky debris continued down the slope in mudflows" (Fiske et al., 1963, p. 74). A few beds of breccia and pumice lapilli accumulated as air-fall pyroclastics. The porphyritic andesite plug filling the main vent is now a massive solfatarized rock consisting of opal with some kaolinite, tridymite, cristobalite, and pyrite. Vertical dikes radiating outward from the summit consist of porphyritic andesite like that of the lava flows and probably represent fissure feeders to some of the flows. A few small, young cones of olivine andesite lava developed on the flanks of the mountain and thin falls of light-colored pumice and ash blanketed the area as recently as 500 years ago. Steam still issues from a crater in the small summit cone, and infrared photos show a recent rise in temperature; but no other recent eruptive activity has been verified.

Mineralogically, the lavas are predominantly pyroxene andesite, containing phenocrysts of progressively zoned plagioclase (An_{45-60} cores to An_{30-45} rims), hypersthene, augite (sometimes with some early Ca-poor augite), rare olivine, and rare brown hornblende. Accessory minerals include magnetite and apatite. Cristobalite or, less commonly, tridymite lines cavities in some porous lavas. Xenoliths of the underlying granodiorite occur in some flows, but they show no effects of assimilation.

Fiske, Hopson, and Waters suggest that the andesite magma of Mount Rainier formed by fusion of crustal rocks. Gneisses and schists derived from "geosynclinal" sediments and volcanic rocks would be partially melted by extreme regional metamorphism at depth, to form a quartz diorite magma closely approximating that erupted to form the pyroxene andesite of Mount Rainier.

The Andesite to Rhyolite Rocks of the Elkhorn Mountains, Montana[1]

One of the largest known ash-flow fields (Smith, 1960) is that of the Upper Cretaceous Elkhorn Mountains Volcanics of west-central Montana. Originally covering (and intruded by) the Boulder batholith, especially on the east and northeast, these volcanic rocks form a thick plateau or shieldlike accumulation covering about 3,000 sq miles and once covering as much as 10,000 sq miles, and accumulated to a thickness of more than 10,000 ft. The volcanic pile is subdivided into: (1) "a lower unit dominantly of andesitic, rhyodacitic, and basaltic pyroclastic" rocks and volcanic sediments, autobrecciated lavas, "mudflows, and a few thin partly welded quartz latitic ash flows; (2) a middle unit characterized by sheets of rhyolitic ash flows, most of which are now welded tuff, and intercallated debris similar to that of the lower unit; and (3) an upper unit dominated by bedded and water-laid tuff and andesitic sedimentary rocks" (Smedes, 1966).

The most important rock types throughout include volcanic breccia and conglomerate, quartz latite tuff, andesitic sedimentary rocks, and andesitic to basaltic lava flows. The volcanic breccia and conglomerate include both monolithologic (fragments are all the same rock type) and polylithologic types. The fragments include coarse-grained to very fine-grained and porphyritic basalt, andesite, rhyodacite, and lithic tuff. The monolithologic breccia and conglomerate are deposited in widespread beds commonly 50 ft to several hundred feet thick. They have a sheetlike form, no sorting or grading, no fine ash or mud matrix, and lack explosive features, suggesting formation by autobrecciation of slowly moving lava and some redistribution by mudflows. They probably also include some talus and landslide debris, agglomerate, and rubbly ash-flow deposits. The polylithologic breccia is not so thick or extensive. It includes talus, mudflow, and landslide debris from heterogeneous source areas and grades to well-rounded, clearly water-laid volcanic conglomerate deposited as alluvial fans and channels.

The ash-flow tuffs are thought to have been emplaced as "an avalanche or flow of hot gas-laden ash," the heat and weight causing collapse, flattening, and welding together "into compact sheetlike masses bearing some resemblance to both lava and tuff" (Smedes, 1966). These ash-flow

[1]Smedes, 1966; Klepper et al., 1957.

tuffs consist of fragments of pumice and other rocks in a matrix of devi-trified glass, fine-grained ash, and euhedral and broken crystals of feld-spar, quartz, hornblende, pyroxene, biotite, and magnetite. The pumice fragments, of direct magmatic origin, are generally 1/2 to 8 in. long and one-half to one-twentieth as thick, flattened parallel to the ash-flow sheet. In general, the deposits with flatter fragments are more thoroughly welded. In the more densely welded parts of the sheets, these fragments have feathery or wispy edges, and some are draped around compact for-eign rock fragments and crystals. In hand specimen or outcrop, the pre-ferred orientation (parallelism) of squashed pumice fragments gives the rock a streaky flowlike appearance. The maximum collapse of pumice fragments and the greatest degree of welding occur in the central part of an ash flow, somewhat below the middle. Less welded parts of an ash flow, gradationally upward and downward from the intensely welded parts, are less streaky and less compact. The unwelded top and bottom of an ash flow tend to be loose, structureless, unsorted tuff resembling an air-fall tuff. The latter, however, usually shows some sorting, grading, and more distinct bedding. The ash flows grade in composition from quartz latite in the lower unit to rhyolite in the middle unit. Plagioclase phenocrysts are largely labradorite (An_{60-48}) in the former to andesine (An_{50-32}) in the latter. Other minerals are sparse. Fragments of "foreign" volcanic rocks tend to be abundant, especially in the basal part of an ash flow.

Bedded tuff and andesitic sedimentary rocks in the area have largely been worked by water. Distinctive features in them include channeling and cross-bedding to even bedding, unsorted to well-sorted grains, and a wide range of grain size and mineral and rock fragment content. Some of these rocks show mud cracks, cross-bedding, ripple marks, and mud-flake breccias. Some contain accretionary lapilli (a core of a host-rock grain with a very fine grained concentric rim of ash), lack any other structure, and are interpreted as airborne material.

Lava flows of basaltic composition make up a small part of the lower and middle units of the volcanics. Most are autobrecciated and some are continuous with intrusive breccias that may represent the feeders to these lava flows. Most are medium to dark gray rocks containing 0.3-to-1-cm-long phenocrysts of plagioclase (andesine and labradorite) in an aphanitic groundmass.

Numerous shallow intrusive bodies of andesite and basalt that are associated with the Elkhorn Mountains Volcanics are very similar to the latter, mineralogically and texturally. They probably "represent the source and feeders of the extrusive rocks" (Smedes, 1966). They include dikes, sills, laccoliths, plugs, and irregular bodies. Where sills were intruded into moist, semiconsolidated sediment, the magma in some instances has be-come intimately mixed with the adjacent sediment (forming "peperite").

Shreds, stringers, and digitate blocks of sediment of all shapes and sizes are strewn for up to 20 ft into the margins of the sills.

The volcanics are interpreted (Smedes, 1966) as having erupted, at least in part, from shallow fractionating magma chambers. In the middle unit, "ash flows were erupted when fractures tapped the chamber after the magma had built up a high level differentiate of rhyolite to quartz latite composition and high volatile content." Following expulsion of this differentiate, fracturing "at various times allowed the escape of basaltic, andesitic, or rhyodacitic magma having low to intermediate concentration of volatiles and produced lava and pyroclastic rocks and shallow intrusive bodies."

Following faulting and folding of the Elkhorn Mountains Volcanics, they were intruded by the granodiorite to quartz monzonite of the Boulder batholith.

General Features of the Basalt, Andesite, Rhyolite Association

Rocks of this association are characteristically developed during and immediately following the moderate to strong orogenic movements involved in deformation of the "eugeosyncline." With the possible exception of basaltic members, they never occur in the oceanic environment. They have been called the basalt, andesite, rhyolite association (Turner and Verhoogen, 1960), the andesitic association (Dickinson, 1962, 1970a; Taylor and White, 1965; Dickinson and Hatherton, 1967), the orogenic suite (Taylor and White, 1965), the hypersthenic rock series (Kuno, 1950, 1958), and the calc-alkaline series (e.g., Green and Ringwood, 1968; Best, 1969). Basaltic members have been called high-alumina basalt (Kuno, 1960).

The association includes the chains of prominent stratovolcanoes and associated rocks as exemplified by Mount Rainier and other volcanoes of the Cascade Range of the western United States; the cordillera of Mexico, Central America, and South America; the Lesser Antilles of the eastern Caribbean; the Indonesian arc; central Japan; and the Aleutian arc. These are rocks of the island arcs or other areas along the continental margins. They develop in "eugeosynclinal" rocks over what is generally thought to be oceanic crust (e.g., Taylor and White, 1965; Hess, 1966a), and over the subduction zone of the sinking oceanic crustal plate (e.g., Dickinson and Hatherton, 1967; Dickinson, 1970a; Stoiber and Carr, 1971; Gilluly, 1971). The dominant rock type is a pyroxene (or basaltic) andesite or an aluminous basalt. Minor amounts of more felsic andesite, dacite, rhyodacite, and even rhyolite are common associates. Eruptions include both lava flows and pyroclastics of all sorts, the latter generally outweighing the former. Flows and pyroclastics characteristically alternate in widely varying proportion, and individual eruptions may include both. Associated ash flows (e.g., from glowing avalanches or nuées ardentes) are moderately common in some areas such as Central America but are minor in volume.

As exemplified by the Elkhorn Mountains Volcanics, another part of the association includes the widespread Cretaceous and Tertiary volcanic deposits of southern Nevada, Utah, and parts of California and Colorado; parts of Idaho, western Montana, and Yellowstone Park; Arizona and New Mexico; Mexico and Central America; the Andes of Chile and Peru; the north island of New Zealand; Sumatra; and southeastern Europe. This group of rocks is equally widespread and develops commonly on "miogeo-synclinal" rocks over continental crust, somewhat farther from the ocean. The tectonic environment is also quite different—being generally associated with block faulting (e.g., Gilbert, 1938; Mackin, 1960; Gibson, 1970). The dominant rocks are ash-flow tuffs, thought to be deposited by huge nuée-ardente or glowing-avalanche type of eruptions. Compositionally, these rocks are much more siliceous than those of the basaltic group of the association. They consist of dacite, rhyodacite, quartz latite, and rhyolite (Smith, 1960). More mafic basalt to andesite lavas and tuffs do occur but they are decidedly subordinate. Most of these ash-flow tuffs appear to be erupted from fissures to form deposits ranging in volume from a few to several thousand cubic kilometers (Smith, 1960). Others appear to be erupted from large calderas (e.g., Lipman, 1966; Ratté and Steven, 1967; Lipman et al., 1970; Elston and Smith, 1970). The lower-central parts of ash flows a few tens of feet thick tend to collapse and fuse together to form welded tuffs. Many ash-flow sheets are compositionally zoned, suggesting compositional zoning in the magma chamber (see, for example, Ewart, 1965; Martin, 1965; Fisher, 1966; Smith and Bailey, 1966; Lipman et al., 1966; Lipman, 1967; Byers et al., 1968; Noble et al., 1969).

Mineralogically, rocks of similar composition occur in both the stratovolcanoes and associated rocks and the ash-flow tuff parts of this association. Basalts of the association show some of the mineralogical characteristics of both tholeiitic and alkaline olivine basalt of the flood basalt association described in a following section. Phenocrysts in the basalts and andesites may consist of plagioclase, pyroxenes (augite and/or hypersthene), olivine, and sometimes corroded grains of quartz, sodic plagioclase, or sanidine. The plagioclase often consists of two different varieties, one zoned and one uniform (e.g., Turner and Verhoogen, 1960). For example, the phenocrysts in a hypersthene andesite from the Aleutian Islands consist of irregular Ca-rich cores (An_{88-92}) with sharply bounded, wide, more sodic rims (An_{60-68}), surrounded in turn by a shell of slightly more calcic composition (An_{65}), the same as the smaller unzoned phenocrysts or groundmass plagioclase (Byers, 1959). Olivine phenocrysts, as in tholeiitic basalt, may show reaction rims of hypersthene; but, by contrast to those in tholeiitic basalt, they commonly contain inclusions of picotite (chrome spinel). Groundmass minerals include olivine, often but not invariably surrounded by a reaction rim of pigeonite or hypersthene (Kuno, 1960), augite, hypersthene, and, less commonly, minor pigeonite. Alkali feldspar is common (Kuno, 1968), and silica minerals may occur either in the groundmass or lining vesicles. Mafic xenoliths are sometimes abundant, as are more or less resorbed xenoliths of granitic and sedimentary rocks (Kuno, 1950).

More felsic members of the basalt, andesite, rhyolite association contain hypersthene as phenocrysts (except in rhyolite); and, in the groundmass, they contain augite especially in the andesites, hornblende phenocrysts in andesite to rhyolite, and biotite phenocrysts in rhyolite (Kuno, 1968). In andesite to dacite, phenocrysts of augite often show resorbed (embayed) outlines and are surrounded by reaction rims of hypersthene (Kuno, 1950).

Chemically, the basalts and basaltic andesites of this association are characterized by high Al_2O_3 and so have been called high-alumina basalts (e.g., Kuno, 1954, 1960; Turner and Verhoogen, 1960). Compared with the flood basalts, discussed in a later section (see Table 4-8), they also tend to be somewhat low in TiO_2. Their Na_2O content is similar to that of most tholeiitic flood basalts but less than that of the alkaline olivine flood basalts, and K_2O is greater than in oceanic tholeiites. In general, their alkali contents are intermediate between those for tholeiitic and alkaline olivine flood basalts having similar silica contents (see Fig. 4-19, below). K_2O content for a given SiO_2 in rocks of the association increases in going across a volcanic chain toward the continent (Kuno, 1959, 1960; Dickinson and Hatherton, 1967; Dickinson, 1968, 1970a). Relatively minor basalts in the basalt, andesite, rhyolite association appear equivalent to these flood basalts. Average (and range of) chemical composition of the basalts (predominantly high-alumina basalt) of the basalt, andesite, rhyolite association is tabulated in Table 4-3. Comparison should be made between this and the chemical composition of flood basalts, tabulated in a later section (Table 4-8).

Chemical characteristics of the more felsic members of the basalt, andesite, rhyolite association are less important for distinguishing between the different volcanic rock associations, but the general trends are of interest. With increasing SiO_2, Al_2O_3 tends to fall slightly; $Fe_2O_3 + FeO$, MgO, and CaO fall; Na_2O rises somewhat and K_2O more so. The behavior of iron is rather variable, but most commonly there is no concentration of iron with respect to MgO and alkalis (Bowen, 1928, p. 110). This is in contrast to basalt of certain other associations and under other conditions (Fenner, 1926, 1929). These have been reviewed by Kuno (1968) and Coats (1968). This lack of iron enrichment has been explained as due to high oxygen pressure in the magma (due to high water content) and resulting

Table 4-3 *Basalt and andesite from basalt, andesite, rhyolite association.* Average and range of chemical composition of basalt.*

	Basalt		Andesite	
	AVERAGE	RANGE	AVERAGE	STANDARD DEVIATION
SiO_2	50.16	45.89–53.00†	58.17	4.06
TiO_2	1.20	0.32– 3.07	0.8	0.35
Al_2O_3	17.61	13.55–23.00	17.26	1.56
Fe_2O_3	3.41	Trace–11.13	3.07	1.38
FeO	6.34	0.63– 9.97	4.18	1.62
MnO	0.17	Trace– 0.52		
MgO	6.22	2.89–11.02	3.24	1.24
CaO	9.99	7.39–12.86	6.93	1.63
Na_2O	2.90	1.58– 4.29	3.21	0.72
K_2O	0.85	Trace– 3.33	1.61	0.75
P_2O_5	0.21	Trace– 1.06	0.21	0.15

*Compiled from the literature, primarily from circum-Pacific occurrences; 141 analyses. Average and standard deviation of Cenozoic andesite from Chayes (1969), 1,775 analyses having been compiled from worldwide occurrences. The andesites were accepted as labeled by the original worker.

†Upper limit of SiO_2 arbitrarily taken here as 53.0 weight percent.

oxidation of the iron and its early separation as magnetite (Osborn, 1959, 1962). Further explanation is given above in the section on differentiation (see especially Fig. 3-7). Most andesites are chemically equivalent to quartz-bearing diorite or granodiorite plutons of the same orogenic environment.

Important points with respect to the origin of the basalt, andesite, rhyolite association:

Complete range of compositions from basalt to rhyolite.

High-alumina basalt is the most common, but not the only, mafic end member.

Basalt to andesite is the most abundant range of rock composition in the mafic stratovolcano part of the association developed typically over oceanic rather than continental crust and over an active subduction zone.

More K_2O-rich magmas erupted over deeper parts of active Benioff zones.

Explosive eruptions and fragmental products are characteristic.

Rhyodacite to rhyolite is the most abundant range of rock composition in the felsic ash-flow tuff part of the association developed typically over continental crust in orogenic regions.

Most commonly lesser or no iron enrichment occurs during differentiation. This in contrast to the tholeiitic flood basalts (described below).

The plagioclase often consists of two varieties, one showing complexly zoned crystals and the other uniform. In many instances, a resorbed crystal core is surrounded by a more calcic plagioclase rim.

Partly resorbed xenoliths of mafic, granitic, and sedimentary rocks are sometimes abundant.

Suggested Origins of Basalt, Andesite, Rhyolite Rocks

The wide range of mineralogy and chemistry in the basalt, andesite, rhyolite association, along with the usual lack of separation of the predominantly mafic stratovolcanoes and associated rocks along the continental margins from the predominantly felsic ash-flow tuffs on the continents, has contributed to an equally wide range of suggested origins. Different workers are not even in agreement whether the magma is predominantly mantle- or crust-derived. Suggestions include:

Differential melting of the upper mantle at different depths or fractional crystallization at different depths in the mantle (than for other basalts of the tholeiitic and alkaline olivine flood basalts described below), presumably along the inclined seismic zone dipping under the continental margin (Kuno, 1959, 1966; Yoder and Tilley, 1962; Kushiro and Kuno, 1963; O'Hara, 1965; Dickinson and Hatherton, 1967; Smith and Carmichael, 1968; Gorshkov, 1969) and under hydrous conditions (Hamilton, 1964, 1966; and 30–40-km depth with $P_{H_2O} < P_{load}$, T. H. Green and Ringwood, 1967).

Melting of oceanic crustal rocks dragged under the continental plate (Coats, 1962; Gilluly, 1969, 1971; Press, 1969; Hamilton, 1969; Green and Ringwood, 1969; Souther, 1970; Dickinson, 1969, 1970a; Cristensen, 1970).

Fractional crystallization of parent basaltic magma (Bowen, 1928; Battey, 1966) under high-water, oxidizing conditions, resulting possibly from the expansion into a large reservoir for slower cooling and more extensive differentiation (than for flood basalts) (Osborn, 1959, 1962, 1969a; Byers, 1961; Waters, 1962; O'Hara, 1965, 1968; Kuno, 1965, 1968; Roeder and Osborn, 1966; T. H. Green and Ringwood 1967b; Best, 1969).

Fractional crystallization of different parent basalt magmas to give different andesites (Katsui, 1961; Kuno, 1969a).

Assimilation by basalt magma (especially high-alumina basalt) of crustal rocks (Tomkeieff, 1949; Kuno, 1950, 1968; Wilcox, 1954; Waters, 1955a, 1962; Buddington, 1959; Byers, 1961; Dickinson, 1962; Kleeman, 1965; Hamilton and Myers, 1967; Coats, 1968; Ewart and Stipp, 1968; Peterman et al., 1970; Vitaliano, 1971; Gilluly, 1971) or of crust-derived granitic magma (Holmes, 1932; Williams, 1942; Macdonald and Katsura, 1965; O'Hara, 1965) or combination fractional crystallization of basaltic magma and assimilation of granitic crustal rocks (Larsen et al., 1940; Kuno, 1960; Aoki and Oji, 1966; Hamilton, 1966; Hedge, 1966; Noble and Hedge, 1969).

Differential melting of crustal rocks by heat from basaltic magmas presumably derived from the mantle (Steiner, 1958; Clark, 1960).

Differential melting of "new" crustal rocks to produce magmas of rhyolitic to andesitic composition (Holmes, 1932; Smith, 1960; Turner and Verhoogen, 1960; Dickinson, 1962; Waters, 1962; Fiske et al., 1963; Lidiak, 1965; Mattson, 1966; McBirney and Weill, 1966; Souther, 1967, 1970; Ewart and Stipp, 1968; Armstrong et al., 1969; McBirney, 1969a, 1969b; Forbes et al., 1969; Doe et al., 1969).

Differential melting of old crystalline basement of the lowermost crust or upper mantle (Peterman et al., 1969; Pichler and Zeil, 1969; Rittmann, 1971).

Magma of a granitic pluton may break through to the surface to form volcanic materials (Billings, 1943; Lapadu-Hargues, 1947; Smith, 1960; Geze, 1962; Ustiev, 1965; Hamilton and Myers, 1967; Souther, 1967; Tabor and Crowder, 1969; Cater, 1969).

Any single origin proposed for the basalt, andesite, rhyolite association as a whole will encounter the differences in chemistry and environment between the dominantly mafic stratovolcano part of the association formed over oceanic crust in the "eugeosynclinal" environment and the dominantly felsic ash-flow tuff part formed over continental crust largely in the "miogeosynclinal" environment. If it is acknowledged that the differences may result from difference in origin, the task is simplified. Support for this separation of different parts of the association is also found in lower Sr^{87}/Sr^{86} for andesitic ash flows (around 0.7035) than for rhyolitic ash flows (mostly 0.704 to 0.710) (Williams and McBirney, 1969).

With this in mind, each of the first five suggested origins noted above, or any combination of them, would be largely compatible with formation of the mafic stratovolcano part of the association, whereas any or some combination of the last four suggested origins would be compatible with formation of the felsic ash-flow tuff part of the association. Formation of basalt and basaltic andesite of the mafic stratovolcano part by partial melting of basalt or peridotite of the descending lithospheric plate seems to be most reasonable, especially where it intersects the seismic "low-velocity zone." This zone at about 50- to 150-km depth is thought to contain a small amount of partial melt (Anderson et al., 1971).

The common occurrence of more sodic cores of plagioclase, partly resorbed and rimmed by more calcic plagioclase, can perhaps be explained by experimental fractional crystallization of quartz diorite magma at high pressures (Green, 1968, 1969). Liquidus plagioclase is An_{44} at 13.5 kbars, An_{51} at 9 kbars, and An_{59} at 0 kbar. Crystallization of plagioclase at high pressure followed by rapid movement toward the surface could result in resorption and crystallization of more calcic rims.

The problem of formation of basalt of somewhat different compositions in the upper mantle will be pursued at greater length in the discussion of the flood basalts. Perhaps the most striking compositional difference of this part of the basalt, andesite, rhyolite association from the other basaltic associations is the greater abundance of the intermediate and felsic rock types such as andesite, rhyodacite, and rhyolite. This seems to require either more extensive fractional crystallization under somewhat different conditions, melting of more felsic basaltic rocks of the oceanic crust rather than ultramafic rocks of the upper mantle, selective assimilation of more felsic crustal rocks, or some combination of these. The lack of iron enrichment, presence of both complexly zoned uniform plagioclase crystals and xenoliths argue for one or a combination of these possibilities. Assimilation of more felsic old crystalline basement rocks can also be inferred by the attainment of high strontium isotope ratios (see pp. 89–94).

Formation of the felsic ash-flow tuff part of the association seems to require, because of its felsic composition and large volume, an origin by partial melting of crustal rocks—that is, an origin like that of the large granitic plutons (generally of similar age and tectonic environment) described below. Many granitic plutons and their associated volcanic rocks, in fact, are thought to be cogenetic (e.g., Billings, 1943, 1956; Mackin, 1960; Ustiev, 1965; Smedes, 1966). Whether both the felsic volcanic rocks and the granitic plutons are derived from the same magma chamber or the granitic plutons represent the magma chamber for the volcanic rocks or they are independent of one another must be determined for individual occurrences. The origin of granitic magma by partial melting (anatexis) of new "eugeosynclinal" crustal rocks will be discussed at some length in the following section on granitic plutons. The high volatile content and high viscosity of the erupting granitic magma account for the rapid vesiculation and development of abundant ash flows and other pyroclastic materials, especially in this part of the basalt, andesite, rhyolite association. Although high, the volatile content of the magma is probably less than that required for saturation, for otherwise a drop in water pressure would cause crystallization before the magma could reach the surface (see, for example, Figs. 2-4 and 4-9). With an initial water pressure significantly less than load pressure, the granitic magma could come quite close to the surface before completely crystallizing. The rapid increase of water-vapor pressure to saturation, coupled with the increase in viscosity, could account for the eruptive characteristics of an ash flow.

Involvement of felsic old crystalline basement rocks, either by assimilation or possibly anatexis, can be inferred from high Sr^{87}/Sr^{86} ratios (e.g., Noble and Hedge, 1969).

GRANITIC BATHOLITHS AND STOCKS

The Sierra Nevada Batholith, California[1]

The rocks constituting the Sierra Nevada batholith form a large part of a nearly continuous belt of plutonic granitic rocks extending from Baja California in the northwestern part of Mexico, north through the Sierra Nevada Range, into northeastern California and adjacent parts of Nevada. Of this, the Sierra Nevada batholith is exposed over an area about 300 miles long and 40 to 60 miles wide (see Fig. 4-6). The batholith was intruded into Paleozoic and Mesozoic sedimentary and volcanic rocks just west of the boundary zone between the cordilleran "miogeosyncline" and "eugeosyncline." These country rocks are preserved in the walls of the batholith, roof pendants, septa, and inclusions. The "eugeosynclinal" rocks west of the batholith consist of marine shale, graywacke, and conglomerate, interbedded with intermediate to mafic volcanic rocks—andesitic tuff and breccia, and basaltic lavas including some spilite pillow lavas. Thinly interbedded chert and lenses of limestone appear in some places. East of the batholith, the Paleozoic to early Mesozoic rocks consist of limestone, quartzite, siltstone, and shale, whereas the Mesozoic rocks are "eugeosynclinal" and much like those to the west. These Mesozoic

[1]Bateman and Wahrhaftig, 1966; Bateman et al., 1963; Moore, 1959; Moore et al., 1961, 1963; Hurley et al., 1965; Loomis, 1966; Putnam and Alfors, 1969; Evernden and Kistler, 1970; Bateman and Dodge, 1970.

Fig. 4-6 *The Sierra Nevada batholith; a composite granitic batholith of Jurassic to Cretaceous age. Intrusive contacts of the batholith (including around roof pendants) are outlined in heavy lines. Fault contacts, contacts of individual plutons within the central best-known part of the batholith, and contacts of satellitic bodies are outlined in lighter lines. Moore's quartz diorite line running the length of the batholith is outlined in gray dashes. Modified from Bateman and Wahrhaftig (1966); Bateman et al. (1963); Moore [1969, Copyright (1959) by the University of Chicago].*

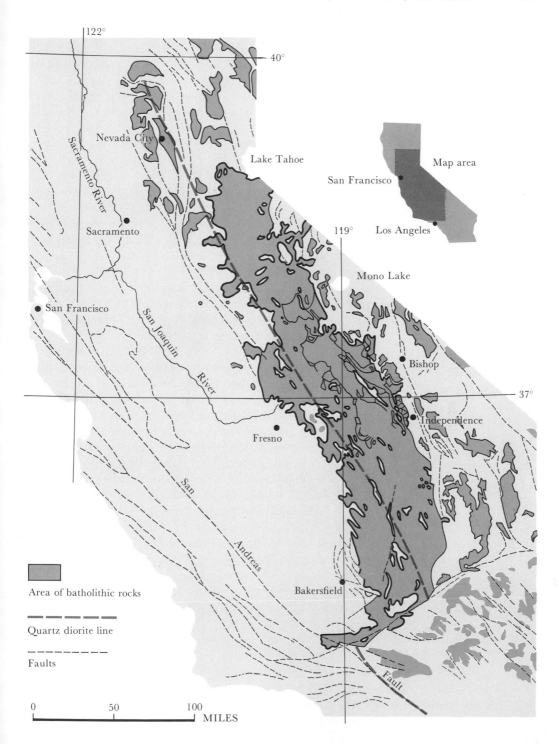

Area of batholithic rocks

Quartz diorite line

Faults

0   50   100
MILES

rocks include graywacke, felsic pyroclastic rocks (both ash-flow tuffs and air-fall tuffs), and interlayered mafic flows. Most of these country rocks were subjected to greenschist facies metamorphism (relatively low grade) before emplacement of the batholith.

Structurally, these country rocks conform to a complexly faulted synclinorium, the axial part of which is occupied by the Sierra Nevada batholith. Thus the stratigraphic units are grossly parallel to the borders of the batholith. The west limb of the "synclinorium" recently has been interpreted as forming by stacking up of oceanic plates being dragged under the continental margin (Hamilton, 1969). The synclinorium and associated steeply dipping strike faults probably "began to take form in Permian or Triassic time" (Bateman and Wahrhaftig, 1966) before emplacement of the batholith. The Nevadan orogeny near the close of the Jurassic caused the principal folds in the rocks west of the batholith and was the strongest disturbance in the region. The main plutons of the Sierra Nevada batholith were intruded before, during, and after this disturbance.

The rocks of the batholith consist predominantly of compositions from "quartz monzonite" (i.e., granite) to granodiorite but include many rocks ranging from quartz diorite to leucocratic granite. The individual plutons that make up the batholith range from numerous small stocks, some less than 1 sq mile in outcrop area, to a few large batholiths, some more than 500 sq miles in area. All the large plutons and some of the smaller ones are elongate north-northwest parallel to the length of the batholith, but most of the smaller ones are rounded, irregular, or elongate in other directions. In general, the major plutons in the western part of the batholith are older than those along the crest of the range, and in some areas such as the Yosemite region, the plutons appear successively younger toward the east, but the pattern of intrusion is not simple. Isotopic ages indicate five major epochs of plutonic igneous activity, 30 million years apart and each about 10 million to 15 million years long: 210 million to 195 million years ago (Middle and Late Triassic), 180 million to 160 million years ago (Early and Middle Jurassic), 148 million to 132 million years ago (Late Jurassic), 121 million to 104 million years ago (Early Cretaceous), and 90 million to 79 million years ago (Late Cretaceous) (Evernden and Kistler, 1970). The use of intrusive relations observed in the field, along with the isotopic ages, shows that the oldest group of plutons lies along the east side of the batholith. The known Late Jurassic plutons are confined to the western part of the batholith. The youngest group of plutons occurs in a belt about 25 miles wide along the crest of the range. Broad chemical and mineralogical changes also occur across the batholith. Grossly, the granitic rocks are more mafic (also have more FeO, MgO, CaO), less silicic, and less potassic (and less K_2O/SiO_2) toward the west (Bateman and Dodge, 1970). Moore (1959) has recognized a "quartz diorite line" running through the batholith, continuing north through Washington, and in fact (Moore et al., 1963) continuing the full length of western North America (see Fig. 4-10). This line divides granitic rocks of dominantly quartz diorite composition on the west from rocks of dominantly "quartz monzonite" (granite) and granodiorite composition on the east. Room for the granitic magma would be provided by concurrent settling of the country rocks downward around the magma. Some plutons even may have become entirely detached from their source and be underlain by country rock.

K-feldspar "phenocrysts" (megacrysts) are restricted to rocks of intermediate to calcic "quartz monzonite" (granite) or potassic granodiorite composition. Because they contain inclusions of many other minerals, the "phenocrysts" are thought to have formed during the later stages of crystallization, but because of dimensional orientation near external contacts and in dikes, they are thought to have formed before complete consolidation. The presence of identical porphyroblasts of K-feldspar in some adjacent wallrocks or inclusions "indicates pressure-temperature conditions like those in the magmas, but does not indicate that the phenocrysts in granitic rock are porphyroblasts" (Bateman and Wahrhaftig, 1966).

Mafic inclusions occur in many of the mafic "quartz monzonites," granodiorites, or quartz diorites. These inclusions range in greatest dimension from a few inches to a few feet, appear elongate in outcrop, and are disc-shaped in three dimensions. In some places, they are rounded or irregular in shape. Progressively flatter inclusions toward the margins of many plutons suggests shaping of soft inclusions by movements of the magma. Mafic inclusions are abundant only in hornblende-bearing granitic rocks, especially near the margins of plutons. Many such concentrically zoned plutons are lower in mafics and free of both hornblende and mafic inclusions in their interiors. This concentric zoning is most obvious in the field from variation in the content of mafic minerals and is best determined in the laboratory by measurement of the specific gravities of systematically collected hand specimens. Zoning may also be lateral or, in some large plutons, patchy or streaky. "Some compositional zoning can be related to contamination by wallrocks, but much zoning, particularly concentric zoning, is independent of the composition of the wallrocks and probably represents differentiation processes within the magma" (Bateman and Wahrhaftig, 1966). Or it may represent assimilation below the levels of exposure.

Joints are of three major types. The earliest sets, cooling joints, can be identified where they are occupied by dikes and dike swarms related to the batholith. A regional system of steeply dipping conjugate joints (oriented at equal angles across a central line), prominent in the eastern Sierra Nevada, crosses boundaries between plutons. The joints are nearly at right angles (NE and NW) and spaced at less than one inch to thousands of feet. Within a given region, the widest spacing tends to be in the coarsest rocks. This regional set formed after consolidation of the batholith, as indicated by the regional extent and continuity across pluton boundaries. The latest joints form gently dipping sheets that are subparallel to topographic surfaces, and are presumably related to unloading during erosion of the landscape.

Development of the Sierra Nevada batholith is envisioned (Bateman and Wahrhaftig, 1966; Bateman and Eaton, 1967) as occurring in the axial zone of the synclinorium, where the sialic upper crust was downfolded and thickened, "possibly as the result of convective overturn in the earth's mantle" (Bateman and Wahrhaftig, 1966, p. 125). Moore (1959) infers that his "quartz diorite line," which runs the length of the batholith, lies close to the edge of the continental crust and that (p. 206) "granitic rocks emplaced east of the line are generated within a thick sialic layer, whereas those emplaced west of the line are developed within the sima or a thinner sialic layer with great thickness of associated geosynclinal sediments and volcanic rocks." The granitic magmas (Bateman and Wahrhaftig, 1966; Bateman and Dodge, 1970) "were generated by the melting of sialic rocks of the upper crust as a result of their being depressed into deeper regions of high temperature" (Bateman and Wahrhaftig, 1966, p. 122), a contribution coming from radioactive disintegration in the thickened downfold of sedimentary rocks. The first melts, formed at temperatures between 600 and 700°C, at a depth of, say, 50 to 60 km, would be high in the constituents of quartz, orthoclase, and albite. Further melting would produce melts richer and richer in calcium, iron, and magnesium, and poorer in water; progressively from perhaps granite, to granodiorite, to quartz diorite. The density of granitic magma is lower than that of solid granitic rock and would permit the magma "to work upward in the manner of a salt intrusion, exploiting lines of structural weakness wherever possible" (Bateman and Wahrhaftig, 1966, p. 123).

One major period of melting presumably occurred in the Late Jurassic, since Late Jurassic sedimentary rocks retain the effect of the regional metamorphism accompanying the rise in temperature; and K-Ar dates have shown crystallization of many of the plutons to have occurred also in the Late Jurassic. Whether the early Late Cretaceous plutons are related to this melting or to a separate episode is uncertain. The initial Sr^{87}/Sr^{86} ratio (see pp. 89–94 at the end of Chap. 3) of the granitic rocks of the central Sierra Nevada has been determined by Hurley et al. (1965) as 0.7073 ± 0.0010. From this they interpret the granitic magmas to have originated from a possible mixture of one-third basalt and two-thirds sial, either by contamination of basalt by sial or by partial melting of that mixture in the "geosyncline."

An eastward increase in measured heat-flow values in the batholith (Lachenbruch, 1968) correlates with calculated heat production from variable radioactivity measured across the batholith and in the country rocks. This also suggests that the granitic magmas were generated by partial melting in "eugeosynclinal" rocks to the west, and "miogeosynclinal" and Precambrian basement rocks to the east (Wollenberg and Smith, 1970).

A contrary explanation for the origin of the granitic magma is by melting along the Benioff seismic zone inclined under the present position of the batholith (Hamilton, 1969) or over a linear zone of high-heat flow like the crest of an oceanic rise (Kistler et al., 1971).

Development of Granitic Plutons through Anatexis in the Shuswap High-grade Regional Metamorphic Complex, British Columbia[1]

The Shuswap metamorphic complex of southeastern British Columbia consists of a series of gently dipping gneissic, schistose, and granitic rocks. The regional metamorphism culminated in the Jurassic Period with development of the sillimanite zone of the upper amphibolite facies. Deformation of the complex in the deep zones of the mountain belt was complex and approximately concurrent with the high-grade metamorphism. At the same time, partial melting (anatexis) accompanying metamorphism of the plagioclase-rich quartzofeldspathic pelitic gneisses and schists developed voluminous granitic melts. The resulting network of concordant to crosscutting veins, lenses, and dikes amounts to more than 20% (and averages about 30%) of an extensive 9,000-ft section of the complex. These dominantly concordant granitic veins range in thickness from less than an inch to a few feet, in general becoming thicker (up to tens of feet thick) and more abundant to higher levels. Although many of the thin concordant veins originated by partial melting in situ, it is clear that much material has moved upward from deeper levels. For one thing, the relative amount of granitic material is too great to have been entirely formed in situ. Also, the similarity in composition and texture of veins in host rocks as diverse as granodiorite, amphibolite, and marble suggests external derivation in part.

Examination of a ternary composition diagram for granitic rocks —$NaAlSi_3O_8$–$KAlSi_3O_8$–SiO_2 (see Fig. 3-1)—but modified to take into account a reasonable water pressure and calcium content (Fig. 4-7) will help illustrate the evolution of the granitic melts. The first melts formed by anatexis would be near (see Weill and Kudo, 1968) the ternary minimum at about 675°C, derived from layers containing quartz, plagioclase, and potassium feldspar. Since the parent gneisses are rich in plagioclase and

[1]Reesor, 1965, 1970; Hyndman, 1968b, 1969, 1971; Little, 1960.

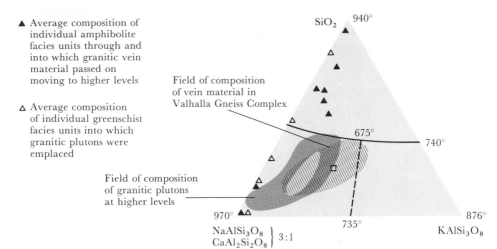

Fig. 4-7   *The ternary system* $NaAlSi_3O_8 - KAlSi_3O_8 - SiO_2$ *extrapolated to 5,000 bars* P_{H_2O} *(from 0 and 2,000 bars), for an average plagioclase composition of* An_{25}. *Plotted from data in Winkler (1967), Tuttle and Bowen (1958).*

contain subequal amounts of quartz and potassium feldspar, increased volumes of melt change composition generally toward the plagioclase corner. The field of composition of the vein material (both formed in situ and injected) in the Valhalla dome of the Shuswap complex is plotted in Fig. 4-7.

As this granitic vein material accumulated in the upper though still high-grade parts of the complex, large volumes of it floated upward or were squeezed off at intervals to still higher levels and into a lower-grade environment. This material was emplaced in the form of steep-sided, sharply bounded granitic plutons of a few to a few tens of square miles in horizontal dimensions, as seen in the lower-grade structural depression north of the Valhalla dome. The compositional field of these plutons is also plotted in Fig. 4-7. As seen from this plot, the pluton compositions are concentrated in the plagioclase-rich part of the field of the vein material in the underlying gneisses. The compositions of the plutons do, however, lap outside this field toward the low potassium feldspar compositions of the overlying metasedimentary and metavolcanic rocks, especially the moderate- to low-quartz compositions of the low-grade rocks into which the plutons were emplaced. This suggests that the rising granitic vein material became somewhat contaminated by the high-grade quartz-rich schists and gneisses through which it had to pass, and was further contaminated by the low-grade pelitic and metavolcanic rocks in which it came to rest. This latest contamination is also apparent in the field, inasmuch as the smaller, highest-level plutons and the border zones of some larger plutons are highest in mafic minerals, plagioclase, and calcium content of plagioclase, and lowest in orthoclase and quartz.

This evolutionary scheme is dependent not only on the compositional environment and grade of metamorphism at the place of melt formation, but on structural environment, distance of migration, and chemical environment at the site of emplacement of the granitic plutons. The importance of these different structural levels is illustrated in Fig. 4-8. In general, those units (15, 17, 19a) that have moved farthest into the lowest-grade rocks at higher levels are richest in plagioclase and mafic minerals. Presumably they are the most contaminated.

As may be seen in the cross section, the anatectic melts have presumably risen a minimum of 5 to 10 km to their site of final emplacement and

Fig. 4-8 *Cross section through upper part of Shuswap metamorphic complex (in Valhalla dome) and overlying low-grade metamorphic rocks (in Nakusp depression) in which granitic plutons were emplaced. Information above the topographic surface is projected up-plunge from outside cross section.* ① *"Hybrid gneiss"—a mixture of older metasedimentary gneisses (quartz-potassium feldspar-plagioclase gneiss, amphibolite, biotite-quartz-plagioclase gneiss, sillimanite schist, and quartzite) and light-colored leucogranite gneiss and pegmatite.* ② *"Veined gneiss"—a mixture of older granodiorite-augen gneiss (hornblende-biotite-quartz-plagioclase gneiss) and younger light-colored granitic material.* ③ *"Mixed gneiss"—a heterogeneous assemblage of amphibolite and metasedimentary layers in hornblende-biotite granite to granodiorite gneiss. Fine dashes indicate concentrations of leucocratic vein material; heavily dashed part indicate more than 50% vein material.*

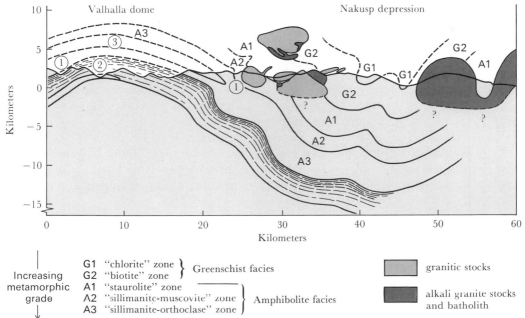

Increasing metamorphic grade	G1 "chlorite" zone ⎫ Greenschist facies G2 "biotite" zone ⎭	granitic stocks
	A1 "staurolite" zone A2 "sillimanite-muscovite" zone ⎫ Amphibolite facies A3 "sillimanite-orthoclase" zone ⎭	alkali granite stocks and batholith

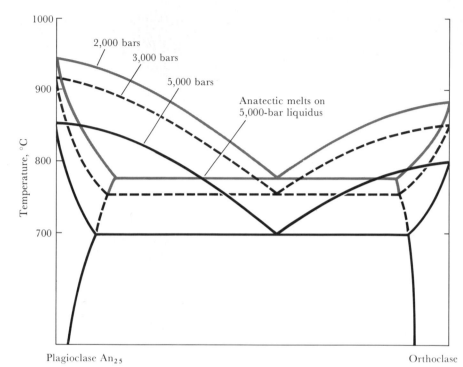

Fig. 4-9 *Diagramatic temperature-composition diagram parallel to the plagioclase-orthoclase side of the SiO_2-plagioclase-orthoclase triangular diagram and through the anatectic melt compositions produced (about 25% quartz). Liquidus-solidus surfaces are shown for 2,000, 3,000, and 5,000 bars P_{H_2O}.*

crystallization. If, as suggested on stratigraphic grounds by Little (1960, p. 99), some 20 km has been eroded from the Valhalla dome since the Middle Jurassic, an approximate pressure of 5 kbars P_{H_2O} may be assigned for the development of the granitic melts.

The initial anatectic melt is formed in quantity at 5,000 bars. This would be 100% melt (i.e., on the liquidus) at, say, 760°C and would have an orthoclase/plagioclase ratio of 40:60 (see Fig. 4-9). The melt may crystallize through any one or a combination of the following (Hyndman, 1969):

1. Melt rises rapidly to a lower-pressure environment at 2,000 bars, remaining at the same temperature and composition until it rests on the solidus, the temperature of which has risen to meet it.
2. Melt remains at same pressure (and depth) and temperature (nearly), assimilating more plagioclase component, crystallizing, and changing composition until it reaches the solidus.
3. Total magma including crystals remains at same pressure (and depth)

and composition and gradually cools and crystallizes until the total mass is on the solidus.

Clearly, for the present example, none of these limiting possibilities is correct by itself. The magmas must have risen at least 5 to 10 km from their point of origin to their point of final crystallization—corresponding to a drop in pressure of at least 1.3 to 2.6 kbars. As may be seen from the diagrammatic temperature-composition diagram, possibility 1 must have been important. Since the magmas clearly became somewhat richer in plagioclase component, as may be seen from Fig. 4-7, possibility 2 must also have played a role. Although there is no direct indication of a fall in temperature of the melt during its rise through cooler rocks and its assimilation of other rocks, both of these effects must have induced some drop in temperature. The amount of drop may well have been small. The amount of minimum rise of magma and degree of assimilation suggests that the magma may have crystallized under about 3,000 bars P_{H_2O} if it was generated under about 5,000 bars P_{H_2O}. If the granitic melt was wet but somewhat undersaturated in water, a higher pressure source would be inferred.

These granitic plutons were emplaced no later than Mid- to Late Cretaceous as indicated by potassium-argon radiometric dates, but following regional metamorphism, as they have intruded and superimposed a contact metamorphism on the regional rocks.

General Features of Granitic Batholiths

The great granitic batholiths, such as the Coast Range batholith of British Columbia, the Idaho batholith, and the Sierra Nevada batholith of California, form immense masses up to tens of miles wide and up to hundreds of miles long. Smaller bodies such as the Boulder batholith of western Montana, and several plutons in southern and eastern California, and in New England have similar characteristics. These bodies characteristically occur within the former "eugeosyncline" and, to some extent, the "miogeosyncline" (e.g., Gilluly, 1965).

Granodiorite and quartz diorite are the dominant rock types, especially in the larger masses, accompanied by lesser amounts of granite ("quartz monzonite"). Quartz monzonite appears to be more prominent in the smaller bodies, especially those on the east side of Moore's quartz diorite line, or in other regions, away from the oceanic side (see Fig. 4-10). Thus, minor associates such as gabbro or ultramafic rocks are not uncommonly confined to small associated intrusives or marginal variants.

As in the Sierra Nevada batholith described above as an example and the Coast Range batholith of British Columbia (e.g., Roddick, 1965; Hutchison, 1967, 1970), most large granitic batholiths are composite bodies,

Fig. 4-10 *The position of the quartz diorite line in northwestern North America and its relation to other features of the continental margin. After Moore et al. (1963).*

themselves made up of small batholiths and stocks of somewhat variable mineralogical compositions, textures, and ages. Despite this variety and resulting complexity, granitic batholiths have many points of similarity, both within themselves and between one another. They tend to be medium grained, hypidiomorphic granular in texture, locally foliated (especially in the earlier and in marginal phases), locally megacryst bearing, and in some cases more mafic rich near the country-rock contacts.

A few minerals are abundant—plagioclase, potassium feldspar, quartz, biotite, and hornblende. *Plagioclase* is generally most abundant, ranging from about 15 to 85% of the rock. It forms medium-grained, subhedral grains that are white in the more sodic varieties (for example, An_{20-35}) to light gray in the more calcic varieties (for example, An_{30-45}). Most grains are zoned; this may be faintly visible in larger grains in hand specimen. *Potassium feldspar* varies from less than 5% to more than 70%. It forms anhedral fine to medium grains, except in coarser, scattered poikilocrysts (poikilitic grains) or megacrysts (large subhedral to euhedral grains). It may be either orthoclase or perthitic microcline and is generally off-white to buff to pink in color. It is most commonly more translucent than the chalky, white plagioclase. Granites in which plagioclase appears only as perthitic streaks in microcline perthite (not as separate grains) have been called *hypersolvus granites* (Tuttle and Bowen, 1958), whereas those in which the plagioclase and potassium feldspar appear as separate grains have been called *subsolvus granites*. As the names imply, hypersolvus granites crystallized above the solvus, whereas subsolvus granites crystallized below the solvus (see Figs. 1-5 and 1-6). Quartz occurs as fine to medium, anhedral, glassy grains, making up less than 10% to more than 50% of the rock. Biotite forms medium-grained, black to dark greenish brown, subhedral flakes ranging up to more than 30% of the rock. Euhedral grains appear in some fairly shallow occurrences. Hornblende forms medium-grained, black to dark green subhedral prisms also ranging up to more than 30% of the rock. Biotite is most prominent in alkali granite to granite, and hornblende is most prominent in granodiorite to quartz diorite or diorite. However, hornblende granites and biotite granodiorites or quartz diorites are not uncommon. Dark clinopyroxene occurs in some of the more calcic rocks, in many cases as a result of contamination of the new magma by calcareous rocks. Hypersthene occurs in the rather dry granites called charnockites. They are typically found in Precambrian metamorphic terrains where their genesis may be somewhat ambiguous—metamorphic (i.e., granitization) or igneous. Muscovite is present in some deep-seated granitic rocks, where its presence is attributed to crystallization under high pressures (see Fig. 4-1 and the overlapping fields of granite minimum melting or crystallization temperature and muscovite stability). It may also develop as a deuteric (late-stage magmatic solutions) or metamorphic alteration of plagioclase. Epidote may develop as an alteration of plagioclase or as a coating on joint surfaces. Sphene is a common accessory mineral that is coarse enough in some rocks to appear in hand specimens as small translucent red-brown grains. The other common accessory minerals, apatite, zircon, magnetite, and ilmenite, are rarely coarse enough to be visible in hand specimen. The normal mafic mineral content of granitic rocks ranges from 5 to 20% in alkali granite to 25 to 50% in diorite. More alkali-rich granites and syenites such as some of those of the White Mountain Magma Series of New Hampshire contain Na- and Fe-rich mafic minerals such as riebeckite, hedenbergite, and fayalite (e.g., Williams and Billings, 1938; Billings and Keevil, 1946; Swift, 1966).

Development of Granitic Melts by Anatexis

As succinctly pointed out by Eskola as long ago as 1933, "granitic magmas have come into existence mainly by two different processes: (1) crystallization differentiation and squeezing out of the residual liquid, . . . and (2) by differential (partial) re-fusion, or anatexis [a term first proposed by Sederholm many years earlier], and squeezing out of the re-fused liquid from older rocks that were heated in one way or another above the melting temperature of the lowest-melting portion of the rock mass mixed with water or other mineralizers" (Eskola, 1933, p. 12).

From heat flow and radioactivity measurements in eastern Australia, Howard and Sass (1964) and R. D. Hyndman et al. (1968) conclude that the temperature at depths of 25 to 40 km in a young continental area will average about 450 to 610°C. This corresponds to a thermal gradient of about 15°C/km. It seems likely that the maximum temperatures reached (due largely to radioactive disintegration) in the "geosynclinal" crust will be significantly higher than this presently "measured" value at some stage and in local areas due to concentrated radioactive "hot spots." Thus temperatures of from about 625 to 725°C should be not uncommon at these depths during some stage of evolution of a "geosynclinal" crust. This would involve thermal gradients of about 18 to 25°C/km. Under these conditions and with water contents approaching saturation (as would be expected), crusted rocks would begin to melt. That such melts may in fact exist today in some evolving "geosynclines" is suggested by recent geomagnetic depth sounding and magnetotelluric measurements (Hyndman and Hyndman, 1968). Considerations such as these, then, strongly support the anatectic origin of at least some of the granitic pegmatites in the abundant veined gneisses described by Sederholm and Holmquist more than 50 years ago.

Such veined gneisses are exceedingly common in deep-seated, high-grade regional metamorphic complexes, including the Skagit gneiss and adjacent complexes of northwestern Washington (Misch, 1952, 1968; Crowder, 1959; Hawkins, 1968), the Shuswap complex of southeastern British Columbia (Jones, 1959; Reesor, 1965, 1970; Hyndman, 1968a), the Precambrian gneisses in the Front Range of Colorado (Lovering and Goddard, 1950), the Wilmington complex of Delaware (Ward, 1959), and the Major paragneiss of New York (Buddington, 1957; Engel and Engel, 1958). Anatexis has probably played an important role in development of many of these veined gneisses, but certainly not all have been developed entirely in this way. Some have formed partly, or in a few cases completely, by injection of vein material or by metasomatism from deeper levels where partial melting is going on. Veined gneisses and criteria for judging their origin are discussed in more detail below. Their characteristics and origin have been reviewed in detail by Mehnert (1968).

Anatectic melts have also been produced experimentally by several workers, especially Winkler and von Platen and their associates. Their experiments, performed largely under a water pressure of 2,000 bars, show that shales and graywackes begin to melt below 700°C to give melts of granitic composition and residua higher in ferromagnesian minerals such as biotite. "The composition and the amount of the melt formed by ana-texis of gneisses depends on, besides H_2O-pressure and temperature, the chemical and therefore mineralogical composition of the gneisses" (Wink-ler, 1967, p. 195). The initial melts formed have compositions lying in the cotectic trough of the quartz-potassium feldspar-plagioclase system (see, for example, Figs. 3-1 and 4-7). The exact composition depends primarily on the anorthite content of the plagioclase, provided the initial rock contained all three of the above components (von Platen, 1965b). The higher the anorthite content of the initial rock, the higher the temperature and the higher the quartz-orthoclase content of the initial melt. Winkler (1967) shows that, beginning with the quartz-orthoclase-albite system (for example, Fig. 3-1), addition of calcium to the system (anorthite to the albite) causes the cotectic trough to move toward the quartz apex—pivoting on the quartz-orthoclase end of the trough. The "minimum" concurrently mi-grates toward the same pivot (away from plagioclase). Winkler's data permit comparison of the initial rock composition (or residual rock) with the melt derived from it by anatexis at 2,000 bars. Similar relationships hold at higher pressures.

During anatexis, therefore, melting begins in quartz-potassium feld-spar-plagioclase−bearing gneisses in which the anorthite content is lowest. Lower melting temperatures are also favored by higher water contents (relative to saturated conditions), as seen in Figs. 3-2 and 3-3, and by higher water pressures. Dry rocks, such as some earlier granites, therefore may melt only at higher temperatures even though they have the requisite mineral content. That most rocks undergoing high-grade regional meta-morphism and anatexis are saturated in water seems likely because through-out progressive regional metamorphism, water is continually driven off. This readily may be seen by comparing the water contents of common low-grade minerals such as chlorite, muscovite, and epidote, with common high-grade minerals such as garnet, biotite, anorthite (component), and sillimanite. The absolute amount of water available at the beginning of melting, however, may be insufficient to produce more than a small amount of melt. Rise of temperature and further melting would undersaturate the melt. As shown experimentally by Wyllie and Tuttle (1961a) and von Platen (1965a), small contents of "mineralizers" such as HCl and HF, which must be present during metamorphism, may also lower minimum melting temperatures or in some instances raise them (HCl on the mini-mum melting of granite: Wyllie and Tuttle, 1964) and alter their compositions slightly. If the granitic melt is undersaturated in water, as suggested by some workers, the water being derived only by melting of 10% muscovite, biotite, or hornblende, the initial melts form between about 705 and 870°C at 2 kbars pressure (Brown, 1970; Brown and Fyfe, 1970; see Fig. 3-3).

Granitic plutons may be found at widely variable depths within the earth's crust, that is, within the epizone, mesozone, and catazone, as reviewed by Buddington (1959). Their characteristics as described below, are largely from Buddington's paper (and written communication, 1970; see also Table 4-4). The *epizone* has been recognized as the uppermost crustal zone. The depth to the base of this zone (as also to the others) must of course be variable, depending on the orogenic environment and on the stage of development. Plutons emplaced in this shallow zone (including most Tertiary stocks and batholiths) may involve a single intrusion or may be composites of several intrusions. Some are associated with ring dikes. Late-stage aphanitic or porphyritic dikes and lamprophyre dikes are common. Distinct pegmatite dikes are rare, but small patches of pegmatitic material may occur locally. Examples of granitic plutons emplaced in the pizone (Buddington, 1959*) include: stocks of the western Cascade Range of Washington and Oregon (including the small batholith under Mount Rainier; Fiske et al., 1963; Erikson, 1969; Cater, 1969); the Boulder (eg., Klepper et al., 1971; Tilling, 1964; Tilling et al., 1968; Biehler and Bonini, 1969), Philipsburg (Hyndman et al., 1972); and Lolo batholiths of western Montana; Alta stock, Utah; Montezuma, Mount Princeton, and Silverton plutons of southwestern Colorado; New Cornelia stock, Arizona; Animas, Hanover, and Organ Mountain plutons, New Mexico; Concepcion del Oro stocks, Mexico; Seagull batholith, Yukon Territory; batholiths of southern Newfoundland; stocks of the White Mountain Magma Series, New Hampshire (e.g., Billings, 1956; Chapman, 1967); Quincy granite, Massachusetts; and Westerly granite and Naragansett Pier batholith, Rhode Island. More recently described epizonal plutons include the Deboville stock, northern Maine (Boone, 1962), and the Rocky Hill stock in California (Putman and Alfors, 1969).

The *mesozone* has been recognized as the next deeper crustal zone, in which are emplaced the majority of the batholiths found in orogenic belts of the world. Generally, the younger intrusions are more alkalic and siliceous. In some, the foliation may be in part systematically oriented at an angle to the outer border. Evidence suggesting upward and outward flowage of magma includes: (1) steep foliation and lineation parallel to the borders; (2) steepening of fold axis and lineation plunges in country rock as the pluton is approached; and (3) crumpling of bedding or dikes

---

*A few more recent references are included for individual plutons.

Table 4-4 *Comparative characteristics of granitic plutons of the epizone, mesozone, and catazone*

	Depth, miles	Contacts	Country Rocks	Pluton
Epizone	0.	Discordant, sharp	No concurrent regional metamorphism; some contact metamorphism; genetically related volcanic rocks common	No planar foliation except near borders; ± chilled borders; miarolitic cavities common
	4.			
Mesozone		Discordant to concordant; sharp or locally gradational contacts	Regional metamorphism to greenschist facies; no related volcanic rocks; well-developed contact metamorphic aureoles around shallower plutons	Most major batholiths; generally composite; planar foliation ± lineation often well developed, generally parallel contact; assimilation may be obvious
	8.			
Catazone		Concordant, only locally discordant	Regional metamorphism to amphibolite to granulite facies; extensive veined migmatites common	Foliation commonly well developed; augen gneisses and porphyroblastic granites, other replacement features common
	12. . . . . ? . . . . ? . . . . ? . . . . (unexposed)			

adjacent to the pluton. In these higher-pressure rocks, miarolitic cavities and chilled borders are absent. Pegmatites and aplites may be common, especially in border zones. The pegmatites may occur in the pluton radially or as inward-dipping, marginal fissures or bordering the pluton as a zone of dike injection. Examples of granitic plutons emplaced in the mesozone (Buddington, 1959) include these pre-Tertiary plutons: Coast Range, Cassiar, and White Creek batholiths, and the stocks of the Nakusp depression, British Columbia (Roddick, 1965; Hutchison, 1970; Hyndman, 1968b); Colville and Wallowa batholiths, Washington; Idaho batholith, Idaho and Montana (Reid, 1959; Hietanen, 1963b; Schmidt, 1964; Taubeneck, 1971); Bald Mountain batholith, Oregon; Sierra Nevada batholith and associated plutons, California; Enchanted Rock batholith, Texas; Snowbank stock, Minnesota; many Acadian granite plutons of New England; LaMotte and Lacorne batholiths of the Canadian Shield in Quebec; Great Slave Lake batholith of the Canadian Shield in the Northwest Territories.

The *catazone* has been recognized as the deepest exposed crustal zone.

Plutons emplaced in this deep zone consist of domes, lens-shaped intrusions emplaced in folded rocks ("phacoliths"), and sheets. They may develop either through metamorphic differentiation and/or injection of magma. Some workers would view many of the "plutons" of the catazone merely as part of the high-grade regional metamorphic complex. However, some such batholiths "may in part crosscut the structural trends of the more rigid members" (Buddington, 1959, p. 715). Examples of granitic plutons of the catazone include: Manawan Lake granodiorites, Saskatchewan; batholiths of the Haliburton-Bancroft area, Ontario; Salmon Lake batholith, New York; the concordant granitic gneiss "phacoliths" of the New York-New Jersey highlands; Killingworth and associated granite domes, Connecticut. These catazonal plutons are most common in the shield areas.

Catazone plutons are largely syntectonic, in the terminology of Eskola (1932a), and most develop at the time of regional metamorphism. Mesozone and epizone plutons roughly can be correlated (but not directly, since they are based on different parameters) with late- and posttectonic plutons. They commonly postdate the peak of regional metamorphism; and, if the metamorphic grade of the surrounding rocks is low enough, they show that they have superimposed a contact metamorphism.

It is tempting to place these granitic plutons of the catazone, mesozone, and epizone in the form of an evolutionary sequence, as has been done by Read (1949, 1951, 1955b) and Hutchison (1970). Read has formulated a "Granite Series" in which the deep-seated, migmatitic, "granitization granites" gradually rise to higher levels with time, becoming separated from their surroundings and evolving into true magmatic, intrusive granites. The degree to which granitization plays a role, even in the deep-seated (catazone) granitic rocks is still argued, but it now generally is considered less important than formerly. This and related genetic problems are pursued further, below.

Radiometric "ages" (K-Ar) in granitic batholiths in western North America have recently been shown to occur in about five main pulses about 30 million years apart (Gabrielse and Reesor, 1964; Evernden and Kistler, 1970).

Means of Emplacement of Granitic Plutons

As a mass of granitic magma rises from its place of origin and during its final emplacement, it must somehow make room for itself. Perhaps the most important means of making room is by forcible emplacement— simple shouldering aside of the country rocks. Mass is conserved by the country rocks squeezing in to fill the "space" left as the pluton passes. Emplacement under shallow conditions may even be accomplished by raising of the roof rocks, as, for example, in the emplacement of a lacco-lith. However, whether the upward movement of the pluton results from stresses arising within the magma itself or from external stresses is largely a matter of conjecture. It may be that deformation of the country rocks (by external means) causes the fluid magma to be squeezed upward through the crust along zones of weakness, much as a bulge of toothpaste is squeezed upward in the tube by externally applied forces. Forcible emplacement is presumably favored by "plastic" rather than brittle country rocks, greater depth resulting in greater plasticity, a constant upward and outward pres-sure, and possibly a compressional environment. Characteristics by which forcible emplacement may be recognized include: (1) regional structure such as formations, bedding, foliation, or fold trends wrapped concordantly around the pluton (e.g., the Bald Rock and Merrimac plutons in north-ern California, Compton, 1955; the Flora Lake stock in Ontario, Heimlich, 1965); (2) deformation of planar structures concentrated adjacent to the pluton (e.g., the White Creek batholith in southeastern British Columbia, Reesor, 1958); (3) development of a foliation in the border zone of the pluton (e.g., the Royal stock in western Montana, Mutch and McGill, 1962; Allen, 1966; and some plutons in the Coast Range batholith, British Columbia, Hutchison, 1970); and (4) all these and development of boudi-nage and lineation in the country rocks such as in the Main Donegal Gran-ite and associated plutons in Ireland (Pitcher and Read, 1960, 1963).

Another important means of emplacement of granitic plutons is by stoping and incorporation of the country rocks. The magma may fracture the country rocks, inject itself into the fractures, and permit the loosened blocks to sink into the magma. This, of course, adds volume to the magma and commonly changes its composition by assimilation. Such stoping is presumably favored by brittle rather than "plastic" country rocks, shal-lower depth resulting in more brittle conditions, zones of weakness such as jointing and bedding planes in the country rock, alternating magma rise with magma relaxation, lifting and stretching of the roof rocks, and possibly a tensional tectonic environment to aid in fracturing and block release. Characteristics by which stoping may be recognized include: (1) structures in country rocks discordant to contacts of the pluton, (2) blocks of country rock visibly pulling away from the contact, and (3) inclusions of country rock (xenoliths) in the pluton. Evidence of assimilation, such as more mafic "contaminated" border-zone rocks, may be a reflection of incorporation of country rocks, their disintegration, and their reac-tion with the magma. Clear examples of stoping are shown by the Marys-ville stock of western Montana (Barrell, 1907; Knopf, 1950), many of the plutons of southwestern Colorado (Lovering and Goddard, 1950), and the Organ Mountain batholith of New Mexico (Dunham, 1935). Related to stoping is cauldron subsidence (stoping along a ring fracture), as is characteristic of ring-dike complexes such as in the White Mountain Magma Series of New Hampshire (e.g., Chapman and Chapman, 1940; Billings, 1943).

A third means of magma emplacement—zone melting—has been proposed by Dickson (1958). He suggests that rise of the magma and emplacement may be aided by melting of roof rocks, with concurrent crystallization of somewhat higher-temperature minerals toward the base of the magma column. The heat provided by the latent heat of crystallization [1] would in part be transferred to the roof region by rising volatile constituents and would contribute to the melting and advance of the magma. Zone melting presumably would be favored by country rocks high in low-melting constituents, slower cooling and easier melting in warmer (deeper-seated) country rocks, and high chemical activity and volatile content of the magma.

Each of the factors involved in magma emplacement is subject to conservation of mass. Hence each must either deform the country rock or remove country rock from one place (e.g., above the magma) and add material elsewhere (e.g., by crystallization of higher-temperature minerals lower in the magma column). However, the volume of magma may change if the heat budget is sufficient to create more melt at the expense of the country rock.

As first suggested by Eskola (1932a), granites probably originate largely by *anatexis* or partial melting during high-grade regional metamorphism (see Fig. 4-11). The granitic melts so formed may be alkali granite in composition if melted from rocks of suitable composition near the low-melting-point temperature (resulting in the low-temperature minerals on Bowen's Reaction Series: quartz, orthoclase, oligoclase, ± biotite). Or they may be more calcic, granite to granodiorite, if melted at higher temperatures (more complete melting) or from more calcic rocks. The residuum from anatexis would be depleted in one or more of the three major minerals in granite. The granitic melts work their way upward along lines of structural weakness to higher levels in the crust, aided by their lower density and deforming forces accompanying metamorphism, coming together to form large bodies of granitic magma. The final location and size of the granitic masses formed will depend on (1) the locus and amount of magma generation; (2) the nature of associated zones of weakness; (3) the temperature, pressure, depth, and composition of the magma; and (4) the deforming forces influencing movement of the magma (Hyndman, 1969). As the magma migrates upward, it commonly assimilates more calcium-rich country rocks resulting in a more typical granodiorite-quartz

---

[1]Latent heat (ΔH) of crystallization is the change in heat content in going from the melt (high H) to the solid (lower H). $\Delta H = T\, dS$, where T = temperature and S = entropy.

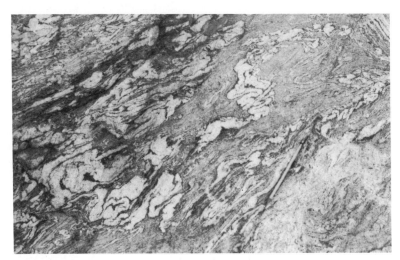

Fig. 4-11 *Pegmatite-aplite streaks probably resulting from anatexis (partial melting) or "sweating out" of granitic fractions during high-grade metamorphism—sillimanite-orthoclase zone of the amphibolite facies. Note especially the dark biotite-rich selvages on the granitic streaks. The mineralogy of the biotite selvages plus the granitic veins essentially equals that of the biotite-plagioclase-quartz gneiss country rock. The biotitic selvages represent the residuum after anatexis. Cirque west of central Kootenai Lake, northern Bitterroot Range, Montana.*

diorite composition. Smaller masses of granite generated near or within less-calcic rocks of the "miogeosyncline" may originate, and remain, more nearly granite in composition. Low Sr^{87}/Sr^{86} ratios in the major granitic batholiths of the "eugeosynclines" are compatible with partial melting within the "eugeosynclinal" graywacke and volcanic rocks (e.g., Peterman et al., 1967; see also pp. 89–94). Also compatible with this origin is the timing of igneous activity and regional metamorphism. Emplacement of common granitic plutons characteristically closely follows regional metamorphism, as noted in the above descriptions of the Sierra Nevada batholith of California and the Shuswap metamorphic complex.

Two other means of *origin of granite* have been supported by many geologists in the past. Both are reasonable alternatives for the formation of small amounts of granitic rock but are considered incapable of forming masses of batholithic dimensions.

The first alternative is *differentiation of a mafic magma* such as of basaltic composition (Bowen, 1928). As clearly shown by Tuttle and Bowen (1958), the end product of crystallization of silica-rich feldspathic magmas approaches the low-temperature point in the cotectic trough of the silica-alkali feldspar system. The final rock, a mixture of alkali feldspars and lesser quartz, should therefore be granitic in composition. By plotting analyses of natural granitic rocks on the same diagram, Tuttle and Bowen have convincingly demonstrated that many granites do fall close to this ternary minimum (see Fig. 3-1). The obvious conclusion is that granitic rocks result from differentiation from a melt, such as basalt. Mafic layered intrusions, discussed below, also show that a basaltic magma crystallizing and differentiating slowly under deep-seated conditions will give rise to a small amount of late-stage granitic melt. Recent experimental studies by von Platen (1965a), involving an addition of varying amounts of Fe and Mg to melts in this system to produce small amounts of ferro-magnesian minerals such as biotite, show that small amounts of mafic materials do not materially affect the crystallization behavior of the quartz and feldspars. Addition of Ca to the granitic melts, however, has a major effect. As Ca enters albite to form plagioclase, the plagioclase field (in place of albite) expands and the ternary minimum moves toward the quartz-orthoclase join, the residual melt consisting largely of quartz and orthoclase. The quartz-orthoclase−rich pegmatite aplite dikes associated with the more calcic granodiorite plutons tend to confirm the experimental work.

However, even if a tremendous volume of basaltic magma evolves in this way, only a small fraction of this crystallizes as granite. Study of the Skaergaard mafic layered intrusion, for comparison, suggests only a fraction of a percent of this basaltic magma crystallized as "granite" (Wager and Deer, 1939). As shown by recent isotopic measurements, even this may have formed at least in part by assimilation of country rock (see p. 230). The amount is not comparable in any way with the huge amounts found in granitic batholiths unless the basaltic magma assimilates a very large amount of "granitic" material. Moreover, the composition of such residual melts is essentially indistinguishable from that formed by partial melting (anatexis) of common quartzofeldspathic metamorphic rocks, for, if the process is reversed, the low-melting fraction has the same composition as the crystallization residuum. The nature of the mafic minerals, however, may provide a clue to permit discrimination of granitic rocks of anatectic origin from those formed by differentiation of a basalt magma contaminated by sialic material in the crust (Chapman and Williams, 1935; Kleeman, 1965). Mafic minerals of the great granitic batholiths are biotite, generally with lesser hornblende or even muscovite, except where assimilation of calcic country rocks has been important. The rocks are essentially peraluminous (alumina in excess of alkalis plus calcium). Mafic minerals (e.g., hornblende, epidote, olivine, and pyroxene) in smaller masses of granitic rock in basaltic provinces and thought to be derived from contaminated basaltic rocks "are generally metaluminous or subaluminous because the parental basic magma has enough calcium to use up any excess alumina in the ingested sediments" (Kleeman, 1965, p. 50). More alkaline minerals such as aegirine-augite or riebeckite may, on the other hand, be indicative of rocks differentiated from alkaline olivine basalt magmas [see section on nepheline syenite and related rocks (pp. 202–207)]. Another argument against derivation of major amounts of granitic magma by differentiation of subcrustal basaltic or peridotitic material results from high-pressure experiments by Yoder and Tilley (1962). They show that basalt is unstable under water pressure in excess of 1,500 bars. Its high-pressure equivalent, eclogite, consists of garnet and a sodium-rich pyroxene. Crystallization of basaltic magmas under high water pressure, therefore, gives these minerals instead of plagioclase and pyroxene. The products of extreme differentiation, though likely to develop, would presumably be quite different from granite. Crystallization of relatively dry basaltic magmas would give rise, as noted above, however, to a small amount of granitic residuum. The possible distance a granite magma can rise, of course, is enhanced with a lower ratio of water-vapor pressure to load pressure, since the corresponding pressure-melting curve is steeper (see pp. 69–70; also Cann, 1970).

The recent development of the concepts of global tectonics, including ocean-floor spreading and subduction of oceanic lithosphere under the margin of an adjacent continent, has given rise to suggestions that the magma to form granitic batholiths is generated by partial melting of this oceanic crustal material (e.g., Dickinson, 1968, 1970a; Hamilton, 1969). This seems to assume, tacitly, that the magmas forming granitic batholiths are the same and have the same origin as those forming the andesites of the basalt, andesite, rhyolite association. The source of these magmas has been inferred as in the vicinity of the Benioff zone (subduction zone).

Presumably magmatic granites can and do form in more than one way. The great granitic batholiths and associated plutons probably result from the rise and consolidation of granitic melts derived by crustal anatexis. Smaller granitic plutons associated with the mafic stratovolcano part of the basalt, andesite, rhyolite association may develop by partial melting in the subducted oceanic crust or by crustal anatexis aided by heat from rising hot water and possibly basaltic melts.

The other alternative is origin of granite by diffusion of fluids through preexisting solid rocks, replacement by the appropriate ions tending to make the rock over into a granite. This solid-state replacement, whatever the specific mechanism proposed, is called *granitization*. The process involves chemical modification associated with activity akin to high-grade metamorphism. The points of view of the "granitizer" have been discussed at length by Perrin and Roubault (1949), Perrin (1954, 1956), Read (1957), Barth (1948), and Raguin (1965). The historical development and basis for arguments in favor of granitization have been concisely reviewed by Marmo (1967). Most proponents of granitization invoke diffusion of ions along grain boundaries in watery solutions. Certainly small-scale replacement along granitic contacts and in inclusions operates in this way as in the reciprocal operations involved in assimilation —an equilibrium being maintained between the granitic magma and the affected rocks, tending to make the latter over into rocks more like granite in composition. However, although marginal "granitization" has clearly been established and is widely recognized, there is little good evidence to suggest that granitization may form bodies of batholithic dimensions. For the opposing point of view, see, for example, Misch (1949a) and Roddick (1965). In fact, as pointed out by Tuttle and Bowen, there is no physio-chemical reason to expect that granitization should preferentially form rocks clustering in the low-temperature trough of the quartz-alkali feldspar diagram. Recent experimental work, however, shows that a hydrous vapor phase in equilibrium with granite magma at 5 to 10-kbars pressure carries 6 to 8% solids of reasonable composition. These could cause local granitization (Luth and Tuttle, 1969).

More extreme advocates of *dry granitization* have invoked diffusion of ions in the solid state without the necessity of interstitial fluids (e.g., Perrin and Roubault, 1939, 1949; Wegmann, 1938; Ramberg, 1944; Reynolds, 1946, 1947). However, as pointed out by Turner and Verhoogen (1960, p. 378), "all experimental evidence to date appears conclusive in showing that solid diffusion is too slow, even at temperatures within the magmatic range, to account for large-scale migration of elements." More recent evidence from diffusion bands in regional metamorphic rocks (Hyndman et al., 1968) gives the same conclusions, especially for the alkalis K and Na. These are precisely the elements that need to diffuse most easily for granitization to be effective.

It must be admitted, however, that granitization ("wet granitization") may well be a valid explanation for the formation of granite in some deep-seated ("catazone") environments.[1] It still may be argued strongly, for example, that the numerous granitic veins in veined gneisses (migmatitic gneisses) or the diffusely bounded, foliated, concordant "gneissic" sheets in high-grade regional metamorphic complexes, have originated by metasomatic replacement, that is, granitization (e.g., Misch, 1968). Although most, but not all, American geologists lean heavily toward a magmatic origin for the large majority of granitic bodies, many European geologists (especially the Scandinavians and the French) lean equally heavily toward an origin by (wet) granitization. A major regional influence is clearly the field environment in which their experience has been gained—predomi-

---

[1] Note that high-grade regional metamorphism and accompanying granitization are confined to the catazone. Diffusion and metasomatism in the mesozone and epizone do not lead to the formation of granite.

(a)

(b)

(c)

Fig. 4-12 *Deep-seated granitic rock—silli-manite-zone contact in northeast corner of Idaho batholith. Probably marginal graniti-zation of Precambrian biotite-plagioclase-quartz paragneiss (metasedimentary rock) by biotite granodiorite of the Cretaceous Idaho batholith. One mile southeast of South Kootenai Lake, northern Bitterroot Range, Montana. (a) Granodiorite of Idaho batho-lith (lower half of photo) grading upslope into streaky, less-"granitized" remnants of biotite-plagioclase-quartz paragneiss (upper half of photo). (b) Light and dark lamina-tions (relict bedding) in granitoid, "partly granitized" biotite-plagioclase-quartz para-gneiss. Possible "marginal mobilization" of granitic material in lower left. Hammer handle is 12 in. long. (c) Very continuous, diffuse "schlieren" of biotite-plagioclase-quartz paragneiss intimately associated with diffuse streaks of granitic material. A pos-sible case of partial granitization.*

nantly granitic plutons of the mesozone and epizone in the United States —predominantly deeper-seated granitic rocks of the catazone on the Baltic Shield and in the French Massif Central.

An acceptable working hypothesis is that high-level "epizonal" granitic rocks are formed, almost entirely, by magmatic processes. "Mesozonal" granites are formed primarily by magmatic processes but solid-state processes such as "granitization," metasomatism, or metamorphism play a role to varying degrees, to be judged on the basis of the individual occurrence. "Catazonal" granites are formed primarily by anatexis, granitization, metasomatism, or metamorphism. The granitic magmas referred to above are formed primarily by anatexis (over continental crust) and to a lesser extent by differentiation of basaltic magmas (over oceanic crust).

Granitization is essentially a metamorphic process, the high-grade

Table 4-5 *Criteria for discrimination between magmatic emplacement and granitization (replacement)*

Evidence	Magmatic Emplacement	Granitization (Replacement)
Contacts with country rocks	Sharp	Gradational or diffuse
Inclusions	Misoriented or rotated	Not misoriented (relict)
Margins of granite	± chilled (finer grained)	Not chilled
Adjacent country rocks	Displaced (forced aside); no relict bedding	Not displaced; relict bedding in granitic rocks
Foliation of granite	Discordant to foliation of country rocks	Passes continuously into regional structure of country rocks
Metamorphism	Contact type (e.g., hornfelses) around borders	Gradational high-grade (e.g., sillimanite zone) metamorphism toward and within granitic rock
Dikes show	Dilation (see Fig. 4-13)	No dilation (replacement; see Fig. 4-14)
Texture	May be granophyric (or graphic)	Not granophyric
Minerals	Paragenesis agrees with Bowen's Reaction Series	Metamorphic minerals may be present (e.g., sillimanite, garnet, etc.)
	May correspond to low-melting fractions	No reason to correspond to low-melting fractions
	Common mica is biotite	May include muscovite (suggests lower temperature, high pressure)

Table 4-5 *Criteria for discrimination between magmatic emplacement and granitization (replacement) (Continued)*

Evidence	Magmatic Emplacement	Granitization (Replacement)
Plagioclase	Strongly zoned (as in volcanic rocks)	Nearly unzoned (as in metamorphic rocks)
	Poor correlation of zoning between adjacent crystals (relative movement of crystals)	Good correlation of zoning between adjacent crystals (no relative movement of crystals)
	Much twinned (as in volcanic rocks)	Little or no twinning (as in metamorphic rocks)
Alkali feldspars	± one only (e.g., perthite; crystallized above solvus)	Two primary (large) alkali feldspars coexist (below solvus)
Isotopic ages	Pb-U age (e.g., on zircon) = K-Ar age (e.g., on hornblende or biotite)	Pb-U age (e.g., zircon) >> K-Ar age (e.g., biotite)
		Ar in biotite affected by metamorphism; U in zircon not affected
"Room problem"	Present if lack of disturbance of country rocks adjacent to large batholith is clearly demonstrable	"Not present" (but incoming ions must take up space; must get rid of outgoing ions; therefore room problem just moved)
Small masses and irregular veins	Connection to magma source	No visible connection to magma source

metamorphism tending to make over the original rock into one of granitic aspect. This solid-state transformation occurs, with the aid of watery solutions, at temperatures and pressures *where the processes of metamorphism merge with those of magmatism*, resulting in a convergence of mineral composition and textures. The resulting rocks are therefore so much alike that it is not too surprising that it is very difficult to be sure of the origin of such rocks. Proponents of either origin generally agree that the other is a reasonable explanation for the origin of some rocks but argue the relative importance and the origin for specific transitional rocks.

Clearly granitic plutons surrounded by hornfelses or by low- to moderate-grade metamorphic rocks (i.e., below the sillimanite zone where metamorphic and magmatic processes begin to merge) have not been formed by either granitization or anatexis at their present level of exposure. Some deep-seated granitic contacts (see, for example, Fig. 4-12), however, are much more difficult to interpret. These may show diffuse, concordant contacts with high-grade regional metamorphic rocks. The same problems arise in the recognition of magmatic and replacement dikes. Some criteria that may aid in discrimination between magmatic emplacement and granitization (replacement) are outlined on Table 4-5.

Where Sr^{87}/Sr^{86} ratios are available for fairly young granitic rocks, high ratios of greater than about 0.710 probably have an origin within the old crystalline basement of the crust but are otherwise ambiguous as to anatexis or granitization. Low ratios of less than about 0.708, however, more likely originate by partial melting in the younger "eugeosynclinal" sedimentary and volcanic pile or in the mantle or oceanic basaltic crust thrust under the continent (see pp. 89–94). A mantle or basaltic crustal origin for the large granitic batholiths is considered unlikely because of the small quantities of granitic melt that could form and of the quantities of more mafic melt that would develop (compare the basalt, andesite, rhyolite association).

Many other criteria have been used in the past to argue either in favor of an origin by magmatic processes or by granitization. In fact, some of the most violent and sometimes entertaining arguments ever aired in geological circles revolved around these possibilities for the origin of granite (e.g., Gilluly et al., 1948). The criteria listed above, although hardly unequivocal, are among the most reliable evidence available. Several such criteria, taken together and used with care, are probably reliable. It must be emphasized, however, that any conclusion must be consistent

Fig. 4-13 *Pegmatite-aplite dikes injected along fractures in calc-silicate gneiss. Note dilation—sharp walls of dikes pulled apart to accommodate granitic material. North of Idaho batholith. On U.S. Highway 12, 10 miles southwest of Lolo Pass, Montana-Idaho.*

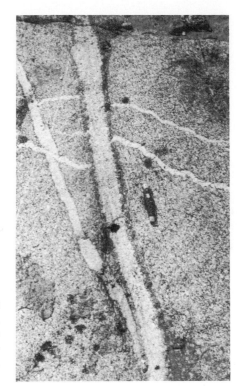

Fig. 4-14 *Granitic dike "sweated out" of homogeneous, foliated biotite quartz diorite, leaving dark, diffusely bounded biotite-rich selvages. Earlier granitic streaks (above knife) appear to continue along original path, rather than offset directly across dike. Knife is 3 in. (7½ cm) long. Half mile west of central Kootenai Lake, northern Bitterroot Range, Montana.*

with other geological information, and recognition of marginal granitization or marginal magmatism does not prove the same origin for the whole body.

Another possible origin for these deep-seated rocks of somewhat ambiguous origin is regional metamorphism of an earlier igneous (i.e., magmatic) body. Such bodies are generally foliated, sometimes streaky, and have a granitoid appearance. They may sometimes be recognized by the presence of relict, clearly magmatic features such as misoriented, angular inclusions. The regional metamorphic foliation, however, will pass continuously through both the inclusion, the granitic rock, and even the country rock. The same considerations may be used in the recognition of regionally metamorphosed dikes—the foliation, discordant to the dike, passes continuously from the country rock through the dike (see Fig. 4-15).

Suggested Origins of the Large Granitic Batholiths

Partial melting of thick "geosynclinal" accumulations of sedimentary and volcanic rocks near the continental margin (Bateman and Wahrhaftig, 1966; Hyndman, 1969; Hyndman and Hyndman, 1968; Bateman and Dodge, 1970; Wollenberg and Smith, 1970)

Fig. 4-15   *Deformed and metamorphosed, zoned pegmatite dike cutting foliated quartz diorite pluton in the Idaho batholith. Foliation in the quartz diorite, parallel to hammer handle (12 in. long), passes continuously through pegmatite dike. Coarse biotite in core of dike parallels hammer handle and outlines foliation. The dike, and presumably also the pluton, was regionally metamorphosed producing the foliation and was folded by slip along this foliation. Note that the foliation parallels the axial planes of the folds in the dike. Half mile west of central Kootenai Lake, northern Bitterroot Range, Montana.*

Partial melting of oceanic basaltic crust or upper mantle (e.g., Hamilton and Myers, 1967) either along the subduction zone inclined under the continental margin (Hamilton, 1969; Gilluly, 1969; Dickinson, 1970a) or over a linear zone of high-heat flow like the crest of an oceanic rise (Kistler et al., 1971)

Arguments in favor of a crustal rather than mantle source for the bulk of the granitic magma have been presented above. That a significant contribution of necessary heat may come from the subduction zone or over an upwelling zone in the mantle may satisfy many of the objections to a continental source for the magma.

ANORTHOSITE PLUTONS

The Michikamau Anorthositic Intrusion of Labrador[1]

Occurring in a wide northeasterly trending belt in the eastern Canadian Shield (especially New York state, Quebec, and Labrador) is a group of

[1]Emslie, 1965, 1968b, 1970, written communication, November, 1970; review by permission of the National Research Council of Canada from the *Canadian Journal of Earth Sciences,* 2, pp. 385–399.

large anorthositic intrusions. Most occur in the Grenville province, meta-morphosed less than 1,000 million years ago, and are deformed and meta-morphosed (Kranck, 1961). Those such as the Michikamau intrusion, lying north of the "Grenville Front," are unmetamorphosed and little deformed. The Michikamau intrusion, emplaced about 1,400 million or 1,500 million years ago, underlies an area of about 800 sq miles (Fig. 4-16). It intruded quartzofeldspathic gneisses that had been regionally metamorphosed to the very high grade granulite facies. Adjacent to the intrusion, these have been thermally metamorphosed to the very high temperature pyroxene hornfels facies.

The Michikamau intrusion shows features reminiscent of many of the gabbroic layered intrusions (described in a later section). It consists of a fine-grained chilled margin of olivine basalt composition, in general overlain by successive coarser-grained units of leucotroctolite (leucocratic olivine-plagioclase rock), anorthosite (> 90% plagioclase), and leucogabbro, with small amounts of late-stage iron-rich dioritic, granodioritic, and syenitic rocks. The overall form of the body is thought to be a composite of two or more intersecting funnels, as interpreted from textural and structural features within and near the intrusion. The chilled margin is structureless near the contact. Gradationally inward, plagioclase lamination (planar orientation of plagioclase grains) and rhythmic layering (some of these repeated layers show graded bedding) become more prominent.

The Marginal Zone of the intrusion is a fine-grained (< 1 mm), equigranular, chilled zone 300 to 400 ft wide. Essential minerals are olivine (Fo_{56-58}), clinopyroxene, and plagioclase (An_{51-56}), with locally abundant deep-brown hornblende and lesser magnetite and ilmenite in fine discontinuous streaks. The hornblende is within what is "regarded as a type of flow banding formed by incorporation of water from the wallrocks" into the otherwise anhydrous margin of the pluton. Plagioclase is equidimensional in contrast to its tabular shape in the Layered Series. The plagioclase content rises rapidly inward to a medium-grained (2 to 5 mm), 1,000 to 3,000-ft zone of olivine gabbro and pyroxene troctolite, apparently formed by crystal accumulation.

Transitionally inward, this Marginal Zone gives way to the Layered Series, a medium-grained (0.5 to 1 cm) leucotroctolite rock forming the bulk of the exposed rocks of the intrusion. The more abundant plagioclase is slightly more calcic (An_{72-50}) and the olivine that typically forms 5 to 20% of the rock is slightly more magnesium rich (Fo_{81-51}). Pyroxene is only a very minor constituent, and the total mafic mineral content is commonly 15 to 25%. Anorthosite forms layers to more than 100 m thick in this unit.

Rocks of the Anorthosite Zone overlie the Layered Series and are still coarser grained and lower in mafic minerals (< 10%), but sporadic pyroxene concentrations (20 to 30% of the rock) do occur. The still more abun-

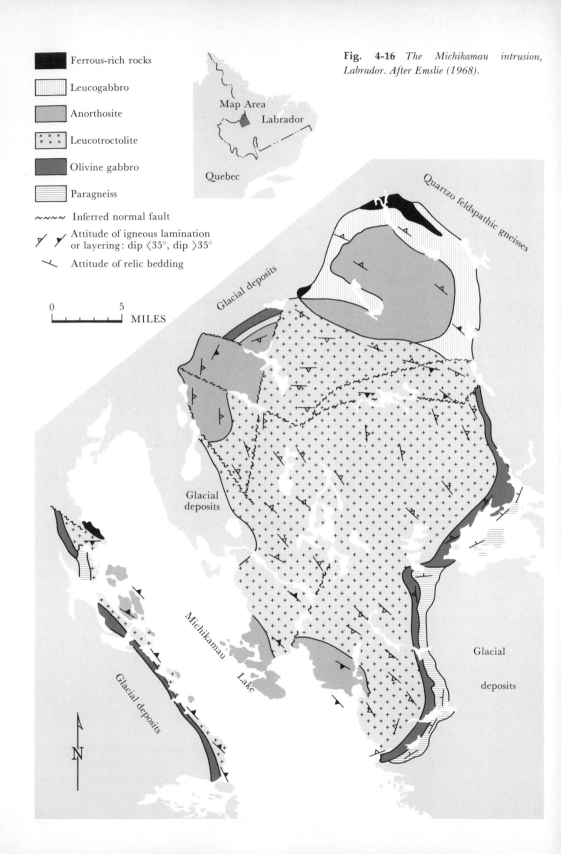

Fig. 4-16 *The Michikamau intrusion, Labrador. After Emslie (1968).*

Ferrous-rich rocks

Leucogabbro

Anorthosite

Leucotroctolite

Olivine gabbro

Paragneiss

~~~~ Inferred normal fault

Attitude of igneous lamination or layering: dip <35°, dip >35°

Attitude of relic bedding

Map Area
Labrador

Quebec

Quartzo feldspathic gneisses

Glacial deposits

0   5
MILES

Glacial deposits

Glacial deposits

Michikamau Lake

Glacial deposits

Glacial deposits

N

dant plagioclase is less calcic (An_{55-44}) and olivine is absent. Except locally, most of the plagioclase forms equidimensional grains.

The Upper Border Gabbros resemble rocks of the underlying Anorthosite Zone but contain a total of about 10 to 35% orthopyroxene, lesser clinopyroxene, and minor magnetite, ilmenite, and apatite. This upper zone is separated from the country rock gneisses by a roof Marginal Zone of medium-grained olivine gabbro.

Latest to crystallize and least voluminous of the rocks in the intrusion is the Transgressive Group, which intrudes, and in places appears chilled, against the Upper Border Gabbros. The rocks range from a mafic-rich ferrodiorite through ferrogranodiorite to syenite. The plagioclase is less abundant, still less calcic, and strongly zoned (andesine cores to oligoclase rims). The ferroaugite and minor olivine (for example, Fo_{18}) show strong enrichment in iron, and perthitic alkali feldspar appears as a minor to major constituent. Quartz is commonly present and Fe-Ti oxides may comprise up to 10% of the rock.

Although inclusions are rare in the Michikamau intrusion, the few that exist are cordierite gneiss clearly derived from a cordierite gneiss country rock or are orthopyroxenite. A few fine-grained gabbro dikes cut the country rock and the Layered Series.

Emslie (1965) considers the following characteristics of the intrusion important in interpretation of its origin: (1) rhythmic layering and planar orientation of the plagioclase point to solidification by bottom accumulation of crystals as in the gabbroic layered intrusions; (2) the anhydrous mineralogy in the chilled margin of basaltic composition indicates that the magma was not in equilibrium with high water pressures; (3) the structureless aspect, inward mineralogical variation, and grain-size variation of the chilled margin suggest that this zone was formed under moderate-to-strong thermal gradients and before crystal accumulation became a prominent process; (4) detailed textures show that much of the plagioclase and olivine accumulated by crystal settling; some other plagioclase, olivine, and clinopyroxene precipitated as overgrowths and between the settled grains; and (5) perthitic intergrowth between subequal amounts of the alkali feldspars in the ferrogranodiorite unit suggests low water-vapor pressures and high temperatures (see, for example, Fig. 1-6).

The parent magma of the Michikamau anorthositic intrusion is thought to consist of an aluminous basalt liquid derived from the mantle and heavily charged with labradorite crystals. Such a situation might arise "if the intrusion were coupled to an active volcanic system," by settling (or perhaps more likely, floating) of plagioclase crystals in the basalt or by continuous subtraction of basalt liquid. Crystals probably nucleated near the cooler roof of the intrusion, accumulated on the bottom with primary dips of about 30° and "the general lack of crushing or fracturing of plagioclase crystals argues against filter press action playing a significant part in the development of the anorthositic rocks." The trend of fractional crystallization "has been toward strong enrichment in iron, moderate concentration of potash, and only slight increase in silica." Only small amounts of syenitic rocks could be produced in the latest stages by this mechanism. Concentrations of Fe–Ti oxides so common in anorthositic rocks are "consistent with a fractionation trend that produces high iron differentiates." Fracturing and collapse of the anorthosite zone and upper gabbros probably permitted squeezing out and injection of the residual iron-rich magma of the Transgressive Group.

General Features of Anorthosite Plutons

Intrusions characterized by anorthosite range in size from a few hundred feet across (e.g., adjacent to the Idaho batholith), to great batholiths covering thousands of square miles (in the Adirondacks of New England and elsewhere in the eastern Canadian Shield). The larger plutons, at least, seem to be confined to high-grade regional rocks of Precambrian metamorphic terranes. Their ages apparently cluster around 1,300 million to 1,400 million years and they lie in broad arcuate belts when plotted on a reconstruction of the continents before continental drift (Herz, 1969). They include those in (1) the eastern Canadian Shield (e.g., Buddington, 1939, 1968; Engel and Engel, 1953; Wheeler, 1942, 1960, 1965, 1968; Rose, 1960; Kranck, 1961; Hargraves, 1962; Papezik, 1965; Emslie, 1965, 1968b; Philpotts, 1966; Jenness, 1966; deWaard and Romey, 1968, 1969; Davis, 1968; Letteney, 1968), (2) central Virginia (Watson and Taber, 1913; Herz, 1968), (3) western Greenland (Sørensen, 1955; Berthelson, 1957; Ayrton, 1963), (4) Laramie Range, southeastern Wyoming (Fowler, 1930; Klugman, 1965, 1968), (5) southern California (Higgs, 1954; Crowell and Walker, 1962), (6) southwestern Norway (Barth and Dons, 1960; Michot and Michot, 1968), (7) Madagascar (Boulanger, 1957, 1959), and (8) India (Subramaniam, 1956; De, 1968).

They consist, to a major extent, of medium-grained to very coarse-grained (up to several inches) gray anorthosite to gabbroic anorthosite, grading especially in the border phases to finer-grained more mafic members such as noritic gabbro or olivine gabbro. Closely associated rocks occurring in smaller amounts in the late stages of the same intrusion or as separate intrusions include quartz syenite, pyroxene syenite (sometimes called mangerite), hypersthene granite (sometimes called charnockite), and ferroaugite granodiorite. Locally, these more felsic members of the anorthositic pluton may intrude the earlier more mafic members. Primary-flow foliation and layering is common in some of the larger intrusions. Intergranular crushing and shearing is common in most but not all anorthosites and is generally thought to correlate with postemplacement regional metamorphism and deformation (e.g., Buddington, 1939). It has also been correlated with deformation during intrusion (e.g., Bowen, 1928; Higgs, 1954). Many, perhaps most, of the larger anorthosite plutons form thick sheets or "sills" in, and may strongly disrupt, the high-grade metamorphic country rocks. High-grade contact metamorphic effects occur except where obscured by later regional metamorphism. The external form appears to be that of a large thick sill, laccolith, or widely flaring funnel. The roof is generally a domical surface or group of domical surfaces. A gravity survey over the Adirondack anorthosite suggests that it, at least, may have two vertical feeder pipes (Simmons, 1964).

Mineralogically, the large masses of anorthosite typically consist of nearly 100% plagioclase in the andesine-labradorite range (usually An_{34-65}). Most are coarse grained and antiperthitic (Berrangé, 1966) and, in many instances, were deposited as "cumulus" crystals that have settled to the bottom of the magma chamber. The finer interstitial grains tend to be more mafic rich, approaching the composition of norite or gabbro. They contain bronzite-hypersthene and titaniferous magnetite or hemoilmenite (e.g., Hargraves, 1962; Anderson and Morin, 1968). Some of the grains are overgrown by later crystallization of plagioclase, pyroxene, or ilmenite-magnetite, to produce very large poikilitic grains (Kranck, 1961; Isachsen and Moxham, 1968). In at least some instances, the plagioclase shows oscillatory zoning—a rather common feature of clearly magmatic, but not metamorphic, rocks (e.g., Boone and Romey, 1966). Garnet is widespread (Berrangé, 1966). With increasing content of augite, hypersthene, or olivine, the anorthosite grades to gabbroic anorthosite, gabbro, norite, and troctolite (olivine-plagioclase rock). In these rocks, the plagioclase is somewhat more calcic (An_{50-70}). Alteration products include hornblende as a common alteration of pyroxenes, biotite, chlorite, and serpentine. The foliation or parallelism of plagioclase grains is parallel to the compositional or grain-size variations that form the gneissic layering, to the contacts between the main compositional units, and approximately parallel to the margins of the pluton. Dikes are rare, but a few of andesine anorthosite have been described cutting more calcic anorthosite (Osborne, 1949; Wheeler, 1960). Sr isotope ratios range from 0.6997 to 0.7061, ratios comparable with those in basalts derived from the mantle (see Table 3-1, p. 93).

The Problem of Emplacement Temperature

Just as with alpine peridotite bodies, the high temperature of crystallization of these rocks represents a difficult problem. Pure plagioclase that has a composition of An_{50} crystallizes above 1250°C, as seen from the plagioclase phase diagram (Fig. 1-3). A fraction of a percent potassium in solid solution in the natural system could lower the temperature a few degrees but not appreciably, as seen from the alkali feldspar diagram (Fig. 1-6). A few thousand bars water pressure would be expected to lower the crystallization temperature a few hundred degrees, but there seems to be no reason to suppose that anorthosites crystallized under the influence of significant water (Emslie, 1965, 1970; Luth and Simmons, 1968). Anorthositic rocks transitional to gabbro, norite, or troctolite could crystallize at slightly lower temperatures, as seen from phase diagrams for basaltic systems (Fig. 3-4). Consider, as a first approximation, the system albite-anorthite-diopside.

Suggested Origins of Anorthosite Plutons

Accumulation of plagioclase crystals separated from a parent magma of gabbroic, noritic, or high-alumina basalt composition (Bowen, 1917; Smith, 1922; Emslie, 1965, 1968a; deWaard and Romey, 1969; Yoder, 1968; Windley, 1967; Luth and Simmons, 1968; deWaard, 1968a; [?] Heath and Fairbairn, 1968; Phinney, 1968; Reynolds et al., 1968); separation by flotation of plagioclase (Bridgewater, 1967; Emslie, 1970).

Accumulation of plagioclase crystals from a parent magma of dioritic, quartz dioritic, or granodioritic composition (Balk, 1931; Barth, 1936; Polkanov, 1937; Philpotts, 1966; Letteney, 1968; Green, 1968, 1969; deWaard and Romey, 1968). Mafic constituents settle out of crystal mush due to shock waves resulting from deformation of the rigid roof of the intrusion (Middlemost, 1968).

Separation of clinopyroxene enriched in Ca, Mg, Al (relative to NaAl) from gabbroic magma at high pressures near or below the base of the continental crust (Emslie and Lindsley, 1969).

Mixing of olivine basalt magma with granitic melt formed by anatexis, followed by rise and crystallization (Kranck, 1968; Philpotts, 1968).

Injection as a largely liquid feldspathic magma of anorthosite or gabbroic anorthosite composition (Miller, 1914, 1931; Fowler, 1930; Kolderup, 1936; Buddington, 1939, 1968; Osborne, 1949; Higgs, 1954; Subramaniam, 1956; Lindsley, 1968; Davis, 1968; De, 1968), perhaps by differential fusion of an assumed bytownite anorthosite layer near the base of the earth's crust (Buddington, 1939; Anderson and Morin, 1968; Michot, 1968).

Injection of a largely liquid magma under at least moderate water pressures (and therefore more reasonable temperatures) (Higgs, 1954; Turner and Verhoogen, 1960; Bryhni, 1966; Yoder, 1968;

Anderson, 1968; Anderson and Morin, 1968; Buddington, 1968; Davis, 1968; Crosby, 1968; Herz, 1969).

Anatexis of calcium-rich clays, quartz diorite, or other rocks, leaving behind an anorthositic plagioclase-rich residue (Michot, 1955; Winkler, 1960; Berg, 1968; Michot and Michot, 1966, 1968; Green, 1968, 1969; deWaard, 1968b).

Assimilation of pelitic rocks by basaltic or andesitic magma (Michot and Michot, 1968; Philpotts, 1966).

Metamorphic and metasomatic processes (introduction of Si and Al from adjacent Al-rich sediments and removal of Ca) on a limestone, followed by partial mobilization and intrusion (Hietanen, 1963a, 1968).

Melting in the lowermost crust or uppermost mantle in response to a unique cataclysmic event such as birth of the earth-moon system or meteorite impact during a high thermal gradient time early in the earth's history (Herz, 1969).

Any origin proposed for anorthosite plutons must account for the following characteristic features:

Developed in high-grade regional metamorphic terranes.

One of the most prominent rock types is generally more than 95% plagioclase, of andesine or labradorite composition.

More mafic noritic, gabbroic, or troctolite phases are closely associated nearer the borders.

More felsic iron-rich granitic phases are associated in the later stages, sometimes at higher levels in the pluton, sometimes as separate adjacent plutons. These tend to show local crosscutting features against the anorthosite.

A primary-flow foliation and layering occurs except where obscured by later deformation and metamorphism.

Much of the plagioclase occurs as coarse phenocrysts, whether in the anorthosite or in associated gabbroic phases.

Absent as lavas.

High-grade contact metamorphism is apparent except in some areas exhibiting a superimposed high-grade regional metamorphism.

As with certain other kinds of igneous rocks, such as granites, more than one mode of origin may give rise to the same kind of rock. The origin which seems to best fit the field, textural, and mineralogical characteristics of this association is that of high-pressure partial melting to produce, or differentiation of, a parent magma of basaltic (high-alumina?) or gabbroic (including noritic) composition. Abundant plagioclase would crystallize slowly in a deep-seated (plutonic) environment, probably near the base of the continental crust (compare the "low or moderately low" pressure diagrams of Fig. 4-21), the plagioclase settling to the bottom of the mass and accumulating in layers. If, as suggested by Emslie (1965), the magma chamber was coupled to an active volcanic system, much of the potential mafic content could be removed from the final system by early crystallization of plagioclase and extrusion of mafic components still in the basaltic melt.

Resembling these large anorthosite plutons in mode of occurrence, but of much smaller size (less than 40 sq miles), are anorthosites associated with the border zones of the Idaho batholith in Idaho (Hietanen, 1963a, 1968) and Montana (Berg, 1968), and possibly those in northern Cape Breton Island, Nova Scotia (Jenness, 1966). The composition of the plagioclase is most commonly An_{45-55}, but those in Idaho are unzoned and characteristically contain unevenly distributed finer-grained, more calcic plagioclase (An_{85}). Those in Montana are strongly normal zoned (more calcic cores) and of only one type. Neither occurrence shows more mafic borders but both have associated mafic rocks. By analogy with the alpine peridotites and serpentinites, these smaller anorthosites could be called "alpine anorthosites."

POSTOROGENIC SUITES

FLOOD BASALTS—THOLEIITIC BASALTS AND ALKALINE BASALTS

The Flood Basalt Associations of the Hawaiian Islands[1]

By far the most completely studied region of flood basalts is that of the Hawaiian Islands. The Hawaiian archipelago of the central Pacific is a chain of islands extending more than 1,600 miles west-northwest of the island of Hawaii (Fig. 4-17) to beyond the island of Midway. The islands toward this west end of the chain tend to be smaller and lower. Becoming

[1]Stearns and Macdonald, 1946; Macdonald, 1949; Powers, 1955; Eaton and Murata, 1960; Macdonald and Katsura, 1964; Peck, Moore, and Kajima, 1964; Murata and Richter, 1966; Richter and Moore, 1966; Murata, 1966; Richter and Murata, 1966; White, 1966; Macdonald, 1969, written communication, 1970.

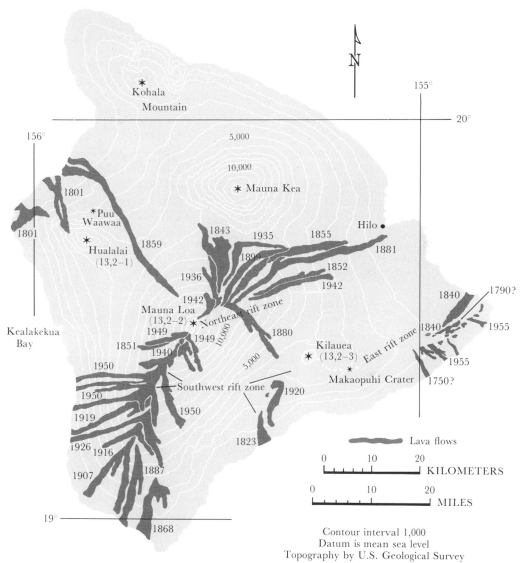

Fig. 4-17 *Map of the island of Hawaii, showing the location of the active volcanoes and their rift zones. After G. A. Macdonald (1955, p. 8).*

extinct progressively, in general, from west to east, the volcanoes are still active only in the southeastern part of the chain. The commonest rock type on the islands is an olivine-bearing tholeiitic basalt approximately saturated in silica. Alkalic basalts, undersaturated in silica and high in alkalis, are confined to the later stages.

The stages during the formation of the Hawaiian volcanoes are now recognized as follows:

Extensive frequent eruption of thin flows of very fluid tholeiitic basalts build a shield volcano. Although most of the basalts are approximately saturated with silica, many contain rounded and embayed phenocrysts of olivine, out of equilibrium with the magma at the time of eruption.

Repeated collapse of the summit area of the shield to form a caldera. Continued eruption of tholeiitic basalts on the flanks of the shield and in the caldera, gradually filling it. On some of the volcanoes, there is no evidence that a caldera was ever formed. Toward the end of this stage, alkalic basalts may erupt alternately with the tholeiitic basalts.

Less frequent but more explosive eruption of alkalic basalt and its differentiation products in the late stages of growth of the volcano. Pyroclastic rocks become more abundant and interflow stream gravels and soil horizons become common. The lavas are more viscous, resulting in shorter, thicker flows and steepening of the upper part of the volcano.

Following a long period of quiescence and erosion, rejuvenated eruption of alkalic and strongly undersaturated olivine basalts, and nepheline-rich "basaltic" rocks (nephelinite).

The quantitatively minor late stages of differentiation of both the tholeiitic and alkalic suites are much more felsic than the basalts. The tholeiitic suite evolves to rhyodacite. The alkalic suite evolves to andesite and sodic trachyte.

The lavas are erupted from narrow zones of fissures intersecting at the summit of the shield volcano. The *tholeiitic magmas* presently erupted at the surface can be followed seismically as they rise slowly over a period of months from their apparent source, 60 to 100 km below (deep in the mantle). Slow bulging of the volcano, as shown by sensitive tiltmeter measurements, suggests filling of a magma chamber high within the volcano prior to eruption. This bulging subsides as the lava is extruded at the surface. The flows generally average about 5 ft in thickness near the summit, thickening with falling temperatures and increased viscosity to about 20 ft near the lower limit. The more fluid types form smooth, billowy surfaces and are called pahoehoe lavas. With increasing viscosity in the more differentiated types, the surfaces of the moving flows break into jagged blocky forms and are called aa lavas. The summit caldera apparently results from collapse of the roof of the magma chamber along steep, arcuate ring fractures. Small "pit craters" and narrow rift-zone grabens result from collapse of the surface over lava in a fissure. Secondary cones of spattered lava, cinders, pumice, and ash are superimposed on the main shield along the rift zones radiating outward from the summit. They range from 5 to 700 ft in height and have steeper slopes up to about 40°. Cinders and ash form the higher and steeper cones. Other ash cones form from steam explosions where aa lava runs into the sea. Pahoehoe lava tends to form pillow lavas without violent explosions, where it flows into the sea.

Most of the flows are gray to black in color. Tholeiitic olivine basalts are most abundant, often containing phenocrysts of olivine and plagioclase, and rarely pyroxene. The olivine is magnesium rich (about Fo_{80}), commonly partly resorbed, and sometimes rimmed by pyroxene. It is absent from the groundmass. The plagioclase is quite calcic (commonly An_{70-80}) in the phenocrysts and more sodic (commonly An_{30-45}) in the groundmass. The groundmass also contains pigeonite (low-Ca clinopyroxene), often augite, and locally hypersthene, along with magnetite. Inclusions in the tholeiitic basalts are rare and appear from their textures and mineralogy to be aggregated groups of phenocrysts from the enclosing magma.

The *alkalic basalts* ($< 1\%$ by volume) erupting in the later stages of evolution of the volcanoes, commonly contain euhedral phenocrysts of magnesium-rich olivine, more iron-rich olivine occurring on the rims of the phenocrysts and in the groundmass. The characteristic pyroxene in these rocks is augite, often a titaniferous variety. In addition to a few plagioclase phenocrysts, the alkalic basalts commonly contain groundmass andesine and alkali feldspar (potash oligoclase), which is apparently absent in the tholeiitic rocks. Nepheline basalts are also known. Angular to subangular inclusions of dunite and other peridotites are abundant in the alkaline basalts. Less common are gabbro and even anorthosite. The minerals and textures of the inclusions are identical to those formed by accumulation of grains in mafic layered intrusions, suggesting that they too were formed by accumulation of crystals from basaltic magma.

Chemically, the tholeiitic and alkalic basalts in Hawaii are also somewhat distinct. The alkalic basalts are almost invariably higher in alkalis for a given silica content than the tholeiitic basalts. They also range to higher total alkalis and lower total silica. The average and range of each of the major oxides in each of the two types of basalts in Hawaii are noted in Table 4-6. It is emphasized, however, that, in naming the rock, the mineralogy takes precedence over the chemistry. As may be clearly seen from Table 4-6, there is considerable overlap for all common oxides, between the two basalt types. Those that show the greatest difference are SiO_2, TiO_2 (average), Na_2O, and K_2O. $(Na_2O + K_2O)/SiO_2$ is almost always greater for the alkali basalts than the tholeiitic basalts.

Table 4-6   *Average (and range of) chemical composition of tholeiitic basalt (200 analyses) and alkaline olivine basalt (36 analyses) on the Hawaiian Islands*

	Tholeiitic basalt, wt.%		Alkaline Olivine Basalt, wt.%	
	AVERAGE:	RANGE:	AVERAGE:	RANGE:
SiO_2	49.4	43.49–52.56	45.4	41.83–47.48
TiO_2	2.5	1.79– 3.69	3.0	1.85– 3.68; 1.01
Al_2O_3	13.9	10.28–16.66	14.7	10.11–17.61
Fe_2O_3	3.0	0.98– 7.90	4.1	1.98– 7.79
FeO	8.5	5.22–11.05	9.2	6.33–10.39
MnO	0.16	0.13– 0.21	0.14	0.16– 0.19
MgO	8.4	5.44–17.30	7.8	4.69–17.87
CaO	10.3	7.86–11.91	10.5	7.75–11.86
Na_2O	2.2	1.45– 2.95	3.0	1.35– 3.78
K_2O	0.4	0.07– 0.70	1.0	0.27– 1.22
P_2O_5	0.3	0.22– 0.56	0.4	0.16– 0.48

SOURCE: Macdonald and Katsura (1964); Macdonald (1969).

The Flood Basalt Association of the Columbia River Plateau, Washington, Oregon, and Idaho[1]

In many respects resembling the flood basalt association of the Hawaiian Islands is that of the Columbia River Plateau in the northwestern United States (Fig. 4-18). The most obvious difference is that the former consists of a group of immense shield volcanoes, whereas the latter forms a vast nearly flat lava plateau deformed by gentle warps. The Columbia River basalts were erupted during Miocene time, over some 55,000 sq miles, to depths ranging up to more than 10,000 ft in the central part of the plateau, thinning to a few thousand feet at the margins. The earliest flows at the bottom of the section may be as old as Eocene, suggesting an average eruption of one flow every 500,000 years (Brown, 1970).

Individual flows range from about 30 to 150 ft thick (average 77 ft) and many appear to have traveled 50 miles to more than 100 miles from their source. Calculations of flow rates based on slope thickness, extent, and rate of cooling of flows, size of the eruptive vent, assumed viscosity, and kind of flow suggest emplacement in a few days (Shaw and Swanson, 1970). Tops of the flows are typically covered with a few inches to more than 15 ft of jagged fragments of lava and clinker (aa type). As in the case of the Hawaiian shield volcanoes, the lava reached the surface through nearly vertical fissures. However, instead of two or three intersecting fissures, these flows erupted through huge swarms of approximately north-south fissures (now represented by dikes) distributed over several hundred square miles (e.g., Taubeneck, 1970). All four of the known swarms are

[1]Waters, 1955b, 1960, 1961; Mackin, 1961; Swanson, 1967; Hoffer, 1967; Lefebvre, 1970; Brown, 1970.

located near the periphery of the plateau. Most of the basalt is thought to have flowed to the west and northwest from the dike swarms in the eastern part of the plateau (e.g., Waters, 1961; Bingham, 1970).

The Columbia River basalts are divisible into two formations, the earlier Picture Gorge basalts and the later Yakima basalts. As may be

Fig. 4-18 *Columbia River flood basalt flows in Washington, Oregon, and Idaho (simplified from Geologic Map of North America, 1965; E. N. Goddard, Chairman of Map Committee). Minor modifications from Wise (1969).*

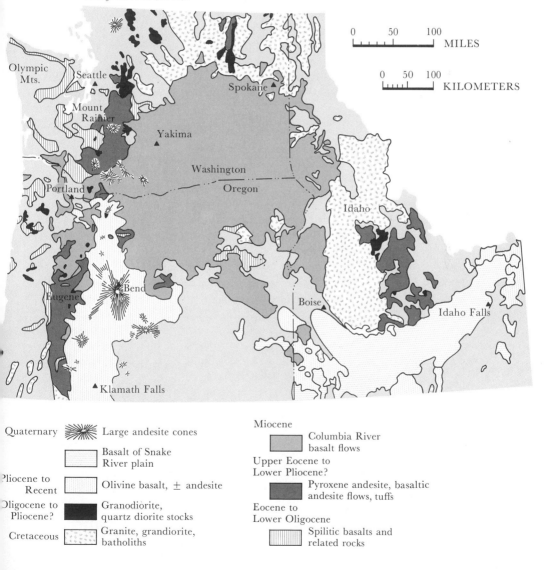

seen in Table 4-7, these earlier basalts are alkalic with tholeiitic affinities, having lower SiO_2, K_2O, TiO_2, and higher MgO and CaO. The later basalts are clearly tholeiitic, with higher SiO_2, K_2O, TiO_2, and lower MgO and CaO. A late variant of the later (Yakima) basalt is clearly alkalic, with low SiO_2 and high TiO_2 and K_2O. These differences are also reflected in the mineralogy. Most flows of the *Picture Gorge basalt* contain only a little brownish glass and contain plagioclase, augite, and a little olivine, both as phenocrysts and in the groundmass. Plagioclase phenocrysts are calcic ($An_{>50}$) and zoned, the augite is in some cases zoned to titaniferous augite, and the olivine is high in iron (Fa_{10-30}) and partly altered to greenish saponite (an iron-rich clay mineral). Chlorophaeite, another iron-rich alteration product, is apple-green in color, but on weathering gives a deep orange-brown color to exposed surfaces of the flows. The silica-poor alkalic mineral analcite is a common accessory mineral. These mineralogical characteristics, especially the titaniferous nature of the augite and the presence of alkalic minerals such as analcite suggest an alkaline olivine basalt. By contrast, flows of the *Yakima basalt* contain more than 20% dark glass and are essentially nonporphyritic. Except for the late alkalic variant, plagioclase in these rocks is less calcic (about An_{45-50}), augite is slightly less calcic and not titaniferous, and olivine is present in only trace amounts (up to 2%). Alteration is nearly absent. Alkalic minerals are absent, but many flows contain excess silica in the form of small ball-like clusters of cristobalite or opal in cavities. The *late variant of the Yakima basalt* is vesicular, also contains zoned plagioclase phenocrysts and groundmass

Table 4-7 *Average chemical compositions of the three main types of Columbia River basalt (Waters, 1961; Swanson, 1969) and Steens and associated Miocene basalts of southeastern Oregon which probably correlate with Columbia River basalts (Walker, 1970; Avent, 1970)*

	Steens Basalt	*(Earlier) Picture Gorge Basalt*	*(Later) Yakima Basalt*	*(Late) Late Yakima Basalt*
SiO_2	48.4	49.3	53.8	50.0
TiO_2	2.1	1.6	2.0	3.2
Al_2O_3	16.3	15.5	13.9	13.5
Fe_2O_3	4.6	3.5	2.6	1.9
FeO	7.0	7.8	9.2	12.5
MnO	0.2	0.2	0.2	0.25
MgO	5.9	6.5	4.3	4.4
CaO	9.0	10.3	7.9	8.3
Na_2O	3.0	2.7	3.0	2.9
K_2O	1.0	0.5	1.5	1.4
No. of analyses	28	16	12	

microlites (labradorite), somewhat less augite, in some cases pigeonite (Ca-poor clinopyroxene), a little olivine, a titaniferous "iron-oxide" mineral, and brown glass. Pahoehoe tops are more common than in the earlier flows.

The Columbia River basalt was largely erupted in a nearly stable tectonic environment, although a low angle erosional unconformity separates the Picture Gorge from the Yakima basalts. In the later stages of eruption, the margins of the plateau were upwarped gradually and the flows deformed into large open folds. Continued eruption filled in the synclines as they formed. Where flows dammed or filled in river channels, shallow lakes spread over the margins of the lava plain and deposited silt, sand, and clay. Accumulations of pillow lavas formed where later lavas entered these lakes. In some instances, pillow lavas sporadically comprise the thin lower part of individual flows (Lefebvre, 1970).

That the trend of these dike swarms lies along that of the post-Oligocene feeder dike swarms within and east of the Basin and Range province to the south and southeast suggests a close relationship of the Columbia River flows and tectonic environment to that of post-Oligocene volcanism elsewhere in the western United States (Taubeneck, 1970).

General Features of the Flood Basalt Associations

The great flood of basalt that have been extruded onto the earth's surface, especially during Tertiary and Recent times, form extensive, gently dipping flows erupted from fissure swarms. In other instances, they form low coalescing shield volcanoes a few miles across (e.g., Kuno, 1969b; Jones, 1970). They include the so-called plateau basalts and the shield volcanoes. On the basis of mineralogy, chemistry, and occurrence (Kennedy, 1933; Tilley, 1950), they may be divided into two major groups: the *tholeiitic* and olivine (*alkaline olivine*) basalt. Both types look rather similar in the field and less so in thin section. However, several distinguishing characteristics are apparent. These characteristics, first recognized by Bailey and Thomas (1924) in their classic study of the island of Mull in Scotland, were later extended by Kennedy (1933) to basalts around the world. Since then, numerous papers have expanded on the distinctions and areas of applicability of the two basic types (see, for example, Turner and Verhoogen, 1960). The recent work of Engel and coworkers first indicated that the continental tholeiites should be distinguished not only from alkaline olivine basalt but also from the "oceanic tholeiites." In each group, basalts are dominant, small amounts of light-colored felsic materials appearing as late differentiates.

The tholeiitic basalts (called subalkaline basalts by Chayes, 1966) are characteristic of most of the great plateau basalts found on each of the continents, over most of the deep-ocean bottoms, being generated along spreading oceanic ridges (e.g., Gilluly, 1971), and on a few large oceanic islands (e.g., Hawaii). There is disagreement as to whether there is (Bass, 1971) or is not (Bonatti and Fisher, 1971) a systematic variation in composition away from oceanic ridge crests. Examples of the continental members include the extensive flows of the Columbia River Plateau in eastern Washington, Oregon, and Idaho (previously described); the Snake River Plain of southern Idaho (Stearns et al., 1938; Malde and Powers, 1962; Tilley and Thompson, 1970); the Paraná basalts of southern Brazil; and the Deccan traps of India. All these rocks are essentially saturated in silica and contain little or no olivine (except in some oceanic areas such as Hawaii). Such olivine as is present is unzoned and bears a reaction rim of Ca-poor pyroxene—hypersthene on the phenocrysts or pigeonite in the groundmass (Tilley, 1950; Kuno, 1960)—showing that it is out of equilibrium with the magma. Plagioclase ranges from anorthite to bytownite in the phenocrysts, to bytownite to labradorite in the groundmass. Late rhyolite differentiates contain quartz. Nodular inclusions of peridotite are absent, although rare gabbroic inclusions may be present.

Alkaline olivine basalts are characteristic of oceanic islands and the upper parts of submarine volcanic cones to about 1,500 to 2,000 m below sea level (Engel and Engel, 1964a; Aumento, 1968). They also occur on the continents, in association with the tholeiitic flood basalts such as in the Britain, Iceland, Greenland region (Turner and Verhoogen, 1960), in the plateau of southern British Columbia (Souther, 1970), and in areas of faulting such as the African Rift Valleys. Olivine is usually abundant in these rocks and may even be seen in hand specimen. The olivine tends to be zoned, and clinopyroxene (most commonly titaniferous augite) crystallizes simultaneously and in equilibrium with it (Tilley, 1950; Wilkinson, 1956). The olivine therefore shows no reaction relation to pyroxene. The alkalic nature of these basalts is emphasized by the presence of interstitial alkali feldspar and in some instances alkalic minerals such as nepheline and by the formation of small amounts of trachyte and phonolite as late differentiates. Plagioclase is also more sodic, ranging from labradorite to andesine in the phenocrysts, to andesine in the groundmass. Quartz is absent throughout. Nodular inclusions of peridotite, and to a lesser extent eclogite, occur in some flows. These are now generally thought to be crystal accumulates from the basaltic magma rather than xenoliths of mantle source material (e.g., White, 1966; Steuber and Murthy, 1966; MacGregor, 1968), but deformational features and presence of orthopyroxene (which is otherwise absent in alkaline olivine basalt) in some inclusions suggest that these may be mantle xenoliths (Talbot et al., 1963).

It is desirable to compare the chemical characteristics of different basalts, more so than between most other kinds of rocks, because of the

Table 4-8 *Average chemical compositions of different kinds of flood basalts (in wt.%). Continental tholeiites (144 analyses); oceanic tholeiites (161 analyses*); alkaline basalts (199 analyses).*

	Continental Tholeiites		Oceanic Tholeiites		Alkaline Olivine Basalt	
	AVERAGE	RANGE	AVERAGE	RANGE	AVERAGE	RANGE
SiO_2	50.7	44.35–54.60	49.3	42.8 –52.56	47.1	41.04–51.4
TiO_2	2.0	0.9 – 3.99	1.8	0.35– 3.69	2.7	0.92– 4.52
Al_2O_3	14.4	12.48–16.32	15.2	7.3 –22.3	15.3	10.11–26.26
Fe_2O_3	3.2	0.95– 7.56	2.4	0.69– 7.90	4.3	0.53–15.85
FeO	9.8	4.18–13.60	8.0	2.87–13.58	8.3	0.48–13.63
MnO	0.2	0.10– 0.3	0.17	0.09– 0.44	0.17	0.06– 0.36
MgO	6.2	3.52–11.16	8.3	4.59–26.0	7.0	2.66–17.87
CaO	9.4	7.45–11.8	10.8	6.69–14.1	9.0	6.81–14.46
Na_2O	2.6	1.8 – 3.47	2.6	0.90– 4.45	3.4	1.35– 4.8
K_2O	1.0	0.19– 1.74	0.24	0.04– 0.70	1.2	0.13– 2.5
P_2O_5		0.09– 0.81	0.21	0.06– 0.56	0.41	0.09– 0.93

*Macdonald and Katsura (1964) and Macdonald (1969) tabulate 200 analyses of Hawaiian tholeiites and the average of all of these was used in Table 4-8. However, so as to not unduly weight the Hawaiian rocks, the average for Table 4-8 was calculated as if there were only 20 Hawaiian tholeiite samples.
SOURCE: Worldwide, compiled from the literature.

difficulties inherent in their very fine-grained to glassy nature. Table 4-8 will permit comparison of continental and oceanic tholeiitic basalts and alkaline olivine basalts. As may be seen from this table, although most basalts have somewhat similar compositions, there do seem to be some real differences between the groups. Note that the groups are primarily separated on the basis of mineralogy and occurrence and to a lesser extent on the chemical composition.

Chemically (Table 4-8 and Fig. 4-19), the tholeiitic basalts are *on the average* lower in Na_2O, K_2O, and TiO_2 and higher in SiO_2 than the alkaline olivine basalts. Chayes' (1964a) comparison of predominantly continental tholeiites with alkaline olivine basalts emphasized the reliability of the low TiO_2 of the tholeiites, and Engel and others (1965) emphasized the very low K_2O and the high SiO_2 of the continental tholeiites. The latter workers also noted that Na/K for oceanic tholeiites is greater than 10, whereas for those on the continents the ratio is generally greater than 3 but less than 6. Engel and others (1965) also note that oceanic tholeiites are lower than alkaline olivine basalts in elements usually considered characteristic of granitic rocks—Ba, La, Nb, Rb, Sr, P, Pb, Th, U, Zr— and they are less oxidized (Fe_2O_3/FeO is usually less than 0.4, versus more than 0.4). The high Al_2O_3 of some oceanic tholeiites suggests that these may be high-alumina basalts (Nicholls, 1965; Ringwood in discussion of Kuno, 1966). TiO_2 and K_2O are very low and Al_2O_3 is very high compared with most other basalts. Comparison with the high-alumina basalts of the basalt, andesite, rhyolite association (Table 4-1) shows that these high-alumina basalts of the deep-ocean basins are very similar in composi-

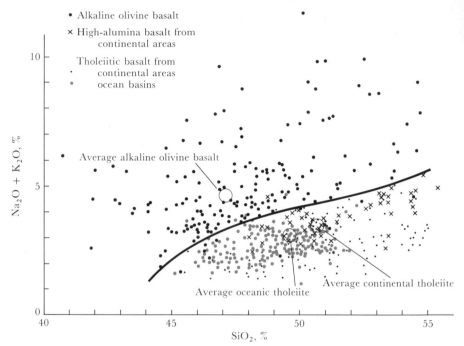

Fig. 4-19 *Alkali-silica variation of basalts. Worldwide occurrences. Rocks that have* SiO_2 *greater than about 53% are normally called andesites. Analyses from the literature. Plot is of the type used by Tilley (1950) and Kuno (1959).*

tion, except for their very low content of K_2O. Miyashiro and coworkers (1969b) have subdivided oceanic tholeiites into low- and high-alumina types, the order of crystallization in the former being olivine→plagioclase→clinopyroxene and in the latter being plagioclase→olivine→clinopyroxene.

Consideration of the foregoing comparison between the different basalt types might lead to the conclusion that the difference is imaginary and all gradations exist between these types. The consistent pattern of basalts of a particular occurrence with a particular mineralogy and chemistry, however, leads most petrologists to conclude that the tholeiitic and alkaline olivine basalt types are distinct and recognizable. At the same time, it is clear that although both types of basalt may be found in a single region—for example, Hawaii, as previously described—the two types are separated spatially, if not in time.

These considerations suggest that, although tholeiitic basalt is the voluminous parent magma to minor rhyolitic differentiates and alkaline olivine basalt is the voluminous parent magma to minor trachytic or phonolitic differentiates, some genetic relationship may exist between them. One may be the parent magma for the other, though of course under different conditions or through a different process than that which forms the minor late-stage differentiates. Alternatively, both types of basalt may develop from the same source but under different conditions or at different stages. A third general possibility is that both develop from different sources but move into the same region because of a single influence such as crustal fracturing.

The origin of basaltic magma is now almost universally accepted as involving melting within the earth's mantle (e.g., Turner and Verhoogen, 1960). The demonstration (Eaton and Murata, 1960) that basaltic magma can be followed seismically from its source in the mantle to its eruption at the surface nicely supports this conclusion. The controversy regarding the origin of basalt arises, however, at the next step. The obvious questions are how and from what is the magma derived, what is its composition, and how does it evolve prior to reaching the surface. If the earth's mantle is essentially dry (strongly undersaturated in water), as is thought to be the case (e.g., Green and Ringwood, 1967b), melting will be at higher temperatures under higher pressures (in this case predominantly rock pressures). As a result, any decrease in pressure such as by crustal and subcrustal fracturing will not only cause partial melting but also will provide a conduit for movement of the magma toward the surface. But tensional failure would presumably give rise to a higher rock pressure after failure than during the tensional stage prior to failure. This problem has prompted Uffen (1959) to suggest that basaltic magmas develop by compressional failure (stress relief), such as along the locus of earthquake foci (Benioff zone) dipping under the continent from major oceanic trenches. Another possibility is frictional melting of uppermost mantle or lowermost crust along the subduction zones, coincident with the Benioff zone (e.g., Oxburgh and Turcotte, 1968b, 1970; Shaw, 1969). Basaltic melts in the vicinity of a midoceanic ridge are presumably associated with tensional stresses accompanying rifting of the lithosphere (e.g., Clarke and Upton, 1971).

Recent experimental work on the basalt-eclogite transition and on their melting temperatures (see Fig. 4-20) suggests that this transition lies below the oceanic Moho,[1] which is at about 12 km, and either below the continental Moho, which is at about 25 to 45 km (e.g., Yoder and Tilley, 1962; Cohen et al., 1967), or above it (e.g., Green and Ringwood, 1967a). Nearly vertical seismic reflections from the Moho (Tuve et al., 1954; Maureau, in Cohen et al., 1967; Cloos, 1969) indicate a fairly sharp transition, probably much sharper than the 15 to 19 km suggested by the experimental basalt-eclogite transition. The Moho is more likely a basalt-peridotite under the oceans and amphibolite- (or granulite-) peridotite

---

[1]Note also that peridotite has been dredged from midoceanic ridges (see p. 107) and apparently thrust up onto and into continental plates (see pp. 111, 113) as discussed under "Alpine peridotite-serpentinite bodies."

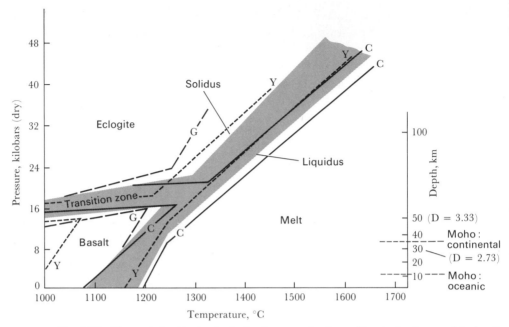

Fig. 4-20 *Phase relationships in the dry basalt system, showing melting temperatures and the basalt-eclogite transition. Heavy phase boundaries averaged from those of Yoder and Tilley (1962, p. 498) for Glenelg eclogite (Y); Green and Ringwood (1967a, p. 789) for quartz tholeiite (G); Cohen et al. (1967, p. 495) for average oceanic tholeiite (C).*

under the continents. This does not eliminate the possibility of some eclogite occurring with peridotite at still-deeper levels (e.g., Clark and Ringwood, 1964; Press, 1969).

Some of the more recently suggested reasonable origins for tholeiitic and alkaline olivine basalt magmas are as follows:

Differentiation of parent tholeiitic basalt magma to form alkaline olivine basalt (Tilley, 1950; Macdonald, 1949, 1969; Powers, 1955; Macdonald and Katsura, 1964; Engel and Engel, 1964a, b; Engel et al., 1965). The mechanism of differentiation could be gravitational segregation of crystals such as clinopyroxene and upward transfer of alkalis in aqueous solution (Macdonald, 1949; Engel and Engel, 1964a, b, 1966; Engel et al., 1965; Richter and Moore, 1966; Hekinian, 1968) or segregation of olivine, pyroxenes, and spinel (Ito and Kennedy, 1968).

High-pressure fractional crystallization of labradorite, subcalcic augite, and a little titaniferous magnetite from low-potash (oceanic) tholeiite and possibly some upward diffusion of alkalis, to form alkaline olivine basalt (Bryan, 1967).

An early stage of partial melting of garnet peridotite could produce tholeiitic basalt, whereas an intermediate stage could give rise to alkaline olivine basalt (O'Hara and Yoder, 1967).

Limited partial melting of garnet peridotite or eclogite to give alkali basalt and extensive fusion to give tholeiitic basalt (Gast, 1968; Philpotts and Schnetzler, 1970a, b).

Higher-pressure (deeper in the mantle) or lower-temperature partial melting of mantle material (possibly eclogite but now generally thought to be garnet peridotite) could produce alkaline olivine basalt, whereas lower-pressure or higher-temperature partial melting could produce tholeiitic magma (Kuno et al., 1957; Kuno, 1959, 1966; Kushiro and Kuno, 1963; O'Hara, 1965; Cann and Vine, 1966; Green and Ringwood, 1967b; Aumento, 1967; Gast, 1968; McGregor, 1968; Miyashiro et al., 1969b; Robinson, 1969; Hawkins, 1970; Leeman and Rogers, 1970; Belousov, 1971). Decrease in activity and crystallization of the magma chamber from the top down could increase pressure and lead to development of alkaline olivine basalt (Macdonald, 1969).

Higher-pressure fractionation of basalt formed by partial melting in the mantle could give alkaline olivine basalt, whereas lower-pressure fractionation of the same basalt could give tholeiitic magma (Powers, 1955; Yoder and Tilley, 1962; O'Hara, 1965, 1968; McBirney and Williams, 1969).

Lower-pressure (shallower) mantle source could produce alkaline olivine basalt, whereas higher-pressure source could produce tholeiitic basalt (Webber, 1966; Peterman and Hedge, 1971).

O'Hara (1965) has elegantly outlined the reasons why the simple concept of partial fusion at depth, followed by rise and extrusion of the basalt without fractionation, is unacceptable:

Such basalt should be extruded well above the temperatures of and should show a long interval of crystallization of one or two phases before cotectic crystallization. But some Hawaiian tholeiitic glasses contain a few phenocrysts of olivine, pyroxene, and plagioclase (Tilley, 1960).

A similar close approximation between the composition and a cotectic crystallization relationship is apparent for alkaline olivine basalt and high-alumina basalt (Yoder and Tilley, 1962; Tilley, Yoder, and Schairer, 1963).

Many Hawaiian basalts have compositions lying in the low-temperature part of the basalt system for 1 atm pressure—"near the cotectic condition olivine + orthopyroxene + augite + feldspar + liquid in temperature and composition" (p. 23). Experimental melting of

Fig. 4-21 *Phase relationships of basalt (hypothetical but in agreement with known data) at various pressures. Triangles represent projections from clinopyroxene apex of tetrahedron. Therefore, clinopyroxene is present in all compositional fields with other minerals. Dry conditions. After O'Hara (1965, pp. 28, 30); minor additions from O'Hara and Yoder (1967) and O'Hara (1968).*

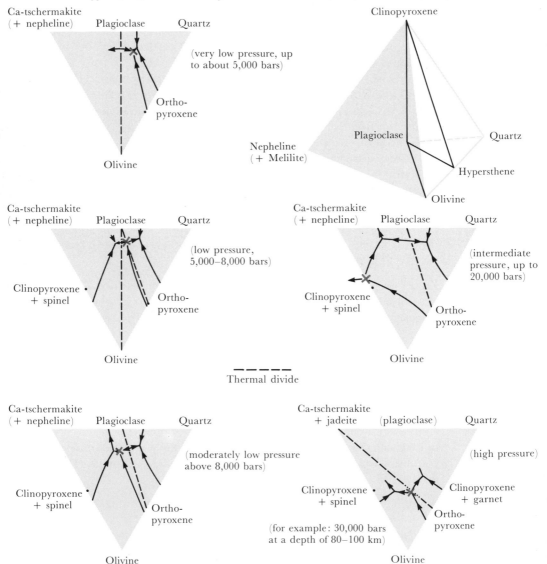

garnet peridotite or eclogite at high pressures yields liquids far from tholeiitic basalt in composition (O'Hara, 1963).

Localized melting followed by rise into cooler rocks implies continuous chilling "and copious removal of the appropriate crystalline phases at all stages" (p. 24) during the 2 to 12 months between magma generation 50 to 60 km below (Powers, 1955; Eaton and Murata, 1960) and eruption at the surface.

Undersaturated (in silica) basalt cannot be derived by the fractional crystallization of a normal (silica-saturated) tholeiitic magma under any conditions because "the composition plane plagioclase-augite-hypersthene represents the most silica-rich thermal divide" possible (p. 26). At high pressures, the thermal divides are poorer in silica (Yoder and Tilley, 1962). To get an alkali basalt from a silica-saturated tholeiitic basalt, the magma would somehow have to cross this high in the temperature contours.

As an alternative to this extreme model of no fractionation following formation of the initial magma by partial fusion, O'Hara (1965, 1968) proposes continual fractional crystallization during rise of the magma to the surface. His scheme involves the shift of cotectic troughs with pressure (this is known to occur experimentally but the details are unknown). Basalt phase relationships may be represented by the tetrahedron clinopyroxene–nepheline (+ melilite)–olivine–quartz as shown in Fig. 4-21 for low pressures.

Partial melting of a peridotite or garnet peridotite mantle begins (O'Hara, 1965) with liquid having the composition of the point X under the conditions represented by any triangular diagram (see Fig. 4-21). Under these conditions, clinopyroxene, olivine, orthopyroxene, and liquid all coexist with an additional phase (plagioclase, spinel, or garnet) depending on the pressure. Note the differing position with pressure but close similarity of configuration of the adjoining cotectics at this initial melting composition. The initial magma thus generated depends on pressure as follows:

Very low pressure: silica-saturated tholeiite (after O'Hara, 1965)
Low pressure: feldspathic olivine tholeiite
Moderately low pressure: alkali olivine basalt
Intermediate pressure: olivine-rich alkali basalt
High pressure: tholeiitic olivine-rich basalt

These will be the compositions of the extruded magmas only if the liquid rises to the surface very rapidly, a situation which seems likely considering the rapid few-month rise from source to eruption. Depending on the depth of initial magma generation, rate of rise, and time and depth at which the liquid is held during its rise, there are many possibilities for change in composition (e.g., O'Hara, 1963, 1965, 1968; Boyd et al., 1964; Cohen et al., 1967; Green and Ringwood, 1967b; Ito and Kennedy, 1968). It may be significant that the "low-velocity zone" in the upper mantle is probably a zone of a small percentage of partial melt and that it is widespread at a depth of about 50 to 150 km (Anderson et al., 1971). Since this corresponds to the depths of inferred magma generation in Hawaii (Eaton and Murata, 1960) and would give rise to the tholeiitic basalts found, it seems likely that most flood basalts may be generated in this zone. In addition to these dominant basalts, late-stage fractional crystallization of the tholeiitic basalt could lead to minor amounts of late quartz-bearing fractions. Late-stage fractional crystallization of the alkali olivine basalt could lead to minor amounts of late nepheline-bearing fractions. Each of these may be seen by following down the appropriate cotectic curve.

O'Hara's (1965, 1968) model may also explain why silica-saturated tholeiitic basalt is so prominently developed over the deep-ocean floor, whereas alkali basalt is largely confined to the upper parts of the higher submarine volcanoes and oceanic islands, a situation noted by Engel et al. (1965).

DIABASE DIKES AND SILLS

The Triassic Diabases of New York, New Jersey, Pennsylvania, and Maryland[1]

Late Triassic diabase bodies, including the 1,000-ft.-thick Palisade sill outcropping along the west bank of the Hudson River in New York, extend in a narrow belt for more than 250 miles from southeastern New York to northern Maryland (Fig. 4-22). These basaltic magmas invaded postorogenic terrestrial arkoses and shales of Triassic age and are related to surface flows farther west. These dark-colored rocks, predominantly sills, are characteristically columnar-jointed, forming steep escarpments and cliffs.

Although largely concordant, the sills are locally discordant. In places, slabs of arkose and shale have broken off the contacts and floated upward in the diabase. The contacts are knife sharp and the otherwise gray diabase there is chilled to black basalt, glassy at the contact with the intruded sedimentary rocks. Grains throughout most of the thickness of each sill are of medium size—coarse enough to be clearly visible with the unaided

[1]F. Walker, 1940, 1961; K. R. Walker, 1970; Poldervaart and Walker, 1962; Pearce, 1970; Hotz, 1952, 1953; Lewis, 1908; Stose and Lewis, 1916.

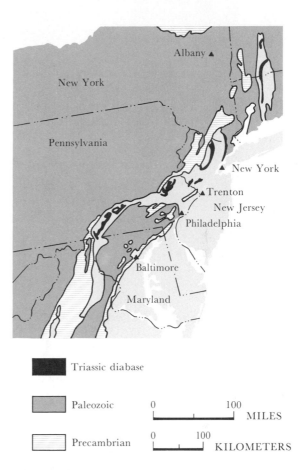

Fig. 4-22 *Distribution of Triassic diabase sills in Connecticut, New York, New Jersey, Pennsylvania, and adjacent states (simplified from Geologic Map of North America, 1965; E. N. Goddard, Chairman of Map Committee).*

Triassic diabase

Paleozoic

Precambrian

```
0                    100
|_____|    MILES

0          100
|_____|_____|    KILOMETERS
```

eye, grains averaging 0.5 to 1 mm being common in the coarse part toward the top of the sill. The minerals are about half light (largely plagioclase) and half dark (largely pyroxene with lesser olivine), the resultant rock having a dark speckled appearance. Thicker bodies such as the Palisade sill are zoned as shown in Fig. 4-23.

The coarsest zone, above the middle of the sill, contains patchy streaks of micropegmatite (pegmatite schlieren) up to 2 ft thick. These coarse-grained streaks consist of quartz and alkali feldspar intergrown in graphic or micrographic textures and coarse blades of pyroxene. Late quartzofeldspathic veins ("white veins" of Walker) up to 3 in. across fill fractures in both the upper and lower chilled zones.

The distinctive subophitic or ophitic texture (see Fig. 4-24) of these and other diabases is attributed to simultaneous crystallization of the plagioclase and pyroxene (Walker, 1957).

Not only does the grain size vary within the thick sills, but also the

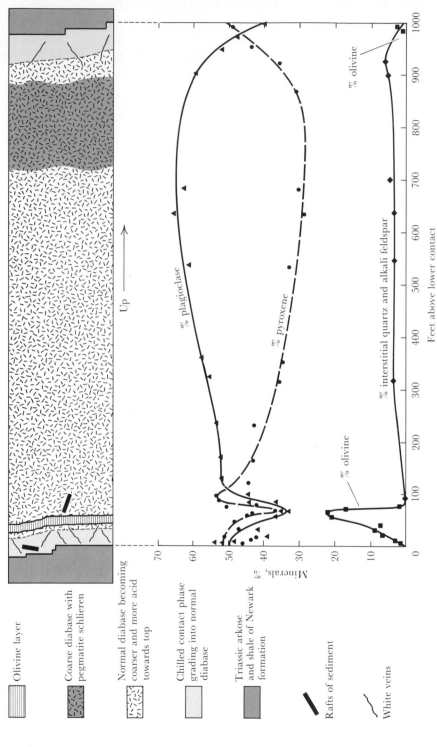

Olivine layer

Coarse diabase with
pegmatite schlieren

Normal diabase becoming
coarser and more acid
towards top

Chilled contact phase
grading into normal
diabase

Triassic arkose
and shale of Newark
formation

Rafts of sediment

White veins

Fig. 4-23 *Diagramatic section across the Palisade sill, New York. Variation of the main mineral constituents, George Washington Bridge section. From Walker (1940).*

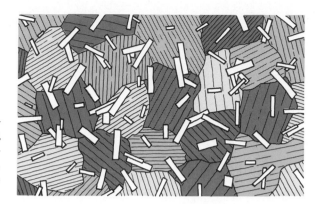

Fig. 4-24 *Subophitic texture typical of diabases. Diagramatic. Note that the smaller, euhedral plagioclase grains are partly surrounded by larger anhedral pyroxene grains. In the less common ophitic texture, the plagioclase grains are completely surrounded.*

mineralogy (see Fig. 4-23). *Olivine* amounts to about 1% in the upper and lower chilled zones and increases to 20 to 25% in the "olivine layer" (where present). The olivine is Fo_{81-79} in the chilled contact, Fo_{77-55} in the olivine layer, and Fo_{20-7} in the late stages. It tends to be euhedral and in some cases is surrounded by a reaction rim of pyroxene. The earlier *pyroxenes* (in the lower third of the Palisade sill) consist of coexisting hypersthene and augite—compare Fig. 1-13. The later pyroxenes (in the upper two-thirds of the Palisade sill) consist of pigeonite and augite— becoming more iron rich as crystallization proceeds. *Plagioclase* in the Palisade sill is intermediate (An_{61-66}) at the chilled contacts (probably representing the original magma), high (An_{64}) near the base, and progressively lower to "low" (An_{37}) near the top—compare Fig. 1-3. *Minor minerals* include ilmenite and titaniferous magnetite, small flakes of brown biotite or grains of dark green hornblende altering from and showing rims on pyroxene and iron oxides, fayalitic olivine, micrographic intergrowths of quartz and alkali feldspar, and fine needles of apatite.

The granophyre forms fine- to coarse-grained lenticular masses up to 100 ft thick, parallel and near to the upper contact of the sills. Miarolitic cavities up to an inch across occur in some places. The granophyres also contain sodic plagioclase, hornblende or biotite or iron-rich clinopyroxene, and minor titaniferous iron oxide and apatite.

Chemically, these diabase sills represent normal tholeiitic basalt (see Table 4-8). Samples from the olivine-rich layer,[1] as would be expected, tend to be somewhat higher in MgO and lower in Al_2O_3 and Na_2O. A few samples of the normal diabase are somewhat higher in Al_2O_3 (up to 16.54%), lower in TiO_2 (down to 0.74%), and lower in MnO (down to 0.03%), suggesting an affinity with high-alumina basalt of the basalt, andesite, rhyolite

---

[1] Recall (for example, p. 165) that olivine may occur in tholeiitic basalts. It may be abundant where, as suggested here, it has settled out to form a concentrate.

association (see Table 4-1). Chemical trends involve an increase upward in alkalis and iron, and a decrease upward in MgO. The granophyre is enriched in SiO_2, alkalis, and iron, and is depleted in MgO and CaO.

The development of variation within these diabase sills is universally attributed to differentiation during crystallization of the magma. Most workers agree that the majority of these sills were injected in a single stage as a nearly complete melt containing a few small early phenocrysts of olivine (as seen in the chilled zones). Compositionally, the magma is basaltic, lying in the forsterite field close to the pyroxene ("diopside") and plagioclase fields of the olivine-diopside-plagioclase system (see Fig. 3-1). Chemical variation in the sill is virtually as predicted from fractional crystallization of olivine and pyroxene from a basalt magma (Pearce, 1970). Sinking of olivine to form the olivine-rich layer was first pointed out by Lewis (1908). However, Poldervaart and Walker (1962) noted that the olivine in this layer occurs in two generations, the coarser less-common grains having accumulated by gravitation, but the smaller (0.02 to 0.1 mm) grains are too small and too iron rich (Fo_{65-70}) for this. They probably formed by injection of a new pulse of magma. Internal chilled phases agree with this newer interpretation. Additional evidence includes reversals in the fractionation trend in the bottom quarter of the intrusion and multicrystal aggregates in the zone of contact between the two injections of basalt (Walker, 1970). Plagioclase and pyroxene began to crystallize soon after olivine, pyroxene presumably settling somewhat as suggested by the higher concentration of pyroxene toward the base of the intrusions (Bowen, 1928, pp. 71–73, and see Fig. 4-23; Pearce, 1970). The commensurate increase upward in plagioclase concentration is presumably due to this settling of pyroxene. The density of the plagioclase is close to that of the magma, resulting in neither settling nor rising.

As these nonvolatile minerals—rich in MgO and CaO and poor in SiO_2, Na_2O, volatiles, and, in the earlier stages, total Fe—crystallize, the magma becomes enriched in the latter elements. Slower cooling and an increase in volatiles develops coarser grains in the central to upper parts of a sill. When the basaltic magma has reached a stage of 75 to 80% crystallization, the composition of the magma has changed sufficiently to begin crystallization of granophyre. At this stage, the residual liquid is concentrated in crystal interstices and some pockets in the upper part of the intrusion, below the solidified roof. Iron enrichment during crystallization was attributed to the presence of volatiles that prevented "iron oxide from going into chemical composition with silica as pyroxene" (Walker, 1940, p. 1092). Amphibole and biotite developed by reaction of pyroxene and iron oxide with the volatile-rich magma.

The basaltic magma that supplied these sills is presumably related to that which appeared to the west as surface flows during the Triassic period in the same region.

General Features of Diabasic Rocks

Diabase[1] sills and, less commonly, dikes are widespread common features in and around regions of flood basalts. Carey (1961) reasons that diabase sills should, and in fact do, occur in nonfolded sedimentary basins of moderate thickness (10,000 to 20,000 ft) because the ascending basaltic magma would do less work by lifting the sediments than by continuing to the surface. Injection into thinner sediments would form cone sheets through fracture. Injection into thicker sediments would form flows, as, for example, in this instance, rise of magma all the way to the surface would involve less work than lifting the sediments. Diabase dikes and sills range in thickness from those a few feet or less to those of 1,000 ft or more, such as the Palisades, New York (e.g., Walker, 1940); the Dillsburg sill, Pennsylvania (Hotz, 1953); the Goose Creek sill, Virginia (Shannon, 1924); the Nipissing sill, Ontario (Hriskevich, 1968); sills in the Ahr Lake area, Quebec (Baragar, 1967); the Endion sill, Duluth, Minnesota (Schwartz and Sandberg, 1940; Ernst, 1960); the Tasmanian dolerites (e.g., Edwards, 1942; Joplin, 1961; McDougall, 1962, 1964; Heier et al., 1965); the Libode sill, South Africa (Walker and Poldervaart, 1949); and the Peneplain sill and Basement sill, Antarctica (Gunn, 1962; Hamilton, 1965b). The general features of diabase sills are largely exemplified by the Palisades sill and related rocks as described above—vertical variation due to differentiation, resulting in an upward increase in SiO_2, alkalis, and iron. Very thick sections of granophyre at the top of some sills have been attributed to updip migration of the granophyre in an inclined sill or to separate intrusion (Ernst, 1960). As noted previously in the section on flood basalts, there are two main types—tholeiitic and alkaline olivine basalts. These could presumably be represented by two comparable types of diabases. Walker (1961) emphasizes that these two varieties do indeed exist—tholeiitic diabase and alkaline olivine diabase (see Table 4-9). Tholeiitic diabase is much more common. Examples of each are described in Wager and Brown (1968).

Suggested Origins of Diabase Dikes and Sills

It is now generally agreed that the basaltic magma forming diabase sills and dikes is the same as that of tholeiitic and alkaline olivine basalt flows (see Table 4-9). The origin of these basaltic magmas has been reviewed above, under the discussion of the flood basalts, and will be pursued no further here. Of immediate concern, however, is the source of the vertical variation found in the thicker diabase sills. Reasonable proposals all in-

---

[1]"Dolerite," a term in wide use outside North America, is essentially synonymous with "diabase" as used in North America. However, the reverse is not true ("diabase" as used in Europe and elsewhere is reserved for certain kinds of altered rocks).

Table 4-9   *Comparison of tholeiitic diabase and alkaline olivine diabase**

	Tholeiitic Diabase (or Quartz Diabase)	Alkaline Olivine Diabase (or Olivine Diabase)
Associated with	Tholeiitic basalt flows	Alkaline olivine basalt flows
Plagioclase	An_{50-80}	An_{50-70}
Pyroxene/plagioclase	40/60	About 30/70
Pyroxene	Augite or subcalcic augite + Mg-pigeonite or hypersthene	Titanaugite
Olivine	± present (up to 25%)	About 10% (up to 70%)
Accessory minerals	More abundant (quartz, potassium feldspar, titaniferous iron oxide, ± pyrite, biotite, hornblende, sphene); micrographic quartz-alkali feldspar as interstitial residuum	Less abundant (iron oxide, zeolites, ± biotite, hornblende, aegirine-augite)
Tendency to gravitational differentiation	Rare in intrusions thinner than 300–400 ft	Rare in intrusions thinner than 100 ft (more prominent)
Settling of	Olivine, pyroxene	Olivine only
Differentiation trend	Rhyolite (granophyre) (high silica, alkalis, iron)	Syenite, zeolites
Chemistry	That of tholeiitic basalt	That of alkaline olivine basalt

*Compare the mineralogical distinctions between tholeiitic basalt and alkaline olivine basalt described in the previous section on flood basalts.
SOURCE: Tabulated from information in Walker (1961).

volve intrusion of an almost completely liquid basaltic magma, chilling of the magma adjacent to the contacts, and vertical variation resulting from differentiation effected by one or more of the following:

> Settling of early formed grains of pyroxene (and olivine if present), with upward migration of residual fluids (Lewis, 1908; Walker, 1940, 1953, 1961; Schwartz and Sandburg, 1940; Edwards, 1942; Turner and Verhoogen, 1960; McDougall, 1961; Gunn, 1962; Pulvertaft, 1965; Drever and Johnston, 1967; Wager and Brown, 1968, pp. 529–537; Hriskevich, 1968; Yagi, 1969; Pearce, 1970)

> Settling of early formed aggregates of pyroxene and plagioclase, with upward migration of residual liquids (Jaeger and Joplin, 1955; Ernst, 1960; Joplin, 1961; Spry, 1961)

> Liquid fractionation involving vertical diffusion in response to pres-

sure-temperature gradients (Lindgren, 1933; Tomkeieff, 1937; Schwartz and Sandberg, 1940; Walker and Poldervaart, 1949; Hotz, 1953; McDougall, 1962; Hamilton, 1965b; Wilshire, 1967)

Settling of early formed pyroxene or olivine and squeezing out of the residual magma (filter pressing) to isolate residual fluids (Tyrrell, 1928; Walker, 1953)

Settling of mafic minerals and possible floating of plagioclase, aided by jostling of crystals during magma movements (Baragar, 1967)

Most crystallization proceeding from the bottom upward (and top downward) without crystal settling, either with convection currents (Hess, 1956, 1960a) or without (Poldervaart and Walker, 1962; Hamilton, 1965b)

Any valid origin of diabase sills must account for the mineralogical, chemical, and textural variation within the sills, including:

Important Data	*Implications*
Chill zones contain only a few small phenocrysts	Magma nearly 100% liquid when injected
Olivine diabase layer (if present) more even on top than bottom	Suggests filling in of depressions by olivine
Progressively lower temperature assemblages higher in intrusion	Intrusion crystallizes from bottom upward
Subophitic to ophitic texture (grains intergrown—see Fig. 4-24)	Grains probably crystallized in place to form multicrystal aggregates, either formed on the bottom or as multicrystal aggregates that sank (compare poikilitic textures of gabbroic layered intrusions—below)
Layering and preferred orientation of grains absent or minor. Interstitial alkali feldspar quartz residuum ("mesostasis") generally gradually increasing upward in the intrusion	Absence of convection currents as important; [?] absence of crystal settling; residuum partly displaced upward during crystallization (pyroxene and plagioclase did not likely sink into *basaltic* magma—i.e., probably grew in place)
Progressive moderate enrichment in iron to higher levels—pyroxenes become higher in Fe^{++}; Fe^{3+} appears as iron oxides throughout but becomes more abundant to higher levels	Moderately low P_{O_2} (low water), so most iron as Fe^{++} until water builds up in later stages at higher levels (see Fig. 3-7)

The scheme that seems best to fit the available data involves injection of a tholeiitic (or, less commonly, alkaline olivine) basalt that is nearly

100% liquid. If the magma composition is such that olivine crystallizes early in significant quantity, the olivine begins settling out to form an olivine-rich layer low in the intrusion. Soon thereafter, crystallization of pyroxene and plagioclase begins on the solid bottom and perhaps within the liquid (to sink as crystal aggregates), the larger pyroxenes partly to largely enveloping the euhedral plagioclase to form subophitic to ophitic textures. Crystallization of these major minerals leaves the interstitial residual liquid enriched in the constituents of alkali feldspar and quartz. These constituents are partly trapped in the interstices between the earlier-crystallized pyroxene and plagioclase and are partly displaced upward to form granophyre at higher levels. The magma is inferred as being low in water (and low in P_{O_2}) until water and other volatiles build up in the later stages at high levels. This increase in volatiles probably accounts for the coarser grains of the pegmatitic diabase and of some of the granophyres. The remaining Fe^{++} is partly oxidized to Fe^{3+} and crystallizes as iron oxides.

LAMPROPHYRE DIKES

Lamprophyres of the Spanish Peaks Area, Southcentral Colorado[1]

The sedimentary rocks enclosing two Tertiary granitic stocks of the Spanish Peaks area have been invaded by a prominent swarm of dikes radiating outward from each. The eastern stock of granite prophyry with a central core of somewhat later granodiorite prophyrys has slightly domed the surrounding rocks. The later, somewhat smaller western stock of pyroxene "syenodiorite" (monzonite) has invaded a syncline without doming the sedimentary rocks and has metamorphosed them for at least 900 ft outward from the contact.

Most, but not all, of the 500 or more radial dikes (see Fig. 4-25) converge on West Spanish Peak stock, the swarm occupying an ellipse extending largely eastward from the focal area. None of the dikes come into contact with the stock and only a few cut its metamorphosed aureole, suggesting that the dike magmas are younger and unrelated to the stock magmas. The nearly vertical radial dikes range from basalt and lamprophyre, to alkali gabbro, diorite, syenodiorite, syenite, and granite; and they were injected in at least seven separate phases.

The lamprophyre dikes, with which we are most concerned here, along with associated, chemically similar syenites, syenodiorites, and syenogabbros, are mostly less than 3 (but up to 14) miles long and well away from the western stock. They are largely later than the dikes of the

[1]Knopf, 1936; Johnson, 1961, 1964, 1968, written communication, November, 1970.

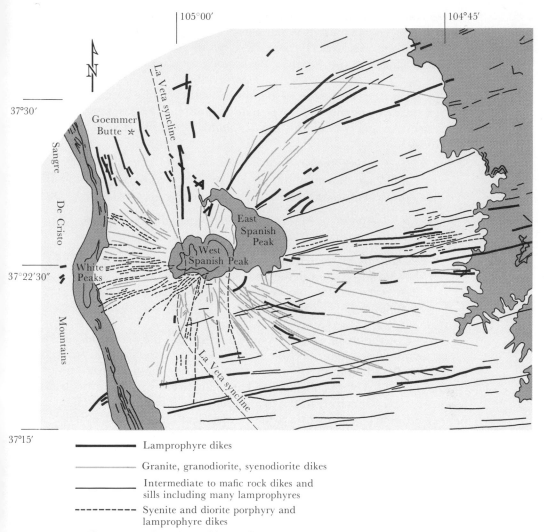

Fig. 4-25 *Dike swarms of lamprophyres and associated rocks surrounding the Spanish Peak stocks, Colorado. After Johnson (1961, plate 1; 1968).*

radial swarm (Johnson, 1961). Most of the dikes are essentially straight, have parallel walls, and range from about 1 to 100 ft thick. Some have metamorphosed the adjacent sedimentary rocks. Lamprophyres, as the term is used by Knopf (1936) and later workers, are mesotype to melanocratic (medium to dark) rocks carrying only ferromagnesian (dark) phenocrysts in an aphanitic or microgranular groundmass. The ferromagnesian minerals in the groundmass tend to be euhedral. A few ultramafic lamprophyres contain small sphereoids ("oceili") of alkali feldspar and analcite with minor ferromagnesian minerals.

Lamprophyres are characterized by their ferromagnesian (mafic) minerals. Those in the Spanish Peaks area include the following types:

Syenodiorite lamprophyres—augite-orthoclase-biotite-plagioclase An_{35} (±olivine or hornblende, + Fe oxide); biotite-orthoclase-augite-plagioclase An_{30-40} (±olivine, ±hornblende, ±glass, ±analcite, + Fe oxide)

Syenogabbro lamprophyres—orthoclase-biotite-olivine-plagioclase An_{55} (± augite, ± hornblende, + Fe oxide); orthoclase-biotite-augite-plagioclase An_{55-65} (± analcite, ± hornblende, ± olivine, + Fe oxide)

Diorite lamprophyres—hornblende-biotite-augite-plagioclase An_{35-45} (±olivine, ±analcite, ±glass, + Fe oxide)

Gabbro lamprophyres—biotite-augite-plagioclase An_{55-60} (± olivine, + Fe oxide); biotite-olivine-analcite (± plagioclase An_{55-60}, ± augite, ± glass, + Fe oxide); biotite-plagioclase An_{55-70} (± analcite, ± augite, ±hornblende, + Fe oxide)

Some of the mafic phenocrysts are fresh, but most are altered. Most, especially olivine, augite, and biotite, are zoned near the rims. In accordance with the definition of a lamprophyre, the light-colored minerals (potassium feldspar, plagioclase, and analcite) are confined to the groundmass. Accessory minerals in the groundmass include apatite needles, magnetite, secondary zeolites or calcite, haüyne (sodalite group), sphene, and rarely glass.

A few of the thicker lamprophyre dikes grade rapidly to coarser-grained cores of alkali gabbro or melasyenite (mafic-rich syenite). The mineralogy of these coarser cores is commonly similar to that of the lamprophyric margins, even to distinctive minerals such as analcite, but in some instances includes one or other extra minor minerals such as hornblende. At least one lamprophyre contains 10 to 20% xenoliths of "white gneiss" derived from an underlying conglomerate.

Chemically, the 27 analyzed lamprophyres from the area range in composition as follows:

SiO_2(41.3−51.6%), TiO_2(0.06−2.16), Al_2O_3(10.9−16.6),
Fe_2O_3(4.12−9.5), FeO (1.3−6.7), MgO (4−12.5), MnO (0.05−0.28),
CaO (6.4−11.9), Na_2O (1.1−4.7), K_2O (0.61−6.6), P_2O_5 (0.44−2.3),
CO_2 (0.05−7.8), BaO (0.13−0.36), SrO (0.1−0.2)

They are therefore compositionally similar to the alkaline olivine basalts (compare Table 4-8), but with a slightly higher K_2O, BaO, and SrO content, a high Fe_2O_3/FeO, and a somewhat lower TiO_2.

Knopf and earlier workers surmised that the lamprophyres and other dikes all originated in the Spanish Peaks stocks. Johnson (1961) points out, however, that there is no reason to believe that intrusion of the stocks should result in radial fractures; that the dikes do not converge on a single source; some dikes are exposed discontinuously along their lengths, indicating that the magma came from below; no dikes contact the west Spanish Peak stock at the surface, and the lamprophyres are rare among the sills and metamorphosed rocks surrounding the stock—also indicating that the magma came from below. Johnson proposes that regional joints are related to orogenic stresses of varying magnitude and direction, resulting in a complex and random pattern. The radial dike pattern would result from selective injection of low-viscosity magmas into those joints that are normal to nearly domical equipotential magma-pressure surfaces.

General Features of Lamprophyre Dikes

The lamprophyres are widespread, distinctive dark-colored, fine-grained to aphanitic dike rocks, recognized in hand specimen by the common presence of mafic phenocrysts. Light-colored phenocrysts are always absent, though they are not uncommon in basalts. Lamprophyres are characterized, in fact, by an abundance of euhedral mafic minerals, in many instances of two generations—earlier, often altered, phenocrysts and later fresh groundmass grains of the same type. Phenocrysts may include biotite, hornblende (including barkevikite), augite or titanaugite, or olivine. Although the groundmass appears dark in hand specimen, it consists predominantly of felsic minerals, including sanidine, plagioclase, analcite, nepheline, or melilite, and secondary minerals such as zeolites, calcite, or talc (after olivine). Albite, in at least some areas, appears not primary but altered from primary plagioclase (Němec, 1966). Xenoliths of quartz, alkali feldspar, granite, gneiss, or metasediments are common, especially in the biotite lamprophyres.

Lamprophyres occur, generally, as narrow (a few feet) dike rocks, often cutting or otherwise associated with granitic intrusions. In some instances, they radiate outward from such intrusions: The Spanish Peaks and Dike Mountain, Colorado (Knopf, 1936; Johnson, 1961); the Shap granite, northern England (e.g., Grantham, 1928); the Ava Stock, Finland (Kaitaro, 1956). In other instances, lamprophyres form subparallel dike swarms in or adjacent to granitic plutons: in the northwestern part of the South Island of New Zealand (Turner and Verhoogen, 1960, p. 252); the Sierra Nevada batholith, California (Moore and Hopson, 1961); the southern uplands of Scotland (Read, 1926; Phillips, 1956). In still other instances, they show no special orientation in granitic plutons, as in southern England (Smith, 1936). In some instances, lamprophyres are associated with volcanic rocks of the alkaline olivine basalt association: the Navajo area of northwestern New Mexico (Williams, 1936); northern Ireland, England, and southern Scotland (e.g., Reynolds, 1931; King, 1937). Most or perhaps all these dominantly volcanic regions containing lamprophyres are underlain, however, by granitic rocks (Turner and Verhoogen, 1960, p. 252).

Chemically, the lamprophyres are like the alkaline olivine basalts (see Table 4-8) and related rocks. A few are higher in Na_2O or K_2O and are compositionally like nepheline basalts or leucite basalts, respectively. Even some of the prominent minerals (e.g., zoned olivine and titanaugite and, locally, analcite or biotite) are identical to those in alkaline olivine basalts and related rocks such as analcite and leucite basalts. The lamprophyres, then, are characteristically low in SiO_2 and high in Na_2O, K_2O, TiO_2, BaO, SrO, and P_2O_5, as are the alkaline olivine basalts. This correlation is fortified by the recognition (e.g., Knopf, 1936; Williams, 1936) of gradations in some dikes from thin dikes or borders of lamprophyre to thick dikes or cores of rocks of the alkaline olivine basalt association.

Although in the past many specific names had been given to different varieties of lamprophyres, current practice is to name them according to the common granitic rock classification (e.g., Johnson, 1961, 1968; Gross and Heinrich, 1966). Thus a lamprophyre consisting of augite phenocrysts in a groundmass of predominantly potassium feldspar would be called a pyroxene syenite lamprophyre. This has all but eliminated the confusion and controversy over terminology.

Suggested Origins of Lamprophyre Dikes

In spite of the textural and, to a lesser extent, mineralogical peculiarities of lamprophyres, their compositional similarities to alkaline olivine basalts and related rocks suggest that some modification of such a magma is involved in the origin of lamprophyres. Most suggested origins use this as a starting point. The origin of such a magma has previously been discussed in connection with the flood basalts and will not be repeated here.

It has been suggested that alkaline olivine basalt magma:

Assimilates granitic rocks, mica schist, or quartzofeldspathic schist (Bederke, 1947; Kaitaro, 1956; Phillips, 1956; Oftedahl, 1957; Turner and Verhoogen, 1960; Cogné, 1962) or the parent magma may be ultrabasic (Williams, 1936).

Is differentiated and enriched in CO_2 and H_2O by an unknown process (Upton, 1965), resulting in delayed crystallization of Mg- and Fe-bearing silicates. This would concentrate the ferromagnesians with the alkalis in the late magmatic (lamprophyric) fluids (Eskola, 1954).

Assimilates concentrations of biotite and hornblende (low in SiO_2 and high in alkalis and water) with simultaneous crystallization of augite and olivine (Beger, 1923; Bowen, 1928).

Differentiates to give soda-rich lamprophyres and potassium-rich basalt magma differentiates to give potassium-rich lamprophyres (Joplin, 1966).

With a fairly high potassium content, crystallizes under pressures greater than about 1,500 bars and under near-equilibrium conditions, giving rise to olivine, then phlogopite, then pyroxene with resorption of olivine (Luth, 1957; Velde, 1967).

Tholeiitic magma is contaminated at depth by an influx (possibly hydro-thermal) of alkalis and volatiles (Němec, 1971). Any satisfactory origin for lamprophyres must account for their typical characteristics, including:

Occurrence in the form of small- to medium-sized dikes. Association either within (but later than) or near granitic plutons.

Environment characterized by members of the alkaline olivine basalt association.

Chemically, and to some extent mineralogically, similar to alkaline olivine basalt and closely related rocks.

Gradation in some dikes, from lamprophyre to a member of the alkaline olivine basalt association. The lamprophyres occur either as thinner dikes or border zones of thicker dikes.

Xenoliths of quartz or (less commonly) alkali feldspar, or quartzo-feldspathic rocks.

Mafic minerals are euhedral and typically occur both as phenocrysts and in the groundmass.

Common extensive alteration, especially of mafic phenocrysts.

Felsic minerals are anhedral and are generally confined to the groundmass.

The origin that can be pieced together from these characteristics is as follows:

An alkaline olivine basalt magma (the origin of which has been described in the section on flood basalts) is injected into newly formed fractures in or around an earlier granitic pluton. Both magma generation and fractures could result from compressive failure as suggested by Uffen (1959; see also p. 173). Assimilation of granitic, gneissic, or sedimentary wallrocks may contribute to the high alkali content and volatiles and hence also to the oxidized state of the iron and to the alteration. The mafic phenocrysts presumably formed early, before injection to the level of the dike and resultant more rapid cooling to form the groundmass. The lack of felsic phenocrysts is typical of alkaline olivine basalts (much more so than of tholeiitic basalts) and must be attributed to the lower SiO_2 content and other chemical characteristics of these magmas. That a few feldspar phenocrysts may appear in alkaline olivine basalt probably does not represent a significant problem, as a few have also been reported in some lamprophyres (Gross and Heinrich, 1966).

POTASSIUM-RICH BASALTIC ROCKS

The Potassium-rich Basaltic Association of the Highwood Mountains, Central Montana[1]

The Tertiary igneous rocks of the Highwood Mountains of central Montana are well known for their distinctive suite of potassium-rich "basaltic" and related rocks. They occur as a group of laccoliths, stocks, dikes, and related extrusive rocks intruding and covering flat-lying Cretaceous sedimentary rocks (Fig. 4-26). In mid-Tertiary time, volcanic eruptions of potassium-rich quartz latite composition built up a volcano several thousand feet high.

Following considerable erosion, renewed volcanism built a new volcano consisting of dark mafic-rich leucite, analcite, and pseudoleucite phonolites (see Fig. 2-1b) of the potassium-rich "basaltic" suite. The basal flows and breccia of the new volcano dip outward from the higher central area. Flows are mostly 10 to 50 ft thick, dark green to black, dense, and apparently fresh. Breccias are more commonly reddish-brown, vesicular, rich in zeolites, and altered. Most rocks contain numerous 2- to 4-mm-long prisms of augite, some olivine (usually altered to red-brown iddingsite), and rounded grains of leucite, analcite, or pseudoleucite.

Radiating outward from the main volcanic center are a few dikes of amphibole phonolite, many of mafic-rich pseudoleucite, augite and biotite phonolite, and a few of syenite porphyry (in order of intrusion). They

[1]Larsen, 1940; Larsen, Hurlbut, Burgess, and Buie, 1941; Hurlbut and Griggs, 1939; Barksdale, 1937; Osborn and Roberts, 1931; Weed and Pirsson, 1895, 1901; Pirsson, 1950; Nash and Wilkinson, 1970.

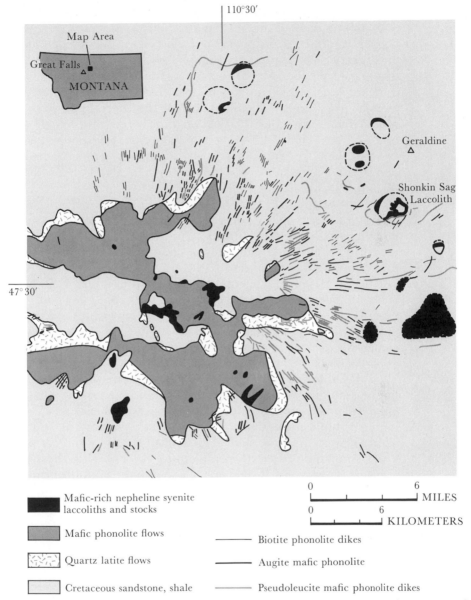

Fig. 4-26 *Potassium-rich basaltic rocks of the Highwood Mountains, Montana. After Larson et al. (1941).*

range from about 2 to 20 ft thick. Although most numerous near the stocks, they are not considered to be offshoots of the stocks. The fractures, into which the dikes were intruded, are thought to be later than intrusion of the stocks (recall the dikes of the Spanish Peaks area). Many dikes show fine-grained, chilled margins; and some are composite, for example, with borders of augite mafic phonolite and a center of biotite, phonolite. Phenocrysts, especially pyroxene or amphibole, near the borders of the many dikes, are well aligned, presumably by flow of the magma.

Six small stocks, not much more than 1 sq mile in area, cut the sedimentary rocks and mafic phonolites of the central part of the Highwood Mountains. Most consist of "shonkinite" and other mafic-rich foyaitic foidites (see Fig. 2-1a)—that is, mafic-rich feldspathoidal (nepheline) syenites. Country rocks have been thermally (contact) metamorphosed and altered by alkali-rich solutions for half a mile from the stocks and for a few inches to a few feet from the dikes.

Nine laccoliths, 1 sq mile to several square miles in area have intruded a thin (250 ft) Upper Cretaceous sandstone in a northwest-trending belt, north and east of the mountains. The laccoliths are all similar and consist primarily of "shonkinite" (augite-sanidine-feldspathoid rock), with an overlying syenite, and dark, fine-grained, chilled margins. The most famous of these is the Shonkin Sag laccolith (see Fig. 4-27). This laccolith is about 250 ft thick in the central part, thinning to about 125 ft at the margins where it merges with the six sills having the same composition as the chilled border phase of the laccolith. One of these merges with the chilled border and others merge with more internal parts of the laccolith crystallization.

Mineralogically, the Shonkin Sag laccolith shows gradations mainly in the amount of individual minerals and in texture (Fig. 4-27). The euhedral mafic minerals form phenocrysts and occur to a lesser extent in the groundmass. These mafic minerals increase in abundance from about 50% at the base (and in the chilled zones) to 65% at the top of the lower shonkinite, then drop rapidly to 20 or 25% in the syenite, before rising rapidly again in the upper shonkinite. Pseudoleucite, a radiating intergrowth of sanidine and cloudy zeolites (altered from earlier leucite or potash analcite) occurs in the chilled margin but apparently became unstable to higher levels where the magma took longer to cool.

Mineralogically, the lavas, dikes, and stocks of the Highwood Mountains are much like the various phases of the Shonkin Sag laccolith. They commonly contain phenocrysts of augite or olivine, and to a lesser extent biotite, pseudoleucite, or analcite. The groundmass commonly consists of barium-rich sanidine, biotite, augite, or analcite, with lesser iron oxide, pseudoleucite, apatite, aegirine, or zeolites. The latest dikes (syenite porphyry) contain phenocrysts of sanidine, augite, with aegirine rims, biotite and locally hornblende, and a groundmass of sanidine, aegirine, and possible analcite.

Chemically, the 36 analyzed mafic phonolites and mafic syenites (including shonkinite) from the Highwood Mountains range in composition

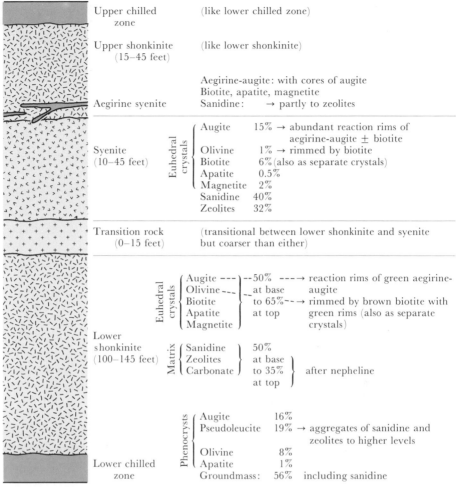

Fig. 4-27 *Generalized section through the Shonkin Sag laccolith. After Hurlbut and Griggs (1939, Figure 3, and from data in text).*

as shown in Table 4-10. Like the lamprophyres they are compositionally similar to the alkaline olivine basalts (compare Table 4-8), except that these potassium-rich basaltic rocks are much higher in K_2O and lower in TiO_2 and FeO than alkaline olivine basalts. K_2O is almost invariably greater than Na_2O by contrast to the alkaline olivine basalts. Chemical trends, as shown by variation diagrams, show that progressive "differentiation" in these rocks is accompanied by an increase in SiO_2, Al_2O_3, K_2O, and Na_2O, and by a decrease in CaO, MgO, and FeO.

Larsen and his coworkers (1941) show by quantitative chemical calculations that settling of early formed phenocrysts of olivine and augite

Table 4-10   *Chemical composition of potassium-rich volcanic and related rocks from the Highwood Mountains, Montana*

	36 Mafic Phonolites and Syenites	Extended Range Including Alkali Syenite Dikes and Alkali Ultramafic Rocks
SiO_2	45.89–52.30	32.20–65.54
TiO_2	0.23– 1.28	0.11– 2.27
Al_2O_3	8.94–18.92	6.08–22.71
Fe_2O_3	0.60– 7.59	0.74– 7.59
FeO	2.67– 8.20	1.15– 8.20
MnO	tr – 0.42	tr – 0.42
MgO	1.81–14.74	0.10–24.92
CaO	4.12–13.64	1.22–18.86
Na_2O	0.71– 4.37	0.32– 8.25
K_2O	2.90–10.33	2.10–10.33
BaO	0.16– 0.70	n.d. – 1.26
SrO	tr – 0.35	tr – 0.38
CO_2	0 – 1.39	0 – 1.39
P_2O_5	0.08– 1.52	tr – 1.52

SOURCE: Larsen et al. (1941); Hurlbut and Griggs (1939); Osborn and Roberts (1931); Weed and Pirsson (1901); Pirsson (1905); Nash and Wilkinson (1970).

in the mafic phonolite (shonkinite) magma, in the proportions in which they occur in the rocks, will form the more felsic members of the association (e.g., alkali syenite). This also agrees with the variation seen from bottom to the top in the laccoliths (see Fig. 4-27). The late biotite phonolites contain abundant inclusions of granitic rocks that show reaction with the magma. Assimilation of granitic material derived from the basement rocks probably accounts for the wide variation in these late rocks. The alkalic ultramafic rocks may have resulted from local assimilation of limestone.

The mafic phonolite of the Highwood Mountains area was thought to be derived at depth (Larsen, 1940; Larsen et al., 1941) from an alkaline olivine basalt magma (described below) by extensive separation of crystals of hypersthene and calcic plagioclase.

Nash and Wilkinson (1970) use the compositions of individual minerals to estimate that during intrusion the vapor pressure was 310 bars (using Barksdale's estimate of 4,500 ft of cover during intrusion and an average density of 2.3) and the temperature 985°C, and at final crystallization the temperature was less than 700°C.

General Features of the Potassium-rich Basaltic Rocks

The potassium-rich volcanic rocks are relatively uncommon but are fairly widespread, continental igneous rocks occurring along the continentward edge of other volcanic associations such as the basalt, andesite, rhyolite association. In most areas, they occur in an environment dominated by crustal faulting, especially rifting. They occur as flows of somewhat restricted areal distributions (a few square miles), tuff breccia, dikes, stocks, and laccoliths. Other examples of the association include the Bearpaw Mountains, Montana (Weed and Pirsson, 1896; Bryant et al., 1960; Schmidt et al., 1961); northern Big Belt Mountains, Montana (Lyons, 1944); Leucite Hills, Wyoming (Cross, 1897; Schultz and Cross, 1912; Yagi and Matsumoto, 1966; Carmichael, 1967); Navajo area, northeastern Arizona (Williams, 1936; Appledorn and Wright, 1957); central Puerto Rico (Jolly, 1971); Mount Vesuvius and related rocks, western Italy (Washington, 1920; Rittmann, 1933, 1962; Tanguy, 1967); western African rift zone, Uganda (Holmes and Harwood, 1932, 1937; Holmes, 1945, 1950; Combe and Holmes, 1945; Higazy, 1954; Sahama, 1960; Bell and Powell, 1969); west Kimberly area, western Australia (Wade and Prider, 1940; Prider, 1960).

The rocks are characterized by a basaltic composition and the presence of potassium-rich minerals such as leucite or pseudoleucite, and to a lesser extent biotite or phlogopite. Either of these distinctive minerals may appear in hand specimens as phenocrysts, along with augite or olivine. Leucite-bearing "basalts," in fact, appear to be the most common rocks in the association. Other distinctive minerals include analcite, nepheline, zeolites, and barkevikitic (high Fe, Ca, Na) amphibole. Basaltic members of the association may contain inclusions of augite peridotite or biotite pyroxenite, whereas more felsic members may carry inclusions of monzonite, trachyte, gneiss, granitic rocks, quartz, or sedimentary rocks. More felsic members of the association (such as phonolites and alkali syenites) are much less widespread, in about the relative proportions found in the Shonkin Sag laccolith, described above.

Chemically, the potassium-rich volcanic rocks are similar to the alkaline olivine basalts but have $K_2O > Na_2O$, are higher in K_2O, BaO, SrO, RbO, and lower in TiO_2 and FeO, much like the more K_2O-rich lamprophyres. Many of the minerals, too, are much like the minerals of such lamprophyres. This extends even to zeolites and calcite as alteration products. In fact, in some areas, biotite-orthoclase lamprophyres (K_2O-rich) can be seen to form the feeder dikes to potassium-rich basaltic flows (e.g., Appledorn and Wright, 1957).

Suggested Origins of the Potassium-rich Basaltic Rocks

Melting of mantle peridotite and overlying crustal rocks, with crystallization of olivine and pyroxene to enrich the melt in K, Ba, Rb, and Zr (Harris, 1957; Yagi and Matsumoto, 1966)

Limited melting of dry oceanic crust along Benioff zone at relatively high pressure such as 36 kbars (Jolly, 1971)

Major melting of low-Si, low-Na (K-feldspar—rich) granulites of the lower crust, possibly with upper mantle rocks to give high potassium melt

Local melting of biotite- or hornblende-rich residues left by anatexis (Waters, 1955a)

High-pressure fractional crystallization of basalt magma in the mantle, removal of hypersthene and calcic plagioclase, and perhaps also diopside and olivine (Larsen, 1940; Larsen et al., 1941; Lyons, 1944) or of hypersthene, clinopyroxene, olivine, and spinel (O'Hara, 1965) with concentration of volatiles and K_2O in the residual melt, with or without lower-pressure crystallization and removal of olivine and augite (O'Hara, 1965—see Fig. 4-21)

Crystallization of soda-rich eclogite and olivine from primary peridotite magma to give ultramafic rocks rich in potassium, plus assimilative reaction with basaltic or granitic rocks (Holmes and Harwood, 1932; Wade and Prider, 1940)

Upward transfer of more volatile components (K, Ba, Ti, Zr, F, Si) in a peridotite magma, followed by possible gravitational separation of olivine (Prider, 1960)

Assimilation of limestone or dolomite by trachytic magma (desilication), sinking of ferromagnesian minerals, upward concentration of potassium by gas transfer, and differential diffusion of sodium into the wallrocks and fumaroles (Rittmann, 1933, 1962)

Sinking and selective reaction of biotite in basaltic magma (Bowen, 1928)

Alkaline olivine basalt magma assimilates (or reacts with) granitic basement rocks (Williams, 1936; Buie, 1941; Turner and Verhoogen, 1960) or with potash salt deposits (Barth, 1962)

Nepheline-titanaugite ± olivine ± biotite (nephelinite) or carbonatite magma assimilates sialic material (Holmes, 1950, 1965; Bell and Powell, 1969)

This wide range of suggested origins reflects the difficulty in obtaining a high potassium content in a rock of a low-silica (basaltic) composition. Important characteristics that must be taken into consideration are:

Occurrence as areally restricted intrusives and flows

Located in continental areas having a fairly thin sedimentary cover over a "granitic" basement

Environment characterized by crustal faulting rather than folding

Located in an environment elsewhere containing rocks of the alkaline olivine basalt association

Often closely associated with or connected to feeders of lamprophyre

Chemically and mineralogically similar to potassium-rich lamprophyres and to alkaline olivine basalts except for $K_2O > Na_2O$

Xenoliths of peridotite, trachyte, gneiss, granitic rock, or sedimentary rock

Demonstration that xenoliths of granite in potassium-rich basalts (Holmes, 1945) and of limestone in basaltic andesite (Brouwer, 1945) are converted to leucite-bearing aggregates

Sr^{87}/Sr^{86} include both low and moderate values (Bell and Powell, 1969)

These characteristics suggest modification of a parent alkaline olivine basalt magma or of another magma that gives rise to alkaline olivine basalt under only slightly different conditions. This modification is not clear but could result from high-pressure crystallization of pyroxenes, olivine, and spinel (or similar minerals), concentration of volatiles and K_2O in the residual melt, and possible assimilative reaction with granitic basement rocks. One alternative is major melting of low-Si, low-Na, K-feldspar−rich granulites of the lower crust, possibly with upper-mantle rocks, to give a high-potassium melt. The melting in dry lower-crust or upper-mantle rock could perhaps result from rifting, possibly above underlying upwelling in convection in the mantle. Once derived, this potassium-rich basaltic rock clearly evolves by settling of early formed olivine and pyroxene, as described by Larsen and coworkers (1941).

NEPHELINE SYENITE AND RELATED SODA-RICH INTRUSIVE ROCKS

The Nepheline Syenites and Related Rocks of the Northern Crazy Mountains, Central Montana[1]

Tertiary igneous rocks intruded into Cretaceous-Paleocene sandstone and shales in the Crazy Mountains consist of an alkali-calcic group overlapping in time with an alkalic group. The alkali-calcic group forms two diorite stocks and associated dikes of diorite to granite composition. The alkalic to subalkalic group characteristic of the northern part of the Crazy Mountains occurs as laccoliths, sills, and dikes of syenite, nepheline syenite,

[1]Wolff, 1938; Simms, 1966.

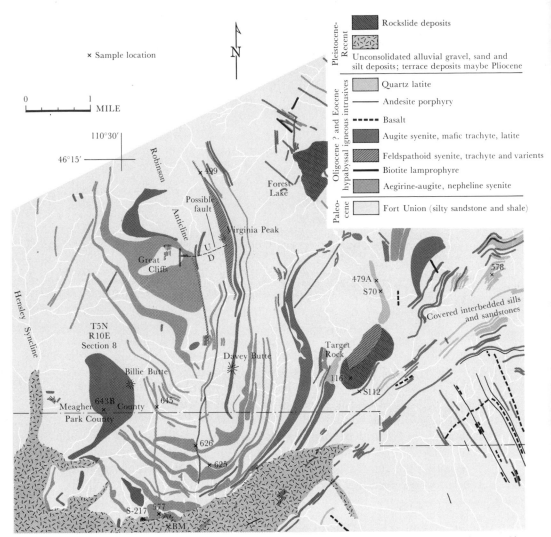

× Sample location

0 _____ 1
|__.__.__.__.__| MILE

110°30′

46°15′

Fig. 4-28 *Sodium-rich intrusive rocks of the northern Crazy Mountains, Montana. After F. E. Simms (personal communication, 1970).*

augite-rich nepheline syenite (mafic-rich foyaite or alkali gabbro), trachyte, leucocratic monzonite, latite, rhyolite, basalt, and lamprophyre (Fig. 4-28). The several laccoliths of augite-rich nepheline syenite, nepheline syenite, and leucocratic monzonite range in thickness from 60 to about 360 ft and up to 2 miles wide and 3 miles long. Most show columnar jointing normal to the upper and lower contacts. The thickest laccolith shows a concentration of augite and olivine toward the base, presumably due to crystal settling. The adjacent shale has been baked to a rock bearing green aegirine (soda iron pyroxene). The numerous dikes form a crude pattern radiating outward from an igneous-poor area 6 or 7 miles across. Most are less than 25 ft thick, but complex dikes range up to 100 ft thick. Sills predominate in areas of steep dip, whereas dikes predominate in areas of low dip, apparently reflecting the relative ease of intrusion in the two directions.

Most of the alkalic rocks are gray to light green to dark green, coarse-grained, and granular. They consist of variable amounts of potassium feldspar, nepheline, sodalite, analcite, aegirine, augite, and lesser amounts of black hornblende, bronze-colored biotite, olivine, magnetite, sphene, and apatite. Secondary minerals include zeolites (natrolite and analcite), yellow cancrinite, and calcite. Common phenocrysts in the more mafic sills and dikes are the mafic minerals, whereas those in the nepheline syenites are anorthoclase (alkali feldspar) and aegirine-augite, and those in the syenites (including "laurvikite") are anorthoclase or orthoclase.

All these minerals are characterized by their high sodium content. The aegirine-augite is commonly rimmed by aegirine and in some cases shows a diopsidic augite (low-Na) core. The biotite in some rocks shows darker rims, and the olivine, reddish oxidized rims. The alkali feldspar ranges from a Ba-rich orthoclase to a Sr-Ba-rich anorthoclase (high-temperature K-rich "albite"). The black hornblende is rich in titanium. Olivine (about Fo_{85-90}) forms several percent of most pyroxene-rich foyaites.

Chemically, the 12 analyzed mafic-rich nepheline syenites from the northern Crazy Mountains range in composition as shown in Table 4-11. They are compositionally very similar to the alkaline olivine basalts (compare Table 4-8), except that these soda-rich "basaltic" rocks are much higher in Na_2O, possibly a little higher in K_2O, and lower in TiO_2 (recall also that the potassium-rich rocks of the Highwood Mountains were lower in TiO_2). Like the alkaline olivine basalts, $Na_2O > K_2O$. Chemical trends as shown by variation diagrams, show that, except at high SiO_2 (more than about 60%), increased SiO_2 is accompanied by an increase in Na_2O and Al_2O_3, fairly even K_2O, and a decrease in CaO, MgO, and FeO. The trends then, are similar to the "differentiation" trends of the potassium-rich "basaltic" rocks and most other basaltic rocks. The alkali-lime index (see p. 43) of 47 for the alkalic rocks (strongly alkalic) compares with an alkali-lime index of 50 (alkalic) for the potassium-rich rocks of the Highwood Mountains. That for the alkali-calcic group of rocks in the Crazy Mountains is about 55.

Wolff (1938) considers the sodium-rich alkalic magmas of the Crazy Mountains to be derived from a diorite magma, such as that of the two stocks in the area. He suggests that differentiation of the diorite magma at depth gave rise to a somewhat mafic syenitic magma, which in turn would give rise to both more-mafic and less-mafic nepheline syenites. Larsen (1940) on the other hand, considers the problem as a part of the evolution

Table 4-11 *Chemical composition of sodium-rich mafic syenites and related rocks from the Crazy Mountains, Montana*

	12 Mafic-rich Nepheline Syenites	Extended Range Including 12 Syenites and Nepheline Syenites and 8 Leucocratic Monzonites
SiO_2	43.18–51.03	43.18–64.33
TiO_2	0.71– 1.40	tr – 1.40
Al_2O_3	8.48–18.22	8.48–20.14
Fe_2O_3	3.65– 7.61*	0.89– 7.61*
FeO	2.67– 5.28	0.35– 5.28
MnO	0.03– 0.28	<0.01– 0.28
MgO	3.41–13.01	0.21–13.01
CaO	6.96–10.63	0.56–10.67
Na_2O	3.44– 6.43	3.44– 8.82
K_2O	0.56– 4.83	0.56– 6.67
BaO	0.22– 0.76	0.22– 0.76
SrO	0.04– 0.37	0.02– 0.37
CO_2	0 – 5.42	0 – 5.42
P_2O_5	0.60– 2.20	<0.06– 2.20

*Also one analysis 11.95.
SOURCE: Wolff (1938); Simms (1966).

of different alkalic magmas in the whole central Montana region—Yellowstone volcanics, Crazy Mountains, Highwood Mountains, and Bearpaw Mountains to the north. He prefers fractional crystallization of basalt magma and removal of hypersthene and calcic plagioclase to give the alkalic magmas of the whole region. The difference between the subprovinces would be due in larger part to greater or lesser assimilation of granitic or other siliceous material—little or none for the potassium-rich rocks of the Highwood Mountains—major for the sodium-rich rocks of the Crazy Mountains. Simms (1966) shows that the alkalic and alkali-calcic (including dioritic) groups of rocks are separate and probably unrelated. Otherwise, he seems to agree with Larsen's (1940) interpretation of the origin of the primary alkalic magma (essentially alkaline olivine basalt) through differentiation and contamination by silicic basement rocks. Given the primary alkalic magma, the various alkalic rocks of the area are thought to be formed by fractional crystallization of olivine and clinopyroxene at an early stage, followed by fractional crystallization of plagioclase to enrich the magma in the components of the nepheline syenite.

General Features of the Nepheline Syenites and Related Rocks

Nepheline syenites and related sodium-rich rocks develop as a relatively uncommon but widespread and distinctive association in continental areas of crustal stability, block faulting, or moderate folding (e.g., Barth, 1956; Bass, 1970) in the foreland of continental thrust zones (Gilluly,

1971). Thus, like the potassium-rich basaltic rocks described in the previous section, and the carbonatites and mafic alakaline rocks described in the next section, they are most prominent along the continentward edge of the volcanic associations of the orogenic environment. Typically, they occur as sharply bounded stock- to small-batholith-sized plutons (up to a few hundred square miles), ring complexes, or cone sheets. Most appear clearly intrusive and show distinct contact aureoles. Other major examples of the association include the Poços de Caldas complex, Itatiaia massif, Ilha de São Sebastião, and other areas in southern Brazil (e.g., Ellert, 1959; Ribeiro Filho, 1966; Freitas, 1947; Amaral et al., 1967); the Messum, Okonjeje, and related igneous complexes in South-West Africa (Korn and Martin, 1954; Simpson, 1954; Martin et al., 1960); the plutons of East Tuva and East Sayan in southern Siberia (Yashina, 1957; Vorobieva, 1960; Kudrin, 1963; Gavrilova and Khryukin, 1964; Lebedev and Bogatikov, 1965; Petersil'ye and Yashina, 1971); the Khibiny, Lovozero, and related plutons of the Kola Peninsula of northwestern Russia (Vlasov et al., 1959; Turner and Verhoogen, 1960, p. 393; Vorobieva, 1960; Attananov et al., 1962; Sørensen, 1970); the Oslo district, Norway (Barth, 1945, 1954; Saether, 1947; Offedahl, 1946, 1948, 1952, 1957; Dons, 1952); the alkalic plutons of southern Greenland (Callisten, 1943, Upton, 1960, 1964; Emeleus, 1964; Sørensen, 1958, 1970; Ferguson, 1970). Volcanic equivalents such as phonolite are typically associated with alkaline olivine basalt as described above, such as in the East Otago volcanic province of New Zealand (e.g., Turner and Verhoogen, 1960, pp. 165–173; Coombs and Wilkinson, 1969).

These rocks consist of nepheline syenite, syenite, nepheline-rich syenite, and monzonite (foyaite and essexite) (see Fig. 2-1a), and their volcanic or subvolcanic equivalents such as phonolite. Associated mafic-rich rocks include mafic-rich syenites, various alkali gabbros (including ijolites: aegirine-augite + nepheline rocks), and nepheline "basalts." The more mafic-rich rocks commonly occur as dikes or as border phases on plutons. The felsic rocks are generally white to buff colored, medium grained, and massive except in the syenites, which contain little other than alkali feldspar. These rocks tend to have a trachytoid texture, because of the parallel arrangement of the tabular feldspars. The adjacent country rocks are often locally affected by alkali metasomatism (called "fenitization"), resulting in the formation of a syenitelike rock containing alkali feldspar, aegirine, and blue soda-ampliboles.

Mineralogically, the nepheline syenites are gray to brownish translucent nepheline-alkali feldspar rocks, commonly with lesser amounts of other feldspathoids, especially white to blue sodalite, analcite, and white to yellow cancrinite. The alkali feldspar is a perthite (thin streaks of exsolved albite in orthoclase or microcline) or, less commonly, anorthoclase (high-temperature potassium-rich "albite"). The typical mafic minerals are the soda-pyroxenes aegirine-augite (Na, Fe, Ca) and aegirine (Na, Fe^{3+}), the soda-amphiboles black barkevikite (Na, Fe, Mg, Ca), greenish-black arfvedsonite (Na, Fe), dark blue to black riebeckite (Na, Fe); and lepidomelane (high Na, Fe biotite). Albite and apatite may be fairly abundant, as to a lesser extent may be zircon, sphene, melanite (black, Ti-rich andradite garnet), perovskite (yellow to brown, $CaTiO_3$), eudialyte [bright red-pink, $(Na, Ca, Fe)_6Zr(SiO_3)_6(OH, Cl)$], astrophyllite [yellow-mica-like, $(K, Na)_2(Fe, Mn)_4(Ti, Zr)(Si_2O_7)_2(OH, F)_2$], ilmenite, titanomagnetite, columbite, tantalite, pyrochlore [$NaCa(Cb, Ta)_2O_6F$], zeolites, and calcite. Plagioclase is usually absent.

Chemically, the nepheline syenites and related rocks range in composition from the oxide minima shown in Table 4-11 for the Crazy Mountains in Montana, to SiO_2 (64.99%), TiO_2 (2.8%), Al_2O_3 (24.55%), Fe_2O_3 (11.95%), FeO (10.38%), MnO (0.82%), MgO (13.01%), CaO (10.67%), Na_2O (15.76%), and K_2O (6.67%). They are therefore high in alkalis with Na > K (almost invariably), have high FeO/MgO, high Zr, Ti, Nb, Cb, Ta, rare earths, P, F, Cl (Turner and Verhoogen, 1960). It is also significant that the nepheline syenites cluster around the minimum melting point of the silica-deficient part of the $SiO_2-NaAlSiO_4-KAlSiO_4$ diagram (see Fig. 3-1). The syenites (lacking both nepheline and quartz) fall near the low-melting saddle on the albite-orthoclase $(NaAlSi_3O_8-KAlSi_3O_8$ join on the same figure.

Suggested Origins of the Nepheline Syenite and Related Rocks

Differentiation of alkaline olivine basalt magma (Barth, 1945; Turner and Verhoogen, 1960; Streckeisen, 1960; Riebeiro Filho, 1966; Nolan, 1966; Mukherjee, 1967; Coombs and Wilkinson, 1969; Sørensen, 1970; Woussen, 1970; Valiquette and Archambault, 1970), ultrabasic magma (Vorobieva, 1960), a nepheline magma (Sood et al., 1970), a diorite-monzonite magma (Wolff, 1938; Khalfin, 1961), a syenite-monzonite magma (Barth, 1945), or a granite magma (Kudrin, 1963)

Differentiation (fractional cyrstallization) of basalt magma, removal of hypersthene and calcic plagioclase, and assimilation of large amounts of granitic material (Larsen, 1940; [?] Simms, 1966)

Differential movement of alkalis and volatiles (and other residual elements such as Zr, Ti, Nb, rare earths, F, Cl) to restricted parts of magma chamber (Saether, 1948; Barth, 1954, 1962; Sørensen, 1958, 1960) with assimilation by alkaline olivine basalt of surrounding gneisses (Upton, 1960) or of limestone roof rocks (Lebedev and Bogatikov, 1965)

Reaction of granophyre at top of diabase sill with silica-poor argillite (for one occurrence only, Barker and Long, 1969)

Separation of an immiscible alkali syenite liquid from a mafic alkaline magma (Philpotts and Hodgson, 1968)

Partial melting of alkaline olivine basalt in the deep crust or upper mantle (Bailey, 1964; Bailey and Schairer, 1966)

Partial melting of Precambrian basement rocks (Barth, 1954)

Differential fusion of appropriate minerals (e.g., hornblende or biotite, or jadeite), resulting from a relief of pressure (Tilley, 1958), or by other means (Bose, 1967)

Partial melting of deep crustal material desilicated by persistent flux of water vapor through the affected rocks (Currie, 1970)

Desilication of granitic or granodioritic magma by reaction with limestone to form lime silicates—removal of SiO_2 in garnet, epidote, etc. (Daly, 1910; Chayes, 1942; Shand, 1945; Barth, 1962; Schuiling, 1964)

Metasomatism of dolomitic limestone (or dark gneisses) by undersaturated solutions from nepheline syenite magma—for banded or gneissic rocks (Grummer and Burr, 1943, 1946; Moyd, 1945; Tilley, 1958; Tilley and Gittins, 1961; Gittins, 1961; Sturt and Ramsay, 1965)

As may be surmised from the above range of suggested origins, the nepheline syenites and related rocks have long been discussed in the geological literature. No single origin clearly stands out as controlling the development of these rocks. Important characteristics for consideration include the following:

Differentiation of alkaline olivine basalt typically gives rise to undersaturated magmas of trachyte to phonolite composition—the volcanic equivalents of syenite and nepheline syenite

Formed in areas of crustal stability on the continents—in the same general environment as the potassium-rich basalts, mafic alkaline rocks and carbonatites

Relatively small size (less than a few hundred square miles)

Common cone-sheet and ring-dike form (shallow depth of intrusion)

Oversaturated rocks (especially granite) may occur either in the early stages of plutons ending in nepheline syenite or in the late stages of plutons containing earlier nepheline syenite

Intrusion of oversaturated granitic magmas into limestone normally results in the formation of skarns (mainly calc-silicate minerals such as garnet, epidote, diopside)

Many nepheline syenite plutons occur away from known areas of limestone (e.g., Streckeisen, 1960)

High-alkali content with Na > K; high FeO/MgO, Zr, Ti, etc.

Most nepheline syenites would plot in the low-melting area of the silica-deficient part of the $SiO_2 - NaAlSiO_4 - KAlSiO_4$ diagram

Sr^{87}/Sr^{86} ratios of the alkalic rocks of the Monteregian Hills, southern Quebec, are low (0.7031 to 0.7052) (Fairbairn et al., 1963)

Clearly, *differentiation* of alkaline olivine basalt magma may give rise to magmas of nepheline syenite and syenite composition. The absence of such alkaline olivine basalt magmas in many areas characterized by the presence of nepheline-bearing syenitic rocks suggests, however, that such differentiation is not the universal control. It could be argued that a large volume of such mafic rock exists at depth; but mafic magmas, although not unknown for such behavior, are more noted for their unrestricted passage to the surface. An alternative that would give rise to magmas of identical composition is *partial melting* of rocks of alkaline olivine basalt composition in the deep crust. This could give rise to syenite to nepheline syenite magmas without the necessity of molten basalt at depth. Differential movement of alkalis, although a possible explanation, is generally accompanied (as seen in normal differentiation) by movement of silica and potassium. However, little experimental work on diffusion is available as an aid to evaluation of the process. Transitions from nepheline- to quartz-bearing syenites may be affected by the oxygen partial pressure in the magma and therefore the oxidation state of the Fe and Ti (Tilley, 1958; Turner and Verhoogen, 1960; Bailey and Schairer, 1966). Substitution of Fe for Al in albite, and formation of pyroxenes could extract silica from a somewhat oversaturated magma, to give a silica-deficient liquid. Partial melting of associated tholeiitic basalt, either before or after partial melting of alkaline olivine basalt, could also account for the close association of oversaturated and undersaturated syenitic rocks. Metasomatism of rocks by solutions from a nepheline syenite magma seems to be demonstrated for some areas, but this still leaves unexplained the origin of the parent nepheline syenite magma.

The writer's preference for the formation of nepheline syenite and related magmas is for differentiation of, or possibly partial melting of, alkaline olivine basalt magma. If differentiation is important, it is presumably strongly influenced by the greater time available for differentaition in the environment of relative crustal stability on the continental side of the orogenic belt. Another factor suggests differentiation as being more important than partial melting. The stable, relatively thin crust in which this association occurs is strikingly different from the orogenic environment in which partial melting of rocks in the lower crust is known to occur. This does not preclude, of course, the possibility of partial melting of rocks in the dry lowermost crust that probably consists of gabbroic material (basaltic composition). Partial melting of such "basaltic" rocks at pressures reasonable for those depths could develop melts on the undersaturated side of the thermal divide (Fig. 4-21)—alkali syenite magmas. The relatively small volume of nepheline syenites and related rocks is also consistent with an origin by differentiation or partial melting of basaltic material. More diagnostic is the low Sr^{87}/Sr^{86} of these sodic rocks, suggesting differentiation of a mantle-derived basalt rather than partial melting of old crystalline basement.

CARBONATITE AND MAFIC ALKALINE ROCKS

The Carbonatite, Mafic Alkaline Rocks, and Felsic Alkaline Rocks at Magnet Cove, Arkansas[1]

The Late Cretaceous alkaline igneous complex at Magnet Cove in central Arkansas forms a roughly circular ring-dike complex a little more than 2 miles across. It intrudes folded and faulted Paleozoic shale, sandstone, conglomerate, chert, and minor limestone, which have been metamorphosed for 1/4 to 1/2 mile from the intrusive contact.

The igneous complex (Fig. 4-29) consists of a core of ijolite (nepheline-diopside rock) and carbonatite, an intermediate ring of trachyte and phonolite, an outer ring of nepheline syenite, and two peripheral bodies of alkalic magnetite pyroxenite ("jacupirangite"). Smaller later dikes cut this main ring-dike complex and the country rocks nearby. They include phonolite ("tinguaite"), nepheline syenite, trachyte, nepheline syenite pegmatite and aplite, carbonatite, and various syenitic lamprophyres (outside the complex). The main ring dikes are vertical or steeply dipping outward. The central core of mafic rocks forms a topographic basin, whereas the outer felsic ring dikes form higher topographic ridges. With the exception of the coarse-grained pegmatites, the rocks are fine to medium grained and massive. Most of the rocks weather to mottled saprolite (clay).

Rocks of the Inner Core

IJOLITE—Nepheline + diopsidic pyroxene, with lesser Ti-rich andradite, phlogopite; ± perovskite ($CaTiO_3$), sodalite; ± sphene, apatite, pyrite and magnetite

[1]Erickson and Blade, 1963; Fryklund et al., 1954; Washington, 1900, 1901; Zartman et al., 1967.

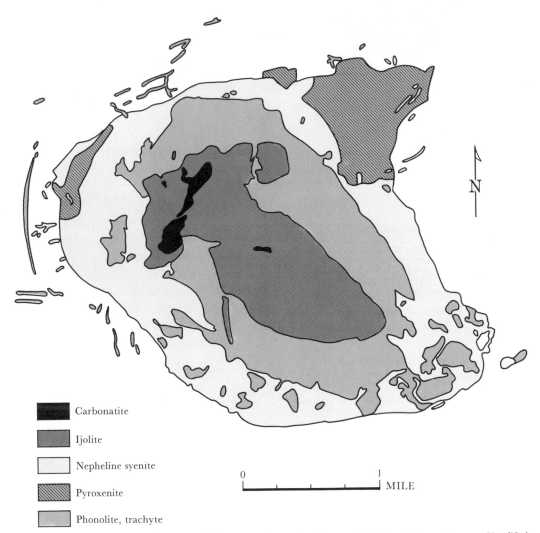

Carbonatite

Ijolite

Nepheline syenite

Pyroxenite

Phonolite, trachyte

0   1 MILE

Fig. 4-29 *Geologic map of the Magnet Cove ring-dike complex, Magnet Cove, Arkansas. Simplified from Erickson and Blade (1963).*

The rocks are gray and greenish gray to mottled white and pink, with green, brown, and black. They range from aphanitic to medium grained and massive, to porphyritic, amygdaloidal, or fine-grained breccia.

CARBONATITE—Calcite (55 to 95% of rock), with lesser apatite, brown monticellite ($CaMgSiO_4$); black, Zr-rich andradite ($\pm Ti$); green phlogopite, pyrite, "magnetite" (high Mg), and perovskite

Inclusions of ijolite a few inches to more than 50 ft across have reaction rims adjacent to the carbonatite. These consist of a magnetite-pyrrhotite-calcite zone (next to the carbonatite), a biotite-rich zone and an idocrase-rich zone.

Rocks of the Intermediate Ring

PHONOLITE TO TRACHYTE—Phenocrysts of diopsidic augite rimmed by aegirine-augite; nepheline; apatite; groundmass of sodic orthoclase; plagioclase, hornblende; replacing early pyroxene and nepheline are calcite, phlogopite, and magnetite

These rocks are light gray to greenish and dark gray, aphanitic and porphyritic to amygdaloidal. Locally, they form a breccia, including fragments of metamorphosed sedimentary rocks, altered pyroxene-rich rocks, and phonolite. It is considered to be an intrusive breccia and the earliest magmatic intrusive in the area.

Rocks of the Outer Ring

NEPHELINE SYENITE—Sodic orthoclase, nepheline; diopside-hedenbergite rimmed by aegirine-diopside, sphene, ±apatite, magnetite, sodalite, garnet, and plagioclase

The rocks are light to dark gray, fine to medium grained.

GARNET-PSEUDOLEUCITE-NEPHELINE SYENITE—Nepheline (partly altered to analcite, calcite, and yellow cancrinite), sodic orthoclase; pseudoleucite (white, euhedral crystals 1/4 to 2 in. across); diopside-hedenbergite (rimmed by green aegirine-diopside); Ti-rich andradite, biotite, ± apatite, fluorite, perovskite, and magnetite

These rocks are light gray to greenish gray, medium grained, and they commonly show conspicuous white crystals of pseudoleucite weathering out into relief. Inclusions of metamorphosed sedimentary rocks and ijolite up to 10 ft across are abundant. Miarolitic cavities up to 3 in. across are locally common.

ALKALIC MAGNETITE PYROXENITE ("JACUPIRANGITE") AND SPHENE PYROXENITE—Diopside-hedenbergite (sometimes rimmed by titanaugite or aegirine-diopside or biotite); magnetite-ilmenite (±rimmed by perovskite); ±perovskite, sphene, apatite, biotite, garnet, pyrite, and pyrrhotite

These rocks are dark gray and fine to medium grained and are composed principally of pyroxene, which appears to have formed an early crystal mush, and lesser magnetite-ilmenite. It is cut by dikes of nepheline syenite and ijolite.

Biotite-garnet ijolite ("melteigite") is a biotite-garnet-nepheline-clinopy-roxene rock with lesser amounts of sodalite, sphene, perovskite, apatite, magnetite-ilmenite, and pyrrhotite.

Dikes cutting the ring complex and outside it include phonolite, trachyte, syenite, nepheline syenite, pegmatite (± with pink eudialyte), and lamprophyre.

Chemically, the three analyzed nepheline syenites, phonolites, and trachytes (excluding the altered phonolites) of the Magnet Cove area are similar to the nepheline syenites and syenites of the nepheline syenite association, ranging possibly to somewhat lower SiO_2 and MgO. The 10 analyzed ijolites and alkalic pyroxenites of the area are similar to the alka-line olivine basalts (compare Table 4-8) but range to somewhat lower SiO_2 and TiO_2, lower Al_2O_3, and higher Na_2O. They are higher in K_2O, MnO, and especially CaO, and lower in MgO. The one analyzed carbonatite, aside from being high in CaO and CO_2, is very high in TiO_2, K_2O, BaO, SrO, and especially P_2O_5. These alkalic rocks typically (but not invariably) show $Na_2O>K_2O$ as in the nepheline syenites and related rocks. They are also high (>0.01%) in several distinctive trace elements: Zr, V, La, Nb (in order of abundance), and to a lesser extent Cr, Cu, Ce. The Zr is concentrated in sphene and garnet; the V in sphene, garnet, and apatite; the Nb in sphene and perovskite.

Veins that cut the igneous rocks and contact zone include a sugary-textured albite-dolomite, microcline-calcite, albite-ankerite, coarse-grained albite-perthite carbonate, calcite-rutile, and coarse-grained calcite (car-bonatite?). More minor types include combinations of the minerals in dolomite-rutile-pyrite-green biotite, quartz-feldspar, quartz-brookite-rutile, apatite, fluorite, and biotite-garnet-apatite.

The oldest rocks of the Magnet Cove complex are considered to be the phonolite and trachyte of the intermediate ring. This is followed by the alkalic pyroxenites, then the alkalic syenites of the outer ring that are chilled against the altered phonolite and form dikes cutting the alkalic pyroxenites. Then came the intrusion of the central mass of ijolite that forms dikes cutting the nepheline syenites and is chilled against the altered phonolite. The carbonatite, which is the last major intrusive, intrudes and contains inclusions of ijolite.

Erickson and Blade (1963) would derive the various rocks of the Magnet Cove complex by fractional crystallization of a mafic-rich phono-lite magma—of the average chemical composition of the whole complex. This mafic-rich phonolite magma is to be derived by fractional crystalliza-tion (apparently of labradorite, pyroxene, and a little olivine) from a re-gional alkaline olivine basalt magma. They point out that their conclusion is supported by the regional occurrence of strongly alkaline rocks, includ-ing alkaline olivine basalt, in an arcuate belt from west Texas to central Mississippi. They further suggest that the mafic phonolite magma was the first liquid released from the magma reservoir and explosively extruded. Differentiation in the magma chamber involved crystallization and floating of pseudoleucite crystals and sinking of pyroxenes and magnetite. Tapping of the lower part of the reservoir could produce the alkalic pyroxenites, followed by emplacement of nepheline and pseudoleucite syenites from the top of the reservoir. This removes most of the felsic constituents, leaving the magma "largely mafic but with ever increasing amounts of volatile constituents, P, Ti, Fe^{+++}, Zr, and rare earths" (Erickson and Blade, 1963, p. 88). Most of this remaining magma crystallizes as ijolite, the highly volatile content giving rise to high fluidity and growth of large crystals (as in pegmatites), monomineralic segregations, and alteration of the nepheline.

The Carbonatite of the East African Volcanoes (Oldoinyo Lengai, Tanganyika, and Napak, Uganda) [1]

One of the most important recent discoveries bearing on the origin of carbonatites was by J. B. Dawson, who in 1960 entered the crater of Oldoinyo Lengai to find carbonatite lava in active eruption. In 1966, he observed that the style of activity changed to violent ash eruptions of sodium carbonate containing crystals and blocks from ijolite. This volcano is one of a chain of several recent carbonatite volcanoes in a north-south rift valley in northern Tanganyika. Oldoinyo Lengai is a well-formed, steep-sided volcanic cone composed primarily of ijolitic pyroclastics. It is about 5 miles in diameter and stands 6,500 ft above the surrounding plains. One of the two summit collapse "craters," 1/3 mile across and 600 ft deep, is active. The main vent on the floor of this crater is surrounded by "scoria cones, recent ejectamenta, and minor lava flows" (Dawson, 1966).

"The following stratigraphic sequence has now been established" (Dawson, 1966, pp. 155–156):

6. Modern sodium carbonate lavas of the northern crater
5. Carbonate ashes of the active crater and soda ash deposit of the summit area
4. Melanephelinite extrusives
3. Black nephelinitic tuffs and agglomerates
2. Grey pyroclastics of the parasitic cones and tuff rings
1. Yellow ijolitic tuffs and agglomerates with interbedded lavas, making up the main mass of the volcano

The finer-grained pyroclastics "are crystal tuffs, consisting of nepheline and pyroxene in a matrix of carbonate, limonite, and zeolites and grade into agglomcrates containing blocks of nephelinite (nepheline with melilite,

---

[1]Dawson, 1962a, b, 1964b, 1966; DuBois et al., 1963; Guest, 1963; King and Sutherland, 1966; King, 1948; Dawson et al., 1968.

sodalite, ± wollastonite), phonolite," ijolite, jacupirangite (magnetite pyroxenite), biotite pyroxenite, fenite (alkali-iron metasomatized rocks) and related rocks. These rocks and their contained minerals are much like those at Magnet Cove. Also present are sodium-carbonate-rich ashes and blocks of olivine basalt and possibly gneiss, in the agglomerates.

Secondary to the pyroclastics in volume are lava flows of phonolite, nephelinite, and pyroxene nephelinite composition. Perhaps most important for our present discussion are *lavas of sodium carbonate* (with lesser calcium carbonate) composition. Those observed by Dawson in 1960 and 1961 were extruded as highly mobile pahoehoe flows and viscous aa flows, behaving in every way like silicate lavas, as previously described in the section on volcanic materials. When first extruded, the lava flows are black in color, but begin to turn white after 25 to 36 hr, becoming light grey-white in 6 to 7 days as water is absorbed.

The great dissected volcano of Napak, Uganda, rests on the Precambrian gneisses some 400 miles north of Oldoinyo Lengai. The original cone, 20 miles across, has been deeply eroded, exposing the central core underlying the lavas of the cone. The bulk of the cone consists of agglomerates and other pyroclastics and minor lavas, most of which are nephelinite to melanephelinite in composition—nepheline, diopside, ilmenite-magnetite, ± olivine, ± melilite (a fine-grained ijolite-like rock). The pyroxene may contain appreciable aegirine, and the accessory minerals include Fe-Ti oxides, perovskite, and apatite.

The central plutonic core of the volcano consists of ijolite and related rocks, with a central plug of carbonatite. The ijolites and pyroxenites are nepheline, pyroxene (diopside to aegirine-augite) rocks with lesser amounts of Fe-Ti oxide, perovskite, apatite, sphene, brown biotite, black Ti-rich andradite garnet. Coarse-grained pegmatitic and nepheline-rich phases are abundant and replacement textures are common. Cancrinite and nepheline syenites form narrow veins and dikes, cutting this almost non-feldspathic complex. *Calcite carbonatite* of the central plug contains minor siderite, magnetite, and pyrochlore. Basement gneisses surrounding the central ijolite-carbonatite core have been fractured and partly transformed into fenites while still preserving the original texture of the rock. The fenites contain several new minerals, including aegirine, soda amphibole, biotite, ± sphene, ± magnetite.

Disagreement exists as to the origin of both the mafic alkaline rocks and the carbonatites. King and Sutherland (1966), from their study of Napak and associated volcanoes, suggest a parent magma of mafic ijolite composition (between ijolite and pyroxenite). Fractional crystallization of pyroxene, ± olivine, ± melilite would form the later, more felsic rocks. This results in enrichment in Fe oxides and in the plutonic members, a strong enrichment of CaO to MgO, resulting in separation of calcite in the final stages of crystallization. This, along with retention of $CO_3^=$ under plutonic conditions, forms carbonatite in the late stages. Dawson (1966), on the other hand, from his study of Oldoinyo Lengai, suggests "that carbonatite magma accumulated during the extrusion of the earlier basaltic rocks, either by differentiation or by preferential gas streaming during migration of material into the zone from which the basalts were trapped."

Progressive reaction of the intruded basement rocks with the alkali-rich (Na, K, Ca) carbonatite magma would produce a range of alkalic rock types from fenite through ijolite to mafic ijolite. High pressures of CO_2 would form because of the reaction and would fragment the upper part of the metasomatic envelope to form the yellow ijolitic tuffs and agglomerates. Further reaction with the surrounding ijolitic rocks forms still more calcic, less-siliceous rocks ("nephelinites"). Finally, carbonatite magma is able to erupt without contamination by the surrounding rocks. This forms the final sodium-carbonate-rich ashes and lavas of Oldoinyo Lengai.

General Features of Carbonatite and Mafic Alkaline Rocks

Carbonatites and mafic alkaline rocks, like the nepheline syenites described in the previous association, develop as somewhat uncommon but widespread and very distinctive rocks in areas of crustal stability, rifting or block faulting, or moderate folding. They have been recognized, however, as an integral part of the tectonic-metamorphic environment of northern Norway (Sturt and Ramsay, 1965). Commonly, the mafic alkaline rocks form small circular or elliptical plutons (up to a few hundred square miles), ring complexes, or cone sheets in which members of the related nepheline syenite association sometimes also occur. The carbonatites (see Fig. 4-30) occur as plugs or irregularly shaped bodies (up to 3 sq miles in area), centrally enclosed in the mafic alkaline rock complex or as inward-dipping cone sheets, outward-dipping rings (up to several hundred feet wide), or crosscutting dikes (generally several inches to tens of feet wide). Carbonatites and mafic alkaline rocks also develop in the volcanic environment (Fig. 4-30)—a volcano of mafic alkaline flows, tuffs, and breccias with carbonatite flows,[1] tuffs, and breccias developing in smaller amounts in the later stages. In many carbonatites, an area of breccia or carbonatite agglomerate suggests a volcanic vent. Although the carbonatite is most commonly later in all these environments, it developed earlier than the mafic alkaline rocks in a few areas such as Fen, Norway (Brogger, 1921; Bowen, 1924, 1926; Barth and Ramberg, 1966), and Alno, Sweden (von Eckermann, 1948, 1966). Kimberlite (mica peridotite), dunite, or serpentinite may appear as associated rocks in a few localities.

[1]The flows are sodium carbonatite at Oldoinyo Lengai (see above) and calcium carbonatite at Fort Portal, western Uganda (Holmes, 1956; von Knorring and Dubois, 1961).

1   Ijolite tuff, agglomerate, carbonatite, phonolite, nephelinite, trachyte, etc.
2   Agglomerate, carbonatite, and lava
3   Zone of brecciation, recrystallization, and replacement
4   Satellite vent from tension dike (cone sheet)
5   Shear dike (ring dike) of carbonatite
6   Shear dike (ring dike) of nepheline syenite
7   Tension dike (cone sheet) of early center

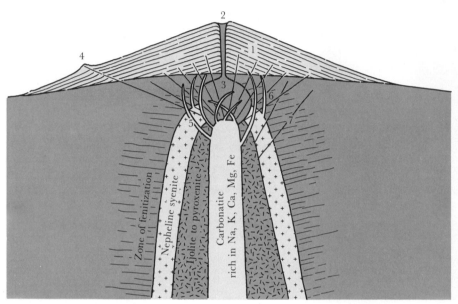

Fig. 4-30 *Schematic diagram of the structural pattern in a carbonatite complex. Modified from Garson (1966).*

Major examples of the carbonatite-mafic alkaline rock association, in addition to these and those described above, include the following in North America: Mountain Pass, southeastern California (Olson et al., 1954); Iron Hill or Powderhorn, Colorado (Larsen, 1942; Temple and Grogan, 1965); Gem Park, Colorado (Parker and Sharp, 1970); the Balcones area of southwestern Texas (Spencer, 1969); Oka, Quebec (Gold, 1966a, 1969; Gold et al., 1967; Davidson, 1963; Deines, 1970; Watkinson, 1970); Monteregian Hills, Quebec (e.g., Gold, 1967; Philpotts, 1970; Currie, 1970; Woussen, 1970; Gandhi, 1970; Valiquette and Archambault, 1970); Manitou Islands, Ontario (Rowe, 1954); Lackner Lake and Firesand, Ontario (Parsons, 1961); Ice River, British Columbia (Campbell, 1961; Rapson, 1966; Currie and Ferguson, 1970). Outside North America, some of the more prominent occurrences include Jacupiranga, Brazil (Melcher, 1966); Serrote, Brazil (Leonardos, 1956); Stjernøy and Sørøy, northern Norway (Heier, 1961, 1964a; Sturt and Ramsay, 1965); Kaiserstuhl, West Germany (Wambecke, 1966; Wimmenauer, 1962, 1966; Ozernaya

Varoka, Afrikanda, and others on the Kola Peninsula, northwest Russia (Afansyev, 1939; Bagdasarov, 1959); those in the Maymecha-Kotuy province, northern Siberian platform (Zhabin and Cherepivskaya, 1965; Egorov, 1970). Africa, especially the southern and eastern parts, has far more carbonatites than any other region. They include Palabora, South Africa (Russell et al., 1954; Heinrich, 1970); Spitzkop, South Africa (Strauss et al., 1951; Holmes, 1958); Chilwa Island and Tundulu, Malawi (Smith, 1953; Garson, 1966); Homa, Kenya (McCall, 1959); Ruri-Sokolo, Kenya (McCall, 1959, 1963; Pulfrey, 1954); Chasweta and associated bodies, Zambia (Bailey, 1960); Tororo and Sukulo, Uganda (Williams, 1952; King, 1965; King and Sutherland, 1966); Fort Portal, Uganda (Holmes and Harwood, 1932; von Knorring and DuBois, 1961); Kerimasi and Monduli-Arusha, Tanganyika (Dawson, 1964b, c); Panda Hill and Sengeri Hill, Tanganyika (Fawley and James, 1955; Fick and van der Heyde, 1959); Okorusu, Kalkfeld, and probably Gross Brukkaros, South-West Africa (Martin et al., 1960; Janse, 1969). Brief descriptions of these and most other known carbonatite-mafic alkaline rock complexes are given by Gittins (1966a) and Heinrich (1966).

The carbonatites are composed of mainly medium-grained carbonates but include very fine- to coarse-grained varieties that, except for their content of distinctive minerals, look much like marble. The majority are largely calcite (sometimes called "sovite"), but others are dolomite (sometimes called "beforsite") or ankerite. Colors range from white to brownish white, and less commonly yellow or brown. Texturally, most carbonatites are massive; but a few show a foliation in the form of a trachytoid texture or layering, most commonly due to parallelism and sometimes streaky concentrations of dark minerals that tend to parallel the country-rock contacts or wrap around inclusions of the country rocks. Dark streaks or bands tend to be high in pale green apatite, magnetite (often Mg rich), green to brown phlogopite, or pale yellow to brownish-red pyrochlore. Apatite and pyrochlore, in fact, are considered diagnostic of carbonatites (Quon, 1966). Other minerals that occur in carbonatites include both the major and minor minerals of the associated mafic alkaline rocks:

Alkali feldspar, nepheline, diopside or aegirine, melanite (Ti-andradite), ± forsterite, biotite, melilite (like anorthite but less SiO_2)

±barite, monazite [(Ce, La, Th) PO_4], pyrochlore, perovskite, ilmenite, eudialyte, bastnaesite [Ce, La, Di(CO_3)F], baddeleyite (ZrO_2), dysanalite (Ca, Ce, Na) (Ti, Nb, Fe)O_3, fluorite, zircon

Two mineralogical types of carbonatites, perhaps end members of a series, have been recognized (Pecora, 1956; Deans, 1966): a magnetite-apatite ± phlogopite type including Magnet Cove, Arkansas; Oka, Quebec;

Manitou Islands, Lackner Lake, and Firesand, Ontario; and a less common rare-earth type including Mountain Pass, California, Iron Hill, Colorado, and the Bearpaw Mountains, Montana (Pecora, 1962).

Most contacts of carbonatite bodies with the country rocks are sharp, but a few are gradational (e.g., Budeda Hill, Uganda, King and Sutherland, 1966), showing replacement of the nepheline and pyroxene of the alkaline rocks by calcite of the carbonatite. Xenoliths of alkaline and other country rocks in various stages of assimilation are found in some carbonatites, such as Tororo, Uganda, and Kaiserstuhl, Germany. Adjacent to the carbonatite contacts, normally within several hundred feet (up to 1 1/2 miles at Tundulu, Malawi), the country rocks (most commonly granite, gneiss, and sandstone) are fenitized (see, for example, Heinrich, 1966, pp. 68−92; McKie, 1966; Bell and Powell, 1970; Heinrich and Moore, 1970). That is, they are transformed by alkali-ferric iron metasomatism into syenite or nepheline syenite. The replacement begins along cleavages, grain boundaries, and fractures, gradually expanding to replace the rock. The original structures of the replaced rock are often preserved, though the minerals are now largely orthoclase and soda pyroxene, with lesser albite and soda amphibole (e.g., Turner and Verhoogen, 1960, p. 400; McKie, 1966). Other elements involved in the fenitization include: Ca, $CO_3^=$, P, F, Cb.

Chemically, the carbonatites are like no other igneous rocks (see Table 4-12). Aside from the extremely high CaO and CO_2 and extremely low SiO_2 contents, the carbonatites are high in TiO_2, Fe_2O_3, MnO, BaO, SrO, P_2O_5, Nb_2O_5, rare earths, and La. SrO > BaO in a large majority, and rare earths Y, Ta, Th, U, and F are high (Gold, 1966b; Heinrich, 1967). The sodium carbonatites are especially characterized by their extremely high content of Na_2O and to a lesser extent K_2O and CaO. The wide compositional variability in the "accessory" mineral content of the carbonatites is reflected in the wide range in composition shown in Table 4-12.

The mafic alkaline rocks, sometimes called alkali "gabbros," and related rocks (though they rarely contain plagioclase) generally include medium-grained ijolite (aegirine-diopside nepheline rocks), mafic-rich and mafic-poor members of the ijolite series, magnetite pyroxenite ("jacupirangite"), biotite pyroxenite, orthoclase-nepheline gabbro ("essexite"), along with fine-grained equivalents such as nephelinite (close to a fine-grained ijolite) and melanephelinite. Most of these rocks are medium to dark colored and high in nepheline and a member of the aegirine-diopside or aegirine-augite series. The pyroxene in the ultramafic rocks is titanaugite. Minor minerals that occur in widely variable amount and kind include: other feldspathoids (sodalite, cancrinite, analcite, melilite, ± hauyne, ± leucite), phlogopite, sphene, apatite, melanite (Ti-andradite), perovskite, magnetite (Ti, Mg rich) or ilmenite, calcite, zeolites, wollastonite, ±soda

Table 4-12 *Chemical composition of carbonatites and (for comparison) sedimentary carbonate rocks*

	Carbonatites (182 Analyses for Most Elements)		Sodium Carbonatites (4 Analyses)	Sedimentary Carbonate Rocks
	RANGE	AVERAGE		
SiO_2	tr −49.65	9.58	tr− 1.18	5.14
TiO_2	0 − 5.20	0.65	0.08− 0.10	0.07
Al_2O_3	tr −18.19	2.90	0.08− 0.09	0.40
Fe_2O_3	0 −38.88	4.33	⎱0.26− 0.32	⎱0.49
FeO	0 −44.60	4.37	⎰	⎰
MnO	tr − 8.60	0.72	0.04− 0.24	0.14
MgO	tr −42.68	6.69	0.14− 2.35	7.79
CaO	0.83 −64.04	34.06	12.74−19.09	42.30
Na_2O	tr − 2.3	1.02	29.0 −30.00	0.03
K_2O	tr −12.8	1.47	6.58− 7.58	0.16
BaO	tr − 8.40	0.40	0.95− 1.05	0.001
SrO	tr − 4.76*	0.81	0.85− 1.24	0.07
CO_2	3.12 −48.02	29.29	30.73−32.40	41.74
P_2O_5	tr −11.56	1.86	0.83− 1.06	0.045
$(Nb, Ta)_2O_5$	0 − 3.7	0.08		0.00002
F	0 −39.72	0.73	1.84− 2.69	0.033
ZrO_2	0.002− 0.003			

*Also $SrO = 15.92, 18.24$ in strontianite-rich ankeritic carbonatites (Garson, 1966).
SOURCE: Turekian and Wedepohl (1961); modified from Gold (1966b).

amphiboles, ± eudialyte. Feldspars are uncommon, and orthopyroxenes typically are absent (Upton, 1967).

Contact zones of the mafic alkaline rocks to granitic or other country rocks are often chilled to finer grain sizes and are in some cases porphyritic or more mafic. The surrounding granitic rocks are commonly but not invariably "fenitized," the alteration being gradational to unfenitized rocks. Stoping of country rocks seems to be a more common means of emplacement than forcible intrusion (e.g., Philpotts, 1970a, b).

Chemically (Table 4-13), the mafic alkaline rocks show most of the distinctive chemical characteristics of the alkaline olivine basalts—the high Na_2O, K_2O, TiO_2, Ba, Sr, Nb, P, Zr, Ta, Y, rare earths. The Na_2O especially, is perhaps even higher here than in the alkaline olivine basalts. It will be noted from Table 4-13 that the mafic and ultramafic alkaline rocks have a wide range of chemical composition, though not as wide as the carbonatites. It should also be noted that the alkali pyroxenites are similar to some lamprophyres, which consist of titanaugite, biotite, serpentinized olivine, titaniferous magnetite, in a nepheline-rich, ± analcite, ± melilite matrix (Upton, 1967).

The Melting Temperature of Calcite and Its Bearing on the Formation of Carbonatites

Until recently, the very high melting temperature of calcite (1339°C, as measured by Smyth and Adams, 1923), provided a convincing argument

Table 4-13 *Chemical composition of mafic and ultramafic alkaline rocks*

| | 105 Ijolites, Nephelinites, and Similar Rocks | | 34 Pyroxenites | 12 Other Ultramafic Rocks (e.g., Mica Peridotites) |
	AVERAGE	RANGE		
SiO_2	41.1	31.55–47.94	31.28–51.12	33.22 –47.33
TiO_2	2.8	0.16– 5.35	0.32– 9.02	1.26 – 6.08
Al_2O_3	13.9	3.97–22.56	1.52–15.84	3.82 – 9.71
Fe_2O_3	5.3	0.02– 9.92	0.39–16.33	0.02 – 9.68
FeO	5.3	1.98–12.00	2.65–12.63	2.70 – 6.47
MnO	0.2	0.02– 0.77	0.05– 0.36	0.10 – 0.52
MgO	6.9	0.50–20.17	3.97–23.15	9.92 –33.84
CaO	15.3	5.3 –24.69	10.26–30.72	2.84 –16.99
Na_2O	5.0	0.41–11.69	0.25– 6.77	0.88 – 1.56
K_2O	3.2	0.14– 8.81	0 – 4.74	0.88 – 6.98
BaO	0.1	0 – 0.59	0 – 0.67	0.03 – 0.32
SrO	0.2	0 – 1.	0 – 0.35	0 – 0.44
CO_2	1.1	0.03– 2.08	0 – 2.08	tr – 4.02
P_2O_5	1.1	0 – 3.11	0 – 4.33	0.21 – 1.18
F	0.06	0.01– 0.14	0 – 0.25	0.01 – 0.27
ZrO_2		0 – 0.09	0 – 0.12	0.001– 0.12

SOURCE: Compiled from the literature.

against the feasibility of a carbonatite magma. This conflicted with the field evidence for a magmatic emplacement, leading some workers to propose "heavy gaseous CO_2" or "hot concentrated CO_2-rich solutions." Alkali-rich carbonates are known to melt at temperatures as low as 750°C, and other workers (e.g., von Eckermann, 1948) suggested an alkali carbonatite magma. This is strongly supported by the alkali carbonatite lava flows at Oldoinyo Lengai, but most carbonatites seem to be largely $CaCO_3$.

Since late in 1958, however, it has become apparent that $CaCO_3$ can be melted at moderate temperatures of 640°C under moderate pressures of $CO_2 + H_2O$ (Paterson, 1958; Wyllie and Tuttle, 1959, 1960). More recently, with MgO and SiO_2 as additional components (with $CaO + CO_2 + H_2O$), melting has been achieved down to 605°C at 1-kbar pressure and about 590°C at 4 kbars (Wyllie, 1965). These temperatures are comparable with emplacement temperatures of carbonatite magmas deduced from contact metamorphic effects (Watson, 1967). The addition of certain other components present in the natural rocks may be expected to lower the crystallization temperatures still farther—for example, FeO, Na_2O, CaF_2, P_2O_5 (Wyllie, 1966b; Biggar, 1967a, 1969).

Suggested Origins of Carbonatite and Mafic Alkaline Rocks

The intimate and almost invariable association between carbonatites and mafic alkaline rocks indicates that they are genetically related and dictates that their origins should be discussed together. Possibilities are that:

Carbonatite magma forms by differentiation from a parent mafic or ultramafic magma of:

> Alkaline olivine basalt composition under stable conditions (Gold, 1967)
>
> Mafic ijolite composition (King and Sutherland, 1960, 1966; Wimmenauer, 1966; King, 1965; Wyllie, 1966b; Heinrich, 1967; Barth and Ramberg, 1966; Dawson, written communication, November, 1970; Watkinson, 1960)
>
> Pyroxenite composition (Davies, 1952; Smith, 1956)
>
> Kimberlite or other alkaline ultramafic magma (Saether, 1957; von Eckermann, 1961, 1967; Garson, 1961, 1966; Sturt and Ramsay, 1965; Gold, 1966a; Franz and Wyllie, 1966, 1967; Powell et al., 1966; Koster van Groos and Wyllie, 1966; Heinrich, 1967; Wyllie, 1966b; Egorov, 1970)
>
> Peridotite composition (Strauss and Truter, 1951)

In many instances the differentiation is considered accomplished by upward migration of volatiles, CaO, and other elements (e.g., Garson, 1966; Gold, 1966a). In other cases the carbonatite is thought to separate as immiscible liquid globules (von Eckermann, 1961, 1967; Koster van Groos and Wyllie, 1963, 1966; Barth and Ramberg, 1966; [?] Deines, 1970; Dawson, written communication, 1970; Egorov, 1970).

Carbonatite hydrothermal solution or aqueous "fluid" forms by differentiation of an alkalic magma (Larsen, 1952; Olson et al., 1954; Pecora, 1956, 1962; [?] Smith, 1956), later modified by alkalic rocks.

Partial melting of garnet peridotite or peridotite under very high pressures in the mantle (see Fig. 4-21) to form an alkalic magma (Johnson, 1966); fractional crystallization enriching the magma in volatiles, alkalis, Rb, Sr, Ba, rare earths, etc., to form kimberlite or carbonatite (O'Hara, 1965).

Partial melting of hydrated alkali peridotite or pyroxenite containing pargasite (amphibole) as a stable phase (Varne, 1968).

Carbonatite magma by relief of pressure on the upper mantle under crustal upwarps (Bailey, 1966) with "differentiation or preferential gas streaming during migration of material into the zone from which the basalts were tapped" (Dawson, 1966).

Carbonatite magma by melting of limestone by an alkali gabbro magma (Brogger, 1921).

Carbonatite body forms by hydrothermal or metasomatic replacement of an earlier nepheline syenite and ijolite complex (Bowen, 1924; Agard, 1960).

Carbonatite body forms by metasomatic replacement (juvenile solutions rich in CO_2 react with differentiate of basaltic magma to form nepheline syenite magma from which solutions are derived) of an earlier dunite or pyroxenite formed by differentiation of the basaltic magma (most Russian geologists; e.g., Kukharenko and Dontsova, 1964; see Gittins, 1966b).

Mafic alkaline rocks form by reaction of primary carbonatite magma with surrounding crustal rocks (von Eckermann, 1948, 1966; Holmes, 1950; Higazy, 1954; Parsons, 1961; Temple and Grogan, 1965; Melcher, 1966; Dawson, 1962a, 1964a, 1966).

Mafic alkaline rocks form by desilication of deep crustal or upper-mantle material by persistent flux of water vapor through them. This would be above a rising convection cell (Currie, 1970).

Mafic alkaline magmas form by partial melting of peridotite under very high pressures in the mantle (O'Hara, 1965; Johnson, 1966; Bailey and Schairer, 1966; Bultitude and Green, 1968; Spencer, 1969; Bass, 1970; Kumarapeli, 1970).

Mafic to felsic alkaline magmas form by differentiation of alkaline olivine basalt magma (many workers).

Any origin proposed for most carbonatite-mafic alkaline rock complexes must account for the following points:

Important Data	*Implications*
Carbonatite almost invariably forms the central member of a mafic alkaline rock (especially ijolite) complex	Carbonatite related to and perhaps differentiated from mafic alkaline rocks
Ring-dike and cone-sheet form typical but not invariable	Most shallow and intrusive
Carbonatites occur as lava and tuffs as well as intrusive types	At least some clearly magmatic
Some carbonatite vents show major explosive activity, such as the tuff cones of northern Tanganyika (Dawson, 1964b)	High volatile content, suggesting volatile buildup during differentiation
Carbonatite forms less than 1% of mass of central mafic alkalic complex at Napak, and central complex forms less than 3% of volume of associated mafic volcanic rocks (King, 1965)	Relative volumes consistent with carbonatite as differentiate but not as parent of mafic alkaline rocks
Country rocks surrounding carbonatites are characteristically fenitized	Affected by the high Na^+, Fe^{3+} of the magmas

Important Data	Implications
Noncarbonate minerals in carbonatites are the minerals that characterize the *later* members of the associated mafic alkaline rocks—e.g., aegirine-rich pyroxene, alkali feldspar, micas, sphene, apatite, magnetite	Consistent with carbonatite differentiate from mafic alkaline rocks
Acicular habit of apatite in carbonatites resembles that crystallized rapidly from an experimental melt, by contrast with the short, stubby habit of apatite coexisting with liquid or vapor (Wyllie, Cox, and Biggar, 1962)	Carbonatite crystallized from a melt (magma)
Many carbonatites show "igneous" textures, such as phenocrysts, primary flow foliation, or partly assimilated xenoliths of country rocks	Carbonatite crystallized from a magma
Carbonatites high in "granitic" trace elements, for example, Sr, Ba, Zr, P, rare earths, F	Extreme differentiate
Carbonatite lavas forming today are rich in alkali carbonates	Carbonatites originally (at time of emplacement) richer in alkalis, or these lavas may be a differentiate from a deeper carbonatite magma
Simplified carbonatite melts (including Ca-rich types) can exist down to about 600°C under moderate to low pressures of CO_2 and H_2O (Wyllie and coworkers)	Carbonatite magmas may exist under reasonable temperatures
Wide immiscibility gaps exist between carbonate and silicate magmas, at least in the systems albite–anorthite–Na_2CO_3–H_2O and Na_2O–Al_2O_3–SiO_2–CO_2 at pressures below 1 kbar (Koster van Groos and Wyllie, 1963, 1966)	Possible separation of carbonatite magma from silicate parent by liquid immiscibility during cooling, rise in CO_2 pressure, or increase in amount of CO_2
High Fe^{3+}/Fe^{++} compared with most rocks	High P_{O_2}, probably due to high water content
Sr^{87}/Sr^{86} in carbonatites is very low and is comparable to or lower than that in basalts, especially alkali basalts (Hamilton and Deans, 1963; Powell, 1966; Powell et al., 1962, 1966)	Carbonatites derived from mantle either directly or indirectly (e.g., from basalts)
Carbonatites most commonly emplaced in shield area of granitic and gneissic rocks, in which limestones in the vicinity are not known	Occurrence of limestone not important to origin of carbonatites
Carbonatites have extremely low C^{13} and O^{18} compositions compared with sedimentary limestones (Taylor et al., 1967); but a complete transition from high to low values exists with progressive metamorphism of limestone (Deines and Gold, 1969)	Carbonatites are formed by assimilation of limestone; but since a complete transition in isotope values exists, this criterion is of limited value

The origin that best seems to fit these characteristics is derivation of the mafic alkaline rocks (e.g., alkaline olivine basalt) by partial melting of peridotite under high pressures in the mantle, rise of this partial melt toward the surface, and separation of carbonatite as an immiscible liquid. The lower crystallization temperature of the carbonatite permits it to remain liquid to the later stages, for example, in the magma chamber to form a central plug or injected to form cone sheets or ring dikes near the surface or even to be extruded as lava or tuffs. The carbonatite magma may be a K-Na-bearing $CaCO_3$-rich melt, from which alkalis may diffuse into the country rock to produce fenite and other Na-rich "igneous" rock. Differentiation within the magma may permit segregation of alkali-rich carbonatite to the upper part of the magma chamber and possibly eruption to form sodium carbonate lavas and ash as at Oldoinyo Lengai.

GABBROIC LAYERED INTRUSIONS

The Muskox Intrusion, Northwest Territories, Canada[1]

The Muskox layered intrusion was not discovered until 1956, but an aggressive program of study (including deep drilling) by the Geological Survey of Canada, beginning in 1959, has made it one of the best-known intrusions of this type. The intrusion has an exposed length of 74 miles, apparently extends at least 75 miles farther north under cover, and has a maximum exposed width of 5 or 6 miles, apparently widening to the north under cover. In form, the intrusion is dikelike, with a wide flaring, funnel-like top (Fig. 4-31) plunging 4° to the north. Thus the top of the funnel is exposed to the north and the thin dikelike bottom of the funnel (the Feeder Dike) is exposed to the south.

The Muskox intrusion (K-Ar ages are 1,150 million to 1,250 million years on biotite in the marginal zone) is intruded into schists, gneisses, and granitic rocks of the Canadian Shield that have K-Ar ages of 1,700 million to 1,900 million years. It is capped by an unconformity overlain by a nearly flat lying sequence of sandstone, dolomite, and tholeiitic basalt flows of Middle Proterozoic age. Since the fault appears to cut the intrusion but not the dolomite and basalt, the basalt is thought to be younger than the intrusion.

The main structural units of the intrusion (Fig. 4-31) are the Feeder Dike, the lower Marginal Zones, and the Layered Series. The *Feeder Dike* is 500 to 1,800 ft wide, vertical, and forms the southern half of the exposed length of the body. It consists of bronzite gabbro chilled against the

[1]Irvine and Smith, 1961, 1969; Smith and Kapp, 1963; Smith, 1962; Bhattacharji and Smith, 1963, 1964; Findlay and Smith, 1965; Chamberlain, 1967; Smith, Irvine, and Findlay, 1967a, b; Pouliot, 1967; Irvine, 1967b, 1970a, b.

country-rock margins but gradational over short distances to an olivine-rich basalt ("picrite") segregation in the center of the Feeder Dike. The *lower Marginal Zones* are generally 400 to 700 ft thick and occur along the lower inward-dipping (generally 20 to 35°) walls of the funnel. They typically grade upward from bronzite gabbro at the contact, through olivine-rich basalt and feldspathic peridotite, to peridotite (essentially an increase in olivine, upward). This is accompanied by an upward increase in MgO/FeO of the mafic minerals (for example, Fo_{70} to Fo_{85} in olivine). Small amounts of pyrrhotite with Cu- and Ni-bearing sulfides occur sporadically along the walls of the intrusion. The *Layered Series* is about 6,400 ft thick and consists of 42 different layers ranging in thickness from 10 to 1,100 ft. Small-scale "rhythmic" layering (small-scale repetitions) is present but uncommon. The saucer-shaped layers are nearly conformable to the overlying Protero-zoic sediments and discordant to the lower Marginal Zones. The layers have sharp contacts and are laterally very continuous. The series ranges upward from dunite at the base, through peridotite, pyroxenites, gabbros, and granophyric gabbro (gabbro containing more than 3% granophyric intergrowths), to pink granophyre at the top (Fig. 4-31). Parts of the sequence are repeated at several places. Fragments of the overlying country

Fig. 4-31 *Restored cross section of the exposed part of the Muskox intrusion. After Irvine and Smith (1967).*

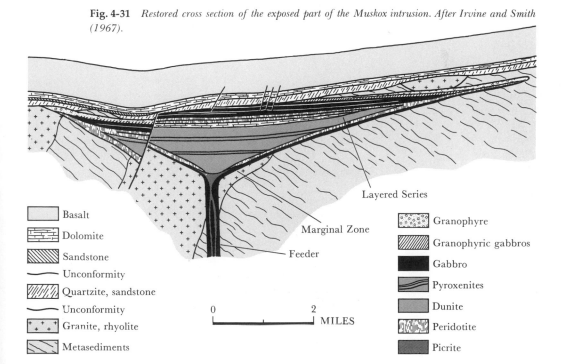

Layered Series

Marginal Zone

Feeder

Basalt

Dolomite

Sandstone

Unconformity

Quartzite, sandstone

Unconformity

Granite, rhyolite

Metasediments

0   2
|_____|_____| MILES

Granophyre

Granophyric gabbros

Gabbro

Pyroxenites

Dunite

Peridotite

Picrite

rocks appear in the roof granophyre. Chromite-rich layers, a few inches thick, occur in two places between peridotite and overlying orthopyroxenite layers.

Most of the rocks in the Layered Series consist of two parts: *cumulus* grains gravitationally settled out of the magma, cemented by *intercumulus* grains formed by crystallization of the trapped magma. Cumulus minerals are olivine (1.2 to 1.5 mm) and minor chromite in the peridotite and dunite; bronzite (0.6 to 0.8 mm) in the orthopyroxenite; augite + bronzite or augite + olivine in the other pyroxenites; and plagioclase + augite + olivine (or bronzite) ±magnetite and ilmenite in the gabbros. Intercumulus minerals in all these rocks include plagioclase, orthopyroxene, clinopyroxene, minor biotite, and, in the pyroxenite, traces of granophyre.

The sequence of crystallization and the proportions of the silicate minerals in most of the rocks are just as predicted by fractional crystallization in the system $CaMgSi_2O_6-Mg_2SiO_4-SiO_2$ (see Fig. 3-4), assuming 20 to 30% parent liquid is trapped as intercumulus material (compare Fig. 4-32). The nearly complete separation of olivine and orthopyroxene (in the peridotites and orthopyroxenites, respectively) is attributed to a peritectic relationship between them (see Fig. 1-11, the olivine being eliminated) and to the olivine being twice the diameter of the orthopyroxene

Fig. 4-32 *Derivation of one type of sequence of Layered Series rocks in the Muskox intrusion. Modal analyses of peridotites and pyroxenites from the intrusion superimposed on a phase diagram for the basaltic rocks (see Fig. 3-4). After Irvine and Smith (1967).*

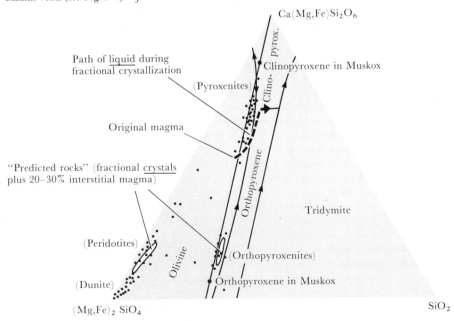

grains, as noted above. On the basis of Stoke's law, the olivine should settle "roughly 10 to 100 times as fast," thereby outdistancing the orthopyroxene that would follow it to the floor of the intrusion. The chromite grains would continue to accumulate after formation of the olivine layer, in the earliest stages of orthopyroxene accumulation. Cu−Ni−Fe sulfides in the "chromitite" may represent accumulation of immiscible liquid droplets.

Studies by Irvine and Smith (1967) of the variation in Ni content of olivine upward through thick dunite layers suggest that each of several 300- to 350-ft dunite layers represents 5 to 10% of its parent liquid. From this they infer that each such repetition must have involved displacement of most of the remaining magma by new parent magma. The displaced magma was probably pushed to the surface as fissure eruptions. This repeated displacement of magma after the early stages of crystallization accounts for the large proportion of ultramafic rocks in this intrusion.

The mineralogy, especially the presence of orthopyroxene, and the chemistry (except for a slightly higher MgO of 9.70%), especially the low TiO_2 (1.06%) and fairly high SiO_2 (50.58%), show clearly that the Muskox intrusion crystallized from a magma of continental tholeiitic basalt type. Olivine in the ultramafic part of the Layered Series is high in magnesium (Fo_{80-85}). That at the top of the Layered Series, in the lower Marginal Zones, and in the core of the Feeder Dike is higher in iron (Fo_{70-80}); and that in a thin zone at the roof and margins of the intrusion is still higher in iron (Fo_{60-70}). Following crystallization, much of the olivine in the peridotite was extensively altered to serpentine and secondary magnetite. Granophyric intergrowths appear in the gabbro up to about 3% and in the upper granophyric gabbros and granophyre up to more than 20% (locally 60 to 70%) of the rock. Biotite forms up to 2% of these same rocks.

The granophyre cap over the intrusion, because of its great volume, the low melting composition of the roof rocks, and the time and temperature during crystallization, is thought to have formed by melting of the roof country rocks (Irvine, 1970b).

General Features of Gabbroic Layered Intrusions

Gabbroic layered intrusions are mafic to ultramafic plutonic rock bodies, characterized as gabbroic because of their typical overall basaltic composition. Their rock types generally range from various peridotites toward the base of the intrusion, upward through gabbro and anorthosite, to ferrogabbro and granophyric gabbro toward the top (see Fig. 4-33). They range in size from that of a stock to a large batholith, the largest known being the Bushveld intrusion of South Africa—about 150 by 300 miles and more than 24,000 ft thick. In shape, most appear to be lopoliths (e.g., Duluth

	Bushveld Intrusion South Africa			Skaergaard, East Greenland and Stillwater, Montana		
	Thickness, ft	Overlying rocks	Irregular masses of melanogranophyre	Thickness, ft	Upper border group	Gabbro and ferrodiorite
	4,900	Upper zone	"Ferrogabbro" or ferrodiorite with melanogranophyre concentrations	2,800	Upper zone	Ferrodiorite or "ferrogabbro" with granophyric concentrations
				2,300	Middle zone	Gabbro
	11,700	Main zone	Gabbro and anorthosite (poor layering)	2,400	Lower zone	Olivine gabbro (Base unexposed)
				Sk./St.	=:?== Gap or overlap uncertain ==:?==	
				2,100	Upper gabbro zone	(Top unexposed) Gabbro
				6,200	Anorthosite zone	Anorthosite / Gabbro / Anorthosite / Gabbro / Anorthosite
	3,400	"Critical series"	Anorthosite, norite, feldspathic pyroxenites, orthopyroxenite (fine layering)	2,200	Lower gabbro zone	Gabbro
				2,700	Norite zone	Anorthosite, norite, feldspathic pyroxenite (fine layering)
	Chromite horizon——Chromite—			1,100		Orthopyroxenite
				Chromite horizon——Chromite—		
	4,000	Basal series	Orthopyroxenite harzburgite, dunite / Peridotites	2,400	Ultramafic zone	Orthopyroxenite, harzburgite, minor dunite / Peridotites
	400	Marginal zone	Fine-grained norite	500	Border zone	Feldspathic orthopyroxenite

Skaergaard intrusion

Stillwater intrusion

Fig. 4-33 *Generalized sections and approximate correlation among three of the thickest and best-described mafic layered intrusions. Note that most of the Bushveld is exposed, as is the upper part of the Skaergaard and the lower part of the Stillwater. Tabulated from data in Wager and Brown (1968), Wager and Deer (1939), and Hess (1960a). The consistency in trend of variation in large mafic layered intrusions is emphasized by this imaginary stacking of the Skaergaard intrusion (Greenland) on top of the Stillwater intrusion (Montana), for comparison with the Bushveld lopolith (South Africa).*

gabbro, Minnesota; Bushveld, South Africa), funnels (e.g., Skaergaard, eastern Greenland), or highly elongate to dikelike (e.g., Muskox, Canada; the Great "Dike" of Southern Rhodesia appears to be part of a large "lopolith" preserved by downfaulting into a rift graben).

In addition to the Muskox intrusion, Northwest Territories, prominent examples of mafic layered intrusions include: Stillwater, Montana (Hess, 1960a; Jones, Peoples, and Howland, 1960; Jackson, 1961, 1963, 1967); Duluth gabbro, Minnesota (Taylor, 1964; Olmsted, 1968; Leighton, 1954; Grout, 1918a, b, c); Bays of Maine complex, Maine (Chapman, 1962); Kiglapait, Labrador (Morse, 1968, 1969; Morse and Davis, 1969; Wheeler, 1942); Bay of Islands, Newfoundland (Smith, 1958; Wager and Brown, 1968; Ingerson, 1935, 1937; Cooper, 1936); Skaergaard, eastern Greenland—the most thoroughly studied of all mafic layered intrusions (Wager and Deer, 1939; Wager and Brown, 1968; Wager, 1960, 1963, 1968; Wager and Mitchell, 1950, 1951; Wager et al., 1957; Deer and Wager, 1939; Carr, 1954; Gay and Muir, 1962; Muir, 1951; Brown, 1957; Brown and Vincent, 1963; Vincent and Phillips, 1954; Hamilton, 1963; Brothers, 1964); Kap Edvard Holm and Kaerven, eastern Greenland (Deer and Abbot, 1965; Wager and Brown, 1968); Rhum, Scotland (Wager and Brown, 1968; Hughes, 1960; Wadsworth, 1961; Brown, 1956); Cuillin complex, Skye, Scotland (Wager and Brown, 1968; Wheedon, 1961, 1965; Harker, 1904); Insch and associated intrusions, northeastern Scotland (Wager and Brown, 1968; Read, 1919; Stewart and Johnson, 1960; Stewart, 1946; Wadsworth et al., 1966); Bushveld, South Africa (Hall, 1932; Wager and Brown, 1968; McDonald, 1967; Cameron, 1963; Jackson, 1967; Willemse, 1959; van Biljon, 1949; Sandberg, 1926; Daly, 1928); Great "Dike," Rhodesia (Wager and Brown, 1968; Jackson, 1967; Hess, 1950; Worst, 1958; Hughes, 1970); Kapalagulu, Tanzania (Tanganyika) (Wager and Brown, 1968; Wadsworth, 1963; van Zyl, 1959); Proshiri-Dake, Japan (Minato et al., 1965); Giles complex, Australia (Nesbitt and Kleeman, 1964; Nesbitt and Talbot, 1966; Goode and Krieg, 1967); Sooke gabbro, British Columbia (Clapp, 1912; Cooke, 1919).

Most mafic layered intrusions are characterized by an overall mineralogical (and chemical) gradation from the bottom to top of the whole intrusion. This gradation, called *cryptic layering*, consists of two aspects (Wager and Brown, 1968): certain minerals in turn begin and end crystallization at specific stages of fractional crystallization, and other minerals belonging to solid-solution series (e.g., olivine, plagioclase) change composition continuously with progressive crystallization. This gradation is reflected in a progressive change in chemistry upward in the intrusion, a change resulting from gravitational settling of early formed grains (Bowen, 1915, 1927). This cryptic layering is sometimes repeated at higher levels due to multiple injection of fresh magma (e.g., in Rhum and Kap Edvard Holm).

Superimposed on this overall vertical variation is a *rhythmic layering* characterized by repeated graded beds. This feature is the main characteristic distinguishing mafic layered intrusions from thick diabase sills (Wager and Brown, 1968). It generally consists of a layer less than 1 in. to more than 100 ft thick, rich in dark, heavier minerals (e.g., pyroxene) at the base, grading to predominantly light minerals (e.g., plagioclase) at the top. This graded bedding is analogous to that found in some sedimentary rocks such as turbidites, the heavier grains having settled through the magma faster, thereby being concentrated at the bottom of the graded layer. Other typically sedimentary structures developed in some of these intrusions include cross-bedding, slump structures, compaction structures around inclusions (e.g., of pyroxenite), and even "trough bedding"— shallow channels directed downward toward the center of the magma chamber. Most of these features were first described in detail by Wager and Deer (1939) in their classic report on the Skaergaard intrusion, eastern Greenland. Some thin layers may be traced for miles along strike; others show marked variation in thickness and sequence of units along strike (e.g., Cameron, 1963; Irvine and Smith, 1967). In some intrusions such as Skaergaard, Rhum, and Kiglapait, a local foliation results from a preferential orientation of tabular olivine or plagioclase grains parallel to the layers (Phillips, 1938; van den Berg, 1946; Brothers, 1964; Morse, 1969).

Mineralogically, the mafic layered intrusions consist of two components, cumulus and intercumulus grains. *Cumulus* grains ("primary precipitate") are those grains that have settled to the floor of the magma chamber to form a crystal mush. The trapped (interstitial) liquid (roughly 35%) may crystallize by further addition to, and enlargement of, the same crystals ("adcumulus" growth) or by nucleation and crystallization of new minerals to form *intercumulus* grains ("interprecipitate") (see Fig. 4-34). The most common cumulus minerals include chromite, olivine, pyroxenes,

Fig. 4-34 *Diagramatic representation of cumulus crystals of plagioclase with (a) intercumulus (and poikilitic) pyroxene; (b) intercumulus pyroxene and overgrowths ("adcumulus") on plagioclase; (c) minor intercumulus pyroxene and overgrowths on plagioclase (giving an almost monomineralic rock). Simplified from Wager and Brown (1968, p. 65).*

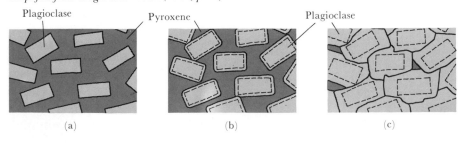

Plagioclase   Pyroxene   Plagioclase

(a)   (b)   (c)

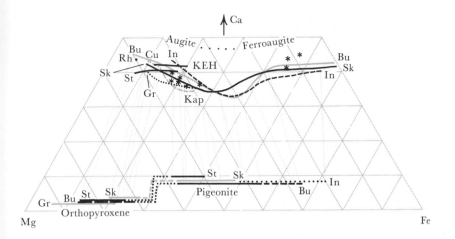

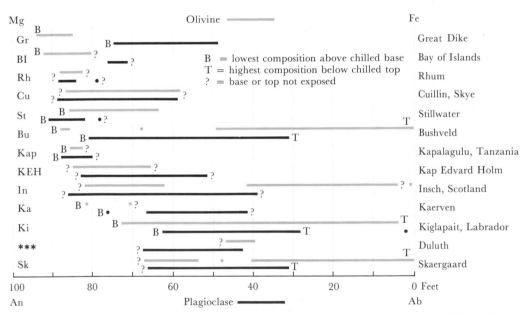

Fig. 4-35 *Compositional variation of the pyroxenes, olivine, and plagioclase in mafic layered intrusions. The earlier-formed minerals (lower in the intrusions) are at the Mg- or Ca-rich end of each line. Letters labeling each pyroxene trend are identified with the olivine and plagioclase trends. Primarily from data in Wager and Brown (1968); more recent additions from Atkins (1969) and Morse (1969).*

and plagioclase. Intercumulus minerals may include any of these and most of the minor later-stage minerals. Any clear understanding of the crystallization behavior of a mafic layered intrusion must include a recognition of both the cumulus and intercumulus components.

Compositional variation within each of the main minerals is shown in Fig. 4-35 (compare also Fig. 1-13). Depending on the original composition of the basaltic magma intruded, the first minerals to crystallize will be high in Mg or Ca. Fractional crystallization proceeds, developing more Fe- or Na-rich grains higher and higher in the magma chamber. The most common order of crystallization appears to be olivine, orthopyroxene + chromite, plagioclase, augite, and pigeonite, followed in turn by the minor minerals magnetite-ilmenite, apatite, and quartz + alkali feldspar (in the latest stages of crystallization). Minor amounts of sulfides, especially pyrrhotite and chalcopyrite, and platinum minerals occur with chromite in the ultramafic zones of some intrusions.

Granophyric gabbro that occurs toward the top of most mafic layered intrusions is characterized by the presence of veins, streaks, and lenses of melanogranophyre (dark rocks rich in quartz and alkali feldspar). This represents the extreme differentiate of the gabbroic (or basaltic) magma and approaches granite in composition. Overlying crusts of "red granite" or granophyre, such as are distinctive at the Duluth gabbro and Bushveld intrusions, have long been thought to represent a product of extreme differentiation. These latter, however, occur in prohibitively large volume, have a much lower iron content than would be expected, and are sharply bounded to the underlying gabbros. It has been suggested, therefore, that these granophyre crusts formed by partial melting of overlying sediments (Brown, 1963; Wager and Brown, 1968; Willemse, 1969; Irvine, 1970b). That even the lesser amount of granophyre in the Skaergaard intrusion may form at least in part by assimilation of country rock is suggested by O^{18}/O^{16}, Sr^{87}/Sr^{86}, and Pb isotopes (Taylor and Epstein, 1963; Hamilton, 1963, 1966). Chayes' (1970) recent calculations of the extent of end-stage alkali enrichment during differentiation also suggest that this may be the case.

Chemically, the magma from which the mafic layered intrusions crystallized is normally taken to be equivalent to the chilled border phase of the intrusion. Where present, this generally turns out to be a tholeiitic basalt or "high-alumina" basalt. A few cases seem to have somewhat alkalic tendencies (e.g., the Duluth gabbro, Minnesota; the Kiglapait intrusion, Labrador); but, with the possible exception of the Iron Mountain layered intrusion, Colorado (Shawe and Parker, 1967), none yet described have clearly crystallized from alkaline olivine basalt. The mineralogy of almost all intrusions also reflects this chemistry. Most striking is the abundant orthopyroxene in the majority of intrusions, a distinctive characteristic of tholeiitic and "high-alumina" basalts. The primary differentiation trends are to increased SiO_2, Na_2O, and FeO, as reflected in both the changing mineralogy and changing composition of individual minerals to higher levels. Enrichment in iron and alkalis is strongly controlled by the oxygen content of the magma, as explained on Fig. 3-7. As shown on that Mg-Fe-alkali diagram, magmas in this postorogenic association undergo pronounced iron enrichment during most of the period of crystallization, presumably due to a low P_{O_2} (i.e., probably a low water content). The low P_{O_2} may derive from a lack of water in the magma from the time of its generation. In the latest stages of crystallization, the trend turns rapidly toward higher alkalis.

Suggested Origins of Gabbroic Layered Intrusions

It is generally agreed that the magma from which crystallized mafic layered intrusions was tholeiitic or "high-alumina" basalt. The origin of these magmas has been described in the section on basalts. Differences of opinion arise primarily in attempting to explain the origin of the rhythmic layering. Unless rather clear circumstances suggest multiple injection, most workers would form the rhythmical layering by a *single injection* of magma, followed by gravitational differentiation in place, and:

> "Short epochs of mild but irregular turbulence" strong enough to temporarily hold plagioclase in suspension without interrupting the downward movement of heavier pyroxene crystals (Hall, 1932; Hess, 1938b; Weedon, 1965).

> Cooling crystal-rich layer adjacent to the roof or walls intermittently and locally descends as a column, spreads over the floor of the magma chamber, and crystals sink at different rates (Hess, 1960a; Wager, 1963, 1968; Wager and Brown, 1968; Morse, 1969).

> Rhythmic crystal nucleation and crystal settling, earlier nuclei having settled out as the crystals reach sufficient size (Wager, 1959, 1968; [?] Wager and Brown, 1968; Wadsworth, 1963; Weedon, 1965; Mathison, 1967; Wilson and Mathison, 1968). Nucleation triggered by intermittent earthquake shocks (Hoffer, 1966).

> Convection current carries crystals to bottom of magma chamber and permits settling at different rates, or lighter grains are winnowed out by currents (Wager and Deer, 1939; Wager, 1953, 1963, 1968; Brown, 1956; Wager and Brown, 1968; Turner and Verhoogen, 1960; Wadsworth, 1963; Brothers, 1964; Morse, 1969).

> Convection cells intermittently shift laterally due to turbulence, or depth of movement rises due to stagnation with bottom crystallization (Turner and Verhoogen, 1960; Jackson, 1961; Cameron, 1963; Taylor, 1964).

> Periodic fluctuation of water pressure (diopside anorthite eutectic— see Fig. 3-4—moves to lower temperature and toward anorthite with increased P_{H_2O}) by periodic fracturing of roof of intrusion to give

alternate precipitation of plagioclase and pyroxene (Yoder, 1954). Fracturing of roof permits eruption of basaltic lava, followed by fresh influx of magma (Brown, 1956).

Abrupt changes in temperature or pressure (McDonald, 1967).

Where repetition exists in the sequence of cryptic layering, such as in the Bushveld, Rhum, and Muskox intrusions, suggested origin of rhythmic layers has involved multiple injection of magma. This repeated injection would be magma:

Of constant composition, injected high in the magma chamber to interrupt convection and crystallization (Cooper, 1936; Irvine and Smith, 1967; Wager, 1968; Wager and Brown, 1968)

Of varying composition from a separate reservoir of differentiating magma (Lombard, 1934; Worst, 1958)

An extreme point of view, proposed for the Bushveld intrusion and the overlying granite, is that of metasomatic (metamorphic) replacement of an original sedimentary complex, the layers representing compositional differences in the preexisting sediments (Sandberg, 1926; van Biljon, 1949; Iannello, 1971).

It is apparent that the cryptic layering must have resulted from fractional crystallization and settling of early formed grains. Bearing in mind that cooling must have been most important next to the roof and walls of the intrusion it also seems likely that crystallization would be most active there. Lack of an obvious mechanism for turbulence exclusive of convection currents in the magma chamber and lack of an obvious source of, or evidence for, significant water pressures may also have a bearing on some of the suggested origins for rhythmic layering. Even with these considerations, it is difficult to choose among the alternatives presented. The writer has a preference for a crystal-rich layer forming near the walls of roof intermittently descending en masse to the floor of the magma chamber, where the grains settle out at different rates.

The immense size and thickness of many mafic layered intrusions, along with their occurrence in tectonically stable regions, associated structural features, and impact phenomena, has recently given rise to a proposal of formation by melting of the upper mantle and ejection of the overlying crustal rocks (D. Alt, personal communication, 1968; French, 1971).

LUNAR BASALTS AND RELATED ROCKS

The return of Apollo 11 from Tranquillity Base (0.67°N, 23.49°E in Mare Tranquillitatis) on July 24, 1969, brought the study of lunar materials from the realm of speculation to the stage of direct observation. The 22 kg of basaltic rocks, microbreccias, and lunar soil were collected from a relatively smooth, level area in the Sea of Tranquillity, one of the large lunar maria. The surface is pitted with numerous craters ranging up to tens of meters in diameter. The lunar soil and breccias appear to be derived largely from the larger rock fragments, but they include glassy particles such as tiny microscopically pitted spheres. Only the basaltic rocks, which were presumably derived from the "bedrock" of the lunar mare, will be here described in any detail.[1]

The lunar basaltic rocks range from very fine-grained vesicular intersertal texture basalts (mainly "type A" rocks) to medium-grained equigranular vuggy ophitic basalts (mainly "type B" rocks). They are predominantly composed of augite, plagioclase, and ilmenite with a minor residuum of generally rhyolitic composition. Pyroxenes in these rocks show a wide range of compositions, tending to cover the exsolution field in the central part of the diagram and extending to extremely iron-rich varieties (compare Fig. 1-12). The mineralogical characteristics are summarized in Table 4-14. The chemical compositions are summarized in Table 4-15. Analyses of the interstitial glassy material correspond to a low-sodium, potassium-rich granitic residuum.

The lunar maria are generally considered to be vast lava lakes that solidified fairly early in the history of the moon. They consist of basaltic rocks characterized as follows (see Table 4-16): Oceanus Procellarum (Ocean of Storms) has much less TiO_2, less Na_2O and K_2O, and in olivine-rich rocks more MgO than Mare Tranquillitatis (Sea of Tranquillity), whereas Mare Tranquillitatis has less SiO_2 and more FeO + CaO than Sinus Medii. The lunar highlands, on the other hand, appear to consist of anorthosite that has much less FeO and TiO_2, less MgO and Na_2O, much more Al_2O_3, and more CaO and SiO_2 than any of these three maria. The lunar maria probably formed by impact of large meteoroids at various times in the moon's history.

The chemical characteristics of the "basaltic" rocks of Tranquillity Base (Table 4-15; especially the high TiO_2 and FeO and the low Na_2O, K_2O) have presumably resulted from fractional crystallization and extreme volatile loss. Crystallization of these lunar "ferrobasalts" appears to have occurred under very low oxygen fugacity ("partial pressure") as indicated from lack of Fe^{3+}, presence of metallic iron-troilite, ilmenite, and probable Ti^{3+}

Further analysis of Apollo 12 basalts from the Ocean of Storms and study of samples of the next Apollo landing in the lunar highlands

[1]Most of the information below has been taken from papers in the Moon issue of *Science*, vol. 167, no. 3918, January 30, 1970, and the Proceedings of the Apollo 11 Lunar Science Conference published in *Geochim. et Cosmochim. Acta,* vol. 34, Supplement, 1970.

Table 4-14 *Mineralogical characteristics of lunar basaltic rocks*

Minerals and Characteristics	%	Intersertal Basalts	%	Ophitic Basalts
Augite	48.9–60.3	With pigeonite cores or intergrowths. Commonly zoned to subcalcic augite and ferroaugite. Minor titanaugite (generally 2–3% TiO_2). See Fig. 4-25.	45.7–55.3	Commonly zoned to subcalcic augite and subcalcic ferroaugite and even to pyroxmangite (a new very high Fe pyroxenelike mineral).
Plagioclase	15.6–29	An_{71-87} but commonly An_{73-81}; some grains zoned, others unzoned; An_{72-75} in residuum.	21.6–37.1	An_{67-95} but commonly An_{83-91}; unknown whether difference from intersertal basalts significant. Some grains zoned, for example, An_{72-89}; others unzoned, for example, An_{88}, An_{83}. High-temperature structural state. Center of large crystals.
Ilmenite	12.8–23.9	Rims on armalcolite phenocrysts. Early crystals (for example: 11% Mg component) lining vesicles. Later tabular crystals in ilmenite-pyroxene-rich areas (for example: 6% Mg component); plates in plagioclase-glass-rich areas (for example: 3–0.3%). → Lower Mg, higher Mn	9.7–20.	Tiny crystals associated with fayalite. Lower Mg, higher Mn
Olivine	0–0.6	$Fo_{41.1-76}$. Some grains zoned, for example: Fo_{71-41}; rare reaction rims of augite.	0–4.8	Scattered phenocrysts, Fo_{58-75}. Some partly resorbed cores rimmed by pyroxene grains. Some grains slightly zoned. Minor fayalite FO_2 in residuum with pyroxferroite and cristobalite.
Cristobalite	0.1–1.1	Low-temperature polymorph; coarse patches in residuum ("mesostasis").	0.3–6.3	Low-temperature polymorph; coarse patches in residuum.

Tridymite		In residuum.	Minor	In residuum in small patches.
K-feldspar	Minor	In residuum.	Minor	In residuum in small patches.
Apatite	Trace-minor	In residuum.	Minor	In residuum.
Troilite	0.21–0.36	FeS in residuum.	0.01–1.3	FeS in residuum.
Native iron	0.013–0.04	Fe in residuum.	0.027–0.18	Fe in residuum.
Armalcolite	Sporadic phenocr.	Opaque $Fe_{0.5}^{++}Mg_{0.5}Ti_2O_5$. Rimmed by Ilmenite with lamellae of chrome spinel.		
Pyroxferroite	0–minor	Yellow $(Fe^{++}, Ca)SiO_3$, intergrown with or rimming the clinopyroxene.		
Chrome spinels			Minor	Fe_2TiO_4-$FeCr_2O_4$-$FeAl_2O_4$
Grain size		Fine.		Fine to medium.
Alteration		None, no hydrous phases.		None, no hydrous phases.
Texture		Intersertal (angular spaces between feldspars filled with glass). Fine and radial intergrowths of augite and plagioclase. Those and "contact metamorphic" textures identical to those in terrestrial lava lakes.		Ophitic, subophitic (plagioclase partly or completely enclosed in large clinopyroxenes). No preferred orientation of grains (similar to that in terrestrial sills or thick flows).
Sequence of crystallization (from textures)		Olivine; ilmenite; pyroxene; plagioclase; rhyolitic residuum (mainly K-feldspar and SiO_2).		Ilmenite; pyroxene; plagioclase; rhyolitic residuum (mainly K-feldspar and SiO_2).
Shock features		None.		None.

Table 4-15 *Average chemical compositions of lunar basaltic rocks (in weight percent)**

	Intersertal Basalt		Ophitic Basalt	
	AVERAGE	RANGE	AVERAGE	RANGE
SiO_2	40.23	39.0 −41.0	40.57	37.8 −42.5
TiO_2	11.85	11.2 −13.2	10.53	8.8 −12.6
Al_2O_3	8.70	7.7 −10.8	10.48	8.9 −12.1
Fe_2O_3	0	0	0.008	0− 0.6
FeO	19.44	18.5 −20.5	18.56	17.3 −19.8
MnO	0.28	0.20− 0.28	0.27	0.24− 0.30
MgO	7.76	7.0 − 8.1	7.07	6.0 − 8.1
CaO	10.49	10.0 −11.0	11.63	11.0 −12.3
Na_2O	0.61	0.49− 0.91	0.58	0.36− 0.85
K_2O	0.30	0.27− 0.36	0.07	0.05− 0.11
Cr_2O_3	0.35	0.32− 0.40	0.30	0.21− 0.47
P_2O_5	0.16	0.13− 0.20	0.06	0.04− 0.12

*Eight intersertal basalts and nine ophitic basalts from Tranquillity Base. Distinction based on texture by Chao et al. (1970b).
SOURCE: Data from Ware and Lovering (1970a, p. 517); Engel and Engel (1970a, p. 527); Maxwell et al. (1970a, p. 530); Wiik and Ojanpera (1970a, p. 531); Peck and Smith (1970a, p. 532); Agrell et al. (1970a, p. 583; 1970b, p. 95); Kushiro et al. (1970a, p. 560); Compston et al. (1970b, p. 1007).

should help sort out whether lunar rocks are inherently different, being developed from a lunar crust or mantle significantly different from that of the earth or whether they are different primarily in degree, being formed under the influence of the moon's lesser gravity, degassing, and lack of an atmosphere. At present there seems no compelling reason why most lunar surface rocks had to be formed from rocks strikingly different from those on earth. In fact, as implied by the available analyses of Apollo 12 rocks, other lunar maria may not be as compositionally extreme as the Sea of Tranquillity (Apollo 11). Lunar basalts do, however, appear to contain more Fe (all Fe^{++}) and Ti, much less Na, and less K, Si, and Al (compare earth basalts, Table 4-8). Anorthosites presumed to compose the highlands would presumably balance this with their low Fe and Ti and high Al, whereas Na, K, and possibly Si remain different.

Although some earth basalts, such as the alkaline olivine basalts, contain high Ti (see Table 4-8; also, for example, Macdonald, 1968; Robinson, 1969; James and Jackson, 1970), their composition is still far from that of the lunar basalts of the Sea of Tranquillity. A basalt hornfels from the somewhat alkalic Duluth gabbro closely resembles some lunar fines except for its higher Fe_2O_3 and Na_2O (Goldich, 1971). Comparisons have also been made with various meteorites (e.g., Mason and Melson, 1970; Ulbrich, 1970; Ringwood and Essene, 1970). Except for the extreme Ti of Sea of Tranquillity samples, reasonable comparisons may be made with some chondrites, eucrites, and basaltic achondrites (see also Mason,

Table 4-16 *Comparisons of chemical compositions of basaltic rocks from three lunar maria and of anorthosites from (?) lunar highlands*

	Oceanus Procellarum (Apollo 12)		Mare Tranquillitatis (Surveyor V, Apollo 11)		Sinus Medii (Surveyor VI)	Lunar Highlands (Surveyor VII; Apollo 11 Presumed Highlands Source)	
	AVERAGE	RANGE	AVERAGE	RANGE			
SiO$_2$	45.33	42.3 –47.1	40.42	37.8 –42.5	46. ± 8.	38.5 ± 8.	45.4 –46.0
TiO$_2$	3.23	2.5 – 4.5	11.11	8.8 –13.2			tr– 0.3
Al$_2$O$_3$	9.79	7.4 –12.1	9.65	7.8 –12.1	12. ±4.	18.2 ± 6.	27.3 –33.8
Fe$_2$O$_3$	0	0	0.004?	0 – 0.6			
FeO	20.08	17.9 –22.1	18.06	17.3 –20.5	6.5 ±2.5	2.7 ± 1.5	2.8– 6.2
MnO	0.26	0.24– 0.27	0.25	0.2 – 0.29			0.1– 0.1
MgO	10.45	6.6 –16.7	7.40	6.0 – 8.1	5. ±5.	6.7 ±5.	1.7– 7.9
CaO	10.16	7.9 –11.8	11.13	10.0 –12.2	8.4 ±3.	8.4 ±3.	14.1 –17.5
Na$_2$O	0.30	0.16– 0.64	0.59	0.36– 0.91	3.	3.	0.3– 0.4
K$_2$O	0.07	0.05– 0.08	0.17	0.05– 0.36			tr– tr
Method of analysis	Wet chemical		X-ray spectrograph, wet chemical		Alpha scattering	Alpha scattering	Electron microprobe average of 7–10 randomly placed analyses
Reference	Kushiro and Haramura (1971); see also LSPET (1970)		Turkevich et al. (1967); references listed for Table 4-15		Turkevich et al. (1968a)	Turkevich et al. (1968a)	Wood et al. (1970a)

1962, 1967; Duke and Silver, 1967; van Schmus and Wood, 1967; van Schmus, 1969; Mueller and Olsen, 1969).

MAJOR SOURCES FOR DISTRIBUTION OF IGNEOUS ROCK ASSOCIATIONS[1]

Geologic Map of Alaska (1957). Compiled by J. T. Dutro, Jr., and T. G. Payne, U.S. Geological Survey, scale 1:2,500,000.

Geologic Map of Canada (1955). Geological Survey of Canada, Map 1045A, scale 1:7,603,200.

Tectonic Map of the Canadian Shield (1965). C. H. Stockwell, Chairman, Canada Geological Survey, map 4-1965, scale 1:5,000,000.

Geologic Map of the United States (1932). Compiled by G. W. Stose, U.S. Geological Survey, scale 1:2,500,000.

Tectonic Map of Mexico (1961). Compiled by Z. de Cserna, *Geol. Soc. Am.*, scale 1:2,500,000.

Tectonic/Geological Map of Greenland (1970). Geological Survey of Greenland, Copenhagen, scale 1:2,500,000.

Geologic Map of North America (1946). Compiled by G. W. Stose, *Geol. Soc. Am.*, scale 1:5,000,000.

Mapa Geológico do Brazil (1960). Departamento Nacional da Produçáo Mineral, Divisão de Geologia e M Mineralogia, escala 1:5,000,000.

Carte geologique de l'Amerique du Sud (1964). Commission de la Carte geologique du Monde, Alberto Ribeiro Lamego, Coordinator.

Carte geologique internationale de l'Europe (1966). Bundesanstalt für Boden forschung et Unesco, Hannover, Parts A3; B3; C1, 2, 3; D1, 2, 3.

Carte Tectonique Internationale de l'Europe (1962). Moscow, 1964, *Congr. Geol. Intern.*, scale 1:2,500,000, 16 Feuille.

Geological Map of Africa (1963). Assoc. of African Geological Surveys, *Intern. Geol. Congr.*, Paris, scale 1:5,000,000.

Geologic Map of Eurasia (1954). Russian Geological Survey, Moscow.

Minato, M., Hokkaido University, Japan, personal communication, 1969.

Geological Map of the World—Australia and Oceania (1963). Bureau of Mineral Resources, Canberra, Australia, *Intern. Geol. Cong.*, Paris, scale 1:5,000,000.

Anderson, A. T., Jr. (1968). Massif-type anorthosite, *in* Origin of Anorthosite and Related Rocks, Y. W. Isachson, ed., *N.Y. State Mus. Sci. Serv. Mem.* 18, pp. 47–55.

Bullard, F. M. (1961). Volcanoes, University of Texas Press, Austin, p. 60.

Rittmann, A. (1962). Volcanoes and Their Activity, John Wiley & Sons, Inc., New York, p. 12.

[1]See maps inside front and back covers.

Catalogue of the Active Volcanoes of the World (1951–1965). *Intern. Assoc. Volcanol.*, Rome, parts I–XVIII.

Hess, H. H. (1955). Serpentines, Orogeny, and Epeirogeny, *Geol. Soc. Am., Spec. Paper* **62,** pp. 391–408.

Fisher, D. J. et al. (eds.) (1963). Symposium on Layered Intrusions, *Am. Mineralogist,* Special Paper **1,** pp. 1–134.

Pecora, W. T. (1956). Carbonatites, a review, *Geol. Soc. Am., Bull.* **67,** pp. 1537–1556.

CHAPTER 5
PROPERTIES
OF
METAMORPHIC
ROCKS

TYPES OF METAMORPHISM AND METAMORPHIC ROCKS

Returning briefly to the overall picture of development of the igneous and metamorphic rocks described in Chap. 1, it is readily seen why the regional metamorphic rocks tend to be so widespread and why they are found in the "geosynclines." As the "geosynclinal" pile of sedimentary and volcanic rocks becomes thicker and thicker by continued accumulation and deformation, the rocks become warmer largely because of radioactive disintegration of the contained U, Th, and K and the heat from below. As a result, the low-temperature (or, as in the volcanic rocks, very high temperature), low-pressure "geosynclinal" rocks gradually depart farther and farther from equilibrium with the environment. The constituent minerals begin to react with one another and with the water and other materials in solution to form new minerals more nearly in equilibrium with the new environment. Metamorphic rocks are formed as the result of the chemical and physical processes involved in these solid-state transformations.

Metamorphism, then, may be defined as all those processes involved in the solid-state transformation of preexisting rocks into those of metamorphic character. Most metamorphism involves recrystallization of the constituent grains without the development of a melt. Metamorphism is thus bounded on the low-temperature end by diagenesis (though the limits are arbitrary) and on the high-temperature end by melting of constituents of suitable composition. The residuum from this partial melting or anatexis remains as a metamorphic rock.

These *regional* (or "dynamothermal") metamorphic rocks have the following distinguishing characteristics. They are characterized by a parallelism of platy minerals, especially the micas (but including flattened grains of quartz) and in some cases by an alignment of prismatic minerals such as hornblende. Such preferred orientation of the minerals presumably results from the intense deformation that must have occurred at the time of metamorphism. The metamorphic zones are widespread (regional) in extent—distances across the zones from one metamorphic grade to the next are measured in miles or even tens of miles. At the same time, these zones are not clearly related spatially to an igneous heat source. Mineralogical and stratigraphic considerations lead us to believe that the pressures at the time of metamorphism are moderate to great—pressures incurred from a few miles below the surface to the base of the earth's crust.

Metamorphic rocks of the other major class, the *contact* (or "thermal") metamorphic rocks, result from emplacement of hot igneous magma into cooler sedimentary or other igneous (e.g., volcanic) rocks. The heating of the intruded rocks results in reaction of the constituent minerals, much as in the case of regional metamorphism. The newly formed minerals again tend to approach equilibrium with the new environment.

As with the regional rocks, the contact metamorphic rocks have some distinctive characteristics. By contrast with regional rocks they tend to be massively textured (though exceptions do exist) and much less deformed. The contact metamorphic zones are limited in extent and are clearly related to the igneous rock (e.g., gabbro) that supplied the heat. Distances across the contact metamorphic zones are measured in feet or hundreds of feet. Pressures at the time of metamorphism must have been low to moderate—commonly within a few miles of the earth's surface.

Rocks of a third class, the *cataclastic* (or "dynamic") rocks, are sometimes included with the metamorphic rocks. These result from crushing or shearing of solid rocks without the addition of significant heat to aid in recrystallization.[1] Such rocks are characterized by a fragmental to streaky texture. Associated with faults or other zones of movement, they form relatively thin tabular zones a few inches to some hundreds of feet thick.

TEXTURES AND STRUCTURES OF METAMORPHIC ROCKS

As noted above, the orientation and arrangement of the minerals in metamorphic rocks differ in different rocks, though the regional metamorphic

[1] In some instances, however, a significant amount of internal frictional heat is generated.

rocks have textural similarities, as do the contact metamorphic rocks. At the same time, a single rock may contain several textures or textural elements. Some of the more important textures are as follows:

Foliation—the parallel arrangement or distribution of minerals. Includes layering as in a gneiss and schistosity as in a schist.

Schistosity—the parallel arrangement of micas or other tabular minerals to give a more or less planar fissility (as in schists and phyllites). With decrease in grain size, this grades to slaty cleavage (as in slates). Stretched or flattened grains, such as quartz in some strongly deformed quartzites, may also form a schistosity.

Gneissosity—the alternation of lighter and darker layers, such as micaceous or hornblende-rich layers with quartzofeldspathic layers (as in gneisses). The term is often used to include metamorphic layering regardless of its origin (as in calc-silicate gneisses).

Lineation—the parallel alignment of linear elements in the rock. Includes, for example, aligned prismatic grains (e.g., hornblende), aggregates of grains, axes of microfolds, and lines of intersection of two or more schistosities.

Preferred orientation—a general term to denote parallelism of tabular or elongate grains, as in schistosity or lineation—equidimensional grains according to their crystal lattice orientations (for example, C axes in much metamorphic quartz).

Hornfelsic or granoblastic—a nondirectional rock texture (as in hornfelses). Planar or prismatic grains, if present, are not oriented. "Hornfelsic" is sometimes reserved for finer-grained and "granoblastic" for coarser-grained textures.

Porphyroblast—large crystals of a mineral (e.g., garnet, andalusite) grown in a solid medium such as a metamorphic rock and surrounded by smaller grains of other minerals. Comparable in appearance to phenocrysts in an igneous rock.

Poikiloblast—a porphyroblast containing numerous inclusions of one or more groundmass minerals enveloped during growth (equivalent to "sieve texture").

Xenoblast—an anhedral porphyroblast (irregular outline).

Idioblast—a euhedral porphyroblast (bounded by its own crystal faces).

Helicitic texture—direction of an earlier foliation is reflected in curved lines of inclusions that are preserved within a porphyroblast. Often

S-shaped as might be formed by rolling of a porphyroblast during growth.

Corona or reaction rim—a new mineral forms as a rim around a mineral that is no longer in its field of stability, e.g., actinolite around augite.

Porphyroclast—coarse, strained, and broken crystals in a finer-grained matrix.

Augen—large "eyes" (porphyroclasts) of feldspar in a finer-grained gneissic matrix (as in augen gneiss).

Cataclastic texture—sheared and crushed rock fabric, not as extreme as mylonitic. Nature of the original rock is recognizable from undestroyed fragments.

Flaser texture—a cataclastic texture in which undestroyed "eyes" of the original rock swim in granulated streaks and laminae.

Mylonitic texture—extremely granulated and streaked-out grains— typically foliated and containing ovoid relict crystals.

Structures and textures in regional metamorphic rocks can conveniently be divided into those formed during regional metamorphism and those formed after regional metamorphism. Those formed *during* regional metamorphism include textures involving parallelism of mineral grains grown during metamorphism—for example, schistosity defined by mica, lineation defined by prismatic hornblende, or preferred orientation of unstrained quartz (as seen in thin section). Also included are most kinds of gneissic layering (excluding bedding) and folds temporarily related to these textures, such as folds with an axial plane schistosity.

Structures formed *after* regional metamorphism involve deformation of the above metamorphic features. The most obvious of these are folds and crenulations that bend the schistosity.

Preferred Orientation of Minerals Formed during Regional Metamorphism

The importance of stress in the development of metamorphic rocks is not well understood. No evidence exists at present to suggest that stress can control the occurrence of a particular mineral or minerals. This contrasts with the concept of "stress minerals" as propounded by Harker (1939, pp. 148–151) and others. As noted by Turner and Verhoogen (1960, pp. 472–477), the thermodynamic effect of nonhydrostatic stress is essentially equivalent to that of a hydrostatic pressure equal to the average of the unequal stresses.

Stress is clearly, however, the controlling factor in the production of schistosity, slaty cleavage, or preferred orientation of the crystal lattices

(a)

(b)

Fig. 5-1 *Small isoclinal recumbent folds in a pelitic schist containing: (a) few quartzofeldspathic gneissic layers; (b) quartzofeldspathic schist. Both cases have schistosity parallel to axial surface of similar style folds. Note almost perfect parallelism of schistosity and layers on limbs but not crests of folds. (a) Half mile east of North Kootenai Lake, northern Bitterroot Range, Montana; (b) Highway 97, 10 miles north of Oliver, British Columbia.*

of minerals such as quartz. Such features are almost always present in deformed regional metamorphic rocks such as schists and gneisses and are rarely present in undeformed contact metamorphic rocks except those affected by an earlier regional metamorphism. Since preferred orientation of mineral grains such as micas in a foliated rock could be formed only at the time of growth of the mineral grains, the schistosity is presumably formed during metamorphism.

Where this schistosity is parallel to the axial surfaces of folds in the bedding or other layering (see Figs. 5-1a and b and 5-2), development of the folds and schistosity must have accompanied metamorphism. Whether folds of this type always accompany development of schistosity is not known. They are present on various scales in many, if not most, regional metamorphic terranes, though in many instances they are found only on close scrutiny of the outcrop (e.g., Jones, 1959; Hyndman, 1968a). Folds of this type are characteristically of similar style (in contrast with the concentric or intermediate styles developed after metamorphism), as shown

Fig. 5-2 *Diagramatic representation of schistosity parallel to axial planes of folds in layering. Folds may be tighter with schistosity at a low angle to bedding on the limbs and at a high angle to bedding at the crest of the fold (left diagram); or the folds may be extremely tight (isoclinal) with schistosity nearly everywhere parallel to bedding as in the right diagram. Such folds may be on any scale from miles across, to the size of an outcrop, to microscopic.*

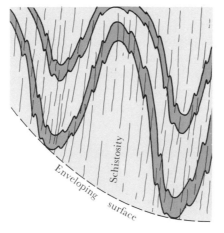

in Fig. 5-2. They form by slip along the schistosity, which parallels the axial surfaces of the resulting folds. Many postmetamorphic folds, by contrast, are of concentric style, forming by slip along the earlier metamorphic schistosity.

Open folds with axial-plane schistosity seem to be more common in low-grade metamorphic rocks; tight isoclinal folds with axial-plane schistosity are more common in high-grade rocks. Note that with increasing tightness of the fold, the original layering is transposed into near parallelism with the schistosity (Figs. 5-1 and 5-2). It commonly, therefore, is referred to as transposition layering. The mapped direction of a lithologically distinctive unit, however, will not be parallel to the "bedding" so measured in outcrop. Stratigraphic estimates of minimum depth for development of slaty cleavage (schistosity) range from about 7,000 to 50,000 ft (Wilson, 1961) or about 600 to 4,000 bars confining pressure.

Evidence for development of slaty cleavage during intense flow of unmetamorphosed shales under high water pressure has been described by Maxwell (1962), Moench (1966), and Braddock (1970). Recent experiments in deformation of wet clays suggest that preferred orientation produced by compaction of shales already buried a few thousand feet is weak, but compaction immediately after deposition produced strong orientation (Engelhardt and Gaida, 1963; Clark, 1970). Orientation in buried shales can form by constant volume distortion. These experiments neglect, however, the growth of micaceous minerals during recrystallization that accompanies deeper burial.

It must be emphasized that the textural and structural (fabric) relationships described above are somewhat idealized and are confined to rocks subjected to only a single deformation during metamorphism or recrystallization. Rocks subjected to more than one deformation show superimposed effects of the multiple deformations. The rocks are correspondingly complex and in some instances are exceedingly complex. Studies of the complexities of deformation are outside the scope of this book. Additional information is available in recent books on structural analysis, for example, Ramsay (1967), Turner and Weiss (1963), and Whitten (1966).

Grain Size of Metamorphic Minerals

The obvious increase in *grain size* of individual minerals with increasing metamorphism results from the smaller surface energy (DeVore, 1959), and therefore lower solubility, of larger grains; the excess free energy, and therefore greater solubility, of smaller grains tends to drive the recrystallization to the more stable coarser grain size.[1] Reasons why the grain size of the same grade of metamorphism is coarser in one region than another are not well understood. Contact metamorphic rocks tend to be finer grained than regional metamorphic rocks, even for the same temperature or metamorphic grade. The coarser grains of the deeper regional rocks are not likely to be promoted by higher pressures. On the other hand, the longer time for reaction and crystal growth would be expected to have an effect, for, as noted above, the lower surface area of coarser grains renders them more stable than finer grains. Still another possibility concerns the rates of reaction. The rapid rise in temperature accompanying igneous intrusion and contact metamorphism would permit more rapid nucleation of new phases as their field of stability is reached. This rapid nucleation would, other things equal, create a larger number of grains and consequently a finer grain size. That the amount of water present (related to pressure) during crystallization also influences increased grain size, presumably by inhibiting nucleation and increasing diffusion, has been shown experimentally by Carter, Christie, and Griggs (1964).

Another factor in control of grain size is exhibited by grain-size reversals in silt-shale graded bedding sequences. The very fine clays and other grains in the shale layers are highly reactive and grow rapidly to form coarser grains. The coarser quartz and feldspar grains in the silt layers are less reactive and grow more slowly to remain as fine grains. Original finer layers may thus become coarser than the original coarser layers, resulting in a reversal of grain size. The coarser (pelitic) layers may therefore appear at the top of a bed.

Porphyroblasts in metamorphic rocks typically develop as a few scattered large grains of a restricted number of minerals. The development of a few scattered large grains (porphyroblasts) instead of numerous small grains is considered to be related to a difficulty of nucleation of the mineral (as in the case of pegmatites). Those minerals that characteristically form porphyroblasts in rocks of pelitic composition are garnet, staurolite, kyanite, andalusite, and cordierite. Sillimanite commonly forms clusters of fine needles. Minerals that seldom form porphyroblasts (except under special circumstances) in pelitic rocks are chlorite, muscovite, biotite, quartz, and the feldspars. If these two groups of minerals are compared with the minerals in the present sedimentary rocks (for example, chlorite,

[1]The surface tension (σ) or surface energy of a grain is the free energy (G) per unit surface area (A), or $(\delta G/\delta A)_{PT} = \sigma_{\text{surface tension}}$. The change in free energy on changing from a fine-grained material to a coarse-grained material would be $\Delta G = \sigma A_{\text{coarse grains}} - \sigma A_{\text{fine grains}}$. Since the surface area of a volume of coarse grains would be much less than that for fine grains, ΔG would be negative so the energy would aid in recrystallization.

illite, montmorillonite, quartz, and albite), it is apparent that the minerals which do *not* form porphyroblasts are chemically and structurally much like the minerals from which they formed. Presumably they nucleate easily. By contrast, those minerals that *do* form porphyroblasts have structures based on isolated tetrahedra (SiO_4) or rings of tetrahedra (Si_6O_{18}). That these structures are radically different from their parent sheet (Si_4O_{10}) or framework (SiO_2) structures may result in a difficulty in nucleation, growth on the few nuclei that do form, and development of large porphyroblasts.

Cases of porphyroblast development in feldspars or in other minerals that normally do not form porphyroblasts can generally be attributed to an unusually high activity of water or other volatiles which could inhibit nucleation. Such conditions may arise, for example, in the vicinity of a crystallizing granitic pluton.

Some porphyroblasts are euhedral and others anhedral. Since a single mineral species may be euhedral in some rocks and may be anhedral or may contain numerous inclusions in others, Rast (1965) suggests that euhedral porphyroblasts result from slow growth, whereas anhedral or poikilitic porphyroblasts result from rapid growth or resolution at a later stage. Rapid growth would not give the growing porphyroblast time to destroy inclusions enveloped during its growth.

The shape of mineral grains in general is considered by many petrologists to be dependent on the interfacial energy (or "surface tension") of the minerals. First suggested by Becke (1913) and developed as a "crystal-

Fig. 5-3a *Boudinage, showing less competent schist layers drawn into troughs between thinner granite boudins.*

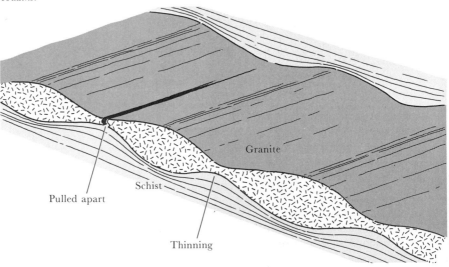

Fig. 5-3b *Boudins of biotite-rich quartzofeldspathic schist enclosed in schlieren-rich zone in granite near the northeastern border of the Idaho batholith. Hammer handle is 12 in. long. Three-fourths mile southwest of South Kootenai Lake, northern Bitterroot Range, Montana.*

loblastic series" of minerals, the concept has been amplified by Buerger (1947), DeVore (1956, 1959), and Kretz (1966a). Absolute surface free energies have been measured for but few minerals (Brace and Walsh, 1962), but those with a high surface energy, such as garnet and staurolite, generally form euhedral grains and are high on the crystalloblastic series. These are also dense minerals containing isolated silica tetrahedra (Barth, Correns, and Eskola, 1939). Others with a low surface energy, such as feldspars and quartz, form anhedral grains and are low on the crystalloblastic series. These are the less-dense minerals consisting of frameworks of tetrahedra. Anhedral, relatively equidimensional minerals, such as quartz and feldspars, have grain boundary intersections near 120°, presumably controlled to a major extent by interfacial tension (Kretz, 1966a).

Layering of Metamorphic Rocks

Lithologic laminations 1 mm to 1 cm or more thick and parallel to the schistosity are a common feature of low-, medium-, and especially high-grade regional metamorphic rocks. This gneissic layering could be original bedding or it could result from a process called metamorphic differentiation. Accompanying deformation during metamorphism, some constituents (e.g., quartz and feldspars) may migrate into preferred zones, leaving behind micas, amphiboles, and other constituents. This process will be discussed in more detail below. For present purposes, although some gneissic layering is probably original sedimentary bedding, in many instances it parallels schistosity that lies at an angle to the original bedding. Some low- to medium-grade examples are illustrated by Talbot and Hobbs (1968).

Deformed metamorphic rocks often show pronounced thinning of layers of rock of a particular composition such as limestone, amphibolite, or pegmatite in slate or schist. These more resistant (competent) layers tend to stretch and thin along subparallel lines of weakness, the less-resistant (incompetent) layers tending to flow into the thinned zones of the more-competent layers. This thinning of more resistant layers is called boudinage, each thicker part being called a boudin (see Figs. 5-3a and b). Boudinage is generally considered to result from stretching, with resultant thinning along lines approximately parallel to the axes of folds that may be related. In other instances, however, the boudins lie perpendicular or oblique to the axes of folds. The geometry, occurrence, and possible genesis of boudinage has been described in some detail by many geologists, including Jones (1959, pp. 104–116) and Whitten (1966, pp. 293–312).

Postmetamorphic Features

Folds formed after regional metamorphism recognizably differ from those formed during regional metamorphism. Instead of the schistosity being planar and oriented parallel to the axial surface of the fold (see Fig. 5-2), the schistosity is deformed by and wraps around most postmetamorphic

Fig. 5-4 *Kink-style postmetamorphic folds in phyllite. Strong schistosity deformed into folds without development of an axial surface schistosity. North side of Box Lake, east of Nakusp, British Columbia.*

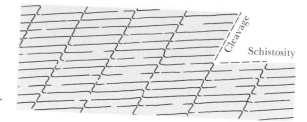

Fig. 5-5a *Deformation of a regional metamorphic schistosity by a later crenulation cleavage.*

folds (for example, Fig. 5-4). By contrast with the similar-style folds generally formed during metamorphism, those formed after tend to be concentric, intermediate between concentric and similar, or kink in style. Some such folds deform a prominent schistosity and exhibit an incipient axial-plane schistosity defined by micaceous minerals. Such folds may be formed during a later, weaker metamorphism, but alternatively the same effect may be produced by folding during the waning stages of the original regional metamorphism.

On a smaller scale, the metamorphic micas, in the form of a slaty

Fig. 5-5b *Photomicrograph of crenulation cleavage in pelitic schist. Schistosity defined by mica formed during regional metamorphism, deformed into crenulations. Two miles west of Wilson Lake, Nakusp area, British Columbia.*

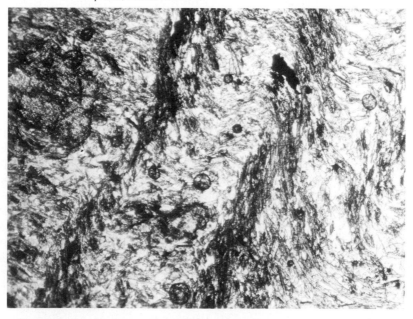

1 mm

cleavage or schistosity, may be deformed into microfolds along a crenulation (or strain-slip) cleavage, as shown in Figs. 5-5a and b. In hand specimen, the crenulation cleavage appears, not as a pervasive surface like schistosity, but as a set of discrete closely spaced surfaces. In many instances, these result in a pronounced planar aspect in the rock. Although this may obscure the schistosity, the orientation of the latter can usually be determined from close examination.

During postcrystallization deformation, minerals other than micas also may be deformed. Of the most common minerals in metamorphic rocks, quartz appears to be most susceptible to deformation. Mild stress results in slight deformation and development of undulose extinction and deformation lamellae seen in the quartz in thin section (Naha, 1959). Further deformation results in segmentation and even granulation of the margins of quartz grains. In hand specimen, this appears as blurring of the boundaries of quartz grains that are otherwise sharply defined.

Methods have also been developed for identification of principal stress (σ_1) using the orientation of deformation lamellae in quartz (Hansen and Borg, 1962; Scott, Hansen, and Twiss, 1965; Carter and Friedman, 1965). These methods have recently been confirmed experimentally (Heard and Carter, 1968). For example, rotation of C axes of quartz by deformation lamellae is toward σ_1, or arrows drawn (in a plane) from the C axis to the pole to deformation lamellae point toward σ_3, or σ_1 is the acute bisectrix of statistically determined orientations of the lamellae. Somewhat similar interpretations may be made for calcite and dolomite (Turner and Weiss, 1963, pp. 408–425). Such measurements are made with the aid of a universal-stage microscope.

Crystallization after the cessation of deformation accompanying regional metamorphism may occur either through the relaxation of deformation in the waning stages of regional metamorphism or through reheating by later igneous intrusions. Quartz in schists and other strongly foliated rocks generally has a strong preferred orientation. Where such rocks have been partly recrystallized, the quartz grains are randomly oriented, though the mica may retain its preferred orientation (e.g., Turner and Weiss, 1963, p. 432). Recrystallization of the bent micas in crenulated phyllites or schists, however, is more easily accomplished. Micas bent around crenulation cleavage in schists adjacent to postmetamorphic granitic plutons are commonly recrystallized into polygonal arcs (e.g., Rast, 1965; Hyndman, 1968a). The crenulations remain, but the individual mica grains have recrystallized to more stable straight grains.

Porphyroblasts containing traces of graphite, quartz, or other grains

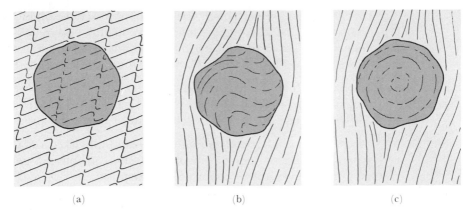

(a)   (b)   (c)

Fig. 5-6 *Inclusions in porphyroblasts. (a) Crenulations in schistosity preserved in porphyroblast; (b) Z-shaped surface preserved in prophyroblast; (c) spiral surface preserved in porphyroblast. (See text for explanation.)*

in the form of crenulations, which are also seen in the micas outside the porphyroblast, can be relegated reliably to a second stage of crystallization (see Fig. 5-6a). The sequence of events leading to this texture presumably involves regional metamorphism to form the schistosity, deformation to form crenulations in the schistosity, and a second metamorphism (e.g., contact) to grow the porphyroblast over the crenulations.

This texture should not be confused with Z- or S-shaped or spiral inclusions in some porphyroblasts such as garnet (see Fig. 5-6b and c). These structures form by growth of the porphyroblast during rolling and deformation accompanying regional metamorphism (e.g., Peacy, 1961; Spry, 1963b; Rosenfeld, 1970). Careful examination of these textures preserved within and around porphyroblasts plays an important role in working out the complicated succession of events (metamorphisms and deformations) developed during an orogenic cycle. Excellent examples of considerations and results are given by Zwart (1960), Rickard (1964), and Rosenfeld (1970).

COMMON KINDS OF METAMORPHIC ROCKS

Metamorphic rock names are not normally placed in a rigorous classification as are igneous rock names, largely because most of the basic names are textural rather than mineralogical in nature. The dozen or so textural names are usually made more specific by mineral-content modifiers such as "muscovite schist" or "andalusite-biotite-quartz hornfels." The modifying minerals are placed in increasing order of abundance in the rock. The following are the most commonly used basic textural names:

Slate—a very fine-grained rock with a well-developed platy rock cleavage. This cleavage results from incipient parallel growth of micaceous minerals, due to metamorphism (generally regional) of fine-grained clastic sediments such as mud, shale, silt, or tuff.

Phyllite—a fine-grained schistose rock resulting from more advanced regional metamorphism than slate. The schistosity surfaces have a lustrous sheen due to the development of new mica and chlorite.

Spotted slate and spotted phyllite—slate and phyllite containing dark spots, the beginnings of porphyroblasts (e.g., biotite) generally resulting from incipient contact metamorphism.

Schist—a strongly schistose, commonly lineated rock having grains coarse enough for hand-specimen identification of most minerals. Micas are commonly abundant. Generally more advanced regional metamorphism than for phyllite.

Gneiss—a medium- to coarse-grained, irregularly "banded" rock (e.g., micaceous layers alternating with quartzofeldspathic layers). It has a fairly poor schistosity because of the preponderance of quartz and feldspar. Equivalent or higher regional metamorphic grade than a schist.

Granulite—an even-grained quartz, feldspar, and to a lesser extent pyroxene and garnet rock poor in micas (not schistose) but typically having a rough foliation due to lighter and darker layers or to the parallelism of flat lenses of quartz or feldspar. Very high grade regional metamorphism.

Amphibolite—a medium- to coarse-grained hornblende-plagioclase rock of metamorphic origin (not the same as hornblendite, which is of igneous origin). Alignment of hornblende prisms to form a lineation is common.

Hornfels—a fine-grained nonschistose rock lacking preferred orientation of the grains. Porphyroblasts are common, giving this contact metamorphic rock a spotted appearance.

Granofels—a "medium- to coarse-grained, granoblastic" rock "without, or with only indistinct, foliation or lineation." Coarser than a hornfels and largely lacking the texture of a gneiss (Goldsmith, 1959).

Skarn—a contact metamorphic and commonly metasomatic (material introduced) rock, commonly composed of red and green calcium-rich silicates such as grossularite, epidote, and diopside. Also called tactite.

Quartzite—a metamorphic rock consisting of recrystallized and interlocking grains of quartz.

Marble—a metamorphic rock consisting of recrystallized and inter-locking grains of calcite or dolomite.

Serpentinite—a serpentine rock formed by metasomatism (especially hydration) of a peridotite and generally containing a little talc or chlorite. Most commonly dark green in color, these rocks weather orange-brown. Discussed above under "Alpine peridotite–serpentinite bodies."

Mylonite—a fine-grained to glassy, flinty looking, banded or streaky rock resulting from extreme granulation of coarser rocks without much chemical reconstitution. Eyes or lenses of the parent rock may persist in a granulated matrix.

Cataclasite—a coarser and less streaky rock than a mylonite. Similar to a mylonite, it is formed by shattering and less extreme granulation.

Phyllonite—a rock macroscopically resembling a phyllite but, like a mylonite, formed by granulation of a coarser rock. A significant degree of chemical reconstitution gives rise to silky films of mica smeared out on the schistosity surfaces.

Eclogite—a medium-grained, commonly green-colored rock, consisting of pale- to medium-green omphacite (jadeite-diopside) and lesser red garnet. Compositionally equivalent to basalt and considered to be an extremely high pressure form resulting from regional metamorphism.

METAMORPHIC ZONES

About 80 years ago, geologists were just beginning to make sense out of the great series of complex and confusing rocks that they called the "crystalline schists" or the "basement complex." Such rocks had been, and to a lesser extent are even now, avoided because of their complexity on a small scale. Early geologists working on these rocks included H. Rosenbusch who in Germany in 1877 described three metamorphic zones around a body of granite, A. Michel-Levy who in France in 1888 distinguished three main stages in the formation of the crystalline schists, and J. J. Sederholm who in Finland in 1891 suggested a connection between depth below the earth's surface and degree of metamorphism. About the same time George Barrow (1893) mapped, in the Scottish Highlands, a widespread series of *zones* of progressive metamorphism, based on the first appearance of a group of distinctive "index minerals" as he approached the granitic bodies in the highest metamorphic grades.

Beginning in unmetamorphosed "geosynclinal" sediments, Barrow mapped the following zones in the pelitic (shaley) members:

First appearance of "digested clastic mica"
(later called the *chlorite* zone)

First appearance of brown *biotite*

First appearance of *garnet* (almandine)

First appearance of *staurolite*

First appearance of *kyanite*

First appearance of *sillimanite* (in coarse gneisses, associated with bodies of granite)

Some of these minerals continue through the higher-grade zones. Biotite, for example, appears in all the higher-grade zones including the sillimanite zone. Staurolite, however, disappears on reaching the kyanite zone.

A line on a map that corresponds to the position of the first appearance of biotite is called the biotite *isograd* (or line of equal grade). Similar isograds can be drawn for other minerals. Isograds are actually surfaces whose traces on (or intersections with) the surface of the earth are drawn on the map.

This work by Barrow formed the nucleus on which our present concept of metamorphic grade is based. Such zones of progressive metamorphism have been interpreted in various ways. Barrow thought they resulted from the heat from the small granite intrusions found in the highest-grade zones. C. E. Tilley (1925), studying the same rocks in an adjacent area, suggested that the temperature of each zone was largely determined by the depth of burial (a function of what we now call the geothermal gradient), modified at deeper levels by heat from the intruded granites. A more general interpretation, not greatly different from what we now believe, stated by Tilley (1924), is that isograds mark rocks originating "under closely similar physical conditions of *temperature and pressure*" (pp. 168–169).

The implication at this stage in our understanding of metamorphism was that the first appearance of a zone mineral indicated a definite metamorphic grade, as long as the rocks were of an appropriate composition for growth of that mineral. It was recognized by Tilley and others, however, that in rocks of a somewhat different composition, (e.g., more Mn rich), a garnet of a different composition (i.e., spessartite) could appear at a lower grade. This was among the earliest indications that strict interpretation of mineral zones in terms of metamorphic grade was too simple a concept. Not known at the time but more clearly appreciated now (e.g., Atherton, 1964) is the fact that virtually all minerals including almandine to a greater or lesser extent vary in composition and appear in rocks of a somewhat different composition. Thus index minerals appear at variable grades, and the concept of an "isograd" is only an idealized approximation.

The appearance of a particular mineral depends on three sets of variables:

1. The conditions under which the rock was formed, i.e., temperature, water pressure, load pressure, etc.
2. The rates of nucleation and reaction involved in formation of the mineral
3. The composition of the rock

The first two factors are discussed below. The composition of the rock is probably the most important factor in production of a given mineral under different metamorphic grades. An isograd mapped within a single, compositionally homogeneous layer probably closely approaches the idealized concept of a "line of equal grade." Correlation of an isograd mapped in one area with that mapped in a widely separated area, even in rocks of approximately similar composition, is hazardous as the rocks may have formed under quite different conditions.

The first real attempt to correlate metamorphic mineral zones between widely separated areas was by Eskola (1914, 1915). Goldschmidt (1911) had a few years earlier described the contact metamorphic mineral assemblages formed at different grades in rocks of various compositions in the Oslo region of Norway. Eskola, working in the Orijärvi region of Finland, similarly found the contact metamorphic assemblages related to the grade and to the rock composition. Although some mineral assemblages were the same between the two areas, others in rocks of equivalent

Table 5-1 *Mineral assemblages for different grades of metamorphism of three common rocks*

Unmetamorphosed Rock	Facies A	Facies B	Facies C	Facies D	Facies E
Shale	*chlorite*	*biotite*	*garnet*	*staurolite*	*sillimanite*
		chlorite	biotite	garnet	garnet
	muscovite	muscovite	muscovite	biotite	biotite
	albite	albite	albite	*plagioclase*	plagioclase
	quartz	quartz	quartz	quartz	quartz
	Slate or phyllite . . . . phyllite or schist . . . . schist or gneiss . . .				
Andesitic volcanic tuff	*actinolite*	actinolite	*hornblende*	hornblende	hornblende
	albite	albite	albite	*plagioclase*	plagioclase
	epidote	epidote	epidote	quartz	quartz
	chlorite	chlorite	quartz		
	quartz	quartz			
	Chlorite or actinolite schist . . . . amphibolite				
Sandy limestone or siliceous dolomite	*dolomite*	*tremolite*	tremolite	*diopside*	diopside
	calcite	calcite	calcite	calcite	calcite
	quartz	quartz	quartz	quartz	quartz
	Marble . . . . . tremolite marble . . . . diopside marble				

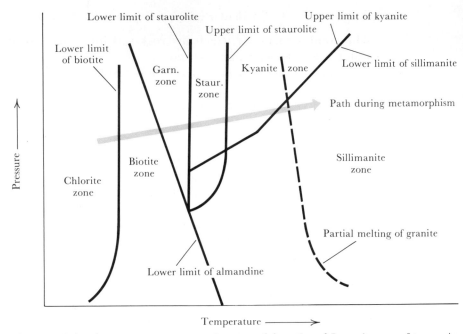

Fig. 5-7 *Inferred temperature-pressure environment of formation of Barrow's zones of progressive metamorphism. Note that most of the zones (except kyanite) are defined on the basis of first appearance of a distinctive mineral, not from temperature and pressure. These conditions are inferences mainly from experimental work.*

composition were quite different. Eskola concluded that the rocks in the two areas were metamorphosed under different conditions such as temperature and pressure. This prompted Eskola (1915, 1920) to propose classification of metamorphic rocks descriptively as an association of metamorphic rocks, each consisting of a mineral assemblage consistently related to the composition of the rock. Each facies could then be related to physical conditions under which it was thought to have formed. As repeatedly emphasized by Turner (e.g., Fyfe, Turner, and Verhoogen, 1958, pp. 8–10), the metamorphic facies are purely descriptive, temperatures and pressures being interpretations subject to change. A metamorphic facies may be defined (Fyfe and Turner, 1966) as

> A set of metamorphic mineral assemblages, repeatedly associated in space and time, such that there is a constant and therefore predictable relation between mineral composition and chemical composition.

The relationship between metamorphic zones and metamorphic facies is simple in principle and can be illustrated with reference to Table 5-1. The mineral zones mapped by George Barrow in shaley rocks included chlorite, biotite, garnet, staurolite, and sillimanite (Fig. 5-7). In the same pelitic rocks, he equally well could have mapped an *albite* zone and a *plagioclase* zone (see Table 5-1), based on the first appearance of those minerals. Alternatively, had he studied metamorphic zones derived from andesitic volcanic tuff, he could have mapped an *actinolite* and a *hornblende* zone or an *albite* and a *plagioclase* zone. Similarly, had he studied metamorphic zones derived from siliceous dolomite, he could have mapped *dolomite, tremolite,* and *diopside* zones. Thus the actual mineral zones mapped depend on the composition of the rock and the individual choice of which initially appearing mineral to map.

If the different rocks illustrated in Table 5-1 (shale, andesitic volcanic tuff, siliceous dolomite) were interbedded, it would be a simple matter to correlate the mineral assemblage derived from the shale with that derived from the volcanic tuff and with that derived from the siliceous dolomite. Such a correlation of mineral assemblages in rocks of different composition for a single grade of metamorphism is the procedure involved in defining a metamorphic facies for that grade.

A metamorphic facies (e.g., Facies A in Table 5-1) encompasses all those mineral assemblages in rocks of all compositions from a single metamorphic grade. The assemblage of minerals, of course, depends on the composition of the individual rock, but in rocks of a single composition the assemblage of minerals is always the same (and predictable) at the same grade. A single metamorphic facies should also conform to the following characteristics (Fyfe, Turner, and Verhoogen, 1958, p. 18):

> The assemblages should occur together and should have formed at the same time (in a mapable metamorphic aureole) and should recur in other areas.

> The mineral assemblage is affected only by the chemical composition of the rock as it *now* exists, regardless of any metasomatism.

> There is no evidence of disequilibrium in the rock: no textural or other evidence of replacement of one mineral by another; no more than about 2 to 6 essential minerals in each assemblage and about 12 essential minerals in each facies.

Again it must be emphasized that metamorphic facies are purely descriptive and are not defined on the basis of assumed physical conditions, regardless of how well those conditions may be determined experimentally.

Reasonable assumptions for the interpretation of metamorphic facies and made by Eskola are:

> Rocks in an individual facies "were formed in the same range of physical conditions" (T, P_{H_2O}, etc.).

Each assemblage represents a group of coexisting minerals that were stable when they were formed.

Granted that these assumptions are probably correct, we may attempt to interpret the physical conditions of formation of each metamorphic facies.

Although metamorphic facies are purely descriptive, they are not suitable in developing a comprehensive classification of individual rock types (Turner, 1968, p. 52). A metamorphic facies shows what minerals should be present in a rock of a given composition but not in what amount. Nor does it indicate texture. As noted above, most metamorphic rock names (e.g., slate, schist, amphibolite) are based on the texture and to a lesser extent on major minerals in the rock. A correlation of such rock types with the metamorphic facies for a given composition of rock can be made only in a general way. Such a correlation is attempted in Table 5-1; but, as noted above, the grain size and texture to a major extent depend (in addition to temperature and pressure) on time, stress, and other factors that apparently have little or no effect on the mineralogy. Examples of meta-morphic-zone and metamorphic-facies maps are shown in Figs. 8-3, 8-6, 8-9, 8-11, and 8-13.

Complications in the simple interpretation of an isograd as a surface of equal conditions of temperature and pressure have recently come to light. Greenwood (1962, 1967) has predicted experimentally and Carmichael (1970) has confirmed in the field that isograds based on reactions involving different proportions of H_2O and CO_2 may intersect. This may be most clearly developed, as shown by Carmichael, where calcareous rocks are intercalated with pelitic and other rocks, especially adjacent to a water source, such as a granitic body emplaced during metamorphism.

CHAPTER 6
CHEMICAL
BEHAVIOR OF
METAMORPHIC
ROCKS

EQUILIBRIUM

As suggested nearly 60 years ago by V. M. Goldschmidt in his classic study of the contact metamorphic rocks near Oslo, the simplicity of the common mineral assemblages in metamorphic rocks results from their having reached equilibrium (or nearly so) under the temperatures and pressures imposed at the time of formation. This tendency of metamorphic reactions —toward equilibrium or minimum free energy and the formation of stable mineral assemblages—is the natural trend in all chemical reactions.[1] Most petrologists agree that general equilibrium is closely attained in most metamorphic rocks, especially under medium to high grades and in regional metamorphism where the time available for reaction is great and catalysts such as deformation are prominent.[2]

[1] $\Delta G = 0$ at equilibrium, i.e., on the phase boundary between reacting phases (free energy of reactants = free energy of products); $\Delta G = \Delta H - T \Delta S = 0$, where ΔG = free energy of reaction, ΔH = heat of reaction, and ΔS = change in entropy.

[2] Time and deformation are minimal in experimental analogies to metamorphic processes, so equilibrium is difficult to obtain in most artificial systems. Other difficulties and considerations for interpretation are reviewed briefly on pages 310–314. These difficulties are much more pronounced with simulation of solid-state reactions (metamorphic) than solid-liquid reactions (igneous), especially with regard to nucleation, equilibrium, and metastability.

How closely metamorphic mineral assemblages approach this equilibrium remains to be seen but depends partly on the scale of observation. Although the trend is clear, the closeness of this approach to equilibrium will govern the reliability of our attempts to assign pressure-temperature data to the rocks. Several pieces of evidence point to equilibrium on the scale that mineral assemblages are studied in thin section:

> Such coexisting minerals show no textural indication of disequilibrium (e.g., rims on grains, relict cores, veinlets of one mineral through another).

> Identical minerals (same composition and structure) form at the same grade from rocks of the same composition.

> Zoning within individual grains (such as is common in plagioclase in igneous rocks) is rare in metamorphic rocks.

> Many (but by no means all) experimental reactions appear to reach equilibrium in a few days whereas hundreds or thousands of years are available in metamorphism.

> The mineral assemblages contain no incompatible phases as determined, for example, by experimental synthesis (e.g., quartz and forsterite, quartz and corundum, graphite and hematite; Zen, 1963), or by crossing tie lines on triangular phase diagrams such as ACF or AKF (see below).

> The mineral assemblages, in accordance with the phase rule discussed below, consist of a small number of minerals.

Most rocks conform to these criteria except for minor, most commonly late-stage, reactions that are generally easily recognized. Such criteria for equilibrium are essentially negative in character as pointed out by Zen (1963). They are necessary characteristics of a system in equilibrium but are not diagnostic. Nonadherence to these criteria implies, of course, nonequilibrium. Additional evidence for nonattainment of equilibrium on certain scales is as follows:

> Thin (a few millimeters) compositionally different layers, however formed, commonly persist to very high grades. In many instances, one or more minerals in a layer are incompatible with minerals in adjacent layers.

> Oxidation states of iron in adjacent thin layers, as indicated by different opaque oxides, are similarly incompatible (Chinner, 1960b).

Many minerals in metamorphic rocks are slightly zoned at the rims (e.g., plagioclase as seen in thin section and garnet as shown by the electron microprobe, Atherton and Edmunds, 1966; Hollister, 1966, 1969a), though some of these can be attributed to minor lower temperature (retrogressive) adjustment.

The scale of near attainment of equilibrium, then, is commonly on the order of individual layers as seen in thin section.

THE PHASE RULE

A chemical system such as a rock consists of a number of components (e.g., chemical oxides) that are combined in the form of discrete phases (i.e., minerals). Under equilibrium conditions these components (c) will combine to the greatest possible extent to form the fewest possible phases (p). The rule governing this minimal number of phases is called the phase rule: $f + p = c + 2$. The phase rule as commonly used is easy to apply to simple systems, such as those involving a single component; but it is very difficult to use for more complex systems, such as are represented by most common rocks. One difficulty generally arises in specifying the number of components. For example, a single element such as Fe^{++} may substitute for Mg in one mineral and may act as a separate component in another, even in the same rock. One way to circumvent the problem is to rephrase the phase rule (since $r = p - c$ or $c - p = -r$, where r = the number of independent equations relating the phases):

$f = 2 + c - p$ f = number of degrees of freedom of the system

$f = 2 - r$ 2 refers to P and T but may be a different number for greater or lesser pressure variables

Although it may still be difficult to specify the number of reactions in some cases, it may be instructive to note the number of degrees of freedom at different points in a chemical system in equilibrium whether it be in an artificial experimental system or a natural rock (see Fig. 6-1). If there are 2 degrees of freedom, the P and T may be varied independently without changing the assemblage. This is by far the most common case in metamorphic rocks in which the phases are in equilibrium. If there is only 1 degree of freedom, either P or T may be varied but not independently (e.g., if P is changed, T is specified) or the system will not remain on the reaction boundary. This case ideally occurs only as a thin surface in space in the earth. If there are 0 degrees of freedom, neither P nor T may vary without moving off the triple point. For almost all practical purposes, this case is nonexistent in nature.

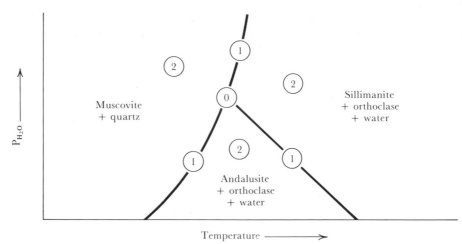

Fig. 6-1 *Number of degrees of freedom in a chemical system in equilibrium. Note that the degrees of freedom are* ⓪ *at a triple point,* ① *on a reaction boundary, and* ② *within a nonreacting equilibrium assemblage.*

V. M. Goldschmidt (1911) noted that any common metamorphic assemblage should exist under a range of pressure-temperature conditions (that is, ② in Fig. 6-1) rather than on a reaction boundary (that is, ①). Thus the equilibrium would be divariant, and the phase rule for such common metamorphic assemblages would reduce to p = c. Thus most commonly the number of minerals in the rock should be equal to the number of chemical components. If the number of minerals is greater, the minerals may not represent an equilibrium assemblage.[1]

GRAPHICAL REPRESENTATION OF PHASES

A graphic representation of metamorphic mineral assemblages in terms of the common chemical components aids visualization of the relationship between chemically equivalent assemblages at different grades, chemically similar assemblages at the same grade, the chemical influence of one phase on another, and the distribution of components between coexisting phases. It is also a kind of graphical application of the phase rule to a group of assemblages.

[1]An alternative phrasing of the phase rule in terms of "open system" or mobile components and "closed system" or inert components has been proposed by Korzhinskii (1950, 1959), Thompson (1959), and Zen (1963). This point of view has been criticized by Weill and Fyfe (1964) and Mueller (1967) as being superfluous. Readers interested in separation of "open" and "closed" system components as in problems involving metasomatism should refer to the pros and cons in these papers.

The common components in metamorphic rocks are SiO_2, Al_2O_3, FeO, MnO, MgO, CaO, Na_2O, and K_2O. This listing neglects minor components such as TiO_2, P_2O_5, and ZrO_2, which occur to a major extent in accessory minerals such as sphene, ilmenite, apatite, and zircon. The common components may be grouped somewhat according to common substitutions in minerals: SiO_2, $(Al, Fe^{3+})_2O_3$, $(Fe, Mn)O$, $MgO*$, CaO, Na_2O, K_2O, H_2O. The remaining eight components are still too many to represent on a simple three- or four-component diagram. Restricting consideration to rocks in which SiO_2 and H_2O are in excess so that silica- or water-deficient minerals cannot form (and most common rocks contain free quartz and were formed under water-saturated conditions) reduces this to six. In order to plot the four most important components $(Al, Fe^{3+})_2O_3$, $(Fe, Mn)O$, MgO, CaO on a tetrahedron, the amounts of these components must be corrected for their presence in alkali-bearing minerals that do not appear on the diagram (K-feldspar, albite), since the alkalis $(K_2O,$ $Na_2O)$ will have to be neglected as being less important. The basic compositions of these minerals in terms of molecular components are:

K-feldspar: $KAlSi_3O_8 = K_2O \cdot Al_2O_3 \cdot 3SiO_2$ (that is, 1 Al_2O_3 to 1 K_2O)

Albite: $NaAlSi_3O_8 = Na_2O \cdot Al_2O_3 \cdot 3SiO_2$ (that is, 1 Al_2O_3 to 1 Na_2O)

Then the four components (ACFM) may be approximated as:

$A = Al_2O_3 + Fe_2O_3 - Na_2O - K_2O$
$C = CaO$
$F = (Fe, Mn)O$
$M = MgO$

ACF and AKF phase diagrams were first used by Eskola (1920, 1939) and AFM projections by Thompson (1957). Both are described in some detail by Winkler (1967) and Turner (1968). The positions of common metamorphic minerals (in terms of molecular proportions) are shown in Fig. 6-2. Similarly F and M could be grouped as a single component (Fe, Mn, Mg)O for some purposes, Na_2O again neglected, and an ACFmK tetrahedron plotted. The components could then be approximated as:

$A = Al_2O_3 + Fe_2O_3 - Na_2O - K_2O$
$C = CaO$
$Fm = (Fe, Mn, Mg)O$
$K = K_2O$ K-feldspar being plotted at the K corner of the tetrahedron

---

*MgO can in many cases be taken as substituting for FeO [(that is: (Fe, Mg)O]. They are kept separate here for the purpose of illustration of the extent of substitution in minerals on an ACFM diagram discussed below.

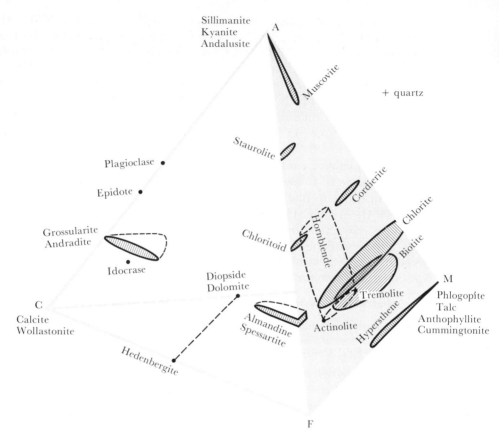

Fig. 6-2   ACFM *phase diagram showing the plotted positions of common metamorphic minerals. Note that only the hornblendes fall within the volume of the tetrahedron. Idocrase, diopside, and dolomite fall on the back* (ACM) *face of the tetrahedron.*

The positions of common metamorphic minerals on an ACFmK diagram are shown in Fig. 6-3.

Use of Phase Diagrams

Phase diagrams such as ACFM, ACFmK, ACFm, AKFm, AFM are intended to show, for a given assemblage, those phases that are in equilibrium in a particular metamorphic facies or zone. For many purposes, the ACFM diagram appears to be most useful as it separates most important metamorphic minerals. For high-potassium rocks in which minerals such as K-feldspar, muscovite, and biotite figure prominently, the ACFmK or AKFm diagrams may be more useful. Still other diagrams may be useful in particular instances; for example, ACFmNa for high-sodium assemblages.

For a given metamorphic assemblage, tie lines are drawn between those mineral phases that appear (e.g., texturally) to be in equilibrium. Only coexisting minerals that are in direct contact and lack any evidence of mutual reaction are so connected. On a four-component tetrahedron, the coexisting phases generally number three or four in accordance with the phase rule as described above. For example, one common high-grade assemblage includes quartz, plagioclase, almandine, biotite, and sillimanite. These are plotted on three different phase diagrams for comparative purposes (see Fig. 6-4). Note that on the simple ACFm diagram, lumping of FeO and MgO as a single component may give crossing tie lines and thus an apparent disequilibrium situation. The position of the whole-rock composition on the diagram may be plotted either by recalculation of the plotted components of a whole-rock chemical analysis to 100% or approx-

Fig. 6-3 ACFmK *phase diagram showing the plotted positions of common metamorphic minerals. None of the minerals fall within or on the back* (ACK) *face of the tetrahedron.*

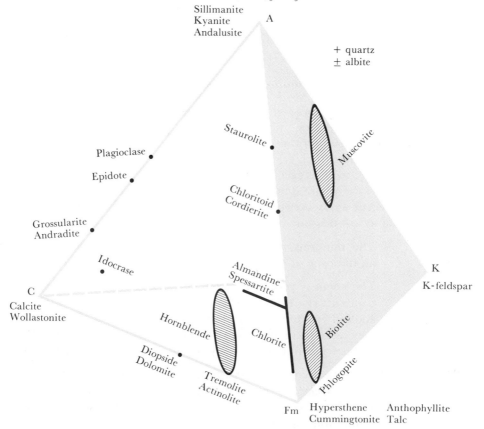

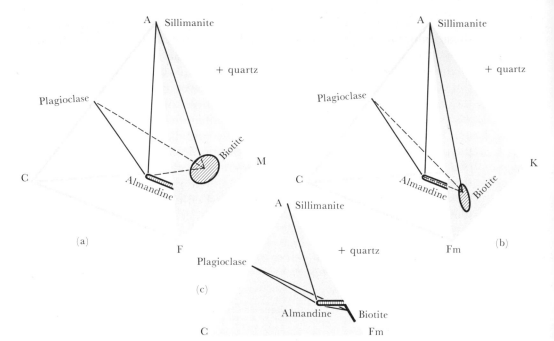

Fig. 6-4 *A single equilibrium assemblage (quartz, plagioclase, almandine, biotite, sillimanite) plotted on three different phase diagrams for comparison.*

imately (and more commonly) by plotting the percentages of the coexisting major phases. The whole-rock composition should of course plot within the subtetrahedron represented by the coexisting phases. Given the approximate rock composition, one can thereby predict the mineral assemblage to be expected under particular metamorphic conditions.

If the phase diagram is representative of the rock and all its components, tie lines between minerals normally will not cross. Several explanations are possible if the tie lines do appear to cross (e.g., as in Fig. 6-4c):

The minerals, although in contact, are not actually in equilibrium (at the time of stabilization of the assemblage)

The corners represented on the diagram are not completely independent; the lumped elements do not substitute completely for one another in the mineral phases present (e.g., although FeO and MgO may be in some instances combined as an Fm corner, FeO-rich minerals such as almandine may accommodate only limited MgO)

Important components for the phases present are not represented on the diagram (for example, Na_2O on an AKFm diagram for an assemblage containing paragonite or MnO on an ACFm diagram containing spessartite).

THE ENVIRONMENT OF METAMORPHISM

Regional metamorphic rocks are largely confined to areas of thick sedimentation at the continental margins. They are most prominent in the thick sedimentary and volcanic rocks along the continental side of the "eugeosyncline." Somewhat different regional metamorphic rocks, interpreted as originating under higher pressures and lower temperatures, occur along the oceanic side of the "eugeosyncline."

As the sediments and volcanic rocks accumulate by prolonged deposition in a "geosyncline," they are gradually subjected to increasing pressure from the weight of overlying sediments and to increasing temperature from heat flow from the earth's mantle and heat generated by radioactivity within themselves. The pressures presumably depend on the rates of sedimentation and volcanism and the total time span available. These rates are probably moderate to high along the continental side of the "eugeosyncline" and adjacent "miogeosyncline," and relatively low along the oceanic side of the "eugeosyncline" (compare Fig. 1-1). After the onset of deformation, a significant pressure contribution may arise through secondary thickening of the depositional pile. Heat flow from the mantle presumably depends on direct conduction through the low-conductivity rocks. Calculations based on measured radioactivities and heat flows in young "geosynclinal" areas suggest that about 40% of the total heat may be derived from this source, about 60% coming from radioactive disintegration (Hyndman et al., 1968). Roy et al. (1968) would reverse the percentages. These two sources of mantle heat probably supply the bulk of the heat to the mass of the sedimentary pile at depth. Heat developed by radioactivity within the depositional pile presumably depends primarily on the content of radioactive elements—primarily U, Th, and K^{40}. All these are higher in felsic rocks, such as are deposited in the "miogeosyncline" and adjacent "eugeosyncline," and are low in mafic volcanic materials, such as are abundant in the oceanic side of the "eugeosyncline." The lower radioactive content in the oceanic side of the "eugeosyncline" may do much to account for the lower temperatures inferred for metamorphic rocks developed in this environment. A small contribution may result from mass transfer of heat in mafic to intermediate magmas, which appear to rise through steep, narrow conduits to the surface. At higher grades of metamorphism a significant contribution may come from the widespread granitic plutons and pegmatites emplaced from below at about the same time (e.g., Takeuchi and Uyeda, 1965).

Turning briefly now to the effect of this changing environment on the minerals themselves; as the temperature rises, for example in a shaley sedimentary rock, the clay minerals and silica that were in equilibrium when formed are removed from their fields of stability as shown in Fig. 6-5. When the rock reaches the field in which micas and quartz are stable,

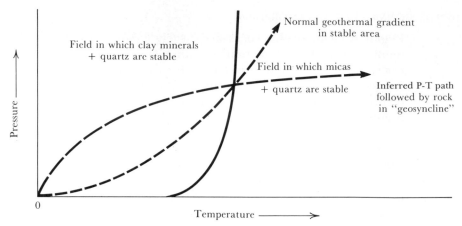

Fig. 6-5 *Hypothetical changing environment of a sedimentary rock buried in a "geosynclinal" deposit.*

these minerals form in preference to the original minerals. Because of slow reaction rates and a certain excess of energy needed to nucleate the new mineral phases, the old phases (clay minerals and quartz) normally persist for a short distance into the stability field of the new phases. Such phases, persisting (or growing) outside their stability field, are called metastable.

THE CONTROLS OF METAMORPHISM

To this point, we have tacitly assumed that metamorphism is caused by temperature and "pressure," without regard to other possible controls. Although this is substantially correct, it is a gross oversimplification. Other factors that influence the development of metamorphic minerals include pressure transmitted in different ways (P_{load}, P_{H_2O}, P_{CO_2}, P_{O_2}, etc.), differential stress, rates of nucleation and crystal growth, and diffusion and metasomatism. The influence of rock composition on the development of metamorphic minerals and zones has been discussed above. Consider in more detail the other factors.

Temperature

Temperatures of regional metamorphism commonly range from about 300 to 400°C to 700 to 800°C. Geothermal gradients based on measured heat flows for a young "geosynclinal" environment average about 15 to 25°C/km, compared to about 10°C/km for stable shield areas (Lee and Uyeda, 1965; Hyndman et al., 1968; Hyndman and Hyndman, 1968; Roy et al., 1968; Wollenberg and Smith, 1970). Earlier estimates range from about 10 to 30°C, and natural geothermal gradients may range equally widely. By extrapolation of this average to a depth of 20 km (about 12 miles or 65,000 ft), corresponding to deep burial in a "geosyncline," the gradient decreasing slightly to depth, we can expect "normal" temperatures of around 380°C (Hyndman et al., 1968), to 500°C (Roy et al., 1968), depending on the estimates. A depth of 25 km would provide about 450 to 625°C. Such temperatures approach the conditions of low- to moderate-grade metamorphism. As discussed below, the heat for regional metamorphism probably results primarily from internal radioactivity and conduction of heat from the mantle.

How the higher temperatures of common regional metamorphic rocks are formed is still a largely unsolved problem. It may be helpful, however, to speculate as follows: to get 700°C with a 20°C/km gradient would require a 35-km-thick "geosyncline." This is probably too thick for simple sedimentation; but, with rising temperatures, loss of strength and drastic thickening of the "geosynclinal" pile by crumpling of the continental margin by an oceanic convection cell, may provide a thickness approaching this value. This would require, of course, a major deformational episode before the climax of high-grade regional metamorphism. This, however, seems to be the case in at least some areas (Sutton, 1965). Read (1955a), Haller (1958), Johnson (1962), Ross (1968), and Carmichael (1968), for example, indicate that the regional metamorphism in particular areas was later than an intensive folding episode. At such temperatures, the high-grade metamorphic rocks deep in the deformed "geosyncline" begin to melt, the resulting granitic magmas rising to higher levels to augment the heat supply there. If this is true, the thermal gradient in high-grade rocks would be expected to be somewhat higher than at lower grades. The geometry and occurrence of such metamorphic highs with steep, closely spaced isograds at the margins has prompted Greenwood (1971) to propose upward transfer of heat largely in rising magmas or heated water drawn convectively into the thermal high.

An alternative to such drastic thickening could involve a locally higher radioactive content for greater heat production or a greater heat flow from the mantle as suggested by Roy et al. (1968). In this connection, it may be noted that the only absolutely known area of presently active metamorphism, the Salton Sea geothermal field, lies along the continental extension of the East Pacific rise (Muffler and White, 1969). Inasmuch as such oceanic rises coincide with positive heat flow anomalies and are considered to be the locus of upwelling convection cells (e.g., Menard, 1958; Wilson, 1963b), it could be argued that metamorphic belts develop over the rising limbs of convection cells. In the same vein, it has been suggested that the voluminous rhyolitic ash flows of the Great Basin of Nevada and Utah are connected with mantle upwelling in the East Pacific rise, which is presumed to underly this region (Menard, 1960, 1964; Cook, 1968; Armstrong et al., 1969; Wilson, 1970). This agrees with the argument of Roy et al. (1968) in favor of the importance of mantle heat flow in raising the temperature of the crust. The influence of mantle heat has also been suggested by Sutton (1965) in his discussion of unusual heat flows and thermal highs in the development of metamorphic belts.

A radically different source of heat for regional metamorphism was proposed by Ambrose (1936). He suggested, on the basis of demonstrably greater movement in the higher-grade zones, that the heat for metamorphism was developed mechanically "by and during intimate regional shearing." Although this would appear to be an extreme point of view, Reitan (1968) has calculated that where "pulses of deformation accomplish significant strain in geologically relatively short periods of time"—about 100,000 to 1,000,000 years—heat generated by friction may be significant. Friction distributed through 3 km of crust may raise the temperature through several kilometers of crust, by about 30 to 80°C. Where the deformation is distributed over longer periods, the rise in temperature is smaller. This frictional heat would increase the thermal gradient in the upper part of the deformational zone and thereby reduce somewhat the depth for attainment of a particular temperature.

As described below, melting at temperatures as low as those encountered in high-grade regional metamorphism requires water-saturated conditions. Such would be encountered within the "geosynclinal" pile but probably not below it, either in the oceanic crust (basalt) or in the deep continental crust (some of which is composed of dehydrated metamorphic rocks from an earlier metamorphic cycle). Although metamorphism is presumably occurring at deeper levels, melting probably occurs in the water-saturated conditions near the base of the "geosynclinal" pile.

Not only does the temperature affect which reactions occur in a given chemical environment, but it also affects the rate of the reactions.[1]

Pressure

Pressures of regional metamorphism are uncertain but for most are estimated to range from about 2,000 or 3,000 bars to 8,000 bars (or atmospheres). The normal load pressure at depth, resulting from the weight of overlying rocks, is about 285 bars/km. At 20 km, we can expect a lithostatic (load) pressure of about $5\frac{1}{2}$ or 6 kbars, and a 35-km depth provided by deformation and thickening of the depositional pile would provide pres-

---

[1]The reaction rate (K) can be expressed as the number of molecules reacting per second (A, which is almost independent of temperature) times an exponential factor dependent on temperature. $K = Ae^{-E/RT}$, where E = size of energy barrier to reaction, R = constant 1.99 cal/mole, and T = temperature (see Fyfe et al., 1958, pp. 55–58).

sures of about 10 kbars. At such moderate to high pressures and the corresponding temperatures described above, the rocks are considered too weak to transmit a significant differential stress applied over a long period (Heard, 1963; Rutland, 1965). The stresses are relieved by rock creep (or flow), resulting in the metamorphic foliation. Rapidly applied stresses, of course, may be transmitted (as in earthquakes), but these are probably too short for nucleation and grain growth. At shallower depths, however, the rocks are more rigid and can resist stresses applied over long periods without continuous deformation. These tectonic overpressures could possibly amount to 2,000 to 3,000 bars (Birch, 1955, p. 116).

Water pressures in pores in the rock are probably equal to the load (solid) pressure in most cases, especially during metamorphism in which most reactions drive off water. This may not be true near the surface, where fractures open to the surface could have pressures corresponding only to the weight of overlying water. Where the water pressure exceeds the load pressure, the latter may be neglected, as the pressure will be transmitted through the fluid. Water pressure, P_{H_2O}, is important in controlling the temperature of a metamorphic reaction. Other fluid pressures, especially P_{CO_2} in carbonate rocks, may be important along with water in some reactions, the total fluid pressure being equal to the sum of the partial pressures ($P_f = P_{H_2O} + P_{CO_2}$). Should the water pressure locally be less than the load pressure, the equilibrium temperature will normally be reduced accordingly (Yoder, 1955b), as expected from thermodynamic reasoning and confirmed by experiment (Greenwood, 1961). As a result, the metamorphic grade, as determined from the mineral assemblage, may appear higher in the drier rock.[1] Such a situation, the case of a metamorphosed granitic pluton surrounded by metasediments, has been described by Barker (1961). Another, the case of orthogneisses (gneisses derived by metamorphism of an igneous rock) surrounded by metasediments, has been described by Buddington (1963).

Metamorphic reactions in which water or carbon dioxide is given off are clearly pressure dependent, as may be seen from the following examples (Fig. 6-6). If a rock is positioned on the reaction boundary, such as at X_1, an increase in P_{H_2O} will, according to Le Châtelier's rule,[2] tend to drive the reaction toward the low-volume side (to X_2). Since water is the only fluid (high-volume) phase, the change will produce muscovite + quartz at the expense of K-feldspar + sillimanite + H_2O. A decrease in P_{H_2O} will, by the same token, produce the high volume, water-bearing phase.

---

[1]This is discussed further, below, in the section on granulite facies metamorphism.

[2]More rigorously $\delta\Delta G/\delta P = \Delta V$, where ΔG = free energy of the reaction (negative for reaction to proceed), ΔV = volume change, and P = pressure. Thus an increase in P will favor the smaller-volume side of the reaction.

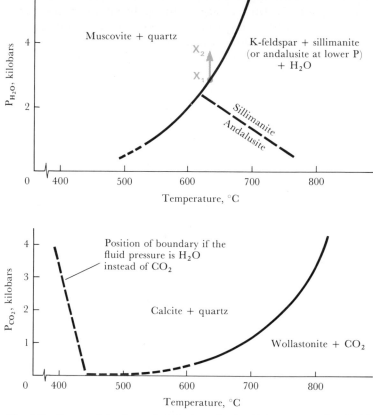

Fig. 6-6   *Pressure dependence of a typical reaction in which water or carbon dioxide is driven off during metamorphism. Muscovite curve from Evans (1965); wollastonite curve from Harker and Tuttle (1956) and Greenwood (1962).*

If the fluid phase on the reaction (Fig. 6-6) muscovite + quartz ↪ K-feldspar + sillimanite + H_2O is not pure water but contains other components, the equilibrium boundary will normally be displaced somewhat to the left as shown by Greenwood (1961). Similarly, if the fluid phase in the wollastonite reaction (Fig. 6-6) is not pure CO_2 but contains some water, the equilibrium boundary is displaced to the left. This latter reaction has been studied experimentally by Harker and Tuttle (1956) and Greenwood (1962), who show that the effect at a few kilobars pressure is rather pronounced, amounting to as much as several hundred degrees.

The effect of water–carbon dioxide mixtures can perhaps best be illustrated with a temperature-composition diagram (see Fig. 6-7). Note that, as noted above, higher water contents of the fluid phase result for the most part in lower equilibrium temperatures. As pointed out on thermodynamic grounds by Greenwood (1962), this reaction must show a temperature maximum at $X_{CO_2} = 0.75$ as this is the composition of the fluid phase produced by the reaction. This can be rationalized on less rigorous grounds by considering that any additional H_2O (or CO_2) over this reaction amount is in excess and could only lower the temperature of the reaction. The reaction would proceed in spite of any excess of one component over that taking part in the reaction.

Application of temperature-fluid composition diagrams such as Fig. 6-7 to interpretation of temperatures in natural systems involves estimation of the composition of the fluid phase. In this particular system, temperatures are nearly constant in situations where the CO_2 content of the fluid phase is greater than $X_{CO_2} \cong 0.25$. This would probably be the case during metamorphism of carbonate-rich metasediments (with limited amounts of water- or OH-bearing phases), the decarbonatization flooding the system with CO_2. The converse would likely be true during metamorphism of pelitic or other water-rich (and carbonate-poor) metasediments, the CO_2 content falling to X_{CO_2} less than 0.25 and resulting in lowered temperatures of reaction. A clear account of the application of this temperature-fluid composition diagram to a contact metamorphic problem has been presented by Greenwood (1967).

The obvious possibilities for difference in slope for reactions in which water versus carbon dioxide is produced, permit the prediction that isograds based on one volatile component such as H_2O

Fig. 6-7 *Temperature-fluid composition diagram for the equilibrium (at 500 and 1,000 bars total fluid pressure P_f): 1 tremolite + 3 calcite + 2 quartz $\leftrightarrows$ 5 diopside + 2 CO_2 + 1 H_2O. After Winkler (1967, Fig. 4, from Metz and Winkler, 1964). See text for explanation.*

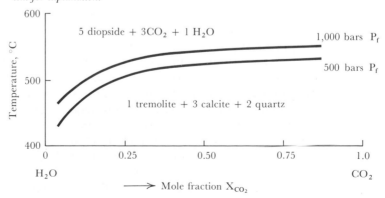

may cross isograds based on another volatile such as CO_2 (Green-wood, 1962). Carmichael (1968) has studied such an environment in which isograds strongly influenced by high content of CO_2 in metasediments are crossed by isograds affected by the influx of abundant water from a nearby granitic pluton.

Water

Water is important in metamorphic reactions because of its influence in transmitting pressure, permitting formation of hydrous minerals, and increasing the rate of reactions such as by breaking down mineral bonds and transporting constituents as outlined below. A great many regional metamorphic rocks begin as water-saturated marine sediments, including volcanic sediments, and undersaturated volcanic flows. With compaction, diagenesis, and low-grade metamorphism, water is continually driven out of the sediments, diffusing upward through the overlying rocks and migrating outward along permeable beds and through fractures. Except in the prehnite-pumpellyite to blueschist facies where some reactions appear to use up water, the sediments presumably remain water saturated throughout this evolution and subsequent high-grade metamorphism. Any intercalated volcanic flows, unless quite permeable, probably remain pretty well undersaturated throughout this low-grade evolution and until somewhat higher grades, where penetrative deformation and water driven off from the surrounding sediments causes saturation of the flows. This deficiency in water is clearly represented in massive flows at low metamorphic grades, by only incipient metamorphism and hydration and by much evidence of relict volcanic minerals in areas where the surrounding sedimentary rocks are clearly metamorphosed. At moderate to high grades, most flows appear to be fully hydrated, the mineralogy being representative of the same grade as the surrounding metasediments. To still higher grades, water is driven off with almost every metamorphic reaction, the rocks remaining water saturated (but contain less total water) to the peak of metamorphism. The higher the temperature, the smaller amount of water (or CO_2) combined in the stable minerals. As the temperature begins to wane, the water vapor would fall off sharply, water remaining primarily as combined water (or OH) and adsorbed on grain surfaces. Mineral reactions would probably come to an abrupt halt. At the same time, expansion accompanying incipient rehydration (retrogressive metamorphism) would seal off pore spaces and further inhibit reaction.

The above comments are intended to apply to conditions of regional metamorphism. Most contact metamorphic rocks, by contrast, begin as sedimentary, volcanic, and metamorphic rocks that are long past the environment of deposition and are probably undersaturated in water. In this situation, the rocks would probably remain undersaturated ($P_{H_2O} < P_{load}$)

until the grade of contact metamorphism reached a grade comparable to that in the rocks undergoing metamorphism. This is perhaps an important reason why low-grade contact metamorphic assemblages show incomplete reaction and disequilibrium. One exception to this undersaturated situation in the lower grades of contact metamorphism is where abundant water is introduced from the igneous intrusion, as in the case of skarns. The situation in which water pressure is less than total fluid pressure has been discussed by Yoder (1955b) and examined experimentally by Greenwood (1962), as noted above.

The interstitial water provided in marine sediments subject to metamorphism is not pure water, of course, but a modification of seawater (see Table 6-1). Comparison of brines with seawater shows that Ca^{++} increases and Mg^{++} and SO_4^{--} decrease from the parent seawater (Kramer, 1969), most of the reduced components probably remaining in the rocks (for example, Mg^{++} into limestone to form dolomite). Among the more important of the former factors are interactions with clay minerals and reduction and dolomitization (Degens, 1965, p. 189). In addition to the relative changes in composition as seen from Table 6-1, the brines are about 2 to 10 times as concentrated as seawater (Degens and Chillingar, 1967). Brine from the Salton Sea geothermal well is about 7.5 times as concentrated as seawater (Muffler and White, 1969).

Table 6-1 gives some idea of the composition of pore waters in the rocks before metamorphism. The constituents in these waters, although progressively modified as mineral reactions occur, probably are available and are used in the reactions of the early stages of metamorphism.

At higher metamorphic grades, the composition of the fluids in the metamorphic environment changes, presumably under the same influences as control in the sedimentary environment—equilibration with the

Table 6-1 *Composition of major components of seawater (3.5 g/liter), subsurface brines (1,860 samples), and evaporite brines (Kramer, 1969), compared with Salton Sea geothermal well (Muffler and White, 1969)*

Component	Seawater, mol. %	All Brines*, mol. %	Evaporite Brines, mol. %	Brine from Salton Sea Geothermal Well
Cl^-	49.	51. −54.	51.	56.2
SO_4^{--}	3.4	0.08− 1.2	1.5	0.0007
HCO_3^-	0.2	0.07− 0.06	0.2	0.03
Na^+	42.	36. −45.	41.	28.1
Ca^{++}	0.9	3.6 − 7.8	4.0	9.0
Mg^{++}	4.8	0.9 − 2.4	2.0	0.03
$Sr\text{-}Ba^{++}$	0.01	0.00− 0.3	0.00	0.07
K^+	0.97†			5.7

*Range is for one standard deviation.
†K^+ from Degens (1965, p. 173).

rocks and movement from elsewhere. Particularly important is the maintenance of a high content of water in pelitic and other rocks containing much combined H_2O or OH and a high content of carbon dioxide in carbonate-rich rocks. The pores will be flooded with these fluids as they are driven off from minerals of appropriate composition with progressive metamorphism, as noted above. Barnes (1970) has described brines high in HCO_3^-, NH_4^+, H_2S, and B, which he believes are products of low-grade metamorphism of marine sediments.

A more direct handle on the composition of fluids in a metamorphic environment has recently been provided by the 1961–1962 drilling of a 5,232-ft-deep well for geothermal power, near the Salton Sea in southern California (White et al., 1963). The pelitic silts and sandy sediments surrounding this and other more recently drilled wells in the area are undergoing active metamorphism in the low grades of the chlorite zone. Recent analyses of this deep high-temperature brine show that it has a very unusual composition (Muffler and White, 1969). Comparison of this geothermal brine with other brines and with sea water (see Table 6-1) shows that the trend begun by the transformation of seawater to brine is continued in the development of this "metamorphic" brine. Cl^-, Ca^{++}, Sr^{++}, Ba^{++}, and K^+ have increased and SO_4^{--}, HCO_3^-, Na^+, and Mg^{++} have decreased. Although this may not be representative of all waters accompanying low-grade metamorphism, it may well approximate the composition of waters accompanying low-grade metamorphism of rocks of the composition found here. This geothermal brine also contains high concentrations of Li, Rb, Fe, Mn, B, As, Cu, Zn, Pb, and Ag (Skinner et al., 1967; Muffler and White, 1969). A knowledge of the composition of brines is important in understanding the final composition of many metamorphic rocks. It helps to explain the mineralogy and compositions of some rocks considered by some geologists to be affected by widespread metasomatism.

Oxygen

Although oxygen is not commonly thought of as being a major phase in rocks undergoing progressive metamorphism, it does have an important influence on the mineralogy of the resulting rocks. Higher activities of oxygen (or higher P_{O_2}) in metamorphic rocks drive more of the iron into the oxidized ferric state, just as for igneous, especially basaltic, magmas (Mueller, 1961; and see Fig. 3-7, p. 80). This Fe^{3+} cannot enter most of the common Fe^{++}-Mg silicates of metamorphic rocks (e.g., biotite, almandine cordierite) but a few Fe^{3+} minerals such as magnetite ($Fe^{++}O.Fe^{3+}{}_2O_3$), epidote ($Ca_2Fe^{3+}Al_2OSiO_4Si_2O_7OH$), and andradite garnet ($Ca_3Fe^{3+}{}_2Si_3O_{12}$) can. Fe is thereby removed from the Fe^{++}-Mg silicates in the rock, forcing them to be richer in Mg (see Fig. 6-8) and possibly Mn (Chinner, 1960b). If the oxygen activity is high enough, certain

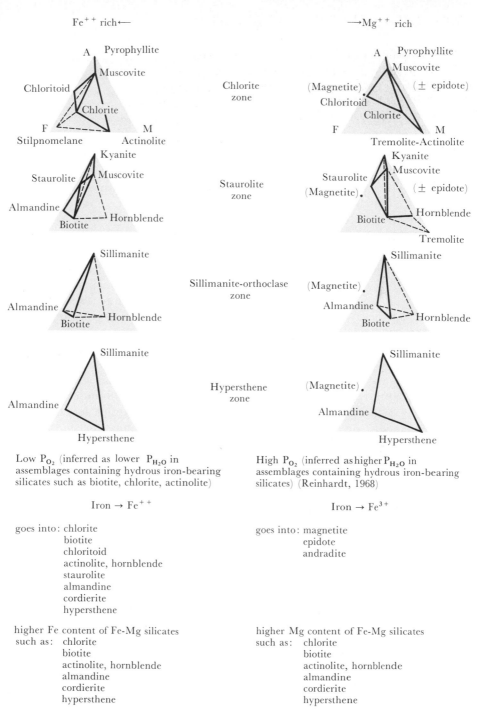

Fe^{++} rich←   →Mg^{++} rich

Chlorite zone

Staurolite zone

Sillimanite-orthoclase zone

Hypersthene zone

Low P$_{O_2}$ (inferred as lower P$_{H_2O}$ in assemblages containing hydrous iron-bearing silicates such as biotite, chlorite, actinolite)

High P$_{O_2}$ (inferred as higher P$_{H_2O}$ in assemblages containing hydrous iron-bearing silicates) (Reinhardt, 1968)

Iron → Fe^{++}

Iron → Fe^{3+}

goes into: chlorite
        biotite
        chloritoid
        actinolite, hornblende
        staurolite
        almandine
        cordierite
        hypersthene

goes into: magnetite
        epidote
        andradite

higher Fe content of Fe-Mg silicates such as: chlorite
        biotite
        actinolite, hornblende
        almandine
        cordierite
        hypersthene

higher Mg content of Fe-Mg silicates such as: chlorite
        biotite
        actinolite, hornblende
        almandine
        cordierite
        hypersthene

Fig. 6.8 *Inferred approximate phase relationships dependent on the oxidation of iron in metamorphic rocks. See text for discussion.*

Fe^{2+}-Mg silicates may even be prevented from forming. The relative sta- bilities of common silicates in pelitic rocks in the vicinity of the staurolite zone and their dependence on oxygen activity, water, and temperature, have been calculated thermodynamically by Fisher (1967). Chinner (1960b), Rinehardt (1968), and Kerrick (1970) have described and dis- cussed striking examples of the influence of oxygen in affecting the com- position of garnet, biotite, muscovite, and cordierite.

Consider biotite as an example (see Fig. 6-8). With low P_{O_2}, iron occurs as Fe^{++}, which can substitute for Mg^{++} in most of the common mafic sili- cates including biotite. The biotite will then take on a high iron content that reflects this high Fe^{++} environment. With higher P_{O_2} (more oxidizing conditions), some of the iron occurs as Fe^{3+}, which cannot substitute for Mg and cannot therefore go into most of the common mafic silicates such as biotite. (Instead it is largely restricted to the few Fe^{3+} minerals magne- tite, epidote, and andradite.) Therefore, the available iron cannot go into biotite, and the biotite is more Mg rich. The same discussion holds equally well for other common mafic silicates including chlorite, muscovite, acti- nolite, hornblende, and hypersthene.

Thompson (1957), Eugster (1959), Chinner (1960b), and Mueller (1960) indicate (on different grounds) that although water may be highly mobile, oxygen is not likely so, suggesting that the oxygen activity is controlled by the original composition of the sediments. Evidence in favor of the immobility of oxygen even in the high-grade metamorphic environ- ment includes the occurrences where "hematite-rich bands [high P_{O_2}] and ilmenite-magnetite-bearing bands [lower P_{O_2}] only an inch or two thick may be found intimately interlayered" (Chinner, 1960b, p. 207). Although volatile, the equilibrium between O_2 and H_2O, coupled with the very low values of P_{O_2}, restrict the movement of O_2 to minute fractions of the movement of H_2O (Mueller, 1967).

CHAPTER 7
PROCESSES IN
METAMORPHIC
ROCKS

RATES OF NUCLEATION AND GRAIN GROWTH

The rate of a chemical reaction depends on the factors affecting two separate steps, nucleation of the new phase and grain growth, each step with its own separate rate (e.g., Fyfe, Turner, and Verhoogen, 1958, chap. III; Rast, 1965). Most metamorphic reactions involve a nucleation step that includes breakdown of single solid phases including dehydration, reaction of two or more solid phases or reaction of a solid with a volatile, first-order (reconstructive) polymorphic transformations (for example, β-quartz-tridymite), and recrystallization involving a change in grain orientation. A few metamorphic changes involve no nucleation step. These include second-order (displacive) polymorphic transformations (for example, α-quartz to β-quartz), and static grain growth not involving a change in grain orientation. Note that these are not the solid-liquid reactions as discussed for igneous rocks but solid-solid reactions in metamorphic rocks.

The important nucleation step requires growth of a mineral nucleus large enough to be stable, the surface area and thus the surface free energy no longer being so large for a given volume as to render the grain unstable. The critical radius has been estimated as of the order of 10^{-6} cm (Fyfe, Turner, and Verhoogen, 1958; Rast, 1965). Nucleation depends on the availability of material, diffusion of the material, and the physical conditions present at the time. These physical conditions, in addition to the temperature, pressure, and other factors necessary to stability of the phase, include crystal surface irregularities, strained parts of a crystal, and the degree of fit of the new and old crystal lattices across the interface between them. In this light it may be noted that the most likely phase to nucleate is not necessarily the most stable one; a metastable phase such as cristobalite may nucleate faster than the stable one such as quartz in its stability field (Turner and Verhoogen, 1960, p. 480).

The other important step in a metamorphic reaction, grain growth, depends on higher temperatures, greater "distance" into the stability field of the mineral (greater negative free energy of the reaction), the presence of fluids to permit rapid movement of constituents to and from the nucleus, the presence of water as a solvent, finer grain size (greater surface area per unit volume), greater concentration of the reaction constituents, and the presence of catalysts. Theory and considerations in reaction rates are reviewed by Fyfe, Turner, and Verhoogen (1958, pp. 53–103), Rast (1965), Kunzler and Goodell (1970). The degree to which the presence of water facilitates reactions is illustrated by the reactions $SiO_2 + 2\,MgO \rightarrow Mg_2$-$SiO_4$, which in the *dry* state is about 26% complete in 4 days at 1000°C; in the presence of water it is 100% complete in minutes at only 450°C— some 100 million to 10 billion times faster (Shaw, 1955, *in* Fyfe et al., 1958, p. 85).

DIFFUSION AND METASOMATISM

A major factor in the rate of crystal growth, as noted above, involves diffusion of reaction constituents to the growing crystal. Diffusion is also of great importance in the process of metamorphic differentiation, whereby constituents segregate into layers, and in metasomatism, whereby chemical constituents are thought to move through the rock (for sometimes considerable distances measured in tens of feet or even miles) to produce changes in the composition of a rock. Although we have tacitly assumed in earlier sections that metamorphism is largely isochemical, mineralogical changes go hand in hand with changes in chemical composition within different parts of the rock. These changes are discussed below.

Increased rates of diffusion are considered to result from higher temperatures and higher concentration gradient or, better, the activity gradient, which depends also on the structure and binding energy (or "solubility") of the source material (Fyfe, Turner, and Verhoogen, 1958, p. 62; Fisher, 1970). Greater strain energy in the crystal lattice may also play a role. The nature of the passages for diffusion depends on whether diffusion is along grain boundaries or fractures (Holser, 1947), considered more important by most authors, or through the crystal lattices themselves. Diffusion in both depends on lattice defects, including site vacancies, dislocations, and the presence of impurity ions having an abnormal charge for that site. Diffusion through the crystal lattice (bulk diffusion) may become more important at high temperatures (Girifalco, 1964, p. 126). Water, whether filling spaces or adsorbed on grain boundaries, is important as a medium through which diffusion may operate and as a solvent in removing constituents from the source. Diffusion in a dry system is commonly many orders of magnitude slower. The nature of the diffusing constituents also plays a significant role—smaller ions, lower charges on the ions, ionic groups (less charged), and presumably "crystal-field effects" influencing transition metal ions (discussed in Chap. 3), all favoring diffusion. Where a constituent that occurs in only trace amounts in the fluid phase (e.g., oxygen, as noted above) is in equilibrium with or buffered by a dominant constituent (e.g., water), transport of that trace constituent is severely limited, as shown by Mueller (1967).

A great deal is known about solid-state diffusion, both from theoretical considerations and experimental work, but little is known about diffusion in an environment containing a chemical "sink."[1] In such a situation, presumably a common one in geology, diffusing constituents react with minerals in the environment to produce relatively stable or insoluble minerals. This removal of diffusing constituents increases the activity gradient, thereby aiding diffusion to the "sinks." Although necessary thermodynamic values are not yet available, it has been shown feasible to calculate diffusion gradients (Fisher, 1970). The insufficiency in the theoretical and experimental base for diffusion may be supplemented by empirical data from the rocks themselves.

During regional metamorphism, the successive breakdown of hydrous and carbonate phases liberates large quantities of H_2O and CO_2 that clearly migrate upward and outward through the overlying and surrounding rocks. Hydrothermal quartz and carbonate veins in lower-grade rocks are presumably derived in part from this source. The particular composition of the vein or fracture filling may result partly from constituents carried along in solution but most commonly bears some resemblance to the enclosing rock, suggesting that the migrating solutions acted partly as a solvent and transporting medium (e.g., Orville, 1963) for local constituents. Most petrologists now believe that with the exception of this H_2O and CO_2, very little migration of common rock-forming constituents occurs in the low to middle grades of metamorphism before

---

[1] A chemical sink is a relatively insoluble phase in which a given element or ion can precipitate, thus being removed from the diffusing state.

anatexis (e.g., Shaw, 1956; Engel and Engel, 1960; Mehnert, 1960; Hiet-anen, 1963, p. C–47; Hyndman et al., 1967).

Regional development of scapolite in amphibolites and other feld-spathic rocks has been attributed to metasomatic introduction of Cl and CO_2 (e.g., Sundius, 1915, *in* Shaw, 1960a). An alternative possibility—isochemical metamorphism (no introduction of material)—depends on the presence of these constituents in interstitial fluids derived from brines as noted above. Similarly, regional development of muscovite from silli-manite or andalusite (e.g., Billings, 1938; Green, 1963) could be attributed to decline in temperature, the K being derived from within the rock. Many described cases of regional metasomatic increase in Fe, Mg, Ca, ± Al, Na in regions containing deep-seated granitic plutons (e.g., Wegmann, 1935; Reynolds, 1946, 1947; Read, 1948; Steven, 1957; Engel and Engel, 1958; Hietanen, 1962) probably should not be classed as metasomatism as used here. Most or all of these cases appear to be possible examples of residue resulting from anatexis as described above in the section on granitic plu-tons. Similarly, many described cases of K ± Na ± Si metasomatism result-ing in the formation of granitic gneisses and plutons (e.g., Barth, 1936; Misch, 1949; Engel and Engel, 1963; Mehnert, 1968, pp. 149–152; Heimlich, 1969) could equally well result from anatexis or metamorphic differentiation and separation of these constituents in place.

A number of cases have been described of diffusion during regional metamorphism, in response to large concentration gradients imposed by juxtaposition of two compositionally different sedimentary or igneous rock types (see Fig. 7-1). For example, diffusion across the boundary between adjacent marble and amphibolite has formed calc-silicate minerals such as diopside, grossularite, idocrase, and scapolite (Turner and Ver-hoogen, 1960, p. 571). Similarly diffusion between carbonate pods and the enclosing dolomite-cemented quartzofeldspathic silt (now orthoclase-hornblende amphibolite) has resulted in reaction rims consisting of diopside, epidote, and grossularite with minor sphene, scapolite, and fluorite (Hyndman et al., 1967). In the latter instance, it was shown that Si and Al, ± Ti, moved readily into the pods and Ca and Mg to a lesser extent moved outward. Under large concentration gradients, Na moved less than 5 mm and K less than 1 mm. Diffusion of K from granite for some 25 mm into amphibolite has been shown by variations in K^{39}/K^{41} ratio (Verbeek and Schreiner, 1967). In these instances, the distance of diffusion appears to be controlled not so much by the ease of diffusion but by the presence of reactive constituents that precipitate the diffusing components. In another example, adjacent aluminous shale and original "calcareous shale" (Hietanen, 1963) formed aluminosilicate-rich schist and anortho-site, respectively, by diffusion of Al into the anorthosite and Ca ± Na out. An alternative explanation suggests diffusion of Al outward to form

Fig. 7-1 *Numerous white granite veins cutting diopside-rich calc-silicate gneiss and showing hornblende-rich reaction selvages along the contacts. Head of hammer is 1 in. across. Cirque southeast of South Kootenai Lake, northern Bitterroot Range, Montana.*

the aluminosilicate-rich schist during metamorphism of an intrusive anorthosite and its pelitic country rocks (Hyndman and Alt, 1971). In either case Al must have moved at least tens of meters during high-grade metamorphism.

During contact metamorphism, the effects of metasomatism are often more apparent. The nature of the metasomatism depends on the composition of the intrusive rock and the composition of the intruded rock, the character of the diffusing elements apparently being controlled primarily by the difference in composition across the contact. The situation is analogous to diffusion across a compositional contact during regional metamorphism but the much greater effects at an intrusive contact probably depend on the large thermal gradient and voluminous movement of hot water (e.g., Greenwood, 1967) across the contact. Just as in the regional situation, the diffusion in most cases tends to erase the compositional difference across the contact.

Adjacent to granitic intrusives (granite to quartz diorite), metasomatism of calcareous rocks (especially limestone or dolomite) appears most pronounced. Here the calcareous country rocks are commonly transformed over a few to many tens of feet into striking coarse-grained brownish red and green calc-silicate rocks called skarns. The resultant minerals include grossularite-andradite garnet, epidote, diopside, hedenbergite, idocrase (vesuvianite), tremolite, hornblende, wollastonite, scapolite, and lesser amounts of hematite, magnetite, quartz, fluorite, fluorapatite, phlogopite, and a host of rare very high temperature lime silicates such as larnite, merwinite, and spurrite. Examples of this type of contact metasomatism have been described by Lacroix (1900); Barrell (1907); Goldschmidt (1911, p. 213); Eskola (1914); Osborne (1931); Wantanabe (1943); Holser (1950); Tilley (1951b); Burnham (1959); Mueller and Condie (1964); Buseck (1966, 1967); and Perry (1969). Most important in this type of diffusion are Fe, Mg, Si, and to a lesser extent Al, F, and B. These elements are abundant in the granitic rocks and are sparse or nearly absent in the intruded limestone. (As noted in the section on assimilation by granitic magma, Ca may migrate into the granitic magma, down its concentration gradient, and up the temperature gradient.)

Adjacent to granitic intrusives, metasomatism of noncalcareous pelitic and quartzofeldspathic country rocks is generally quite limited. Although the metamorphism may reach a high grade, the composition of the country rocks commonly appears unchanged. Exceptions have been described, however, where pelitic rocks have been changed to tourmaline- or axinite-rich rocks by the influx of B from an adjacent granitic pluton (e.g., Agrell, 1941; Parker, 1961; Taylor and Schiller, 1966). As noted by Goldschmidt and Peters (1932, *in* Turner and Verhoogen, 1960, p. 575), however, not all tourmaline-bearing rocks are metasomatized. Small amounts of disseminated tourmaline may result from the boron (0.1% B_2O_3) already present in argillaceous marine sediments.

The development of scapolite on a regional scale in pelitic, sandy, or calcareous rocks has been attributed by a number of workers to Na and Cl metasomatism (e.g., Sundius, 1915, *in* Shaw, 1960a; Barth, 1930; Buddington, 1939; Laitakari, 1947; Edwards and Baker, 1953). It seems reasonable to suppose, however, that under certain circumstances Na and Cl could be retained to higher grades in brines trapped in marine sediments (von Knorring and Kennedy, 1958; White, 1959; Shaw, 1960b; Hietanen, 1967). The development of porphyroblastic albite or oligoclase schists or augen gneisses adjacent to granitic plutons has in many cases been attributed to Na-metasomatism. Turner (1948, p. 116) has noted, however, that "non-metasomatic metamorphic derivatives of basic igneous rocks and of graywackes commonly contain albite porphyroblasts." Alteration of an albite rock to form poikiloblasts of microcline within about 6 in. of a granitic pegmatite shows K-metasomatism over that distance (Green, 1963a).

Adjacent to mafic or ultramafic intrusions, noncalcareous sediments may be transformed into calc-silicate rocks by outward diffusion of Ca (e.g., Turner, 1933; Coleman, 1961, 1967b; Challis, 1965a). Also, metasomatic introduction of Cl and CO_2 from such intrusions may form scapolite in various calcareous rocks including amphibolite, metagabbro, and calcareous shale (Sundius, 1915, *in* Turner and Verhoogen, 1960, p. 576). Extreme metasomatism of pelitic rocks by Na-rich solutions, during the later stages of crystallization of some diabase sills, produces fine-grained quartz-albite rocks called "adinoles" (Agrell, 1939; Davies, 1956). A less extreme Na-metasomatism of arkose by diabase has been described by Butler (1961).

METAMORPHIC DIFFERENTIATION

All metamorphic processes, as noted above, tend toward conditions of equilibrium. Most rocks seem to approach a state of equilibrium closely, at least under peak conditions of metamorphism. Evidence for this contention has been presented at the beginning of this chapter. The natural tendency of most such processes, including those of diffusion and metasomatism just described, is to homogenization of the rocks on a scale larger than that of individual grains.

Certain features in metamorphic rocks, however, suggest that some processes lead not to homogenization but to differentiation (i.e., segregation of constituents) within metamorphic rocks. Perhaps the most common such feature is the development of layers of contrasting mineralogical (and chemical) composition where no such layers can have existed in the original rock (see Figs. 7-2a and b) or where the original layering has been accentuated. Where the new layers are oriented at an angle to the original sedimentary beds or to other original compositional contrasts, formation of the new layers during metamorphism must constitute a tendency toward

Fig. 7-2a *Diagramatic representation of new quartzofeldspathic layers developed parallel to schistosity and axial surfaces of folds, in a series of silt and shale beds. New quartzofeldspathic layers commonly 1 mm to several millimeters thick.*

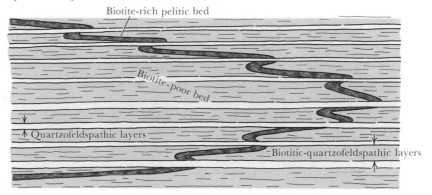

(a)

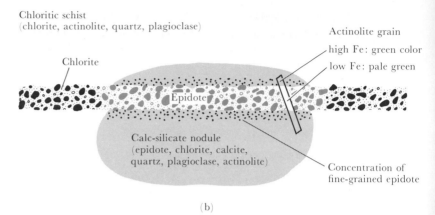

Chloritic schist
(chlorite, actinolite, quartz, plagioclase)

Chlorite

Actinolite grain

high Fe: green color

low Fe: pale green

Epidote

Calc-silicate nodule
(epidote, chlorite, calcite,
quartz, plagioclase, actinolite)

Concentration of
fine-grained epidote

(b)

Fig. 7-2b *Relationship of mineralogy of new composite layer to mineralogy of enclosing rock (Altoona Lakes, Flint Creek Range, Montana). New layer is about 0.1 mm thick. Hornblende hornfels facies (sample X-19b, Stuart, 1966).*

equilibrium but away from homogenization. This is metamorphic differentiation. These new layers appear to develop by diffusion of selected constituents from the immediately adjacent rock. Most are too thin (Fig. 7-2a) or are compositionally inappropriate for magmatic injection or show striking compositional correlation with the immediately adjacent rock (Fig. 7-2b). The resulting layered metamorphic rocks are called gneisses.

As noted in the above discussion of diffusion and metasomatism, the movement of constituents depends on a concentration (or activity) gradient.[1] Some other factor must outweigh the effect of this gradient to create compositional contrasts rather than eliminate them. A large temperature gradient between the rock and the position of the new layers hardly seems likely, considering the small scale of the layers and the nature of regional metamorphism. Another possible factor is a large pressure gradient resulting from the same stresses that form the schistosity of regional metamorphic rocks. Formation of schistosity and accompanying folds implies deformation and thus unequal stresses in the rocks at the time of metamorphism. As implied by Fig. 7-2a, new layers formed in gneisses by metamorphic differentiation are commonly parallel to the schistosity. Similarly, in lower-grade nearer-surface rocks, fractures may develop in which the pressure in the fracture may be lower than the pressure on the grains within the rock (e.g., as in Fig. 7-2b). Or constituents may migrate to lower-pressure "strain shadows" bordering a porphyroblast (see Fig. 7-3).

---

[1]Rate of diffusion = diffusion coefficient × activity gradient (proportional to concentration gradient). Amount of diffusion = rate of diffusion × area × time/distance.

Gresens (1966) has outlined in some detail the thermodynamic effect of a structurally produced pressure gradient on diffusion of chemical constituents in rocks. As in any system, the natural tendency is to lower the free energy (or chemical potential) of the system. The excess energy imposed by deformation dictates that some components will migrate from high- to low-pressure sites and vice versa. A component that expands on moving from the solid to a fluid phase (most components), for example, would prefer a lower-pressure fluid phase and would tend to concentrate within it. (Note that this would be expected also from Le Châtelier's rule.) However, some components will be affected more than others, and the movement of one component affects the free energy and thus the behavior of other components. As the component migrates to a new (e.g., low-pressure) site, its concentration there will increase, thus decreasing the tendency for migration. Crystallization of the component at the new site, however, decreases its concentration in the fluid phase and aids migration. Returning to the examples illustrated, metamorphic differentiation in high-grade,

Fig. 7-3 *Large porphyroblast of almandine garnet in kyanite-biotite-plagioclase schist. Almost equally large porphyroblasts of kyanite are less clearly visible in upper left and lower edge near left. Equant garnet porphyroblast shows triangular wedges of biotite forming "strain shadows" at its right and left edges. The 3/4 -in. aureole of white plagioclase around garnet is probably a case of "metamorphic differentiation" associated with migration of constituents of biotite into "strain shadows," leaving the plagioclase concentrated in the aureole. Adjacent to kyanite-andalusite-sillimanite "triple point" area on Goat Mountain, 18 miles south of Avery, Idaho.*

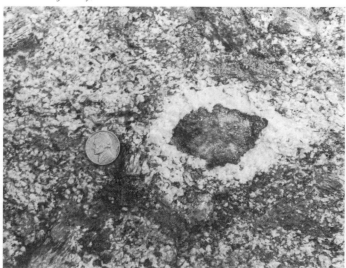

deep-seated rocks (for example, Fig. 7-2a) probably involves migration of some components to lower-pressure surfaces parallel to the schistosity (e.g., to form quartz and feldspars) and migration of other components with less of this tendency to the higher-pressure sites (e.g., to form mafic minerals such as biotite). Metamorphic differentiation in lower-grade, shallower rocks able to sustain an open fracture (for example, Fig. 7-2b illustrates a possible example), probably involves migration of selected components to the lower-pressure site of the fracture.

Some examples of metamorphic differentiation include the following pairs of assemblages:

Rock	Veins or Laminae	Reference
Chlorite-epidote-muscovite schist	Quartz-albite ± calcite	Turner, 1941; Reed, 1958
Quartz-muscovite-chlorite-phyllite	Quartz ± chlorite	Williams, Turner, & Gilbert, 1954, p. 214; McNamara, 1965
Albite-epidote-chlorite schist	Epidote	Turner, 1948, p. 145
Mica (20%)–quartz (30%)–carbonate (50%) phyllite	Carbonate (10%)–quartz (20%)–mica (70%)	Talbot & Hobbs, 1968
Quartz-mica schist (crenulated)	Quartz concentrated at crests of crenulations	Nicholson, 1966
Garnet amphibolite	Omphacite	Essene & Fyfe, 1967
Kyanite-chlorite ± muscovite-iron oxide schist	Quartz-kyanite	Read, 1933
Quartz-plagioclase-K-feldspar ± clinopyroxene-minor magnetite	Quartz-magnetite ± microcline	Hagner & Collins, 1967
Quartz-plagioclase-biotite	Quartz-plagioclase	Crowder, 1959
Quartz-plagioclase-biotite-minor orthoclase ± magnetite gneiss	Quartz-plagioclase-orthoclase	White, 1966
Biotite-cordierite-sillimanite gneiss	Quartz-feldspar	Mehnert, in Barth, 1962, p. 359
Plagioclase-hornblende ± quartz ± biotite amphibolite	Plagioclase, minor quartz ± biotite	Reitan, 1956
Quartz-plagioclase-hornblende amphibolite	Quartz-plagioclase ± hornblende	Kretz, 1966b
Plagioclase-hypersthene-hornblende ± quartz ± magnetite ± apatite	Quartz-plagioclase	Ward, 1959

Metamorphic differentiation is used here in a more restricted sense than that used by many writers. Eskola (1932b), Turner (1948), and Barth

(1962), for example, included such features as growth of porphyroblasts and reaction between adjacent, compositionally contrasting layers, as well as formation of veins or laminae in originally more homogeneous rocks. The last conforms especially to the "solution principle" of Eskola (1932) and metamorphic differentiation is here restricted to this general range of activity, in agreement with most common usage in practice.

MIGMATITES AND THE PROBLEM OF THEIR ORIGIN

Among the most controversial of common igneous and metamorphic rocks are the migmatites or veined gneisses. The term "migmatite" connotes "mixed rocks" consisting of both granitic and metamorphic parts. It includes rocks ranging from igneous intrusion breccias (for example, Fig. 3-9), to deep-seated veined gneisses (for example, Figs. 4-11, 7-4). The former are not particularly controversial. It is the "veined gneisses" to which the term "migmatite" is most commonly and perhaps most usefully applied. These rocks have generated vigorous controversy since the arguments of J. J. Sederholm and P. J. Holmquist some 50 to 60 years ago. Sederholm maintained that the granitic veins were emplaced from external· sources below as granitic magma or "emations" (for example, 1967, 1926). Holmquist, however, insisted that the granitic veins were derived from

Fig. 7-4 *Small-scale, somewhat contorted migmatite or veined gneiss—granitic streaks in biotite-plagioclase-quartz gneiss. These granitic streaks were probably formed by metamorphic differentiation or by anatexis and appear subsequently deformed. Half mile west of Central Kootenai Lake, northern Bitterroot Range, Montana.*

within the rocks themselves, the material between the veins being the source for the veins, which then could develop into granitic magma (for example, 1908, 1921). Thus the former used the term "arterite," whereas the latter used "veinite," by analogy with arteries and veins as conveyors of blood into or out of the tissues. Numerous other terms have been used in the description of migmatites (e.g., "phlebite," "metatect," "diadysite," "agmatite," "embrechite," "nebulite"), some descriptive and some genetic These will not be discussed here but are reviewed in Sederholm (1923, 1926, or the 1967 collection of his selected works), Barth, (1962, pp. 358–364), King (1968), and Mehnert (1968).

More recently, the discussion has revolved around four major possibilities, discussed below. Rational arguments have been presented in favor of each of these possibilities, and each appears to be valid for individual occurrences. However, different workers often disagree on the origin of the granitic material in an individual occurrence. The problem then is to discriminate among these possibilities. Recent general papers on the problem include Dietrich (1960) and King (1965).

1. Injection of magma to form granitic veins (e.g., Osborne, 1936; Hietanen, 1938; Rogues, 1941; Buddington, 1948; Lovering and Goddard, 1950; Sutton and Watson, 1951; Dietrich, 1954, 1960; Hewitt, 1957; Haller, 1958a; Reesor, 1965; Page, 1968).
    This possibility is favored where:
    (a) The amount of granitic material is too great to have formed by metamorphic differentiation or anatexis.
    (b) The "veins" are thick or show emplacement by dilation of the country rocks.
    (c) The veins occur in limestone or some other rock of extreme composition, from which they would be unlikely to form by the other possibilities.
    (d) Geothermometry (see below) of the granitic material suggests temperatures in the magmatic range.
    (e) A contact metamorphic shell occurs adjacent to the granitic "vein."
    (f) Chilled margins occur on the granitic "vein."
    (g) The "veins" occur at an angle to the schistosity in the metamorphic part of the rock.
2. Metasomatic introduction of K, Na, and/or other elements to form granitic veins (e.g., Read, 1931; Wegmann, 1935; Barth, 1936; Buddington, 1948, 1957; Ramberg, 1949, 1952; Misch, 1949; Engel and Engel, 1953; Ward, 1959; King, 1965; Brown, 1966; Butler, 1969; Leveson and Seyfert, 1969).
    This is favored where:
    (a) The characteristics of the vein material do not appear to be

igneous, and mineralogy of the veins and of the metamorphic layers appear to be incompatible (and thus unlikely to form by metamorphic differentiation).

(b)   Geothermometry of the vein material suggests temperatures too low to be in the magmatic range, or geothermometry of the vein suggests high temperatures with a nonmagmatic mineralogy, in a lower-temperature rock.

(c)   Relict textures are suggestive of replacement of preexisting minerals or structures.

(d)   "Dust"-free grains in a rock consisting of "dust"-rich grains, or unbroken or undeformed grains of a certain mineral or group of minerals in a cataclastic or otherwise strongly deformed rock consisting of broken or deformed grains, probably have been introduced (Dietrich, 1960).

(e)   Veins have developed without disturbance of the enclosing rock (this is a very subjective matter and should be handled with the utmost of care). Compare Fig. 4-14 and the criteria for replacement veins listed in the section on granitic plutons.

3.   Metamorphic differentiation (e.g., Greenly, 1923; Eskola, 1932b; Hietanen, 1938; Ramberg, 1949; Harry, 1954; Reitan, 1956; MacKenzie, 1957; Crowder, 1959; Loberg, 1963; Evans, 1966; Bowes and Park, 1966; Kretz, 1966b, 1967; White, 1966; Gresens, 1967; Hawkins, 1968; Talbot, 1968; Ghaly, 1969; Hughes, 1970).

This is favored where:

(a)   The sum of the vein material plus that of the immediately adjacent material is equivalent to that of the original rock, interpreted as the rock still farther from the vein (see, for example, Fig. 4-14).

(b)   The characteristics of the vein material do not appear to be igneous, and the mineralogy of the veins and of the metamorphic layers is compatible and suggests the same temperature.

(c)   Geothermometry of the vein material suggests temperatures too low for the magmatic range.

(d)   Tracing of the layering laterally is possible into lower-grade, more homogeneous or less-deformed rocks.

(e)   High K/Rb and high Ba/K in veins, the opposite of that for pegmatites resulting from fractionation from a granitic magma (White, 1966).

4.   Anatexis or partial melting to form granitic veins (e.g., Eskola, 1933; Rogues, 1941; Mehnert, 1951, 1968, pp. 244–284; Haller, 1958; Dietrich, 1959; Koschmann, 1960; Smulikowski, 1960; Wyllie and Tuttle, 1961b; Zwart, 1963; Reesor, 1965; von Platten, 1965b; Elliot, 1966; Lundgren, 1966; Hyndman, 1968b, 1969; Fonteilles and

Guitard, 1968; Thompson and Norton, 1968; Hawkins, 1968; Hughes, 1970; Mikhaleva and Skuridin, 1971).

This is favored where:

(a)   The sum of the vein material plus that of the immediately adjacent material is equivalent to that of the original rock (as in 3a, above), especially the occurrence of biotite-rich selvages around K-feldspar-rich granitic veins.

(b)   The characteristics of the "vein" material appear igneous, and the mineralogy of the "veins" is compatible with that expected (e.g., from experimental work) from partial melting of the original rock (see, for example, von Platten, 1965a).

(c)   Geothermometry of the rocks suggests temperatures in the magmatic range.

(d)   Granitic streaks or patches are isolated in three dimensions.

As may be judged from the above criteria, the origin of migmatitic layering may be determined in some cases with ease and in other cases with considerable difficulty or not at all. Layering in high-grade gneisses also may be, of course, not migmatitic but relict sedimentary layering (see Fig. 7-5). Although in many cases their parentage is clear, in other cases the

Fig. 7-5 *Gneissic layering in high-grade pelitic gneisses, cut by granitic pegmatite dike. Layering in this instance is probably original sedimentary bedding. Note the very uneven distribution of light-colored layers and the lack of mafic selvages. Hammer in lower left has 13½-in.-long handle. Highway 2, about 15 miles north of Wenatchee, Washington.*

Fig. 7-6 *Relict sedimentary layering in high-grade diopsidic quartzite. Nearly vertical bedding is emphasized by leaching of thinner calcite-bearing layers. Trace of schistosity is nearly horizontal, parallel to axial surface of folds. North side of Scalping Knife Mountain, 10½ miles south of Nakusp, British Columbia.*

problem is not so easy. Some criteria favoring relict sedimentary or other surface layering (including volcanic tuffs and flows) are reviewed by Dietrich (1960) and include:

1. The granitic or felsic layers contain fossils, pebbles, or other composite grains; grain overgrowths and other grain-surface features; cross-bedding, scour and fill structures, mud cracks, graded bedding, and related features.
2. The layered sequence contains adjacent layers of diverse compositions, rather than just two alternating compositions.
3. The "granitic" layers contain minerals that rarely if ever occur in igneous rocks, especially graphite and possibly aluminum-rich silicates such as kyanite, sillimanite, andalusite, staurolite, almandine, corundum, and carbonates. However, at least some of these latter silicates may occur in contaminated igneous rocks or by metamorphic differentiation. Or the proportions are unknown from igneous rocks, such as in rocks that are dominantly quartz (see, for example, Fig. 7-6).

4. Chemical characteristics peculiar to sedimentary rocks (e.g., high ratio of Al_2O_3 to $Na_2O + K_2O + CaO$), although this can equally well be seen in the mineralogy and is subject to the same qualifications; and possibly high concentrations of certain trace elements (for example, Sc, Cu, Pb, Au, Co, Ni, and Cr).

Many migmatites show strong contrasts between the "granitic" and metamorphic portions. These in general are more amenable to interpretation of the processes of origin discussed above, than are the also abundant deep-seated, very diffuse migmatites. These diffuse, streaky rocks in many instances appear in hand specimen to be medium- to coarse-grained, massive or nearly massive granites. In clean exposures the size of an outcrop, however, hazy, streaky layers or folded layers may be quite apparent (see Fig. 7-7). These rocks are so thoroughly reconstituted that it is commonly extremely difficult or even impossible to be sure whether the layers are metamorphosed sedimentary layers or metamorphosed migmatites formed by one or more of the above processes, such as metamorphic differentia-

Fig. 7-7 *Coarser K-feldspar-rich layers (pink in outcrop) a few inches thick are indistinctly visible in nearly massive granite. Layers show similar style folding. These layers could be relict sedimentary layers, the whole rock being a product of granitization (though the pink layers could be a "low-melting" assemblage). Or the rock could be an igneous granite in which coarser granitic veins have been emplaced or were formed by metamorphic differentiation, the veins (and granite) later being folded by slip parallel to the axial planes of the folds (parallel to 13-in.-long hammer handle). Road cut on Red Lodge–Cooke City (Montana) Highway, 9 miles west-southwest of Beartooth Pass.*

Coarser K-feldspar-rich layer

tion, or even a product of differentiation accompanying flow streaking in a body of granitic magma.

Regional relationships within and around migmatite complexes are also important in their interpretation (e.g., see Haller, 1956b, 1958; Eckelmann and Poldervaart, 1957; Reesor, 1965). In most instances, the migmatites develop as concentric layers or shells, the more granitic types being overlain by migmatites containing larger proportions of more clearly metasedimentary material, and in turn by metasedimentary rocks containing little or no granitic material (e.g., Eckelmann and Poldervaart, 1957). Each of these shells is thousands to tens of thousands of feet thick. In some instances, at least, a clear succession has been described—a deep anatectic complex in which granitic melts are formed through partial melting within the rocks themselves, overlain by an "injection complex" in which the granitic melts so formed are injected to the higher levels (e.g., Haller, 1958). In some instances, the whole migmatite complex may rise to higher levels and intrude the overlying rocks (Wegmann, 1935; Haller, 1956b, 1958, 1962). The high-grade migmatite complex and its associated strong similar-style deformation is in many cases separated from the low-grade overlying metamorphic and sedimentary rocks with their typical concentric folds and faults (i.e., the "infrastructure" separated from the "suprastructure"; see Fig. 7-8) by a relatively thin zone having a steep metamorphic gradient with or without complex shearing (Wegmann, 1935; Haller, 1956b, 1958, 1962; Misch and Hazzard, 1962; denTex and Vogel, 1962; Zwart, 1963; Reesor, 1965; Armstrong and Hansen, 1966; Hyndman, 1968a; Fonteilles and Guitard, 1968; Campbell, 1970; Fyson, 1971). The deep-seated migmatites and associated rocks are considered to have been quite hot and plastic, flowing beneath the overlying low-grade rocks, which would be cooler and more rigid and thus would deform in a different manner. In some instances at least, the most prominent penetrative structures (e.g., axial surfaces of folds) in the deep-seated rocks are at approximately right angles to those in the overlying low-grade rocks (Haller, 1958; Zwart, 1963; Reesor, 1965; Hyndman, 1968a, b; Fyson, 1971). It has been suggested that the contrasting structures at the different levels could have developed at about the same time (Zwart, 1963; Reesor, 1965).

Mantled gneiss domes (Eskola, 1949) have been described as dome-shaped cores of old crystalline rocks that are interpreted as having been metamorphosed, remobilized, or partly melted and as having intruded their overlying stratified mantle rocks. The usefulness of this concept derives from the distinction between (and basis for recognition of) a high-grade infrastructure, which is separated tectonically from a lower-grade suprastructure, and an old igneous or metamorphic complex unconformably overlain by younger strata, the whole being later metamorphosed, or

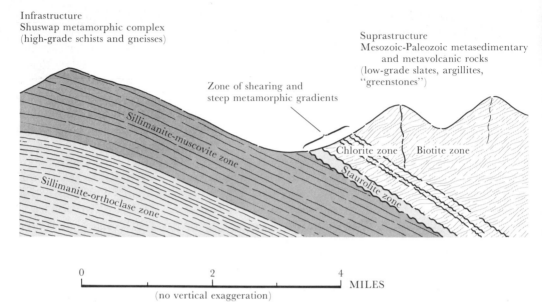

Fig. 7-8 *Cross section of boundary between infrastructure (high-grade Shuswap metamorphic complex) and suprastructure (overlying low-grade metasedimentary and metavolcanic rocks). Detailed structures are diagramatic. Nakusp area, southeastern British Columbia (Hyndman, 1968). Compare Fig. 8-9, p. 349).*

a concordant igneous pluton having intruded older overlying strata, the whole being later metamorphosed. Remobilization of older igneous or metamorphic core complexes has been documented for domes in Finland (Eskola, 1949; Wetherill et al., 1962; Kouvo and Tilton, 1966), Vermont (Faul et al., 1963), the Baltimore gneiss, Maryland (Tilton et al., 1958; Hopson, 1964; Wetherill et al., 1968), southeastern California (Lanphere et al., 1964), and southern Idaho (Armstrong, 1968a). Other crystalline gneiss domes in which the core complex may not be significantly older than the stratified lower-grade cover include the Oliverian domes in New Hampshire (Eskola, 1949; Naylor, 1969) and the Shuswap domes in southeastern British Columbia (Jones, 1959; Reesor, 1965, 1970; Hyndman, 1968a; Campbell, 1970; Fyles, 1970; McMillan, 1970).

RECOGNITION OF THE PARENT ROCK TYPE AND THE ORIGIN OF AMPHIBOLITE

Determination of conditions existing before metamorphism depends to a major extent on recognition of the type of rock prior to metamorphism. Most important is whether the parent rock is igneous or sedimentary; the answer is in many instances simple, but in some very difficult. If the metamorphic rock can be traced directly along strike to the unmetamorphosed equivalent, there is no problem. However, in most instances, this' is not feasible. Otherwise, the answer may lie in the occurrence, textures, and mineralogy (or rock chemistry) of the metamorphic equivalent. The most

Table 7-1 *Possible origins of amphibolite and criteria used for their determination*

Possible Origin	Criteria Used for Determination
Metamorphism of basalt or gabbro sill, dike, or pluton	May show contacts discordant to layering in country rocks ± chilled margins; layering may be due to metamorphic differentiation; may show relict igneous textures; zoning in plagioclase phenocrysts, subophitic or ophitic textures; relict minerals such as augite, hypersthene, or olivine; hornblende and plagioclase subequal in abundance, ± almandine, ± epidote, ± sphene, ± apatite, ± ilmenite, and other opaques; may contain minor amounts of biotite, quartz, or diopside; the chemical trend of different samples (e.g., on MgO/MgO + FeO) follows that for differentiation of diabase or gabbro; Cr, Ni may be high (250 ppm), as may Ti, Cu; Rb/Sr is low (0.03 to 0.33), La/Ce is low (< 0.4). (See, e.g., Wiseman, 1934; Flawn and King, 1953; Williams et al., 1954, pp. 241–243; Walker et al., 1960; Engel and Engel, 1962; Leake, 1963; Gomes et al., 1964; Elliot and Cowan, 1966; Gates, 1967; Tobisch, 1968.)
Metamorphism of basalt flow	May show discordant lower contact (unconformity); may show relict volcanic flow textures; zoning in plagioclase, phenocrysts, amygdules, monolithologic breccia, pillow structures; hornblende and plagioclase subequal in abundance, ± almandine, ± epidote, ± sphene, ± apatite, ± opaques; chemically equivalent to tholeiitic basalt (generally); low oxidation ratio $2Fe_2O_3 \times 100/Fe_2O_3 + FeO$ (30.); Cr, Ni, Ti high. (See, e.g., Williams et al., 1964; Ambrose and Burns, 1956; Engel and Engel, 1962; Leake, 1963, 1964; Gomes et al., 1964; Holdaway, 1965; Elliot and Cowan, 1966.)
Metamorphism of basalt tuff (or tuff contaminated by carbonate)	Constant composition, persistence along strike, layered, relict lapilli or agglomeratic textures, sequential pattern (but tuffs of basalt composition forming thin persistent layers are rare in unmetamorphosed sedimentary sequences); high oxidation ratio ($Fe_2O_3 \times 100/2Fe_2O_3 + Fe$) (av. 68); chemistry does not follow pelitic rock-carbonate trend. (See, e.g., Evans and Leake, 1960; Engel and Engel, 1962; Holdaway, 1965; Elliot and Cowan, 1966; van de Kamp, 1970.)

Table 7-1   *Possible origins of amphibolite and criteria used for their determination   (Continued)*

Possible Origin	Criteria Used for Determination
Metamorphism of shaley limestone or calcareous shale	Layered but not monotonously the same for tens of miles along strike and in successive beds for thousands of feet across strike (layering can probably also develop on a regional scale by metamorphic differentiation; Walker et al., 1960, p. 155, however, disagree); concordantly interlayered with marble, pelitic schist, and other clearly metasedimentary layers; widely varying mineral percentages; hornblende generally more abundant than plagioclase; biotite, quartz, diopside, or epidote may be abundant, sphene and apatite may be present, and almandine is generally absent; the chemical trend of different samples lies at a large angle to the differentiation trend of basalt on CaO: MgO/MgO + FeO or on Al + alkalis: Mg/Mg + Fe: CaO; K>Na; lower in Ni, Cr, Ti, ± Sc ± Cu, negative correlation of Cr, Ni with Mg/Mg + Fe, positive correlation of Ti with Mg/Mg + Fe, ± higher in Pb, Ba, Au. (See, e.g., Engel and Engel, 1957; Flawn and King, 1953; Williams et al., 1954; Poldervaart, 1953; Eckelmann and Poldervaart, 1957; Evans and Leake, 1960; Walker et al., 1960; Heier, 1962; Leake, 1963, 1964.)
Metasomatic replacement of marble	Transition to parent carbonate rock from a rock of suitable composition; may have relict carbonate; marble parent makes sense in the parent stratigraphic sequence (Adams and Barlow, 1910; Buddington, 1939; Engel and Engel, 1962).
Metasomatic replacement of pyroxene granulite (metasediment)	Pyroxene granulite adjacent to metagabbro; Sr 1000–1450 ppm in gabbroic plagioclase (An_{56-67}), Sr 250–900 (and one 1100) ppm in metasomatic amphibolite plagioclase (An_{40-82}). (See, e.g., Skiba and Butler, 1963.)

vexing of the parent rock problems is in the origin of amphibolite—rocks consisting primarily of hornblende and plagioclase, both in major amounts. Parent rocks of equivalent chemical composition include basalt flows, tuffs, sills, and dikes; gabbroic rocks, and various combinations of shaley limestone or calcareous shale. Rocks of equivalent composition may also form by metasomatism of limestone or other rocks. Possible origins and criteria used for their distinction are tabulated in Table 7-1.

Another problem rock is low-mafic quartzofeldspathic gneiss. Several criteria, especially those reviewed by Dietrich (1960) were listed above under the section "Migmatites and the Problem of Their Origin." Criteria suggesting sedimentary parentage included sedimentary features such as pebbles, cross-bedding, and graded bedding; diverse composition of alternating layers; presence of nonigneous minerals such as graphite, sillimanite, and almandine. The shape of zircons, because of their refractory nature, has been used by Poldervaart (1955, 1956). Irregular and subrounded zircons have been considered to be derived from a sedimentary rock (Poldervaart, 1955; Verspyck, 1961), the result being a paragneiss; euhedral and subhedral zircons, rounded ones being uncommon, have been considered to be derived from an igneous rock (Poldervaart, 1956; Kalsbeek and Zwart, 1967), the result being an orthogneiss. Rounding of euhedral zircons apparently may also result from high-grade metamorphism (Kalsbeek and Zwart, 1967).

Most other metamorphic rocks are relatively easily assigned to a parent rock type: pelitic slate, schist, gneiss, or hornfels from shales, graywackes, or other argillaceous sedimentary rocks; marble and skarn from limestones, quartzite from quartz sandstone; serpentinite from peridotite; and iron formation from ferruginous sediments.

REACTIONS BETWEEN METAMORPHIC MINERALS

Nature of Metamorphic Reactions

Metamorphic reactions, like other chemical reactions, depend on the temperature, pressure, and composition of the system, and on the rates of diffusion, nucleation, and crystal growth. Each of these factors has been discussed above.

It is apparent, from studies of the compositions of minerals that coexist in equilibrium with one another, that the composition of a mineral is controlled primarily by the composition of the rock and the nature (composition and structure) of the coexisting minerals but also by factors such as the temperature and pressure of the environment. These latter factors help to control the nature of the coexisting minerals and in some instances they also change the distribution of the elements within the same set of coexisting minerals. Because mineral reactions occur in response to changes in these factors—temperature, pressure, and composition—and because at least one mineral commonly appears or disappears during the reaction, the composition of many or even all of the minerals in the rock also generally changes. To what extent this occurs depends on the specific reaction and ranges from not at all to a radical change in composition of a mineral. In most instances, a mineral changes composition very slightly.

Most commonly, metamorphic mineral reactions are written as "idealized" chemical reactions illustrating the appearance or disappearance of a mineral of an idealized (end-member) composition. Although such simplified chemical reactions serve a useful purpose, and much use will be made of them below, it should be realized that they normally do not describe adequately the chemical changes occurring in the rock.

An illustrative example of how the other constituents of a high-grade pelitic gneiss participate in the simple polymorphic reaction kyanite ⇌ sillimanite is provided by Carmichael (1969). Making the reasonable assumption that rocks of a high metamorphic grade have passed through all the lower-grade zones and thus that the sequence of metamorphic zones mapped in space corresponds to the sequence of metamorphic changes in time, Carmichael reasons that it should be possible to record, in the texture of the minerals, each stage of the reaction that produced the new zone. Using evidence from the size and distribution of aluminum-bearing porphyroblasts, he suggests that aluminum is relatively immobile. In such rocks:

> The sillimanite does not appear to grow directly from kyanite.
>
> Sillimanite needles are embedded in biotite and quartz.
>
> Kyanite is rimmed by muscovite, which in some cases contains sillimanite needles.
>
> Biotite is embayed by plagioclase.

Based on these textures and assuming Al constant, Carmichael suggests that the following reactions make the sillimanite isograd:

$$2 \text{ muscovite} + \text{albite} + 3(\text{Mg, Fe})^{++} + H_2O \rightleftharpoons$$
$$\text{biotite} + 3 \text{ sillimanite} + 3 \text{ quartz} + K^+ + Na^+ + 4H^+$$
$$3 \text{ kyanite} + 3 \text{ quartz} + 2K^+ + 3H_2O \rightleftharpoons 2 \text{ muscovite} + 2H^+$$
$$\text{Biotite} + Na^+ + 6H^+ \rightleftharpoons \text{albite} + K^+ + 3(\text{Mg, Fe})^{++} + 4H_2O$$

If these reactions occur at essentially the same time (i.e., all the reactants on the left go to form all the products on the right), a cursory glance will show that most of the reactants cancel most of the products. The resultant reaction is:

$$3 \text{ kyanite} \rightleftharpoons 3 \text{ sillimanite}$$

which is the simple "idealized" reaction derived by comparing the mineralogy on either side of the isograd. As Carmichael (1969) has shown, however, this simple reaction is not the reaction actually occurring in the rock. He also notes that the activation energy (i.e., for nucleation and growth) is therefore apparently lower for the group of three actual reactions than it is for direct formation of sillimanite from kyanite. The other minerals (i.e., muscovite, albite, quartz, and biotite) act as catalysts, being used up in the destruction of kyanite and reformed in the development of sillimanite. This simple example illustrates also that even though a particular mineral species such as biotite may persist throughout many mineral zones, the individual grains may have been recrystallized more than once and that inclusions of grains within a porphyroblast may be not relics but products of the reaction along with the mineral forming the porphyroblast.

Carmichael (1969) shows also how several other more complex metamorphic reactions similarly involve minerals that do not appear in the overall change in mineralogy of the rock. The reaction:

9 staurolite + muscovite + 5 quartz $\leftrightarrow$

17 sillimanite + 2 garnet + biotite + 9H$_2$O

for example, also appears to involve recrystallization and temporary utilization of the components of albite in addition to changes in the minerals of the net reaction. Thus, although the above reaction is essentially correct for the net change across the sillimanite isograd for some areas, it does not show the actual path of the reaction. Similarly, the changing composition of individual minerals that appear on both sides of the isograd has been suggested by several workers (e.g., Tilley, 1923; Chinner, 1960, 1965; Turner and Verhoogen, 1960, p. 535; Guidotti, 1963; Fawcett, 1964; Evans and Guidotti, 1966; Loomis, 1966; Brown, 1967; Carmichael, 1969) and is generally not shown in idealized reactions of this type. Nevertheless such "idealized" reactions do serve a useful purpose, as long as it is remembered that they tell only part of the story.

Coexisting Minerals and Equilibrium between Them

An assemblage of metamorphic minerals in equilibrium with one another and with the fluid phase is not a static system. These coexisting phases are continually reacting with one another so as to maintain the equilibrium—ions continually moving from the grain surfaces into the fluid as other ions move from the fluid to the grains—a statistical balance being preserved. This equilibrium provides not only a total charge balance within the crystal but also controls the concentration of each element in each phase. Thermodynamically this can be expressed as "the chemical potential of each element is the same in each phase which is in equilibrium" (e.g., Turner and Verhoogen, 1960, p. 21; McIntire, 1963). This means, for example, that the chemical potential of Fe^{++} in staurolite Fe$_2$Al$_9$O$_6$(SiO$_4$)$_4$(O, OH)$_2$, in which it is a major component, is the same as in coexisting muscovite KAl$_2$(AlSi$_3$O)$_{10}$(OH)$_2$, in which it occurs as a trace component.

If the element exists at dilute concentrations (i.e., minor or trace element) the distribution follows a straight line governed by the Berthelot-Nernst distribution law (e.g., McIntire, 1963). The slope of the distribution line (concentration of A in phase 1/concentration of A in phase 2) is the partition coefficient or distribution coefficient K_D. This slope, K_D, depends especially on temperature, and many attempts have been made to correlate different values of K_D with temperature for a number of mineral pairs. This procedure has met with some success, particularly in the following:

Element in Coexisting Minerals		Reference
Fe, Mg:	Biotite, garnet	Engel & Engel, 1960; Albee, 1965a; Hietanen, 1969; Saxena, 1969a; Saxena & Hollander, 1969; Froese, 1970
Mg:	Orthopyroxene, clinopyroxene	Kretz, 1961; Howie, 1965; Saxena, 1969b; Katz, 1970
Fe:	Orthopyroxene, clinopyroxene	Binns, 1962; Katz, 1970
Ca:	Amphibole, plagioclase	Perchuk, 1966
Sr:	K-feldspar, plagioclase	Virgo, 1968
Mn, Ti, V, Co, B:	Hornblende, biotite; garnet, biotite	Engel & Engel, 1960; Carr & Turekian, 1961; Moxam, 1965
Tetrahedral Al:	Biotite, muscovite	Snelling, 1957; Lambert, 1959; Butler, 1967; Guidotti, 1969

K_D is considered to be pressure dependent in Ca in scapolite/plagioclase (Zharikov, 1966) and in garnet/plagioclase (Kretz, 1964; Saxena, 1968). A regular distribution of an element between coexisting minerals is generally considered to indicate equilibrium distribution of that element (e.g., Kretz, 1959; McIntire, 1963).

CONTROLS OVER THE COMPOSITION OF A METAMORPHIC MINERAL

As described above, the composition of metamorphic minerals changes with progressive metamorphism. Most important in controlling the composition of a mineral, however, is the composition of the system in which it develops. For a slightly more complex example, garnet (Ca, Mg, Fe^{++}, $Mn)_3(Al, Fe^{3+}, Cr)_2(SiO_4)_3$ from a high-Ca rock such as marble is rich in Ca (grossularite-andradite); one from a high-Al, moderately high Fe^{++} rock such as schist, gneiss, or hornfels, is rich in Al and Fe^{++} (almandite); one from a high-Mn metasediment is rich in Mn (spessartite); one from a high-Mg rock such as eclogite or peridotite is rich in Mg (pyrope).

Of somewhat lesser importance in the control of the composition of a mineral is the grade of metamorphism. This results from the presence of different phases in equilibrium with the mineral at higher grade and from a difference in distribution of elements between the coexisting minerals at higher grade. A number of minerals show distinct changes in composition (exclusive of rock composition) with changing grade (see Table 7-2).

DETERMINATION OF THE CONDITIONS OF METAMORPHISM

One of the main objectives of the study of metamorphic rocks is to determine their temperature and pressure of formation. There are several possible approaches to the problem. The *relative* conditions of one rock compared with another can be determined by field and petrographic work. The *"absolute"* conditions of formation can be determined (almost invariably with much poorer resolution than relative conditions) from the geological environment, by experimental synthesis of or reactions between several specific mineral species such as occur in the rock, or by thermodynamic calculation of the stability field of such a mineral. Clearly, to best define the metamorphic conditions for specific metamorphic rocks or for metamorphic facies in general, we should use both approaches—relative and "absolute." The difficulties in determination of "absolute" conditions of metamorphism are very great. Only the more obvious difficulties are outlined on the following pages.

Relative conditions of metamorphism can be determined in the field and the details sorted out petrographically, by considerations including grain size, proximity to a heat source such as an igneous pluton, and zone minerals or metamorphic facies that are indications of approximate grade (relative, not specific temperatures).

A number of metamorphic facies are now recognized (Fig. 7-9). These facies may be grouped under those that form under relatively shallow contact metamorphic conditions (low pressure), the hornfelses and related rocks, and those that form under deep-seated regional metamorphic conditions (high pressure). Further subdivision is possible on the basis of relative temperature, low- and high-pressure facies being placed in juxtaposition on the basis of similar mineralogy. These metamorphic facies and their relative conditions of formation are now well established. Each facies is recognized by a characteristic assemblage of minerals in rocks of a particular composition; for example, hornblende-plagioclase ($An_{>15}$) in rocks of basaltic chemical composition is called the amphibolite facies. Several facies may be further divided into subfacies based on mineral assemblages with a narrower temperature span, or into mineral zones in rocks of a particular composition. For example, the amphibolite facies may contain a staurolite and a sillimanite zone in rocks of pelitic composition. Regional metamorphic zones have been mapped in hundreds

Table 7-2 *Change in composition of metamorphic minerals with increase in metamorphic grade*

Mineral	Change with Increasing Metamorphic Grade	Reference
Potassium feldspar $(K,Na)AlSi_3O_8$	Less ordered (Al, Si ordering): Microcline (triclinic) in greenschist facies and staurolite zone, to orthoclase (monoclinic) in upper sillimanite zone and granulite facies, to sanidine in sanidinite facies Higher Na: Commonly orthoclase perthite in granulite facies, locally microcline	Heier, 1957; Hart, 1964; Hanson & Gast, 1967; Heald, 1950; Evans & Guidotti, 1966 Turner & Verhoogen, 1960, p. 555
Plagioclase $NaAlSi_3O_8 \cdot CaAl_2Si_2O_8$	Higher Ca: An_{0-7} in greenschist and albite-epidote amphibolite facies, $An_{>15}$ at higher grades Higher K: Commonly antiperthitic in granulite facies	deWaard, 1959; Billings, 1937; Lyons, 1955; Engel & Engel, 1960b; Chatterjee, 1961; Giraud, 1968; Hyndman, 1968; Vogel et al., 1968; Suwa, 1968; Crawford, 1966
Scapolite $3NaAlSi_3O_8 \cdot NaCl \cdot 3CaAl_2Si_2O_8 \cdot CaCO_3$	Higher Ca: From about Me_{40} in greenschist and albite-epidote amphibolite facies, to about Me_{40-50} in the staurolite zone, to about Me_{65} in the sillimanite zone to Me_{65-85} in the granulite facies	Hietanen, 1967
Chlorite $Mg_5(Mg,Al)(Al,Si)Si_3O_{10}(OH)_8^-$ $(Mg,Fe)_4Al_2Al_2Si_2O_{10}(OH)_8$	Lower Fe: From Fe-rich chlorite in chlorite zone to Mg rich in garnet and staurolite zones (FeO/MgO from 0.5 to 0.8)	Brown, 1967; Turner & Verhoogen, 1960, pp. 533, 539; Miyashiro, 1957; Chatterjee, 1962
Biotite $K(Mg,Fe,Mn,Al)_3Si_3AlO_{10}(OH)_8$	Higher tetrahedral $Al + Ti_{0.5}$ in place of $Si + (Mg,Fe)_{0.5}$ Higher MgO/FeO, higher MnO except in garnet-bearing rocks where it tends to decrease Higher Ba, Cr, V, Co/Fe, Ni/Fe Green colors (low grade) disappear and red colors (high grade and higher in Ti and Fe^{++}/Fe^{3+}) appear (browns at almost all grades)	Miyashiro, 1953, 1956, 1958; DeVore, 1955; Snelling, 1957; Hayama, 1959; Lambert, 1959; Engel & Engel, 1960b; Oki, 1961; Butler, 1965; Evans & Guidotti, 1966

Mineral	Description	References
Muscovite $KAl_2Si_3AlO_{10}(OH)_8$	Higher interlayer ordering and symmetry: 1M or 3T mixed-layer illite-montmorillonite in lowest grade (zeolite facies to lower greenschist facies), to 2M muscovites in biotite zone and higher grades. Higher Al: higher Na through greenschist facies followed by lower Na through amphibolite facies Lower Si, Fe^{++}, Fe^{3+}	Yoder and Eugster, 1955; Miyashiro, 1958; Lambert, 1959, Butler, 1965, 1967; Evans & Guidotti, 1966; Guidotti, 1963, 1969; Maxwell & Hower, 1967; Brown, 1967
Hornblende $(Ca,Na)_2 (Mg,Fe,Al)_5Si_6 (Si,Al)_2$-$O_{22}(OH)_2$	Tremolite or actinolite in chlorite or biotite zones Higher Na + K (0.45–0.58 in amphibolite facies, 0.67–0.85 in granulite facies) Higher Ti (if available) and lower Mn^{++} (0.04–0.05 in amphibolite facies, 0.005–0.03 in granulite facies) Slightly higher MgO/FeO: tetrahedral Al higher (1.5–1.65 in amphibolite facies, 1.64–2.05 in granulite facies) Green or blue-green colors in thin section (garnet zone), to green or olive, to greenish-brown or brown (sillimanite zone and granulite facies)	Wiseman, 1934; Sundius, 1946; Compton, 1958; Shido, 1958; Miyashiro, 1958, 1961; Engel & Engel, 1962a, 1964; Eskola, 1952; Howie, 1955; Suwa, 1968; Bard, 1970
Garnet $(Fe^{++}, Mn^{++}, Mg)_3Al_2(SiO_4)_3$-$Ca_3(Al, Fe^{3+})_2(SiO_4)_3$	Lower Mn: Spessartite in Mn-rich rocks of chlorite zone Higher FeO + MgO: 28% in garnet zone to 34% in sillimanite zone to higher in granulite facies Lower CaO + MnO: 9% in garnet zone to 5% in sillimanite zone	Nandi, 1967; Miyashiro, 1953; Lambert, 1959; Howie, 1955; Parras, 1958; Engel & Engel, 1960; Howie & Subramaniam, 1957; Sturt, 1962; Atherton, 1965; Oliver, 1968; Suwa, 1968

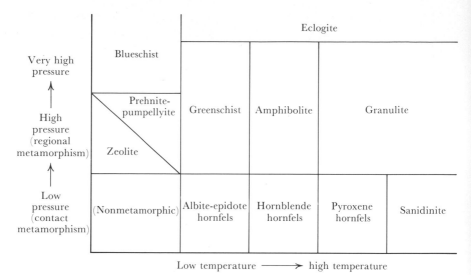

Fig. 7-9 *Relative temperature and pressure of the metamorphic facies as determined from field and petrographic relationships.*

of areas since the time of Barrow and the *relative* conditions of formation of most common minerals are now moderately well known. It remains for a series of absolute temperature determinations to place this grid of relative temperatures and pressures in the correct position at the correct scale on the petrogenic (pressure-temperature) grid. Relative pressures may be determined from stratigraphic relationships or from intersecting boundaries on a pressure-temperature grid (Carmichael et al., 1971).

Limits may be placed on the "absolute" pressure of formation of a rock by careful estimates of the stratigraphic depth at the time of metamorphism (e.g., Bloxam, 1956; Schulling, 1958; Coleman, 1961; Zwart, 1963). Some of the problems and conflicts with experimental estimates of pressure have been reviewed by Rutland (1965). Similarly, crude limits may be placed on the temperature of formation by measurements or estimates of the heat flow or thermal gradient, in conjunction with the estimated depth. Or metamorphic temperature limits may be determined according to the methods of Jaeger (1957, 1959), by estimates of thermal conductivity of the rock in conjunction with estimates of the temperature of the intrusion, its heat content, initial temperature, and the size and shape of the intrusion. Winkler (1965; 1967, pp. 80–83) and Turner (1968, pp. 18–22) review the method and some of the most common results. In particular, Winkler plots the thermal profile in the country rock, with distance from the intrusion, for intrusions of several common types. Turner plots the thermal profile changing with time in the country rock, for granite and diabase intruded into dry and into water-saturated sediments (see also Kudryatsev et al., 1971). Hori (1964, *in* Turner, 1968, p. 21) and Kesler and Heath (1968) show that outward transfer of magmatic water from a granitic (but probably not a dry basaltic) magma should also be taken into account so that the predicted size of the aureole is thereby enlarged and the predicted temperatures are higher. In this connection, it should be noted that although basaltic (or diabase or gabbroic) intrusions are intruded at higher temperatures, they commonly have smaller metamorphic aureoles, presumably because of the deficiency of contained water, which could carry heat into and could aid reactions within the country rocks. It is also important to note that the temperature in the country rock at the contact is always much lower (generally by about 100 to 400°C) than that of the intrusive. This temperature drop results from removal of heat in the heat of vaporization of water in the country rocks and in dehydration of water- or hydroxyl-bearing minerals.

An estimate of the maximum temperature adjacent to the contacts of granite and gabbro intrusions may be prepared from the data of Jaeger (1957). If the probable limits of the facies of contact metamorphism (as discussed below) are plotted on the same diagram (Fig. 7-10), it is immediately apparent that the temperatures adjacent to a granite intrusion are sufficient to produce minerals of the hornblende hornfels facies with the albite-epidote hornfels facies farther away. Adjacent to a gabbro intrusion,

Fig. 7-10 *Expected approximate maximum temperatures in wet country rocks (e.g., porous rocks containing water of any origin) immediately adjacent to granitic and gabbroic rock contacts, from the data of Jaeger (1957). Assumes heat transfer by conduction (heat in moving fluids neglected), magma is 100% melt on intrusion, country rocks are wet, and geothermal gradient is 20°C/km. Illustrates the effect of magma composition and depth of intrusion on grade of metamorphism.*

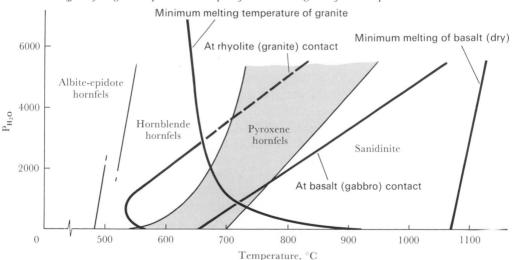

temperatures may reach those of the pyroxene hornfels facies. Rocks of intermediate composition, such as diorite, can be interpolated between the curves for granite and gabbro. Temperatures higher than these approximate limits may be reached in xenoliths within the intrusion. There the temperature in the country rock may closely approach the temperature of the magma.

The most straightforward method for determination of the conditions of formation of a metamorphic rock is by comparison with stability fields of minerals determined by experimental synthesis or reactions between a number of phases. With this in mind a considerable amount of experimental work at high temperatures and pressures has been attempted in the past 20 years. Synthesis of a mineral phase involves both the growth and destruction of the mineral or group of minerals under a range of conditions sufficient to define a reaction boundary such as described in "The Controls of Metamorphism" (pp. 270–280). Most commonly the "hydrothermal bombs" used for such a synthesis are thick-walled cylindrical tubes of high-strength steel permanently closed at one end and sealed at the other end when the charge of chemical reagents or minerals is enclosed. The pressure is applied in the form of water vapor, CO_2, or any other gas, depending on which phase participates in the reaction or comes closest to matching the appropriate natural system. The bomb under pressure is heated to the desired temperature in a furnace and the conditions are maintained for a time ranging from hours to months—however long it takes for the system to reach equilibrium under the applied conditions. Some reactions never do reach equilibrium in a reasonable length of time. Then the bomb is suddenly cooled or "quenched" and the pressure lowered, for study of the resultant phases. The hydrothermal bombs were designed by Tuttle (1948, 1949), have been modified since to some extent, and are now capable of sustained use at about 750°C and 5,000 bars. Higher temperatures are possible at lower pressures. Much higher pressures are attained with different types of equipment such as the triaxial press, Bridgeman anvil ("squeezer"), or piston and cylinder, but the accuracy of measurement falls off significantly. Techniques are discussed in numerous papers, including reviews by Roy and Tuttle (1956), Wyllie (1966), Turner (1968, pp. 87–94), and Fyfe (1969a, b).

The experimental work has resulted in major advances in igneous and metamorphic petrology but it should be realized that it is not a panacea. Experimental papers like any other must be reviewed with a critical eye and the results used with caution. Syntheses of and reactions between assemblages of many silicate minerals under metamorphic conditions has given some indication of the T, P_{H_2O}, and in some cases P_{CO_2}, conditions involved in their formation. However, several points should be considered to aid in the evaluation of experimental or thermodynamic data to the natural system (e.g., Fyfe, Turner, and Verhoogen, 1958, pp. 150–151, 175–185,

1961; Fyfe, 1960, 1962 personal communication; Rosenfeld, 1961; Greenwood, 1963; MacKenzie, 1965; Turner, 1968, pp. 85–171; Zen, 1969;
Holdaway, 1971):

> The reaction forming the mineral or mineral assemblage should
> be reversed. If the mineral is merely synthesized, the data set only
> an upper limit to the low-temperature boundary of the mineral and
> a lower limit to the high-temperature boundary. If the mineral is
> merely destroyed, the data set an upper limit to the high-temperature
> boundary of the mineral. If the reaction is reversed, equilibrium
> is established, but the possibility of a metastable equilibrium should
> be considered. This possibility may be eliminated if the same ex-
> perimental curve is obtained by variation of the experimental time
> by several orders of magnitude (Fyfe and Godwin, 1962). Note that
> much greater time and presence of catalysts should lead to a closer
> approach to stability in the natural system.

> Reactions involving only synthesis are of questionable application
> if the starting materials are artificial and highly reactive (e.g., gels,
> glass, and to some extent even oxides). These often lead to forma-
> tion of metastable products.

> The possibility of retrograde reaction during quenching should
> be considered.

> The resultant phase diagram should be compatible with thermo-
> dynamic principles and with Le Châtelier's rule. For example, the
> high-volume assemblage (especially one containing water) should
> be on the low-pressure side of the equilibrium boundary. Basic prin-
> ciples with examples are reviewed by Turner (1968, pp. 94–109).

> If the composition of the experimental system is not the same as
> the composition of the appropriate rock (the usual case), the ex-
> perimental data can only approximate or be used to set limits on the
> conditions of formation of the rock. Consider also the composition
> of volatiles in the experimental system (for example, H_2O, CO_2) and
> those likely to be in the natural system (for example, H_2O, CO_2, Cl_2,
> F_2, O_2, B, etc.).

> Stability limits for a single mineral set limits on conditions of forma-
> tion of the rock (the presence of an additional phase can only *reduce*
> the stability field of the mineral, *provided* the synthetic mineral has
> the same composition and structure as the natural mineral). For
> example, the common presence of quartz generally reduces the sta-
> bility field of a mineral (e.g., muscovite).

> The method of applying the temperature and particularly the pres-

sure should be such that each is constant over the span of the sample and position of measurement of the variable. The methods of determining the temperature and pressure during the experiment and of identifying the product should be reliable for the type of equipment used and for the nature of the product.

The experimental results should be compatible with the geological data.

The application of experimental results should be compatible with the equilibrium mineralogy in the rock and with the reaction as determined from the mineralogy and textures in the rock.

The addition of an impurity in a mineral may either reduce or extend the stability field. Fe^{++} as an impurity in Mg-bearing minerals almost invariably reduces the stability field. Al^{3+} as an impurity most commonly increases the stability field.

Many of the above difficulties and considerations for evaluation of experimental data also apply to stability boundaries thermodynamically. Thermodynamic data[1] are an inherently reliable way to determine stability boundaries. Presently, however, thermodynamic values (of entropy S, molar volume V, heat of formation ΔH) are known for oxides and a number of other compounds, primarily at 1 bar (e.g. Latimer, 1952; Kelley, 1960, 1962; Robie, 1966). Given V or a point on the equilibrium boundary at elevated pressure, the rest of the boundary may be calculated. Unfortunately, few thermodynamic data are yet available at elevated pressures. Recently, however, it has been shown feasible to calculate thermodynamic values from experimental equilibrium data (Fisher and Zen, 1971). Examples of thermodynamic calculation of equilibrium boundaries for a few systems are given by Turner (1968, pp. 94–109), Fyfe, Turner, and Verhoogen (1958, pp. 115–128, 149–185), Miyashiro (1960), Greenwood (1961), Ramberg (1964), Orville and Greenwood (1965), Anderson (1970).

In order to help overcome the problems arising from the difficulties and deficiency of data in experimental and thermodynamic work, we may turn to the natural system of the rocks themselves, for relative stability data to combine with these data on absolute conditions. And despite the problems in experimental and thermodynamic work, a sizable amount of very useful and important information—in fact, many of the most important recent advances in the field—have been produced from these sources. Relative stability data are available in the form of numerous de-

---

[1]Those interested in this approach but lacking a background in thermodynamic principles may acquire a basic understanding from Krauskopf, 1967; Kern and Weisbrod, 1967.

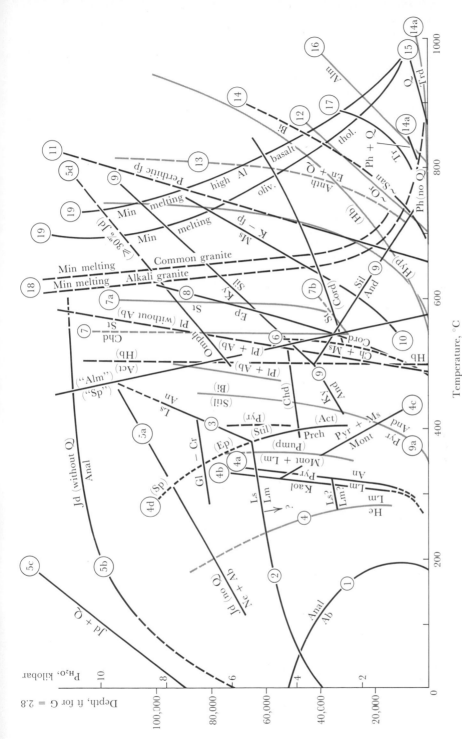

Fig. 7-11 P_{H_2O}-T diagram for the stability limits of common minerals in quartz-bearing, noncalcareous rocks. Numbered boundaries refer to experimentally determined reactions discussed in the text. Other boundaries (minerals in parentheses) are positioned relative to those (see text). Abbreviations used are: Ab = albite; Act = actinolite; Alm = "almandine"; An = anorthite; And = andalusite; Anal = analcite; Anth = anthophyllite; Bi = biotite; Ch = Mg-chlorite; Chd = chloritoid; Cord = cordierite; Cr = crossite; En = enstatite; Ep = epidote; Fp = feldspar; Gl = glaucophane; Hb = hornblende; He = heulandite; Hyp = hypersthene; Jd = jadeite; Kaol = kaolinite; K-fp = K-feldspar; Ky = kyanite; Lm = laumontite; Ls = lawsonite; Mont = montmorillonite; Ms = muscovite; Ne = nepheline; Or = orthoclase; Ph = phlogopite; Pl = plagioclase; Preh = prehnite; Pump = pumpellyite; Pyr = pyrophyllite; Q = quartz; San = sanidine; Sil = sillimanite; Sp = "spessartite"; St = staurolite; Stil = stilpnomelane; Tr = tremolite; Trd = tridymite.

313

scriptions of the assemblages of minerals with progressive metamorphism
and in the compilation of much of this information in the form of phase
diagrams (for example, ACF, AKF, AFM; see Figs. 6-2 to 6-4, 8-4 to 8-19)
for each of the presently recognized metamorphic facies and even zones
(e.g., Turner, 1948; Fyfe, Turner, and Verhoogen, 1958, pp. 199–236;
Turner and Verhoogen, 1960, pp. 510–560; Barth, 1962, pp. 311–337;
Winkler, 1965, 1967, pp. 27–46, 64–172; Turner, 1968, pp. 190–348). A
number of these are reviewed below in the discussion of the different
metamorphic rock associations and the individual facies of which they
are composed. To put these associations and facies in an approximate
pressure-temperature framework to facilitate discussion, a correlation of
P-T conditions for common quartz-bearing assemblages may be prepared,
using mutually compatible experimental and petrographic data. In essence, what has been done here is to prepare a P-T grid of relative pressures and temperatures, and then using what appears to be some of the
more reliable experimental or thermodynamic results, to stretch or collapse this grid in different parts or directions so that compatibility is
achieved. Most important, if both experimental and petrographic data
are correct, they should be mutually compatible.

Review of Selected Experimental or Thermodynamic Work

The following experimental work appears as numbered stability boundaries on Fig. 7-11—P_{H_2O}-T for quartz-bearing assemblages.

(1)   Analcite + quartz $\leftrightarrow$ albite + H_2O
    $NaAlSi_2O_6 \cdot H_2O$ SiO_2 $NaAlSi_3O_8$
Campbell and Fyfe (1965); see also Turner (1968, p. 158).

(2)   Laumontite $\leftrightarrow$ lawsonite + 2 quartz + $2H_2O$
    $CaAl_2Si_4O_{12} \cdot 4H_2O$ $CaAl_2Si_2O_7(OH)_2$ SiO_2
Crawford and Fyfe (1965); see also Turner (1968, p. 154), Liou
(1969), and Thompson (1970).

(3)   Lawsonite $\leftrightarrow$ anorthite + $2H_2O$
    $CaAl_2Si_2O_7(OH)_2 \cdot H_2O$ $CaAl_2Si_2O_8$
Crawford and Fyfe (1965); see also Turner (1968, p. 155), Newton
and Kennedy (1963).

(4)   Heulandite $\leftrightarrow$ laumontite + 3 quartz + H_2O
    $CaAl_2Si_7O_{18} \cdot 6H_2O$ $CaAl_2Si_4O_{12} \cdot 4H_2O$ SiO_2
Coombs et al. (1959), Nitsch (in Winkler, 1967, p. 159).

(4a)   Laumontite $\leftrightarrow$ anorthite + 2 quartz + $4H_2O$
    $CaAl_2Si_4O_{12} \cdot 4H_2O$ $CaAl_2Si_2O_8$ SiO_2
Thompson (1970).

(4b)   Kaolinite + 4 quartz ↔ 2 pyrophyllite + $2H_2O$
$Al_4Si_4O_{10}(OH)_8$ SiO_2 $Al_2Si_4O_{10}(OH)_2$

(4c) 2 "montmorillonite" ↔ pyrophyllite + 2 muscovite
$KAl_4(Si_{3.5}Al_{0.5})(OH)_4 \cdot nH_2O$ $Al_2Si_4O_{10}(OH)_2$ $KAl_2(AlSi_3O_{10})(OH)_2$

Velde (1969).

(4d) 5 prehnite ↔
$CaAl_2Si_3O_{10}(OH)_2$

2 zoisite   + 2 grossular + 3 quartz + $4H_2O$
$Ca_2Al_3Si_3O_{12}(OH)$ $Ca_3Al_2Si_3O_{12}$ SiO_2

Liou (1971); Strens (1968) reports 380°C at 2,000 bars for natural (5% Fe_2O_3) prehnite equilibrium.

(5a) Jadeite ↔ nepheline + albite
$NaAlSi_2O_6$ $NaAlSiO_4$ $NaAlSi_3O_8$

(5b) Jadeite + H_2O ↔ analcite
$NaAlSi_2O_6$ $NaAlSi_2O_6 \cdot H_2O$

(5c) Jadeite + quartz ↔ albite
$NaAlSi_2O_6$ SiO_2 $NaAlSi_3O_8$

(5d) Omphacite + quartz ↔ albite + diopside
$NaAlSi_2O_6 \cdot CaMgSi_2O_6$ SiO_2 $NaAlSi_3O_8$ $CaMgSi_2O_6$

Boettcher and Wyllie (1968), Wikström (1970); see also Bell and Kalb (1969), Kushiro (1969), Fyfe and Valpy (1959), and Robertson et al. (1957).

(6a)   Chlorite +   quartz ↔
$(Mg, Fe)_5Al(AlSi_3O_{10})(OH)_8$ SiO_2

cordierite + $\begin{cases} \text{anthophyllite} \\ \text{or gedrite} \end{cases}$ + H_2O
$(Mg, Fe)_2Al_4Si_5O_{18}$ $(Mg, Fe, Al)_7(Si, Al)_8O_{22}(OH)_2$

(6b)   Chlorite + muscovite + quartz ↔
$(Mg, Fe)_5Al(AlSi_3O_{10})(OH)_8$ $KAl_2(AlSi_3O_{10})(OH)_2$ SiO_2

cordierite + biotite + H_2O
$(Mg, Fe)_2Al_4Si_5O_{18}$ $K(Mg, Fe)_3(AlSi_3O_{10})(OH)_2$

Reaction 6a: Fawcett and Yoder (1966) and Akella and Winkler (1966). Reaction 6b: Schreyer and Yoder (1964), Schreyer (1965), and Hirschberg (*in* Winkler, 1967, p. 180).

(7)   Chloritoid   +   Al-silicate ↔
$(Fe, Mg)_2(Al, Fe^{3+})Al_3O_2(SiO_4)_2(OH)_4$ (e.g., pyrophyllite)

staurolite   + 3 quartz + H_2O
$(Fe, Mg)_2(Al, Fe^{3+})_9O_6(SiO_4)_4(O, OH)_2$ SiO_2

Hoschek (1967); see also Ganguly (1968, 1969) and Richardson (1969).

(7a)   Staurolite   +   quartz $\hookrightarrow$

$Fe_2Al_9O_6(SiO_4)_4(OH)_2$ SiO_2

almandine + sillimanite + H_2O

$Fe_3Al_2(SiO_4)_3$ Al_2SiO_5

(7b)   Staurolite   +   quartz $\hookrightarrow$

$Fe_2Al_9O_6(SiO_4)_4(OH)_2$ SiO_2

Fe-cordierite + sillimanite + H_2O

$Fe_2Al_4Si_5O_{18}$ Al_2SiO_5

Modified to 100°C lower from Richardson (1968a).

(8)   4 clinozoisite   +   quartz $\hookrightarrow$

$Ca_2(Al, Fe^{3+})_3Si_3O_{12}(OH)$ SiO_2

5 anorthite + grossularite (+ hematite) + H_2O

$CaAl_2Si_2O_8$ $(Ca, Fe^{++})_3Al_2(SiO_4)_3$

Modified to 100°C lower from Holdaway (1966) to better correspond
to the relative temperature controls imposed by experimental and
petrographic data for other minerals. The common presence of significant Na in many natural rock systems would probably facilitate
(i.e., lower the temperature of) formation of plagioclase by combination with the anorthite molecule.

(9) Kyanite $\hookrightarrow$ andalusite $\hookrightarrow$ sillimanite

Al_2SiO_5 Al_2SiO_5 Al_2SiO_5

(9a) Pyrophyllite $\hookrightarrow$ andalusite + 3 quartz + H_2O

$Al_2Si_4O_{10}(OH)_2$ Al_2SiO_5 SiO_2

Holdaway (1971), Richardson, Gilbert, and Bell (1969), Gilbert,
Bell, and Richardson (1969), supplemented by the data of Newton
(1966b) and Weill (1963, 1966). Holdaway suggests that the higher-
temperature andalusite stability of earlier studies results from extreme boundary overstepping and resultant formation of fibrolite,
the fine-grained "disordered" form of sillimanite. See also Albee
and Chodos (1969), Hollister (1969b), Bell (1963), Khitarov et al.
(1963), Holm and Kleppa (1966), Anderson and Kleppa (1969).

(10)   Muscovite   +   quartz $\hookrightarrow$ K-feldspar +$\begin{cases} \text{andalusite} \\ \text{or sillimanite} \end{cases}$ + H_2O

$KAl_2(AlSi_3O_{10})(OH)_2$ SiO_2 $KAlSi_3O_8$ Al_2SiO_5

Evans (1965b); see also Weill (1966), Yoder and Eugster (1955),
Velde (1964, 1966), Crowley and Roy (1964).

(11) Homogeneous alkali feldspar $\hookrightarrow$ perthite (or antiperthite)

$(K, Na)AlSi_3O_8$ $(K, Na)AlSi_3O_8 + (Na, K)AlSi_3O_8$

Bowen and Tuttle (1950), Yoder, Stewart, and Smith (1957), Orville
(1963), and Morse (1969) for the maximum on the solvus.

(12) Orthoclase $\leftrightarrow$ sanidine
$\quad\quad$ $KAlSi_3O_8$ $\quad\quad\quad$ $KAlSi_3O_8$
Smith et al. (1965).

(13) Anthophyllite $\leftrightarrow$ 7 enstatite + quartz + H_2O
$\quad\quad$ $Mg_7Si_8O_{22}(OH)_2$ $\quad\quad$ $MgSiO_3$ $\quad\quad$ SiO_2
Greenwood (1963), Fyfe (1962), Zen (1971).

(14) $\quad\quad\quad$ Biotite $\quad\quad$ + $\quad\quad$ quartz $\leftrightarrow$
$\quad\quad$ $K(Fe^{++}, Mg)_3AlSi_3O_{10}(OH)_2$ $\quad$ SiO_2
$\quad\quad\quad\quad\quad\quad\quad$ sanidine + magnetite $\pm$ hematite + H_2O
$\quad\quad\quad\quad\quad\quad\quad$ $KAlSi_3O_8$ $\quad\quad$ Fe_3O_4 $\quad\quad$ Fe_2O_3

Eugster and Wones (1962) and Wones and Eugster (1965) for $Fe^{++}/(Fe^{++} + Mg)$ of 0.5 to 0.6 (Butler, 1965) and $Fe^{3+}/(Fe^{3+} + Fe^{++})$ of about 0.05 to 0.25 in biotite.

(14a) $\quad$ 2 phlogopite $\leftrightarrow$ kalsilite + leucite + 3 forsterite + H_2O
$\quad\quad$ $KMg_3AlSi_3O_{10}(OH)_2$ $\quad$ $KAlSi_2O_6$ $\quad$ $KAlSi_2O_6$ $\quad$ Mg_2SiO_4

(14b) $\quad$ Phlogopite + 3 quartz $\leftrightarrow$ sanidine + enstatite + H_2O
$\quad\quad$ $KMg_3AlSi_3O_{10}(OH)_2$ $\quad$ SiO_2 $\quad\quad$ $KAlSi_3O_8$ $\quad$ $MgSiO_3$
Wones and Dodge (1968); see also Wones (1967), Yoder and Eugster (1954).

(15) Quartz $\leftrightarrow$ tridymite
$\quad\quad$ SiO_2 $\quad\quad\quad$ SiO_2
Tuttle and England (1955).

(16) 5 almandine $\leftrightarrow$ 2 Fe-cordierite + 5 fayalite + hercynite
$\quad\quad$ $Fe_3Al_2(SiO_4)_3$ $\quad\quad$ $Fe_2Al_4Si_5O_{18}$ $\quad\quad$ Fe_2SiO_4 $\quad\quad$ $FeAl_2O_4$
Hsu (1968); see also Yoder (1955a), Coes (1955).

(17) $\quad$ 2 tremolite $\leftrightarrow$ 3 enstatite + 4 diopside + 2 quartz + $2H_2O$
$\quad\quad$ $Ca_2Mg_5Si_8O_{22}(OH)_2$ $\quad\quad$ $Mg_2(SiO_3)_2$ $\quad$ $CaMg(SiO_3)_2$ $\quad$ SiO_2
Boyd (1954, 1959); see also Ernst (1966).

(18) Alkali granite $\leftrightarrow$ incipient melting
$\quad\quad$ Common granite $\leftrightarrow$ incipient melting (subequal amounts of
$\quad\quad\quad\quad\quad\quad\quad\quad\quad\quad$ quartz, orthoclase, plagioclase An_{25})
Tuttle and Bowen (1958), Luth et al. (1964), von Platen and Höller (1966, also reported in Winkler, 1967, p. 206), Merrill et al. (1970), Piwinskii (1968).

(19) Basalt $\leftrightarrow$ incipient melting
Yoder and Tilley (1962). Positions of these boundaries above 8 kbars have been modified after Lambert and Wyllie (1968) and Bryhni et al. (1970). Other significant mineral boundaries, for which little or

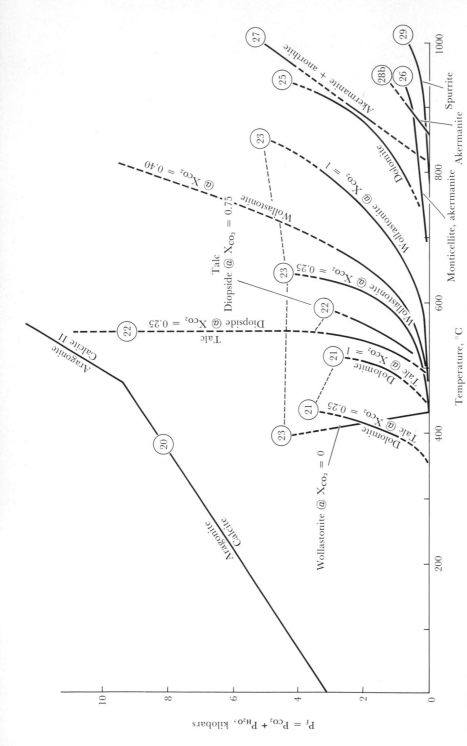

Fig. 7-12 P_{fluid} ($P_{CO_2} + P_{H_2O}$)-T diagram for the stability limits of common minerals in quartz-bearing calcareous rocks. Numbered boundaries refer to experimentally determined reactions discussed in the text. Compare Figs. 6-6 and 6-7.

no experimental data are available or for which the experimental data appear inconsistent with other information, have been plotted (labeled in parentheses) in accordance with petrographic observations. The positions of these boundaries have been sketched as smooth curves (or in the case of reactions not involving a fluid phase, straight lines) compatible with the stabilities of the other minerals and the experimental work discussed above.

Ca- and Mg-rich minerals involving P_{CO_2} are plotted separately on Fig. 7-12. Again the experimental work appears on the figure as numbered boundaries:

(20) Aragonite $\leftrightarrow$ calcite
        $CaCO_3$ $CaCO_3$

Modified to 0.17 kbar lower for 1 mole % $SrCO_3$ from Boettcher and Wyllie (1967), Crawford and Fyfe (1964), Goldsmith and Newton (1969); see also Froese and Winkler (1966), Froese (1970), and Vance (1968).

(21a) 3 dolomite + 4 quartz + $1H_2O$ $\leftrightarrow$
          $CaMg(CO_3)_2$ SiO_2

$$1 \text{ talc} \quad + \quad 3 \text{ calcite} + 3CO_2$$
$$Mg_3Si_4O_{10}(OH)_2 \qquad CaCO_3$$

(21b) 5 dolomite + 8 quartz + $1H_2O$ $\leftrightarrow$
          $CaMg(CO_3)_2$ SiO_2

$$1 \text{ tremolite} \quad + \quad 3 \text{ calcite} + 7CO_2$$
$$Ca_2Mg_5Si_8O_{22}(OH)_2 \qquad CaCO_3$$

Metz and Winkler (1963; also reported in Winkler, 1967, pp. 24–25) and Metz and Trommsdorff (1968). Values of X_{CO_2} less than 0.25 seem unlikely in carbonate-rich rocks in which CO_2 is being formed in quanitity during metamorphism.

(22a)   Tremolite   +   3 calcite + 2 quartz $\leftrightarrow$
        $Ca_2Mg_5Si_8O_{22}(OH)_2$ $CaCO_3$ SiO_2

$$5 \text{ diopside} + 3CO_2 + H_2O$$
$$CaMg(SiO_3)_2$$

Metz and Winkler (1964), Metz (1966), and Metz and Trommsdorff (1968).

(22b)   Tremolite   +   11 dolomite $\leftrightarrow$
        $Ca_2Mg_5Si_8O_{22}(OH)_2$ $CaMg(CO_3)_2$

$$8 \text{ forsterite} + 13 \text{ calcite} + 9CO_2 + 1H_2O$$
$$Mg_2SiO_4 \qquad CaCO_3$$

Metz (1967); lies within about 10°C of the above diopside-forming reaction.

(23) Calcite + quartz ↔ wollastonite + CO_2

$CaCO_3$ SiO_2 $CaSiO_3$

Harker and Tuttle (1955) and Greenwood (1962, 1967).

(24) Brucite ↔ periclase + H_2O

$Mg(OH)_2$ MgO

Fyfe (1958) and Johannes and Metz (1968); see also Kennedy (1956), MacDonald (1955).

(25) Dolomite ↔ periclase + calcite + CO_2

$CaMg(CO_3)_2$ MgO $CaCO_3$

Harker and Tuttle (1955), Graf and Goldsmith (1955).

(26a) Calcite + forsterite ↔ monticellite + periclase + CO_2

$CaCO_3$ Mg_2SiO_4 $CaMgSiO_4$ MgO

(26b) 2 calcite + forsterite + diopside ↔ 3 monticellite + $2CO_2$

$CaCO_3$ Mg_2SiO_4 $CaMg(SiO_3)_2$ $CaMgSiO_4$

Walter (1963b); see also Turner (1968, p. 135).

(27) Grossularite   +   clinopyroxene   ↔

$Ca_3Al_2(SiO_4)_3$ $(2CaMgSi_2O_6 + CaAl_2SiO_6)$

2 akermanite + 2 anorthite

$Ca_2MgSi_2O_7$ $CaAl_2Si_2O_7$

Yoder (1969a).

(28a) Diopside + calcite ↔ akermanite + CO_2

$CaMg(SiO_3)_2$ $CaCO_3$ $Ca_2MgSi_2O_7$

(28b) Diopside + 3 monticellite ↔ 2 akermanite + forsterite

$CaMg(SiO_3)_2$ $CaMgSiO_4$ $Ca_2MgSi_2O_7$ Mg_2SiO_4

Walter (1963a); 28a is coincident with reaction 26.

(29) 3 calcite + 2 wollastonite ↔ spurrite + CO_2

$CaCO_3$ $CaSiO_3$ $Ca_2SiO_4 \cdot CaCO_3$

Tuttle and Harker (1957).

Limitations of the Present P-T Diagrams

Aside from the limitations of experimental and thermodynamic work described above, the stability boundaries plotted do not mark the position for growth or disappearance of that phase in specific rocks, of course, because this depends on the composition of the rock and on other factors enumerated below. The plotted stability limits are intended to encompass the outside limits for all common rocks—quartz-bearing rocks in Fig. 7-11 and calcareous rocks in Fig. 7-12. They are not applicable to rocks of unusual composition, such as those abnormally high in Mn, or to those occurring under "unusual" conditions, such as metamorphism of a granite or a fresh basalt, in which P_{H_2O} (or P_{fluid}) may be much less than P_{load}.

Other limitations that must be kept in mind in using a diagram such as this include the problems and difficulties of experimental work outlined above. For these reasons the diagram is but semiquantitative and must be so treated. However, the major restriction of internal compatibility of many pieces of petrographic and reliable experimental data tends to strongly support the overall reliability and general applicability for most natural assemblages. Several points may be noted:

> Several mineral reactions occur under about the same conditions, especially in the zone between about 500 and 550°C, across the important boundary between the albite-epidote hornfels—greenschist facies and the hornblende hornfels—amphibolite facies.

> The usual mole fraction of CO_2 (X_{CO_2}) for the breakdown of carbonate-bearing minerals (such as dolomite and calcite in the presence of quartz and the appearance of diopside) seems nearly constant between about 0.25 and 0.5 and is perhaps lower than that expected. This may be caused by a large continual flux of water through rocks undergoing metamorphism, the source of the water being other metamorphic rocks deeper in the section.

> No mineralogical criteria are available for distinguishing the hornfels facies from the low-pressure regional metamorphic facies. In fact, since the distinction is a spatial and textural one, dependent on the occurrence of syntectonic crystallization, time, and depth at the time of metamorphism, the largely pressure-dependent boundary is variable in position.

DETERMINATION OF THE AGE OF METAMORPHIC ROCKS
AND THE TIME OF METAMORPHISM

The problem of age determination in metamorphic rocks involves both the age of the rocks and the age of metamorphism. The determination of the age of the parent rocks from which the metamorphic rocks were derived rests largely on conventional grounds such as identification of contained fossils (if any) and correlation with stratigraphic units or other features of known age. Direct determination of the age of sedimentation by radiometric means may be attempted by potassium-argon or rubidium-strontium determinations on glauconite, though the resulting "ages" are generally somewhat low, or on pyroclastic grains or layers in the sedimentary rocks. Maximum age limits on sedimentation may be obtained by determinations on minerals of clearly detrital origin (any nonrecrystallized minerals such as detrital mica). Minimum age limits on sedimentation may be obtained by determinations on minerals of clearly metamorphic or later igneous origin.

Determination of the age of metamorphism may also be obtained on conventional grounds, using as a maximum the age obtained for sedimentation and as a minimum the age of overlying unmetamorphosed sedimentary rocks (especially if in sharp, unconformable contact), unmetamorphosed later intrusive rocks, and postmetamorphic structures or other features. The most widely used radiometric age determinations are those of potassium argon, but the Rb–Sr method is now coming into wide use. In high-grade rocks, both of these methods generally date the time of metamorphism;[1] or more correctly for K–Ar in particular, they place a minimum age on this metamorphism as explained below. Although radiometric ages are very useful, they are only as reliable as the associated supporting field and petrographic information from the rocks.

The most commonly used methods utilize potassium and argon (K–Ar), rubidium and strontium (Rb–Sr), uranium (or thorium) and lead (U–Pb or Th–Pb), fission track, and lead alpha (Pb–α), in approximate order of present use. These methods are based on the occurrence of a radioactive isotope or element (the parent) that spontaneously and continuously decays (or disintegrates) to form one or more radiogenic isotopes (the daughters). Different isotopes decay at different but predictable rates such that there is a constant time for each isotope for the decay of half of the original atoms—the half-life of the isotope. The proportion of parent atoms disintegrating is constant at any given time. Thus half of the parent atoms disintegrate in one half-life (leaving one-half), half of the remaining atoms disintegrate in one more half-life (leaving one-quarter), and so on. Those isotopes that have long half-lives are decaying slowly and have small decay constants. More specifically:

$$\lambda = \frac{0.693}{\text{half-life}} \qquad \text{where } \lambda = \text{decay constant for isotope}$$

The general relationship for the radiometric age of a sample is given by:

$$t = \frac{1}{\lambda} \log_e (1 + \frac{D}{P}) \qquad \begin{array}{l} \text{where } t = \text{time of decay (or "age" of sample)} \\ D = \text{concentration of newly formed} \\ \text{(daughter) atoms} \\ P = \text{concentration of original (parent)} \\ \text{atoms now present} \end{array}$$

[1]The time of metamorphism is presumably recorded in the minerals as their time of final crystallization. This is in general probably close to the time of maximum temperature because a drop in temperature would rapidly lower the vapor pressure and therefore inhibit recrystallization.

Therefore, the half-life (or the decay constant λ) for each parent isotope to be considered needs to be known (these are known to various degrees of accuracy), and the concentration of parent (P) and daughter (D) atoms in the sample must be measured. Several measurement techniques have been used, but most methods (K–Ar, Rb–Sr, U–Pb) use a mass spectrometer whereby the different isotopes of each element may be separated because of their slight difference in mass. The Pb–α method involves measurement of the amount of U or Th by their α-particle activity, using a proportional radiation counter or related method. The Pb (assumed to be all radiogenic daughter material) is determined by chemical analysis. The fission-track method involves the counting of tracks formed by fission fragments produced by decay of U atoms over the life of the mineral. The tracks are enlarged by etching, then counted under an ordinary microscope. Bombardment of the same sample with a known flux of thermal neutrons from a reactor produces a new group of tracks, which are counted to give the concentration of U. The main methods and some of their important characteristics are summarized in Table 7-3. Discussion of techniques and examples are covered by Faul (1954), Aldrich and Wetherill (1958), E. I. Hamilton (1965), Moorbath (1965), and McDougall (1966).

In general, the "age" obtained dates the time of formation of the mineral dated or the time at which diffusion of the daughter product ceased to be important. The different methods have somewhat different problems and pitfalls of interpretation. Ideally, if two or more methods (e.g., K–Ar, Rb–Sr) give concordant ages (same ages), the age is probably accurate. Most discordant ages err on the side of being too young (e.g., loss of daughter element). Most single ages can thus be considered to be minimal.

K–Ar is the most widely used of the radiometric methods. As with the other methods, caution must be used in interpreting the "date" obtained on a mineral. The radiometric "clock" is started after crystallization or recrystallization of the mineral. If, however, the mineral is not rapidly cooled, as in the case of a deep-seated granitic pluton or a regional metamorphic rock, the radiometric "clock" is not effectively started until the temperatures cool sufficiently for argon diffusion out of the mineral to be inhibited. This temperature for major Ar loss in biotite depends, of course, on the time available, but it has been estimated as about 300°C (Evernden et al., 1960, by experimental diffusion in phlogopite) or about 250 to 400°C (Hart, 1964; Hanson and Gast, 1967, by temperature estimates and K–Ar dates in a wide-contact metamorphic aureole). Greenschist facies metamorphism is normally sufficient to reset biotite K–Ar and Rb–Sr ages (Hart et al., 1968; Hart and Davis, 1969).

The time for cooling through this gradational "isotherm" may be millions or even tens of millions of years depending on the size of the body, its depth, and the rate of unroofing (Reesor, 1961a, b; Jäger, 1962; Harper,

Table 7-3 *Comparison of main methods of radiometric age determination*

	K–Ar	Rb–Sr
Parent → daughter (only the most important listed)	$K^{40} \to Ar^{40}$	$Rb^{87} \to Sr^{87}$
Half-life (years)	1.19×10^{10} (11,900 million years)	4.7×10^{10} (47,000 million years)
Useful age: Minimum range (years)	< 50,000 and possibly as low as 2,500 (Evernden & Curtis, 1965)	10,000,000
Maximum range (years)	None	None
Most useful materials	Biotite, muscovite, hornblende, sanidine, nondevitrified volcanic glass, whole rock, slates if no remaining detrital grains (e.g., Goldich et al., 1957; Harper, 1964)	Biotite, muscovite, K-feldspars, felsic igneous and metamorphic whole rocks
Minimum amount of sample	< 1 gm (mica)	
Minerals most retentive of daughter element during heating	Hornblende > slate ≥ muscovite > biotite > orthoclase or microcline	Muscovite > K-feldspars > biotite
Additional samples for increased accuracy	More reliable if same apparent age on two or more minerals in same rock	Isochron lines by plotting Sr^{87}/Sr^{86} for series of two or more minerals in one rock, or two or more rocks. Slope of whole rock isochron proportional to age of metamorphism

Table 7-3 *Comparison of main methods of radiometric age determination (Cont'd.)*

U–Pb	Pb–α	Fission Track
$U^{238} \rightarrow Pb^{206}$	$U + Th \rightarrow \alpha$ particles	$U^{238} \rightarrow$ fission fragment
4.5×10^9 (4,500 million years)		1.012×10^{18} ($1. \times 10^{12}$ million years)
10,000,000		20 years (Brill et al., 1964; Fleischer et al., 1964)
None	None	None
Zircon, monazite, uraninite, pitchblende, K-feldspar	Zircon, monazite, xenotime, uraninite, pitchblende, K-feldspar	In young rocks: high U minerals (e.g., zircon, quartz, glass, apatite, hornblende); in old rocks: micas, hornblende, feldspars (Fleischer & Price, 1964b)
	100 mg (zircon)	A few grains
Zircon ($Pb^{207}/Pb^{206} > Pb^{207}/U^{235} > Pb^{206}/U^{238}$)	Zircon, monazite, xenotime	Zircon, quartz, glass hornblende > ? sphene > micas > apatite > autunite (Fleischer & Price, 1964; Naeser, 1967)
Plot Pb^{206}/U^{238} against Pb^{207}/U^{235} for two or more mineral samples and for series of known concordant ages. Intersection of the sample line with the concordia line gives age	Much more reliable if same apparent age on more than one mineral containing different concentrations of U	More reliable if same apparent age on two or more minerals in same rock

1964, 1967; Armstrong, 1966; Mauger et al, 1968; Condie and Heimlich, 1969; O'nions et al., 1969; Zartman et al., 1970). With a normal thermal gradient of 20°C/km, a deep regional metamorphic complex would have to be unroofed to less than 15 km before the temperature could hope to fall below 300°C. Sillimanite-zone regional metamorphic complexes appear to be particularly susceptible to such delayed cooling through this critical isotherm for limitation of argon diffusion. The time of intrusion of a small, shallow intrusive such as a dike could be obtained by dating the fine-grained, quickly cooled border zone of the immediately adjacent thermally metamorphosed contact rocks. The radiometric "clock" will also be reset by any recrystallization of the mineral, such as by reheating during metamorphism (a temperature of 300°C corresponds to the lowest grades attained in most types of metamorphism as illustrated on the P-T diagrams developed in the last section). These considerations lead to the conclusion that K–Ar dates normally represent minima for the time of (re)crystallization of the mineral. Other influences include alteration (e.g., biotite to chlorite) and shearing of the mineral. Weathering is thought to have little effect on K–Ar ages (Kulp and Engels, 1963). Orthoclase and microcline may lose a major proportion of radiogenic argon even without significant reheating or shearing, so are considered poor bets for dating. By contrast, pyroxenes and volcanic glass may take on excess radiogenic argon (e.g., Gerling et al., 1961; Hart and Dodd, 1962; Damon et al., 1967; Dalrymple and Moore, 1968; Noble and Naughton, 1968) and also are unsuitable for dating. The possibility of solution of excess Ar^{40} in silicate melts has been verified experimentally by Kirsten (1968) and Fyfe and coworkers (1969). However, the reservation may not apply to pyroxene or other minerals formed near the surface and never deeply buried (Richard, *in* Evernden and Curtis, 1965). Although considered rare, even biotite may take on excess radiogenic argon (Wanless et al., 1969; Giletti, 1971).

Rb–Sr is also becoming widely used in dating, but the problems are somewhat different than those associated with K–Ar. All of a group of samples are assumed to have the same age and initial Sr^{87}/Sr^{86} ratio. Although a few instances have been described in which samples of the whole rock have acted as open systems (e.g., in small dikes, xenoliths) and Sr lost (e.g., Lanphere et al., 1964; Peterman, 1966; Arriens et al., 1966), most appear to have acted as closed systems, the Sr^{87} (and Sr^{86}) distributing itself between coexisting minerals throughout the rock, at least over a span the size of a hand specimen (e.g., Compston and Jeffery, 1959; Giletti et al., 1961; Bofinger and Compston, 1967). Loss of Sr^{87} from biotite by diffusion appears to be important above about 350 to 450°C (Hanson and Gast, 1967), by temperature estimates and Rb–Sr dates in a contact aureole. In other instances, later metamorphism of dry rocks such as granite may homogenize some parts but not others. By contrast with the K–Ar situation, weathering may significantly affect the apparent age by exchange of Sr^{87} with common Sr in groundwater, giving ages that are too young (Peterman, 1966; Bottino and Fullagar, 1968). Leaching of Sr or Rb is also possible under hydrothermal conditions and may give an apparent age that is either too young or too old. These effects of weathering or hydrothermal activity are particularly important in young rocks because of their low content of radiogenic Sr^{87}.

U–Pb techniques, although in longer use than the K–Ar or Rb–Sr methods, were until recently less widely used in metamorphic studies. U–Pb ages may be affected by either Pb loss (or U gain), which appears to be most common, or by U loss (or Pb gain). Such a Pb loss may be episodic (Ahrens, 1955a, b; Wetherill, 1956a, b), as caused, for example, by metamorphism (e.g., Catanzaro and Kulp, 1964; Moorbath et al., 1969), or may involve continuous diffusion after formation of the mineral (Tilton, 1960), in either case giving ages that are too young. During metamorphism, isotopic exchange may occur between two rocks, such as an intrusive rock and the country rocks or between a mineral such as K-feldspar and the whole rock (e.g., Doe and Hart, 1963; Doe et al., 1965; Moorbath and Welke, 1968). Zircon ages appear to be unaffected by metamorphism up to the sillimanite zone, being affected at 500 to 600°C on the basis of heat-flow estimates (Davis et al., 1968). Loss or gain of Pb or U may also occur during hydrothermal activity or weathering, secondary and altered minerals being particularly susceptible to Rn loss (an intermediate member of the U–Pb fractionation series), less Pb being formed (Giletti and Kulp, 1955; Stieff and Stern, 1956).

Pb–α dating involves essentially all the considerations just noted for U–Pb dating and suffers from the additional disadvantage of being unable to discriminate between common nonradiogenic lead and the radiogenic lead formed by U + Th decay. This common lead may crystallize in some minerals (e.g., substituting for Ca in plagioclase or K in potassium feldspar) but not to any major extent in zircon, xenotime, monazite, and some other accessory minerals in granites or pegmatites. Pb contamination may also be introduced by certain laboratory procedures such as use of a magnetic separator (E. I. Hamilton, 1965, pp. 160–161).

Fission-track methods are a relatively recent development (Price and Walker, 1962a, b, c). The methods are relatively simple and inexpensive, and they hold considerable promise in age dating (Price and Walker, 1963; Fleischer et al., 1964; Fleischer and Price, 1964a, b; E. I. Hamilton, 1965; Naeser, 1967; Naeser and McKee, 1970). The method is based on the spontaneous fission of U^{238}. The resultant rapidly moving fission fragment resulting from this "ion explosion" leaves a damage trail marking its path through the crystal (Fleischer et al., 1965). The number of tracks

Table 7-4 *Comparison between relative "dates" (or "ages") determined by most commonly used methods and minerals, using available information to 1970*

Probable sequence of relative "dates" in a single rock	U–Pb	Rb–Sr	K–Ar	Fission Track	Reference
Greatest "age" (best retention of daughter product)	Zircon	Whole rock, K-feldspar	Hornblende	Zircon, quartz, glass, hornblende, epidote	Aldrich et al., 1958, 1959; Tilton et al., 1958; Faul et al., 1963; Steiger, 1964; Fleischer & Price, 1964; Doe et al., 1965; Livingston et al., 1967; Davis et al., 1968; Hart & Davis, 1969; Naeser & Dodge, 1969; Menzer, 1970
	K-feldspar	Muscovite	Albite, plagioclase? Slate muscovite		
Intermediate		Biotite	Biotite	Sphene Micas	Tilton et al., 1958; Hart, 1963, 1964; Kulp, 1963; Fleischer & Price, 1964b; Wescott, 1966; Hanson & Gast, 1967; Naeser, 1967; Naeser & Dodge, 1969; Menzer, 1970
Least "age" (least retention of daughter product)			K-feldspar	Apatite Autunite	

is proportional to the age of the sample and amount of U^{238} present. Depending on the mineral, the fission-track method appears either more or less susceptible to later thermal events (or slow cooling) than the comparable radiometric methods described above. Minerals with high melting points (e.g., zircon, hornblende, glass) retain fission tracks to high temperatures whereas in others (e.g., apatite) the tracks are annealed out by heating at relatively low temperatures (Fleischer and Price, 1964a, b). Wagner (1968), and Naeser and Dodge (1969) suggest 150°C or less for retention of fission tracks in apatite, 400°C or less for sphene, and 600°C or less for epidote.

The relative susceptibility to daughter-product loss of the various minerals and methods used in radiometric dating is important both in the choice and the evaluation of "absolute" ages. Although the relative susceptibility depends on thermal, deformational, and other conditions affecting the mineral, a typical order is apparent (Table 7-4).

CHAPTER 8
THE METAMORPHIC ROCK ASSOCIATIONS

DISTRIBUTION OF THE REGIONAL METAMORPHIC ROCKS

Regional metamorphic rocks are in most cases confined to "geosynclines" or former "geosynclines" (see Fig. 8-1). Metamorphic rocks developed in an environment of "abnormally low" geothermal gradient (high pressure/temperature), are the blueschist, many eclogites, and to some extent the zeolitic facies (see Fig. 8-2). These typically develop in the "eugeosynclinal" or thick graywacke-volcanic environment, bordering an oceanic trench. Examples include the Franciscan blueschist-eclogite rocks of western California and the zeolitic rocks of southern New Zealand.

Regional metamorphic rocks that developed in an environment of more "normal" geothermal gradient include the greenschist, amphibolite, and granulite facies. These commonly develop along the "miogeosynclinal" side of the "eugeosyncline" and into the "miogeosyncline"—the same environment as that of numerous granitic plutons and the major granitic batholiths. Examples include most of the major regional metamorphic belts of the world, such as the Cordilleran igneous-metamorphic belt of eastern California, Nevada, Idaho, and western to central British Columbia and the Appalachian metamorphic belt of northern Georgia through the western Carolinas, central Virginia and Maryland, to much

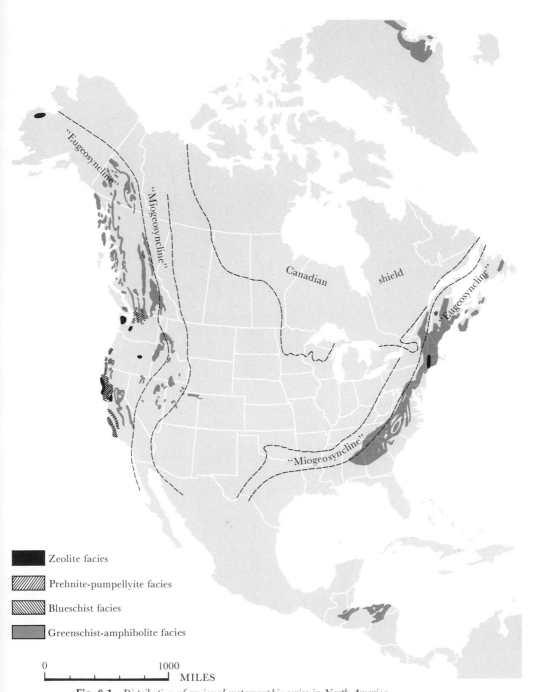

Fig. 8-1   *Distribution of regional metamorphic series in North America.*

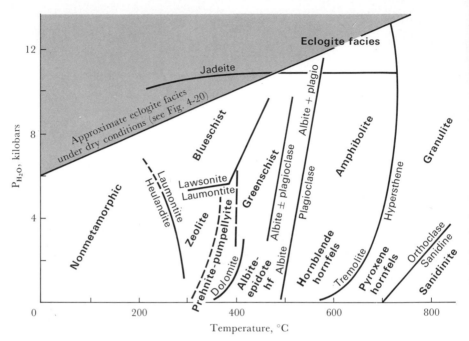

Fig. 8-2 *Metamorphic facies boundaries largely as defined by Fyfe, Turner, and Verhoogen (1958); Winkler (1965); Fyfe and Turner (1966); and Turner (1968) and positioned according to the stability limits shown on Figs. 7-11 and 7-12. Metamorphic facies are designated by boldface type.*

of New England, part of New Brunswick and Nova Scotia, to central Newfoundland.

Regional metamorphic rocks developed in an environment of "abnormally high" geothermal gradient include facies transitional from the greenschist and amphibolite facies of normal regional metamorphism to the albite-epidote, hornblende, and pyroxene hornfels facies of contact metamorphism. These transitional facies have been called Abukuma type (e.g., Miyashiro, 1958, 1961; Winkler, 1965, 1967, p. 125) or Buchan type (Read, 1952; Turner, 1968, p. 28) from high-temperature metamorphism in the Abukuma Plateau of Japan and the Buchan area of eastern Scotland. The conditions of formation are not identical in both areas but both develop in close association with granitic plutons, from which some of the necessary additional heat was presumably derived. Miyashiro (1961) and Zwart (1971) have noted that regional metamorphic belts occur in pairs, the higher-pressure/temperature type developed on the oceanic ("eugeosynclinal") side and the lower-pressure/temperature type developed on the continental ("miogeosynclinal") side (see also Fig. 8-1).

DESCRIPTION AND DEVELOPMENT OF THE REGIONAL METAMORPHIC ROCKS

THE ZEOLITE FACIES AND PREHNITE-PUMPELLYITE FACIES

The Zeolite Facies and Prehnite-Pumpellyite Facies of Western Washington and Adjacent British Columbia[1]

The "eugeosynclinal" environment of Mesozoic marine sedimentary rocks and some pillow basaltic rocks of the Olympic Peninsula and surrounding areas (see Fig. 8-3) appears to be overlain by more than 10 km of Cenozoic basaltic and sedimentary rocks (equivalent to about 2,600 bars lithostatic pressure). Graywackes and slates of the peninsula are exposed in the core of a broad anticline plunging southeast. They show graded bedding in graywacke-conglomerate beds tens of meters thick, to shale-graywacke interbeds less than a centimeter thick, apparently representing a turbidite sequence. In the field, most of the rocks appear to retain their sedimentary characteristics. In thin section, they are seen to be extensively recrystallized, the original detrital grains being primarily oligoclase-andesine, quartz, actinolite, biotite, Fe-chlorite, muscovite, and magnetite, and the final metamorphic assemblage being primarily albite, quartz, prehnite, calcite, $Mg-Fe-Al$ chlorite, sphene, Fe-rich colorless mica, and a trace of pumpellyite (Hawkins, 1967). To the east and north, descriptions suggest a higher content of basaltic and andesitic volcaniclastic rocks and flows. Prehnite and pumpellyite are common as amygdule fillings (Surdam, 1969). Plagioclase is generally albitized, commonly with segregation of pumpellyite (Misch, 1966; Surdam, 1969). Chemically, the rocks are equivalent to granodiorite-quartz diorite, reflecting the probable continental source of the detritus. Shearing is widespread, the medium-grained rocks having become semischists (schistose rock formed by granulation of coarser grains and incipient development of schistosity: Williams, Turner, and Gilbert, 1954, p. 216) and the finer-grained rocks having become slates and phyllites.

The newly formed metamorphic minerals are diagnostic of the prehnite-pumpellyite facies: albite-prehnite ± calcite being formed from oligoclase-andesine. Although these assemblages are widespread and independent of degree of deformation or structural position, assemblages in some layers appear not to have reached equilibrium. Vein fillings of prehnite, quartz, and calcite are also widespread, as to a lesser extent is pumpellyite (Hawkins, 1967; Misch, 1966).

[1]Weaver, 1916, 1937; Misch, 1966; Hawkins, 1967; Surdam, 1966, 1969; Vance, 1968a; Monger, 1970; Carlisle, 1971; Stewart and Page, 1971.

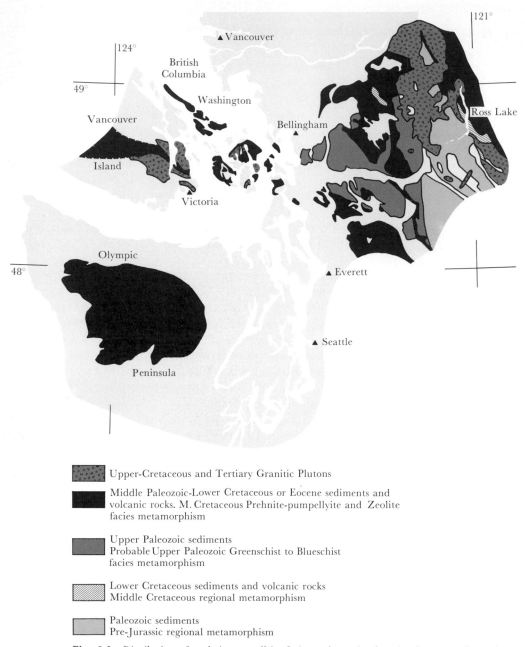

Upper-Cretaceous and Tertiary Granitic Plutons

Middle Paleozoic-Lower Cretaceous or Eocene sediments and volcanic rocks. M. Cretaceous Prehnite-pumpellyite and Zeolite facies metamorphism

Upper Paleozoic sediments
Probable Upper Paleozoic Greenschist to Blueschist facies metamorphism

Lower Cretaceous sediments and volcanic rocks
Middle Cretaceous regional metamorphism

Paleozoic sediments
Pre-Jurassic regional metamorphism

Fig. 8-3 *Distribution of prehnite-pumpellyite facies and associated regional metamorphic rocks in northwestern Washington and adjacent British Columbia. Compiled from Hawkins (1967); Misch (1966); Monger (1970); Surdam (1966, 1969); Vance (1968); Stewart and Page (1971).*

Zeolite and prehnite-pumpellyite facies metavolcanic rocks to the north on southern Vancouver Island contain assemblages that incorporate the zeolite wairakite ($CaAl_2Si_4O_{12} \cdot 2H_2O$), along with laumontite, prehnite, pumpellyite, epidote, chlorite, quartz, phengite, Ca-natrolite, and calcite (Surdam, 1966, 1969).

Hawkins (1967) considers that the zeolite and prehnite-pumpellyite facies rocks of this region developed by burial metamorphism. Conditions are thought to involve 2.5 to 3.5 kbars pressure and 300 to 350°C, or a high thermal gradient of 30 to 40°C/km.

General Features of the Zeolite Facies and Prehnite-Pumpellyite Facies

Burial of "eugeosynclinal" graywackes, shales, and commonly mafic volcanic rocks, to depths of tens of thousands of feet, results in the increased pressures and temperatures of "burial metamorphism." The environment grades downward through the zones of diagenesis, into the zones containing zeolites (especially laumontite), and finally into the zone containing distinctive minerals such as prehnite and pumpellyite. At still deeper levels, the rocks grade into either the greenschist facies or the blueschist facies, depending on the environment (i.e., on the geothermal gradient). The depth of the beginning (top) of the zeolite facies, as recognized on the basis of laumontite occurrence, is in the vicinity of 3,000 to 15,000 ft, depending on the area, and ranging down to about 10,000 to 40,000 or more feet (e.g., Hay, 1966, pp. 70–72; Coombs, 1965; Coombs et al., 1959; Dickinson et al., 1969; Levi, 1970). Prehnite and pumpellyite generally underlie or somewhat overlap this lower limit for laumontite, and range down to mineral assemblages of the greenschist facies. Although there is a large amount of variance in these depth estimates for different areas, the relative order of appearance and disappearance of the mineral phases is quite constant (see Table 8-1). Where the line between diagenesis and metamorphism should be drawn in this sequence is not important. Whether the zone boundaries should be based on mineralogical changes as is done here and by most metamorphic petrologists (e.g., Fyfe et al., 1958, p. 215; Turner, 1968, pp. 263–268; Winkler, 1965, pp. 153–158), or in part on textural changes such as the development of schistosity (e.g., Packham and Crook, 1960), depends on the point of view, but as mineralogical criteria form the basis for metamorphic zonation and interpretation of temperatures and pressures, they will also be used here. It should be recognized, however, that relict minerals persist into these facies—disequilibrium assemblages are common (e.g., Dickinson, 1962; Levi, 1970). In many cases depositional textures are preserved.

Zeolite facies assemblages have been described in some detail from southeastern New Zealand (Coombs, 1954, 1960, 1961; Coombs et al., 1959), from northern New Zealand (Brothers, 1956), New South Wales,

Table 8-1 *Order of appearance of distinctive mineral phases from volcanic-rich detritus in low-grade environments from diagenesis to the zeolite and prehnite-pumpellyite facies*

	Volcanic Material
Diagenesis	Glass and other volcanic material (plagioclase, pyroxene, etc.)
	Mordenite, clinoptilolite, ± heulandite, stilbite, analcite
	Analcite, ± mordenite, clinoptilolite, heulandite, ± albite
Zeolite facies	Laumontite, albite, ± wairakite
Prehnite-pumpellyite facies	Pumpellyite, prehnite, albite

SOURCE: Hay (1966).

Australia (Packham and Crook, 1960; Wilkinson and Whetten, 1964; Whetten, 1965; Smith, 1969), the Fiji Islands (Crook, 1963), Japan (Miyashiro and Shido, 1970), western Alaska (Hoare et al., 1964), west-central Washington (Wise, 1959; Fiske et al., 1963), northwestern Washington and southwestern British Columbia (described above), central Oregon (Thayer and Brown, 1960; Brown and Thayer, 1963; Dickinson, 1962), the northern coast of California (Bailey et al., 1964, p. 91), the Great Valley of California (Ojakangas, 1968; Dickinson et al., 1969), the Rocky Mountain foothills of Alberta (Carrigy and Mellon, 1964), southern Connecticut (Heald, 1956), Georgia (Ross, 1958), Puerto Rico (Otálora, 1964), Chile (Levi, 1970), northern Caucasus Mountains of eastern Europe (Rengarten, 1950), southern Siberia (west of Lake Baikal: Buryanova, 1960; Koporulin, 1961, 1964), and east-central Siberia (east of Lena River: Kossovskaya and Shutov, 1961; Zaporoshteva et al., 1961). An ACFmK phase diagram illustrating coexisting phases and possible equilibrium assemblages (corners of subtriangle volumes) for the zeolite facies is shown in Fig. 8-4.

Possible reactions transitional to the prehnite-pumpellyite facies include (Coombs et al., 1959; Seki, 1961; Martini and Vuagnat, 1965; Hawkins, 1967):

$$\text{Laumontite} + \text{calcite} \leftrightharpoons \text{prehnite} + SiO_2 + 3H_2O + CO_2$$
$$2\ \text{laumontite} \leftrightharpoons \text{prehnite} + \text{pyrophyllite} + SiO_2 + 6H_2O$$
$$\text{Laumontite} + \text{calcite} + \text{chlorite} \leftrightharpoons \text{pumpellyite} + SiO_2 + H_2O + CO_2$$
$$\text{Laumontite} + \text{prehnite} + \text{chlorite} \leftrightharpoons \text{pumpellyite} + SiO_2 + H_2O$$
$$\text{Chlorite} + 4\ \text{prehnite} + H_2O \leftrightharpoons 2\ \text{pumpellyite} + SiO_2$$

Prehnite-pumpellyite facies assemblages have also been described in detail from New Zealand (Brothers, 1956; Coombs, 1954, 1960, 1961; Coombs et al., 1959; Elliot, 1968), New South Wales, Australia (Smith, 1968, 1969; Chappell, 1968), the Fiji Islands (Crook, 1963), New Guinea (Crook, 1961), Celebes (deRoever, 1947), Japan (Seki, 1961, 1965; Hashimoto, 1966; Miyashiro and Shido, 1970), Olympic Peninsula and adjacent areas of western Washington (Misch, 1966; Vance, 1966, 1968; Hawkins, 1967), central Oregon (Thayer and Brown, 1960; Brown and Thayer, 1963; Dickinson, 1962), Panama (Cuénod and Martini, 1967), Chile (Levi, 1970), Haiti (Burbank, 1927), Puerto Rico (Otálora, 1964), the Virgin Islands (Hekinian, 1971), northern Michigan (Jolly and Smith,

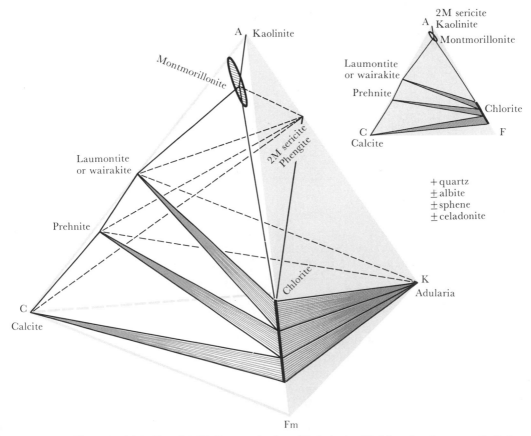

Fig. 8-4 ACFmK *and ACF diagrams for the zeolite facies (modified from Coombs, 1961). Kaolinite, montmorillonite, celadonite, and saponite* (No⁻Mg *vermiculite) are sedimentary minerals that remain stable in this facies.*

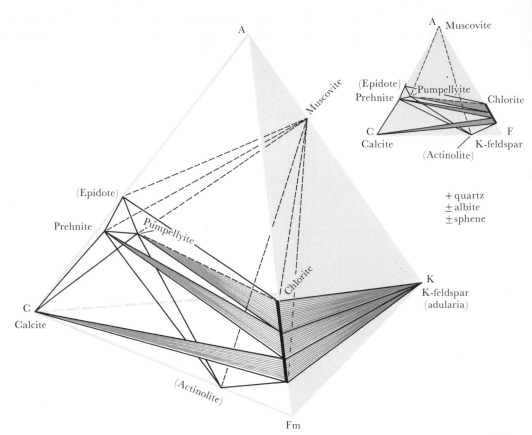

Fig. 8-5 ACFmK *and* ACF *diagrams for the prehnite-pumpellyite facies (modified primarily from Coombs, 1961; Turner, 1968, p. 267; Coombs et al., 1959, 1970). Stilpnomelane and actinolite may appear in the higher-temperature parts of the facies, which may be called the pumpellyite-actinolite metagraywacke facies (Hashimoto, 1966).*

1970), northern Maine (Coombs et al., 1970), and the western Alps (Martini and Vuagnat, 1965; Martini, 1968). An ACFmK phase diagram illustrating coexisting phases and possible equilibrium assemblages for the prehnite-pumpellyite facies is shown in Fig. 8-5. Although the common polymorph of $CaCO_3$ in this and the zeolite facies is calcite, Vance (1968) found aragonite along shear zones in limestone where it has been considered to grow metastably in the sheared environment (Boettcher and Wyllie, 1967).

Pumpellyite is Mg-epidote:
$$Ca_4(Mg, Fe^{++})(Al, Fe^{3+})_5O(OH)_3(Si_2O_7)_2(SiO_4)_2 \cdot 2H_2O$$

[Epidote is $Ca_4 \ldots Al_6 \ldots O_2(OH)_2(Si_2O_7)_2(SiO_4)_2$]

Tie lines apparently intersecting the calcite-pumpellyite or prehnite-actinolite lines may not actually intersect because of independent behavior of Fe^{3+} and Al^{3+} in pumpellyite and epidote or because prehnite-chlorite tie lines disappear as the pumpellyite-actinolite tie line appears. Possible reactions transitional to the greenschist facies include (e.g., Coombs et al., 1970):

4 calcite + chlorite + 8 quartz $\leftrightharpoons$

$$\text{prehnite + actinolite} + 2H_2O + 4CO_2$$

5 prehnite + chlorite $\leftrightharpoons$ 2 pumpellyite + actinolite

Areas in which the zeolite and prehnite-pumpellyite facies assemblages have been recorded are all within the "eugeosynclinal" or other plagioclase- or volcanic detritus-bearing rocks such as the Great Valley sequence of California (Dickinson et al., 1969). Whether this means that the conditions for development of these facies are found only in this environment or whether they may also develop in "miogeosynclinal" rocks lacking such Ca-rich assemblages is uncertain. The diagnostic assemblages—those containing laumontite, prehnite, or pumpellyite, especially—are essentially restricted because of composition to calcareous feldspar or volcanic-bearing (i.e., "eugeosynclinal") materials. Diagnostic assemblages have not been recognized in rocks of other composition (e.g., pelitic) that are widespread in the "miogeosynclinal" deposits. Assemblages such as chlorite-muscovite-albite-quartz, however, which are intercalated with diagnostic assemblages in the "eugeosyncline," do occur in "miogeosynclinal" rocks. Textures in these rocks are typically nonschistose or nearly so. In some occurrences, prehnite occurs as tiny inclusions in albitized plagioclase, small porphyroblasts, amygdule fillings, or as diffuse light-colored spots a few millimeters across (Brown and Thayer, 1963; Smith, 1969; Surdam, 1969; Coombs et al., 1970).

The very wide range of pressure (as determined by stratigraphic depth and noted above) necessary for the initial formation of laumontite suggests that high pressure is not crucial for formation of the zeolite facies. The development of laumontite as a product of "hydrothermal" alteration around igneous intrusions (e.g., Coombs et al., 1959; Fiske et al., 1963) suggests the importance of temperature in its formation rather than pressure. Both these considerations agree with the experimental synthesis of laumontite (see curve 4, Fig. 7-11), which suggests a boundary near 275°C that is nearly independent of pressure. Whether the laumontite in the active hydrothermal area of Wairakei, New Zealand, is stable at the measured temperatures of 195 to 220°C (Coombs et al., 1959) or relict from a higher temperature is not clear.

THE GREENSCHIST FACIES

The Greenschist Facies of New England and the Maritime Provinces[1]

The regional metamorphism of New England follows a wide belt trending north to northeast from Connecticut, through Massachusetts and New Hampshire, through southern Maine, and into southeastern Nova Scotia (Fig. 8-6). The high-grade sillimanite zone extends for about half of this distance (400 km or 250 miles), the grade falling off to the chlorite zone in about 50 to 100 km to the east and west. The metamorphism is located almost entirely within the "eugeosynclinal" environment, but "eugeosynclinal" rocks of similar age (e.g., Cambrian-Ordovician-Silurian) in northern Maine are in part almost unmetamorphosed. The preorogenic sedimentary rocks include some 30,000 to 40,000 ft of lower Paleozoic marine clastic sedimentary rocks—shale, muddy sandstone, quartzite, with some limestone and calcareous sandstones and shales—along with some mafic to felsic volcanic rocks. These rocks were folded, faulted, and regionally metamorphosed in the Middle Devonian ("Acadian Orogeny"). A weak (mostly chlorite zone) metamorphism may have affected the western parts of the belt in the Ordovician ("Taconian Orogeny"). Granitic plutons emplaced in the early stages of metamorphism are typically concordant and have been metamorphosed (e.g., Heald, 1950), injected during metamorphism (Page, 1968), or formed by anatexis during metamorphism (Thompson and Norton, 1968). Those intruded after the climax of this orogenic period tend to be discordant and have contact metamorphosed the regional metamorphic rocks.

The low-grade—greenschist facies—metamorphic rocks that were formed include slates, muscovite- and biotite-rich phyllites, and fine-grained quartz-mica schists from the shales and graywackes, actinolite-bearing marbles from impure limestone and dolomite, and fine chlorite-epidote-albite schists from the basalts. Most of the metamorphic rocks are strongly foliated except where the metamorphism outlasted deformation or where the parent rock was massive and dry (e.g., a granitic pluton). The highest-grade zones (i.e., sillimanite) form a wide belt concentrated in the stratigraphically younger rocks (Silurian and Devonian) and are bounded by a relatively narrow zone of a steep thermal gradient (e.g., southeast New Hampshire) to a relatively wide low-grade zone (i.e., chlorite-biotite). Considering the low-grade greenschist facies rocks as having resulted from the normal buildup of heat by radioactive decay within the

---

[1]New England: Heald, 1950; Lyons, 1955; Billings, 1956; Zen, 1960; Goldsmith, 1962; Crawford, 1966; Rodgers, 1967; Thompson and Norton, 1968; Albee, 1968; Cady, 1969. The Maritime Provinces of Canada: e.g., Taylor and Schiller, 1966; Rodgers, 1967; Taylor, 1969.

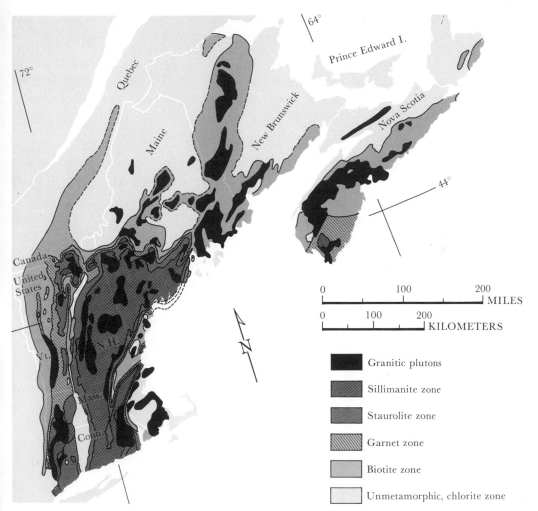

Fig. 8-6 *Metamorphic facies map of New England and the Maritime Provinces. Compiled from various sources, including especially Thompson and Norton (1968), Goldsmith (1962), Billings (1956), Taylor and Schiller (1966), Taylor (1969), Espenshade and Boudette (1964), publications of the Geological Survey of Canada, Maine Geological Survey, and Vermont Geological Survey.*

thick pile of crustal rocks, Hamilton and Myers (1967) suggest that the higher-grade amphibolite facies rocks (especially the sillimanite zone) developed because of the additional heat brought to higher levels by rising anatectic melts. They also advance the intriguing hypothesis that the wide expanse of sillimanite zone is due, not to the presence of a shallow sub-surface batholith, but to the former presence of an overlying batholithic sheet, now largely eroded away. The high-grade rocks would have developed beneath this batholith. The presently exposed plutons in the lower-grade rocks off to the flanks and northward in Maine and New Brunswick would be peripheral plutons where no extensive batholith existed before. It is also possible that the higher-grade rocks reflect a somewhat later (Devonian), more restricted, higher-grade metamorphism than the earlier (Ordovician) widespread low-grade metamorphism (Albee, 1968) or were brought in as large recumbent folds from the "eugeosyncline" to the southeast (Thompson et al., 1968).

The three pelitic rock zones of the greenschist facies—chlorite, biotite, and garnet—are represented in rocks of widely varying composition by a large number of different mineral assemblages. A few assemblages from representative areas will serve to illustrate.

Chlorite zone:

Muscovite-chlorite-quartz-albite   (±epidote,   hematite,   magnetite)

Muscovite-chlorite-quartz-microcline (or chloritoid or paragonite or spessartite)

Calcite-chlorite-epidote-albite

Biotite zone:

Muscovite-biotite-quartz

Muscovite-biotite-chlorite-quartz-albite (magnetite)

Muscovite-biotite-chlorite-quartz-albite-oligoclase-carbonate[1]

Actinolite-biotite-chlorite-quartz-oligoclase epidote[1]

Garnet (almandine) zone:

Muscovite-biotite-quartz-albite-almandine-chlorite

Muscovite-biotite-quartz-oligoclase-carbonate- (±almandine)[2]

Muscovite-biotite-quartz-andesine-calcite-ankerite-chlorite[2]

Hornblende-biotite-quartz-albite-clinozoisite-chlorite

The question of equilibrium among the minerals in these rocks has been discussed by Zen (1960). He shows that the following criteria are

[1]Oligoclase may appear at this low grade in the presence of very high Ca (carbonate or epidote).
[2]Oligoclase or andesine may appear in the presence of very high Ca (carbonate).

met: (1) each coexisting assemblage of minerals obeys the phase rule; (2) clearly incompatible minerals are not found together; (3) no evidence was found to suggest intersecting tie lines on triangular phase diagrams (for example, ACF, AFM); (4) there is no evidence of disequilibrium textures; and (5) recrystallization and therefore probably an approach to equilibrium is shown by growth of porphyroblasts, development of a slaty cleavage, and the same composition for chlorite in both textures.

Although the strong metamorphism accompanying the Acadian Orogeny in the Devonian is most prominent, a later somewhat lower grade metamorphism appears to have accompanied the Alleghenian (or Appalachian) Orogeny in the Pennsylvanian or Permian, at least in southeastern New England (Rodgers, 1967). The metamorphism of a single orogeny need not have occurred at the same time everywhere, however, and there is some indication that the Acadian Orogeny may have been somewhat later (into the Carboniferous) to the southwest.

Retrograde metamorphism (reaction to lower-grade assemblages) such as in north-central Vermont, may have resulted from water driven off by continued prograde metamorphism at depth (Cady, 1969, p. 29). Garnet has been replaced by chlorite and kyanite by muscovite.

General Features of the Greenschist Facies

Greenschist facies regional metamorphic rocks comprise the most voluminous and most widespread of the metamorphic rocks. They appear to develop under conditions of a "normal" geothermal gradient in a "geosynclinal" environment, evolving at deeper levels and higher grades than the zeolite or prehnite-pumpellyite facies. These rocks are extensively developed in the general regions containing the major granitic batholiths, such as the Sierra Nevada, Idaho, and Coast Range batholiths. They are also prominent in other regions in which granitic plutons are less prominent or are largely confined to the peripheries of the high-grade metamorphic belt, such as those adjacent to the Shuswap complex of southeastern British Columbia and the Ocoee series of western North Carolina and eastern Tennessee. These "low-grade" regional metamorphic rocks are generally separated from large granitic batholiths by "high-grade" regional metamorphic rocks of the amphibolite facies. Smaller, postmetamorphic granitic plutons may be emplaced in rocks of the greenschist facies, resulting in narrow zones of superimposed contact metamorphism. The facies is also well represented in many areas of the Precambrian shields.

Typical rocks formed in the greenschist facies include slate, phyllite, and fine-grained schist from pelitic (i.e., shaley) rocks, greenschists from basalts and andesites, magnesian schists (e.g., serpentinites) from peridotites, along with siltstones, and fine-grained quartzite and marble. Most of these rocks, especially the pelitic derivatives such as slates, phyllites, and schists, are characterized by a strong schistosity defined by preferred orientation of flakes of mica and chlorite. Chlorite zone rocks are generally very fine grained—such as argillite and slate—but in some areas are as coarse as medium-grained schists. With increasing grade in a single area, the rocks become coarser—such as phyllite or schist—and metamorphic differentiation may develop segregation layers. The metamorphic schistosity may be later deformed by a strain-slip or crenulation cleavage accompanied by a lineation lying in the schistosity.

The greenschist facies comprises the chlorite, biotite, and garnet zones of Barrow's classic locality (Barrow, 1893, 1912; Tilley, 1925). The garnet zone is also known as the epidote amphibolite or albite-epidote amphibolite facies (e.g., Eskola, 1939; Barth, 1962, p. 323; Lambert, 1965; Fyfe and Turner, 1966) and the greenschist-amphibolite transition facies (Turner, 1968, p. 186). Characteristic minerals include the green minerals chlorite, actinolite, and epidote, along with albite in metavolcanic rocks, and chlorite, muscovite, and biotite, along with albite and quartz in pelitic and most other metasedimentary rocks. The low-temperature limit of the greenschist facies is distinguished by the absence of zeolites, prehnite, and pumpellyite, minerals characteristic of lower grades at low pressure and by the absence of lawsonite, jadeite, omphacite, aragonite, and glaucophane (Turner, 1968, p. 268) of the low-temperature–high-pressure blueschist facies. The high-temperature limit of the greenschist facies is recognized by the final disappearance of albite (An_{0-5}), which coexists with oligoclase (An_{18-24}), above the upper biotite zone, and the appearance of oligoclase ($An_{>18}$) or zoned plagioclase in the general range of oligoclase (Crawford, 1966; Turner and Verhoogen, 1960, p. 533). This problem of the changing composition of plagioclase with increasing grade is discussed in more detail below. This upper limit is also marked by being prior to the appearance of the distinctive minerals diopside and staurolite.

Coexisting mineral phases and equilibrium assemblages for the greenschist facies are illustrated on ACFmK and ACF diagrams in Fig. 8-7. Four-phase assemblages plus quartz, represented by the corners of any of the subtetrahedra within either diagram, make up possible assemblages in greenschist facies rocks. Albite and graphite are common additional phases. Apparent crossing tie lines, even on a single face of the ACFmK (or ACF) diagram are, as noted above, in most cases explained by lumping of components at a single corner in the tetrahedron. For example, calcite-muscovite and epidote-microcline are both on the ACK face, but A represents Al^{3+} in muscovite and $Al^{3+} + Fe^{3+}$ in epidote. Thus the lines do not really intersect in space. Example of a common assemblage derived from a pelitic sediment is quartz + albite + muscovite + chlorite ± microcline

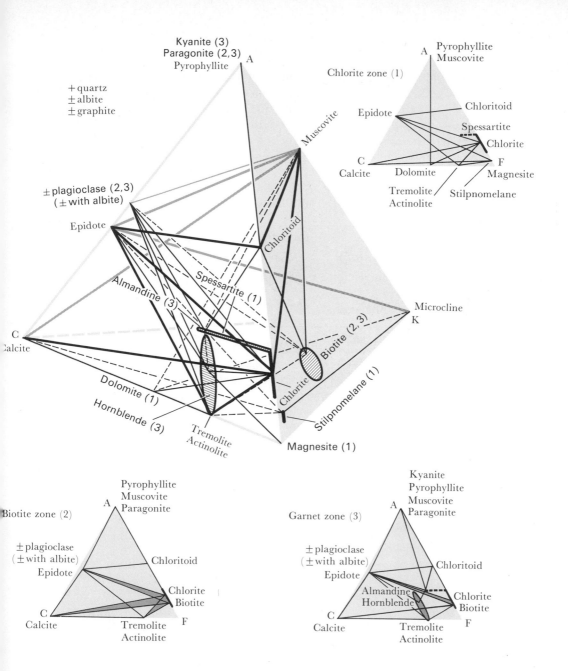

Fig. 8-7 *Greenschist facies ACFmK and ACF diagrams. Heavy lines are tie lines between minerals that coexist throughout the facies. Minerals in lightface and joined by thin tie lines coexist in individual zones. Numbers in parentheses refer to zones in which the mineral occurs. Restrictive-zone minerals are given in parentheses: (1) chlorite zone (dolomite, magnesite, stilpnomelane, spessartite); (2) biotite zone (biotite, paragonite, ± plagioclase); (3) garnet zone (biotite, paragonite, ± plagioclase, hornblende, almandine, kyanite).*

345

± epidote. That of a common assemblage derived from a mafic volcanic rock is albite + epidote + actinolite + chlorite + calcite + minor quartz. A few other uncommon minerals may appear under unusual chemical conditions—for example, spessartite or piemontite in high-Mn rocks or margarite in silica-deficient rocks.

Possible reactions transitional to and within the greenschist facies include those listed below (Carr, 1963; Winkler, 1964, 1965, 1967; Metz and Winkler, 1963; Zen, 1960; Zen and Albee, 1964; Ernst, 1963b; Miyashiro, 1958; Crawford, 1966; Thompson and Norton, 1968).

Below (lower grade than) the chlorite zone:

Kaolinite + 2 quartz ↬ pyrophyllite + H_2O
2 kaolinite + albite ↬ pyrophyllite + paragonite + $2H_2O$

Below the biotite zone:

Dolomite + quartz + H_2O ↬ tremolite + calcite + CO_2
Chlorite + quartz + paragonite ↬ albite + chloritoid
10 calcite + 3 chlorite ± 21 quartz ↬
　　　　　　　3 actinolite + 2 epidote + $10CO_2$ + $8H_2O$
3 muscovite + 5 chlorite ↬
　　　　　　　3 biotite + 4 Al-rich chlorite + 7 quartz + $4H_2O$
5 muscovite + 3 chlorite + 8 albite ↬
　　　　　　　5 biotite + 8 paragonite + 9 quartz + $4H_2O$
13 microcline + 3 chlorite ↬
　　　　　　　7 biotite + 6 muscovite + 12 quartz + $5H_2O$
8 phengite + chlorite ↬ 3 biotite + 5 muscovite + 7 quartz + $4H_2O$

Below the garnet zone:

Chlorite + tremolite/actinolite + epidote + quartz ↬
　　　　　　　hornblende + H_2O
Fe–Mg–Al chlorite + quartz ↬ almandine + Mg chlorite
6 chlorite + muscovite + 15 quartz ↬
　　　　　　　13 almandine + biotite + $36H_2O$
3 chlorite + paragonite + quartz ↬ 7 almandine + albite + $19H_2O$
Chlorite + 3 muscovite + 7 quartz + 6 calcite + 3 albite ↬
　　　　　　　7 plagioclase + 3 biotite + $8H_2O$ + $6CO_2$

Change of Plagioclase Composition with Increasing Metamorphic Grade

As already noted above, plagioclase in most rocks of the greenschist facies is nearly pure albite, whereas at higher grades such as in the amphibolite facies, the calcium content suddenly jumps to oligoclase or andesine (e.g., Lyons, 1955; deWaard, 1959). X-ray studies of low-temperature plagioclase show that sodic plagioclases (between about An_2 and An_{18}) are unmixed into two phases of about An_{0-1} and An_{25-28} (Laves, 1954; Gay and Smith, 1955; Brown, 1960; Ribbe, 1960). Calcium-bearing plagioclase, more sodic than An_{18-20}, is very uncommon (Hunahashi et al., 1968). The perthitelike product of this exsolution in sodic plagioclase is called peristerite. The peristerite solvus was inferred earlier on the basis of the composition of plagioclase coexisting with a calcium-rich phase such as epidote (Lyons, 1966; Christie, 1959; Rutland, 1961, 1962; Noble, 1962; Brown, 1962). More recently, coexisting albite and plagioclase have been identified and measured with the electron probe microanalyzer and with the microscope and universal stage (Evans, 1964; Crawford, 1966, 1970). The albite and plagioclase occur as discrete grains, the oligoclase first appearing as tiny grains along the margins of albite grains, gradually enlarging until it completely envelops the albite. Plagioclase below the garnet isograd is not visibly zoned but that above is reverse zoned (more calcic rims). Crawford's careful study of coexisting albite, plagioclase, and more calcium-rich phases through a series of metamorphic zones (inferred as increasing temperature) has developed a semiquantitative picture of the peristerite solvus (see Fig. 8-8).

With increasing grade, albite and plagioclase appear to behave as follows. In the chlorite and biotite zones, albite in calcium-bearing rocks coexists with a Ca-bearing phase such as calcite or epidote. In the high-

Fig. 8-8 *Approximate temperature-composition diagram for plagioclase + carbonate. Peristerite solvus peaks at top of greenschist facies. Composition of plagioclase coexisting with carbonate is somewhat different for plagioclase coexisting with epidote or some other high-Ca phase. After Crawford (1966).*

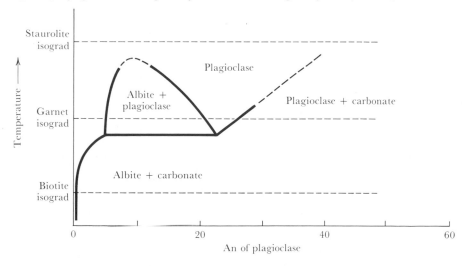

temperature part of the biotite zone, albite may coexist with more calcic plagioclase (say, An_{23}; see Fig. 8-8) or in a more calcic rock plagioclase (for example, An_{23}) may appear without albite or in a still more Ca-rich rock, plagioclase may coexist with a Ca-rich phase. At higher temperatures in the garnet zone, the situation is similar: albite + plagioclase (for example, An_{15}) or one plagioclase (for example, An between about 15 and 34) or plagioclase (for example, An_{34}) with a Ca-rich phase, depending on Ca content of the rock. At still higher temperatures in the amphibolite zone: plagioclase between An_0 and, say, $An_{40 \text{ or higher}}$, or $An_{40 \text{ or higher}}$ with a Ca-rich phase.

In a similar study in contact metamorphic rocks, Crawford (1970) finds that the peristerite solvus under these lower-pressure conditions is somewhat narrower. This agrees with experimental work by Yoder et al. (1957) who found that the alkali feldspar solvus also expands with increasing pressure (see p. 19).

THE AMPHIBOLITE FACIES

The Amphibolite Facies of the Shuswap Metamorphic Complex of Southeastern British Columbia[1]

The high-grade regional metamorphic rocks of the Shuswap metamorphic complex occupy a 50-mile-wide belt trending north-northwest for some 300 miles in southeastern British Columbia (Fig. 8-9). Farther south it appears to include rocks of the Okanogan gneiss dome ("Colville batholith") in north-central Washington. Most of the complex itself consists of sillimanite-zone schists, gneisses, and migmatites. The metamorphic grade falls off to the east and west; but much of the border of the complex is fault bounded so that the sillimanite zone rocks abut greenschist facies rocks, either directly or with only a very narrow intervening sheet of staurolite or kyanite-zone rocks. The premetamorphic sedimentary and volcanic rocks include a thick section of Paleozoic and late Precambrian shales, muddy sandstones, quartzites, with some limestone and calcareous sandstones and shales—along with some minor mafic volcanic rocks.

These rocks were isoclinally recumbent folded and regionally metamorphosed to the upper amphibolite facies in the Jurassic ("Nevadan Orogeny"), as the metamorphism affects Triassic and possibly Lower Jurassic rocks. They were in turn intruded and contact metamorphosed by granitic plutons at least as old as Lower Cretaceous. With few excep-

[1]Reesor, 1961b, 1965, 1970; Hyndman, 1968a,b, 1969, 1971; Jones, 1959; Campbell, 1970; Campbell and Campbell, 1970; Fyles, 1970; Fyson, 1970; McMillan, 1970; Wheeler, 1965; Froese, 1970; Moore and Froese, 1971; Fox and Rinehart, 1971.

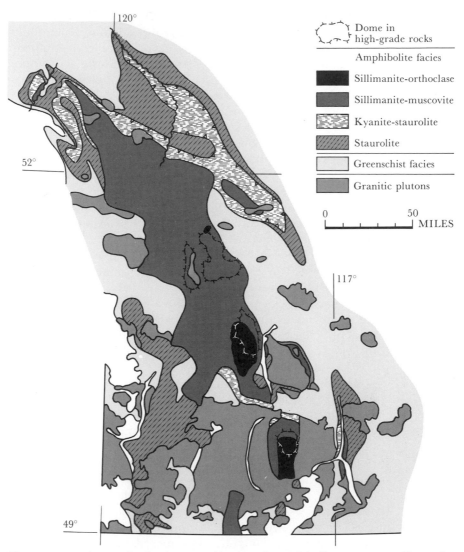

Fig. 8-9 *Map of zones of regional metamorphism in and around the Shuswap metamorphic complex, southeastern British Columbia. After Reesor (1970), Campbell (1970), Monger and Hutchison (1971), and Hyndman (1968).*

tions, these postorogenic granitic plutons are massive, mostly discordant, and have been emplaced in the lower-grade greenschist facies rocks outside the high-grade complex. Synorogenic granitic rocks are locally abundant in large gneiss domes forming core zones along the eastern side of the Shuswap complex (see Fig. 8-9). These generally consist of concordant, strongly foliated, migmatitic, pegmatite-sheeted biotite-quartz-feldspar gneisses derived from pelitic to quartzofeldspathic metasedimentary rocks. Also present are great lenticular sheets of hornblende-biotite granodiorite augen gneiss, probably derived from granodiorite. Enveloping this core zone is a mantling zone of varied high-grade metasedimentary rocks such as pelitic schists, quartzite, quartzitic gneiss, calc-silicate gneiss, and marble. Migmatites similar to those in the core zone develop in places within these metasedimentary rocks. Overlying this is a fringe zone that, in addition to containing metasedimentary rocks like those in the mantling zone, is dominated by the development of extensive sheets of pegmatitic material. These low-mafic granitic rocks form concordant layers up to tens and even hundreds of feet thick. The fringe zone is characteristic of much of the Shuswap complex. In these rocks, foliation including schistosity defined by parallel flakes of mica, and gneissic layering defined by alternating pelitic and quartzofeldspathic laminae a few grains to a few centimeters thick, is almost everywhere gently dipping. It contains a strong mineral lineation defined by elongate hornblende or biotite, stretched quartz or crenulated micas, generally trending east-west. All these infrastructure rocks, core-zone gneisses, mantling-zone metasedimentary rocks, and fringe-zone metasedimentary rocks with pegmatite sheets, lie in the upper amphibolite facies, predominantly the sillimanite zone. Together they constitute the Shuswap metamorphic complex (Fig. 8-9; see also Fig. 7-8, p. 298).

Overlying the infrastructure and bounded to it through a sheared zone a few hundred yards to a few miles wide in the lower amphibolite facies (staurolite to kyanite zones) is the lower-grade suprastructure in the greenschist facies. The tectonic thinning and steepening of metamorphic zones during separation of the plastic, relatively homogeneous infrastructure from the more brittle heterogeneous suprastructure accounts for the narrowness, and in some areas absence, of medium-grade metamorphic rocks along the margins of the high-grade complex (Fig. 8-9). Suprastructure rocks surrounding the complex include pelitic phyllites and argillites, finer-grained pelitic and quartzose schists, andesitic metavolcanic rocks, and marble, all intruded by concordant to discordant pegmatite and aplite dikes and by granitic stocks and small batholiths. Isograds of regional metamorphism in both the infrastructure and suprastructure are essentially parallel to the stratigraphic boundaries except on a regional scale. Large-scale anticlinoria and synclinoria outline both major stratigraphic units and metamorphic facies.

Folding accompanying regional metamorphism in both the high- and low-grade rocks was isoclinal and recumbent, the metamorphic schistosity lying parallel to the axial surfaces of the folds and the east-west mineral lineation lying parallel to the fold axes. Small-scale folds tend to be sheared off near crests and are thus normally seen only on close inspection. Later folds deform both the metamorphic foliation and lineation and the early folds. Core zones rose diapirically, as did the whole infrastructure with respect to the suprastructure. Later crenulation cleavage and other small-scale structures are best developed in the lower-grade rocks because of their finer grain size and better-developed schistosity. Some of the later folds, local recrystallization of bent micas, and contact metamorphism appear related to the intrusion of the slightly later granitic plutons.

A wide range of rock compositions produces an equally wide range of mineral assemblages, including the ones listed below (Reesor, 1965; Hyndman, 1968b).

Staurolite zone:

Quartz-plagioclase-biotite-garnet

Quartz ± plagioclase-biotite-muscovite-staurolite (± garnet
± chloritoid)

Hornblende-plagioclase-biotite-epidote (sphene ± apatite)

Amphibole-plagioclase-biotite-epidote-diopside-apatite

Sillimanite-muscovite zone:

Quartz-plagioclase-biotite-muscovite-sillimanite ± garnet

Quartz-plagioclase-biotite ± muscovite-orthoclase (± apatite)

Quartz-plagioclase-diopside-calcite-scapolite-graphite (sphene)

Hornblende-plagioclase-quartz (± biotite ± apatite ± sphene)

Sillimanite-orthoclase zone:

Quartz-plagioclase-biotite-orthoclase-sillimanite ± garnet (± apatite
± sphene)

Hornblende-plagioclase ± quartz ± biotite ± garnet ± pyroxene

General Features of the Amphibolite Facies

Amphibolite facies regional metamorphic rocks, although not so widespread as greenschist facies rocks, occur in most orogenic belts where they underly the lower-grade greenschist facies. As with these overlying rocks, the amphibolite facies appears to develop under conditions of a "normal" or possibly somewhat steepening geothermal gradient in the deepest levels of the "geosynclinal" environment. They occur in the cores of high-grade metamorphic complexes such as the Shuswap and Wolverine complexes of eastern British Columbia, the infrastructure of the eastern Great Basin (e.g., Armstrong and Hansen, 1966; Howard, 1966; Misch and Hazzard, 1962) and throughout the eastern Appalachians from New Hampshire to northern Georgia (e.g., Lyons, 1955; Billings, 1956; Barker, 1961, 1962; Woodland, 1963; Green, 1963; Sriramadas, 1966; Ward, 1959; Espenshade and Potter, 1960; Overstreet and Bell, 1965). The amphibolite facies also occurs in narrow to wide belts adjacent to major granitic batholiths such as the Coast Range batholith (Hutchison, 1967, 1970; Roddick, 1965; Hutchison and Roddick, 1971), the Idaho Batholith (e.g., Hietanen, 1956, 1961, 1963c, 1969; Reid and Greenwood, 1968; Chase, 1968; Nold, 1968), and the Sierra Nevada batholith (Durrell, 1940; Baird, 1962; Bateman et al., 1963; Moore, 1963; Kistler, 1966). In this environment, the granitic plutons of the granitic batholith association appear to be slightly younger (say a few million to possibly 100 million years) than the regional metamorphism and may well be derived from it as noted above. Smaller granitic plutons most commonly tend to be emplaced not within large belts of amphibolite facies rocks but in the greenschist facies of the surrounding (or overlying) suprastructure.

Typical rocks formed in the amphibolite facies include micaceous schists and gneisses from pelitic rocks, quartzofeldspathic schists and gneisses from feldspathic sandstones, amphibolites from basalts and andesites, and calc-silicate rocks from calcareous sandstones and shales, along with medium- to coarse-grained quartzite and marble. The pelitic and semipelitic schists and gneisses generally have a strong but rough schistosity (not as smooth as the lower-grade equivalents), as do isolated examples of each of the other high-grade rocks. This schistosity is defined by the preferred orientation of biotite or muscovite in most micaceous rocks and by flattened quartz or other grains in many other rocks. Amphibolites more commonly show a lineation defined by aligned prisms of hornblende. Quartzites and some other rocks may show a lineation due to the elongation of quartz or some other grains. Although bedding may be preserved in some rocks such as calc-silicate rocks, much of the layering in gneisses and other amphibolite facies rocks presumably results from metamorphic differentiation. The problem of origin of layering in metamorphic rocks was reviewed in an earlier section.

Isoclinal folds containing an axial-plane schistosity were presumably formed at the time of formation of the schistosity—that is, at the time of metamorphism. Although such folds are widespread, they are difficult to spot on the scale of an outcrop or a hand specimen, except on close examination. Their distribution also tends to be uneven. Where large-scale folds are present, though not necessarily readily apparent, the small folds tend to concentrate in the axial or crestal zone of the large fold, and to be sparse or absent on the limbs. Later postmetamorphic folds and crenulations deform all of the metamorphic structures—the schistosity, lineation, and metamorphic folds. Gneiss domes, the development of granitic material, and the relationship between high-grade infrastructure and low-grade suprastructure are in many instances similar to that described above for the Shuswap metamorphic complex. Other illustrative examples are described by Haller (1958, 1962), Armstrong and Hansen (1966), and Eskola (1949).

Amphibolite facies rocks characteristically occur in the infrastructure, as in the Shuswap complex. The boundary between infrastructure and suprastructure generally occurs in the lower amphibolite facies (e.g., staurolite zone) or upper greenschist facies.

Mineral assemblages of the amphibolite facies are illustrated on ACFmK and ACF phase diagrams of coexisting phases and possible equilibrium assemblages in Fig. 8-10. As in the greenschist facies, four-phase assemblages plus quartz, represented by the corners of any of the subtetrahedra within the ACFmK diagram, make up possible assemblages. Graphite and other accessories such as common black Fe-tourmaline (schorlite; olive green in thin section), apatite, sphene, zircon, and iron oxides are common additional phases. Apparent crossing tie lines or five-phase assemblages, all of which plot on the ACFmK diagram, may usually be explained by the lumping of components at the corners where two elements do not actually act as a single component. A common example is the separate behavior of Fe^{++} and Mg in almandine and biotite. Possible reactions transitional to and within the amphibolite facies include those listed here (e.g., Ramberg, 1952; Hoschek, 1967; Metz and Winkler, 1964; Hietanen, 1961; Billings, 1937; Francis, 1956a; Crawford, 1966; Thompson and Norton, 1968; Carmichael, 1969; Evans, 1965b; Guidotti, 1963; Kretz 1959).

Below (lower grade than) the staurolite zone:

$$5 \text{ chloritoid} + 2 \text{ quartz} \leftrightarrow 2 \text{ staurolite} + \text{almandine} + 3H_2O$$
$$31 \text{ chloritoid} + 5 \text{ muscovite} + \text{quartz} \leftrightarrow$$
$$8 \text{ staurolite} + 5 \text{ biotite} + 27H_2O$$
$$4 \text{ chloritoid} + 5 \text{ pyrophyllite} \leftrightarrow 4 \text{ staurolite} + 32 \text{ quartz} + 16H_2O$$
$$3 \text{ chlorite} + 7 \text{ muscovite} + \text{quartz} \leftrightarrow 4 \text{ kyanite} + \text{biotite} + 3H_2O$$
$$\text{Tremolite} + 3 \text{ calcite} + 2 \text{ quartz} \leftrightarrow 5 \text{ diopside} + 3CO_2 + H_2O$$
$$5 \text{ vesuvianite} + 4 \text{ epidote} + 11 \text{ quartz} \leftrightarrow$$
$$10 \text{ diopside} + 16 \text{ grossularite} + 12H_2O$$
$$4 \text{ epidote} + \text{muscovite} + 4 \text{ quartz} \leftrightarrow$$
$$8 \text{ anorthite} + 2 \text{ microcline} + 4H_2O$$

(component in plag.)

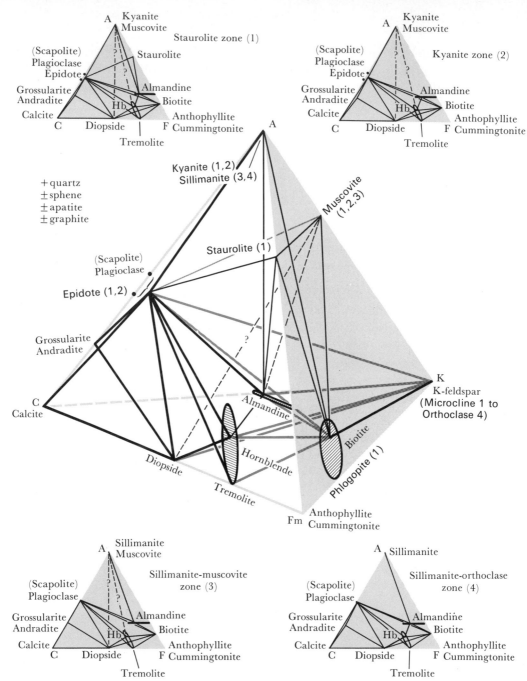

Fig. 8-10 *Amphibolite facies ACFmK and ACF diagrams. Minerals joined by heavy tie lines occur throughout the facies. Minerals in lightface and joined by thin tie lines occur in particular zones (e.g., 1, 2, 3, 4) as indicated in parentheses. Restrictive-zone minerals are given in parentheses: (1) staurolite zone (epidote, staurolite, kyanite, muscovite); (2) kyanite zone (epidote, kyanite, muscovite); (3) sillimanite-muscovite zone (sillimanite, muscovite); (4) sillimanite-orthoclase zone (sillimanite, sillimanite-orthoclase).*

12 epidote + 7 quartz + chlorite + 3 muscovite + 3 albite $\hookleftarrow$
$$27 \text{ plagioclase} + 3 \text{ biotite} + 14H_2O$$
(An$_{88}$ component)

Below the sillimanite-muscovite zone:

3 staurolite + 4 quartz $\hookleftarrow$ 2 almandine + 4 kyanite + 6H$_2$O
(or sillimanite)

6 staurolite + 7 quartz + 4 muscovite $\hookleftarrow$
$$31 \text{ sillimanite} + 4 \text{ biotite} + 3H_2O$$

2 epidote $\hookleftarrow$ anorthite + hematite + grossularite/andradite + quartz
(component)   [Ca$_3$(Al, Fe^{3+})$_2$(SiO$_4$)$_3$]   + H$_2$O

Kyanite $\hookleftarrow$ sillimanite

Below the sillimanite-orthoclase zone:

Muscovite + quartz $\hookleftarrow$ orthoclase + sillimanite + H$_2$O
Muscovite + quartz + Na-rich plagioclase $\hookleftarrow$ Na-bearing orthoclase
+ more-calcic plagioclase + sillimanite + H$_2$O
Muscovite + biotite + 3 quartz $\hookleftarrow$ 2 orthoclase + almandine + 2H$_2$O
Muscovite + garnet $\hookleftarrow$ biotite + 2 sillimanite + quartz

THE GRANULITE FACIES

The Granulite Facies of the Westport-Gananoque Area of Southeastern Ontario[1]

The high-grade regional metamorphic rocks of the "Grenville Group" of the Canadian Shield underly a large part of southeastern Ontario, southwestern Quebec, and the Adirondacks of New York (Fig. 8-11). The rocks include quartzofeldspathic gneisses and granulites including pyroxene-bearing types, quartzite, and marble. They appear to be the metamorphosed equivalent of a layered sequence of sedimentary rocks perhaps 15,000 ft thick. Radiometric K–Ar ages approximating the time of metamorphism range from about 800 to 1,100 million years (Stockwell, 1961, 1964). Metamorphism in the "Grenville Group" of this area grades from northeastward-trending belts of the granulite facies to the adjacent, somewhat lower temperature rocks gradational to the hornblende hornfels facies or to the sillimanite zone of the upper amphibolite facies. The metamorphic rocks have been deformed into upright, isoclinal, similar-style folds generally plunging gently northeast. They are overlain to the northeast by unmetamorphosed lower Paleozoic sandstone, shale, and limestone.

Rocks in the granulite facies include low-mafic hypersthene granulites and hypersthene-biotite granulites, and gneisses containing brownish-yellow perthite and antiperthite (the former is K-feldspar containing

[1]Wynne-Edwards, 1967a,b; Wynne-Edwards and Hay, 1963; Reinhardt, 1968.

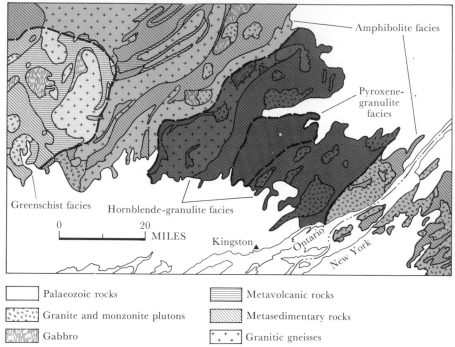

Fig. 8-11   *Map of zones of regional metamorphism in the "Grenville Series" of the Westport-Gananoque and adjacent areas of southeastern Ontario. After Wynne-Edwards (1967a).*

40 to 50% exsolved plagioclase; the latter is plagioclase containing exsolved K-feldspar). The granulites are almost structureless quartzofeldspathic rocks, nearly lacking both schistosity and layering. The gneisses,
on the other hand, nearly lack a schistosity but are compositionally layered. Minerals in the granulites and gneisses are the same except for abundance. Both rock types are "permeated by coarse-grained pink granitic
material, but [in the granulites it] is irregularly distributed in pods and
dykelets owing to the absence of foliation" (Wynne-Edwards, 1967a, p. 29).
These rocks that contain coarse quartz-feldspar mafic-free granitic layers
are called lit-par-lit gneisses. Several large concordant granitic plutons and
a few bodies of massive diorite, gabbro, and anorthosite were also emplaced
into the metamorphic rocks.

   The associated lower-grade rocks include abundant cordierite-
sillimanite and cordierite-garnet gneisses, which appear to have "crystallized in an environment intermediate between those of the pyroxene
granulite subfacies and the hornblende hornfels facies" (Wynne-Edwards,
1967a, p. 55). In these rocks, antiperthite is uncommon and hypersthene
is absent.

Mineral assemblages that are apparently stable in the granulite facies of this area include the following (Wynne-Edwards, 1967a; Wynne-Edwards and Hay, 1963; Reinhardt, 1969):

Quartz-plagioclase $\pm$ perthite-diopside-scapolite-calcite $\pm$ grossularite
       (An_{25-35})

Quartz-plagioclase-biotite $\pm$ diopside-hypersthene $\pm$ Fe-ore $\pm$ rutile

Quartz-perthite-biotite $\pm$ sillimanite $\pm$ garnet $\pm$ Fe-ore $\pm$ spinel
                                                                      $\pm$ apatite

Quartz-plagioclase-perthite-cordierite-hypersthene-garnet[1]

Quartz-plagioclase-perthite-hornblende + Fe-ore $\pm$ apatite

Plagioclase-biotite-hornblende + Fe-ore $\pm$ apatite

Perthite-biotite-sillimanite-garnet + Fe-ore

Several distinctive layers form useful marker horizons for structural and stratigraphic mapping in the metasedimentary rocks. Garnet gneiss includes both granitic and biotitic varieties but both are characterized by low Ca and high FeO/MgO. Biotite gneiss commonly contains oligoclase as the only feldspar (no K-feldspar). Diopsidic gneiss is dark green and well layered and generally also contains biotite, brown (high-temperature) hornblende, and plagioclase. Amphibolite is dark brown to black and consists of brown hornblende and plagioclase (An_{22-32}). Lit-par-lit gneisses are thought to result from partial melting or metasomatism and granitization. The large homogeneous granitic bodies appear to have developed at structurally favored sites.

Part of the area has been subjected to later lower-grade metamorphism around the postregional metamorphic granitic plutons. Hypersthene has been partly altered to biotite and chlorite; the diopsidic pyroxene to green hornblende, chlorite, and biotite; the garnet to green mica; ilmenite to sphene; plagioclase to sericite.

The metamorphic grade in the Westport-Gananoque area corresponds to the pyroxene granulite facies in the central part, where it forms biotite-free, hypersthene-bearing gneisses. Gradationally outward from this, biotite becomes more abundant than hypersthene and hornblende appears in somewhat lower grade rocks corresponding to the hornblende granulite facies.

Rocks of equivalent metamorphic grade and stratigraphic sequence

---

[1]These cordierite-bearing rocks occur in the lower-grade part of the area (possibly gradational to the hornblende hornfels facies).

(though inverted) occur across the St. Lawrence River to the southeast in the northwestern Adirondacks and have been studied by Engel and Engel, 1953, 1958, 1962a, b, 1963; Engel, Engel, and Havens, 1964; Buddington, 1963, 1965, 1966; deWaard, 1965a, 1967; deWaard and Romey, 1969.

General Features of the Granulite Facies

Granulite facies regional metamorphic rocks are widespread but apparently are not very voluminous in the exposed Precambrian shield areas. They probably develop in the deepest parts of the continental crust, grading upward to the lower-grade amphibolite facies, which develops at somewhat shallower depths. Small to large bodies of diorite, gabbro, or anorthosite in many instances are associated. In addition to the Westport-Gananoque and Adirondack areas noted above, these highest-grade "normal" regional metamorphic rocks occur elsewhere in the late Precambrian Grenville "Series" to the north and east (e.g. Kretz, 1959; Kranck, 1961; Harry, 1961; Roach and Duffell, 1968; Katz, 1968, 1969); just north of the Grenville in Labrador (Emslie, 1970); northern New Jersey (Vogel et al., 1968); the Wilmington complex, Delaware (Ward, 1959; Clavan et al., 1954); southwestern Minnesota (Himmelberg and Phinney, 1967); in the Pony and Cherry Creek Groups of southwestern Montana (Reid, 1957; Heinrich, 1960); northwestern Washington (Misch, 1971); and the Santa Lucia Range and the San Gabriel Mountains of coastal and southern California (Compton, 1960; Reiche, 1937; Hsu, 1955). Outside North America, prominent examples of the granulite facies include those in Greenland (Kranck, 1935; Ramberg, 1948a, b); Scotland (Davidson, 1943; Sutton and Watson, 1950; Muir and Tilley, 1958; O'Hara, 1960, 1961; Dearnley, 1963; Howie, 1964); Norway (Bugge, 1940; Heier, 1960; Touret, 1968, 1971); Sweden (Quensel, 1951; Saxena, 1968); Finland (Hietanen, 1947; Simonen, 1948; Eskola, 1952; Parras, 1958); U.S.S.R. (Lebedev, 1939; Moor, 1940; Drugova and Bugrova, 1965; Lutz, 1968); India (Holland, 1893, 1900; Howie, 1955; Howie and Subramaniam, 1957; Subramaniam, 1959, 1960; Leelanandam, 1968); Ceylon (Adams, 1929; Cooray, 1962; Hapuarachchi, 1967, 1968; Searle, 1968); Australia (Prider, 1945; Wilson, 1959, 1968; Binns, 1964); Antarctica (Klimov et al., 1968; Suwa, 1968); Africa (Groves, 1935; Pulfrey, 1946; Hepworth, 1968; McIver and Gevers, 1968; Giraud, 1968; Macdonald, 1968); Spain (den Tex and Vogel, 1962); and Guyana (British Guiana: Cannon, 1966).

Rocks of the granulite facies are generally moderately coarse to coarse grained and have a greater proportion of granular—quartz and feldspar—constituents than most rocks in the lower grades. The granulite facies consists of two zones, a lower-temperature hornblende-orthopyroxene zone (gradational to the sillimanite zone of the amphibolite facies) and a higher-temperature orthopyroxene zone. Rocks in the lower-temperature zone may contain biotite and hornblende and thus include micaceous schists and gneisses, whereas those in the higher-temperature zone are beyond the stability limits of the micas and amphiboles and are characterized by a virtually anhydrous mineralogy. This orthopyroxene zone is clearly recognized in only a few areas: Westport, Ontario (described above); the Wilmington complex of Delaware (Ward, 1959); near Madras, southern India (e.g., Holland, 1893, 1900; Howie, 1955; Subramaniam, 1959); Uganda, East Africa (Groves, 1935; Macdonald, 1968); the Fraser Range of Western Australia (Wilson, 1968); the Willyama complex of New South Wales (Binns, 1964); and possibly in the San Gabriel Mountains of southern California (Hsu, 1955), on Langøy, northern Norway (Heier, 1960), and in southwestern Ceylon (Hapuarchchi, 1968). Both zones may show a poorly developed to well-developed schistosity defined by small flattened lenses of quartz. They may also exhibit a gneissic layering defined by alternating layers of quartz- and feldspar-rich or quartzofeldspathic and proxene-bearing (or other mafic minerals if present) layers.

Individual rock types include hypersthene granites (also called charnockites), hypersthene granodiorites, quartz-hypersthene gabbros (quartz norite), scapolite-diopside-quartzofeldspathic gneisses, garnet-cordierite-K-feldspar-quartz gneisses, and marble.

Quartzofeldspathic pyroxene-bearing gneisses are common, greasy to waxy looking, and are medium grayish-green to brown in color because of the color of the plagioclase. Quartz is in some instances bluish gray to greenish. Rocks rich in plagioclase and pyroxenes are in many instances low in K-feldspar (e.g., perthite) or quartz, whereas those rich in K-feldspar and biotite are more commonly low in plagioclase (or antiperthite).

Coexisting mineral phases and assemblages of the granulite facies are illustrated on ACFmK and ACF diagrams in Fig. 8-12. Possible reactions transitional to and within the granulite facies include (e.g., Ramberg, 1949, 1952, pp. 68, 152; Boyd, 1954; Kretz, 1959; deWaard, 1965a, b, 1966; Muller, 1966; Hapuarchchi, 1967, 1968):

Anthophyllite (or cummingtonite) $\leftrightarrow$ hypersthene + quartz + H_2O
Tremolite $\leftrightarrow$ 3 enstatite + 4 diopside + 2 quartz + $2H_2O$
Biotite + sillimanite + 2 quartz $\leftrightarrow$ almandine + orthoclase + H_2O
Hornblende + almandine + 5 quartz $\leftrightarrow$
$$7 \text{ hypersthene} + 3 \text{ plagioclase} + H_2O$$
$$(An_{76} \text{ component})$$
Hornblende + 2 biotite + 17 quartz $\leftrightarrow$
$$15 \text{ hypersthene} + 4 \text{ orthoclase} + 3 \text{ plagioclase} + 5H_2O$$

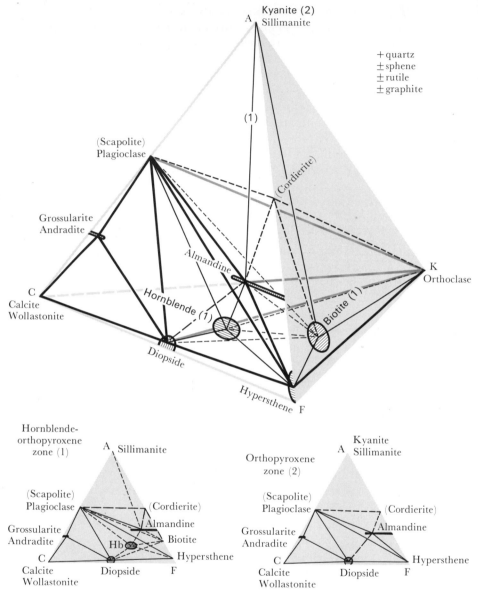

Fig. 8-12 *Granulite facies ACFmK and ACF diagrams. Minerals joined by heavy tie lines occur throughout the facies. Minerals in lightface and joined by thin tie lines occur in particular zones (i.e., 1, 2) as indicated in parentheses. Restrictive-zone minerals and assemblages are given in parentheses: (1) Hornblende-orthopyroxene zone (hornblende, biotite, sphene, sillimanite-almandine); (2) orthopyroxene zone (diopside-hypersthene-almandine) at higher P_{load}/P_{H_2O}; the diopside-almandine join replaces the hypersthene-plagioclase join (deWaard, 1965); ( ) cordierite may appear, especially in biotite-bearing types, probably at lower P(?).*

$$\text{Hornblende} + \text{quartz} \leftrightarrow 3 \text{ hypersthene} + 2 \text{ anorthite} + H_2O$$
(component in plagioclase)

$$\text{Biotite} + 3 \text{ quartz} \leftrightarrow 3 \text{ hypersthene} + \text{orthoclase} + H_2O$$

Origin of Granulite Facies Rocks

Both metamorphic and igneous origins have been long considered and strongly argued for hypersthene-bearing plutonic rocks in various areas. Each appears valid for individual occurrences. Evidence favoring each is reviewed below:

Metamorphism of Supracrustal Rocks (Sedimentary and Volcanic)[1]

This origin is favored where:

1. The rocks or interbedded layers are of undoubted metasedimentary composition, such as marble, calc-silicate rocks, quartzite, high-quartz rocks (for example, > 60% quartz), graphite-bearing rocks, quartz-banded iron ores, or rocks high in Al_2O_3 (for example, > 17%) or high in sillimanite + garnet.
2. The rocks show a wide variety of compositions of layers or of thickness of layers such as are not expected from metamorphic differentiation of a nearly homogeneous igneous rock.
3. Concentrations of dark or accessory minerals occur at the contacts between some layers (original heavy mineral concentrations?).
4. Zircon grains are rounded, and in many the length is not parallel to the C axis; plagioclase grains are untwinned.
5. Correlation may be made with less metamorphosed equivalents at the same stratigraphic horizon elsewhere in the region.
6. Relict amygdules may be recognized in amphibolites.
7. Hornblende crystallized before hypersthene (texturally) and hypersthene in the mafic-rich granulites is more Fe rich than that in the granitic granulites.

Metamorphism of an Earlier Igneous Intrusion[2]

This origin is favored where:

1. Relict ophitic or other plutonic igneous texture is preserved.

[1]Ramberg, 1948a, 1949; Eskola, 1952; Hsu, 1955; Howie and Subramaniam, 1957; Parras, 1958; Ward, 1959; Engel and Engel, 1960; Heier, 1960; Compton, 1960; Cooray, 1962; Wynne-Edwards, 1967a,b; Roach and Duffell, 1968; Reinhardt, 1968; Giraud, 1968; Searle, 1968; Viswanathan and Murthy, 1968; McIver and Gevers, 1968; Hepworth, 1968; deWaard, 1968; Macdonald, 1968; Katz, 1969.
[2]Tyrrell, 1929, p. 317; Groves, 1935, 1951; Prider, 1945; Quensel, 1951; Buddington, 1952; Howie, 1955; Compton, 1960; O'Hara, 1961; Dearnley, 1963; Singh, 1966; Wynne-Edwards, 1967b, 1969; deWaard, 1968b.

2. "Metadolerites" (metadiabases) grade to lower-grade equivalents of recognizable plutonic igneous parentage.
3. Granulite facies metasedimentary rocks occur as inclusions in and are deformed around a more massive pluton.
4. Chemical analyses fall along reasonably smooth granite to gabbro (calc-alkaline) trend on variation diagrams (both triangular and Larsen type).
5. The more felsic rocks ("late differentiates") are strongly enriched in FeO/MgO.
6. The rocks contain unzoned plagioclase and other minerals.
7. Biotite crystallized before hornblende and in turn before hypersthene (texturally), the reverse of Bowen's Reaction Series.

Metamorphism of Earlier "Basement" Gneisses (Initially "Dry") [1]

Igneous Origin (Intruded as a Dry Magma Perhaps Resulting from Anatexis of the Enclosing Rocks) [2]

This origin is favored where:

1. The plutonic rocks are sharply bounded to lower-grade rocks.
2. The rocks form large, homogeneous, massive bodies.
3. The rocks show intrusive relationships to the surrounding rocks (e.g., veins, dikes, or contact crosscutting foliation of country rocks).
4. Texture is massive except adjacent to country rock contacts.
5. Angular to rounded inclusions of metasedimentary or other country rocks are present.
6. Pegmatitic phases are present.
7. Rhythmic graded layering in mafic or ultramafic rocks.
8. Well-twinned or zoned plagioclase or ophitic texture.
9. Zircons are largely euhedral and doubly terminated, and have unimodal length/width ratios. See also discussion of the anorthosite pluton association in Chap. 4.

Water in the Granulite Facies

As noted above, water is driven off almost continuously during progressive metamorphism (except in the zeolite and prehnite-pumpellyite facies). Wet sediments are water saturated, and saturation is maintained by nu-

[1]Wynne-Edwards, 1967b, 1969.
[2]Dunn, 1942; Rama Rao, 1945; Pulfrey, 1946; Eskola, 1952; Hsu, 1955; Parras, 1958; Subramaniam, 1959; Compton, 1960; Wynne-Edwards, 1967a; Viswanathan and Murthy, 1968; deWaard and Romey, 1969; Katz, 1969.

merous dehydration reactions that drive water into the intergranular spaces (compare Fig. 7-11 and the related text discussion). These dehydration reactions in metasedimentary rocks culminate within the granulite facies with the breakdown of hornblende and biotite. Although the remaining minerals are anhydrous, the rock may continue to be water saturated as long as the temperature is maintained or continues to rise. Thus conditions of water saturation can exist during granulite facies metamorphism (e.g., Fyfe, Turner, and Verhoogen, 1958, pp. 233–234; Katz, 1969).

During metamorphism of intially dry rocks such as plutonic igneous rocks, however, rather different conditions prevail. $P_{H_2O} < P_{load}$ throughout metamorphism of these rocks unless water is injected into them from surrounding dehydrating metasediments (e.g., along fractures produced during deformation). Under such water-deficient conditions, the boundary for a dehydration reaction (e.g., biotite $+ 3$ quartz $\leftrightarrow 3$ hypersthene $+$ orthoclase $+ H_2O$) will lie at a lower temperature. The new steep boundary (almost parallel to the pressure axis) intersects the normal water-saturated curve at the remaining water pressure. Thus the water-deficient rocks could develop diagnostic minerals such as hypersthene at a lower temperature (lower "grade") than in the adjacent water-saturated metasedimentary rocks. This problem of metamorphism of water-deficient rocks has been discussed from a theoretical point of view by Yoder (1955b) and deWaard (1966b) and from an experimental point of view by Greenwood (1961).

It seems possible therefore that certain metaigneous rocks with granulite facies mineralogy could have developed at lower temperatures than normally expected for the granulite facies. Examples have been described by deWaard (1968b), Wynne-Edwards (1969), and Touret (1971).

Granulite Facies and the Lower Crust

Although a few exceptions exist (e.g., Compton, 1960) almost all granulite facies complexes occur in Precambrian rocks, most of them in the Precambrian shield areas. It is clear that such shield areas underly sedimentary rocks of the continental interiors and even most "miogeosynclinal" rocks near the continental margins. It is also generally agreed that the Precambrian shields were once "geosynclinal" belts of sediments and volcanic rocks, which have been regionally metamorphosed, dehydrated, and partially melted (anatexis) deep in the crust. At the deepest levels in the continental crust, well below the minimum depths for anatexis (i.e., sillimanite zone of the amphibolite facies), not only must the rocks be thoroughly dehydrated, but they must have also lost their minimum melting constituents (alkali granite). That the granulite facies is widespread in the lower continental crust is not amenable to direct observation but has been suggested by different lines of evidence. These include the correlation of the top of the granulite facies with the Conrad seismic discontinuity (e.g., Belousov, 1966, 1969; Closs, 1969) and with the base of the high electrical conductivity layer (Hyndman and Hyndman, 1968).

THE BLUESCHIST FACIES[1]

The Blueschist Facies of the Franciscan Assemblage of Western California[2]

The distinctive high-pressure regional metamorphic rocks of western California extend more than 500 miles in a northwesterly direction, especially in the northern part of the state (Fig. 8-13). Lower-grade rocks of the prehnite-pumpellyite and zeolite facies occur in adjacent belts to the west. The glaucophane schists and associated metamorphic rocks are within the "eugeosynclinal" environment on the oceanic side of the more "normal" greenschist and amphibolite facies regional metamorphic rocks associated with the Sierra Nevada batholith. They occur adjacent to (below) a regional thrust fault containing an extensive sheet of serpentinite (see Fig. 8-13).

Premetamorphic graywacke with less shale, chert, and mafic metavolcanic rocks comprise the 50,000 or more feet of the Franciscan assemblage. The abundant graywackes contain plagioclase, volcanic rock fragments, quartz, and chert. K-feldspar is virtually absent. These rocks, deposited in the Upper Jurassic to Cretaceous, were metamorphosed almost continuously through the Upper Jurassic and Lower Cretaceous (Lee et al., 1969; Suppe, 1969; Bailey, written communication, January, 1971), very soon after deposition.

Granitic plutons, such as are common in the more "normal" regional metamorphic rocks described above, are absent in this blueschist facies environment. Masses of serpentine (alpine peridotite-serpentinite bodies) are abundant in some areas and rather uncommon in others. Most are

[1]Blueschist facies (Bailey, 1962) is defined as those rocks forming in the stability field of lawsonite. Thus lawsonite is confined to and diagnostic of the facies but blue amphiboles such as glaucophane are not.
[2]Bailey et al., 1964, 1970; Bailey and Blake, 1969; Blake et al., 1969; Bloxam, 1966; Ernst et al., 1970; Blake, written communication, December, 1970.

Fig. 8-13 *Map of regional metamorphic zones in the blueschist (3), prehnite-pumpellyite (2), and zeolite (1) facies within the Franciscan formation of western California. Metamorphism decreases in grade westward from the underside of the regional thrust fault. After Bailey, Blake, and Jones (1970, Fig. 5).*

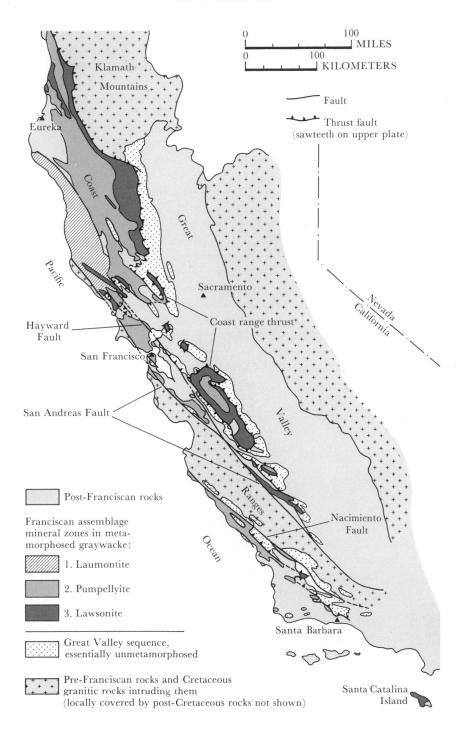

0   100
├────────┤ MILES
0   100
├────────┤ KILOMETERS

——— Fault

⌣⌣⌣ Thrust fault
(sawteeth on upper plate)

Klamath
Mountains

Eureka

Coast

Pacific

Great

Sacramento

Coast range thrust

Hayward
Fault

San Francisco

San Andreas Fault

Valley

Nevada
California

Ranges

Ocean

Nacimiento
Fault

Santa Barbara

Santa Catalina
Island

Post-Franciscan rocks

Franciscan assemblage
mineral zones in meta-
morphosed graywacke:

1. Laumontite

2. Pumpellyite

3. Lawsonite

Great Valley sequence,
essentially unmetamorphosed

Pre-Franciscan rocks and Cretaceous
granitic rocks intruding them
(locally covered by post-Cretaceous rocks not shown)

probably the basal part of the Mesozoic oceanic crust deformed along with the "geosynclinal" assemblage in the Late Cretaceous (Bailey, Blake, and Jones, 1970) but a few were emplaced "cold" along faults or as diapiric bodies.

Metamorphism of the graywackes resulted in jadeite- and/or lawsonite-bearing metagraywacke with little glaucophane, whereas metamorphism of mafic igneous rocks resulted in blueschists rich in blue amphibole (Ernst, 1965; Bloxam, 1966). Albite-bearing metagraywacke is converted to jadeite-bearing metagraywacke, in some instances over only a few inches of distance. Locally there is no apparent relationship of this metamorphism to distance from serpentinite bodies, but regionally the lawsonite zone (blueschist facies) is 1 to 4 km thick, lying immediately under the Coast Range thrust (see Fig. 8-13). The pumpellyite zone underlies this to the west. Although the jadeitic graywacke resembles less-metamorphosed Franciscan graywacke in lacking a schistosity, it may generally be distinguished by a greenish-gray waxy appearance on fresh surfaces and because grain boundaries appear to be less distinct. Where the jadeite is coarse enough to see, it occurs as tiny radiating clusters. Graywackes (specific gravity commonly 2.55 to 2.65) having specific gravities greater than 2.71 have been found almost invariably to be jadeitized (Bailey et al., 1964).

Gradational conversion of "greenstone" (metavolcanic rocks) or metagraywackes to blueschist has occurred outward from a few serpentinite contacts for tens of feet. Elsewhere, "greenstones" are locally veined and replaced by blue amphibole, and jadeitic graywacke locally contains a little glaucophane. In a few areas blue amphibole- (generally glaucophane-) rich rocks underly tens of square miles. Most of these are schistose parallel to bedding and lineated but a few are massive. Some schists are tightly folded and cut by veins of quartz and albite. Relict features such as diabasic textures and pillow structures are preserved in some rocks.

Tectonically emplaced blocks a few feet to tens of feet across include eclogite, glaucophane schist, and related rocks, in some but not all instances enclosed in serpentine and emplaced most commonly in the pumpellyite zone but also in the blueschist facies. Many of these blocks are coated by a few feet of chlorite-actinolite schist. Others show smooth, slickensided surfaces. The eclogites and related rocks do not occur in normal depositional or metamorphic contact with the blueschists.

Distinctive minerals in the blueschist facies rocks of the region include white or pale green lawsonite, pale green or colorless jadeite, dark blue or bluish-black glaucophane or crossite, and aragonite. Mineral assemblages from a few representative areas are listed below (Ernst, 1965; Ernst and Seki, 1967; Ghent, 1965; Davis and Pabst, 1960; Brothers, 1954; Coleman and Lee, 1962; McKee, 1962):

Quartz- "phengite"- chlorite- albite ±   lawsonite-   calcite-   magnetite ± stilpnomelane-sphene

±quartz ± "phengite" ± chlorite ± albite-   lawsonite-   calcite-   crossite ±   stilpnomelane-sphene

±quartz ± "phengite"-   chlorite-   albite ± pumpellyite ± calcite   -sphene

Quartz   -aegirine   -crossite - stilpnomelane

      "Phengite"   -pumpellyite-lawsonite   .   -glaucophane   -sphene

                    Pumpellyite   -glaucophane

Quartz- "phengite"- chlorite- albite-   lawsonite-   aragonite   -stilpnomelane-sphene

Quartz   -jadeite-lawsonite   -glaucophane

      "Phengite"- garnet   -aragonite-glaucophane   -tourmaline

          - Garnet   -clinozoisite-   aragonite-glaucophane   -pyrite

Quartz   -jadeite-lawsonite-   aragonite

Attempts have recently been made to subdivide the blueschist facies into zones (e.g., McKee, 1962; Ernst, 1965; Ernst and Seki, 1967). Rocks in the low-grade zone contain calcite and albite, whereas in the higher-grade zone these disappear, and aragonite and jadeitic pyroxene, respectively, take their place. Lawsonite is found in both zones. Glaucophane, aegerine-augite, probably garnet, and rutile in place of sphene, are also confined to the high-grade zone, whereas crossite may appear in both zones.

Sodium-rich minerals such as the blue amphiboles and jadeitic pyroxenes at first sight suggest that many blueschists are abnormally rich in sodium. Chemical analyses of many blueschists indicate, however, that such rocks are chemically equivalent to typical unmetamorphosed "eugeo-synclinal" rocks (e.g., Turner and Verhoogen, 1960, p. 544; Coleman and Lee, 1963; Lee et al., 1963; Ernst, 1963b, 1965; Bailey et al., 1964; Ghent, 1965; Ernst et al., 1970). Possible exceptions have been noted.

The presence of the high-pressure minerals aragonite, jadeitic pyroxene, and glaucophane-crossite suggests an abnormally low thermal gradient (high pressure, low temperature). No consensus has been reached as to whether the requisite pressures can be reached by rapid deep burial in the "eugeosynclinal" pile (e.g., Brown, Fyfe, and Turner, 1962; Bailey et al., 1964; Essene and Fyfe, 1967) or whether, in addition to stratigraphic depth, tectonic overpressures are necessary to obtain these conditions (e.g., Coleman and Lee, 1962; McKee, 1962; Coleman, 1967; Blake et al., 1967, 1969; Bailey and Blake, 1969).

General Features of the Blueschist Facies

Regional metamorphic rocks of the blueschist facies are much less common than those of the higher-thermal-gradient greenschist facies. They develop under conditions of an "abnormally" low thermal gradient in the "eugeo-synclincal" environment along tectonically active continental margins. There they form the high-pressure oceanward member of a common high-pressure–low-pressure pair of metamorphic belts (Miyashiro, 1961; Zwart, 1967; Landis and Coombs, 1967; Dewey and Bird, 1970). They appear to form under high-pressure conditions, though in many cases at least, at higher present structural levels than the prehnite-pumpellyite facies. Most, if not all, blueschist facies metamorphism is confined to post-Paleozoic orogenic belts (deRoever, 1956).

Blueschist metamorphism outside California extends into coastal Oregon (Dott, 1965), with minor occurrences in northwestern Washington (Misch, 1966, 1971) and central and northern British Columbia (Arm-strong, 1949; Paterson, 1971; Monger and Hutchison, 1971). Elsewhere in the circum-Pacific belt, blueschist metamorphic rocks have been described from Japan (e.g., Suzuki, 1930; Suzuki and Suzuki, 1959; Banno, 1958, 1964; Seki, 1958, 1960; Iwasaki, 1963; Ernst, 1964; Ernst et al., 1970), Celebes (e.g., deRoever, 1947, 1953), New Caledonia (Joplin, 1937), New Zealand (Landis and Coombs, 1967), and Venezuela (Dengo, 1950, 1953; Shagam, 1960). Still other occurrences have been described from Burma (e.g., Lacroix, 1930; Yoder, 1950), Turkey (Schurmann, 1956), Italy (Hoff-man, 1970), Corsica (e.g., Brouwer and Egeler, 1952; Egeler, 1956), the Italian, Swiss, and French Alps (Bearth, 1959, 1966; van der Plas, 1959; Ellenberger, 1960; Zwart, 1967; Chatterjee, 1971), southeastern Spain (deRoever and Nijhuis, 1964), and in the Urals of the U.S.S.R. (Dobretsov et al., 1965).

Common rocks formed in the blueschist facies include fine- to me-dium-grained schistose or massive glaucophane-crossite rocks, jadeite-bearing metagraywackes and associated shales, sandstones, and conglom-erates. Glaucophane- (or crossite-) rich rocks are distinctive because of their dark blue color, though this may be in some cases so dark as to appear nearly black. Porphyroblasts of white lawsonite or albite may be common. Jadeite-bearing metagraywackes may be distinguished on careful examina-tion by their gray to greenish gray, somewhat waxy appearance and high density as described above for the Californian occurrences. Piemontite-quartz schists and phyllites may be recognized by the distinctive pink color imparted by the piemontite. Associated cherts, metabasalts, and serpen-tinites are much like those in California.

The blueschist facies appears to be separable into two distinct miner-alogical zones, a lawsonite-albite zone (probably lower pressure) and a lawsonite-jadeitic pyroxene zone (probably higher pressure). Lawsonite $[CaAl_2Si_2O_7(OH)_2 \cdot H_2O]$ is regarded as diagnostic of the facies. Jadeitic pyroxene and aragonite are confined to this and the higher-pressure eclogite facies; soda amphiboles to this and adjacent higher-temperature, high-pressure parts of the greenschist facies; stilpnomelane to this and the chlorite zone of the greenschist facies; pumpellyite to this and the lower-pressure prehnite-pumpellyite facies. Biotite is absent.

Coexisting mineral phases and equilibrium assemblages for the blueschist facies are illustrated on ACFmNa and ACF diagrams in Fig. 8-14 (note the use of Na$_2$O as a component instead of K$_2$O because of the importance of Na-bearing minerals). As with ACFmK diagrams for metamorphic facies described above, possible mineral assemblages include four-phase assemblages plus quartz, represented by the corner of any of the subtetrahedra within the ACFmNa diagram. Accessory minerals include sphene, pyrite, hematite, "magnetite," and to a lesser extent rutile, apatite, zircon, and pyrrhotite; but only sphene appears to be common.

Possible reactions transitional to and within the blueschist facies include (e.g., deRoever, 1955; Brown et al., 1962; Coombs, 1960; Seki et al., 1960; Coleman and Lee, 1963; Ernst, 1963b; 1965; Coombs et al., 1970):

Anorthite + 2H$_2$O ↪ lawsonite
(component in plagioclase)
Albite ↪ jadeite + quartz
Analcite ↪ jadeite + H$_2$O
2 albite + antigorite ↪ glaucophane + H$_2$O
        (component in chlorite)
13 albite + 3 chlorite + quartz ↪
                5 glaucophane + 3 paragonite + 4H$_2$O
10 albite + 3 chlorite + 3 calcite + 7 quartz ↪
                3 lawsonite + 5 glaucophane + 3CO$_2$ + H$_2$O
Prehnite + H$_2$O + CO$_2$ ↪ lawsonite + calcite + CO$_2$
Prehnite + laumontite + CO$_2$ ↪
                2 lawsonite + calcite + 3SiO$_2$ + H$_2$O
12 pumpellyite + 9 Al-chlorite component + 38SiO$_2$ + 50H$_2$O ↪
                48 lawsonite + 8 chlorite component
Calcite ↪ aragonite

Chemistry of Metabasalts in the Blueschist Facies

The abundance of Na-rich mafic minerals in rocks of the blueschist facies, especially in the metabasalts, has suggested to some workers that Na-metasomatism may be responsible for the development of glaucophane schists. As noted in the above section on California occurrences, however, most workers consider the glaucophane schists and related metabasalts to be chemically equivalent to normal basalts. An examination of analyses of 42 metabasalts from worldwide occurrences of the blueschist facies shows most to fall in the range of chemistry of the low-K oceanic tholeiites (compare

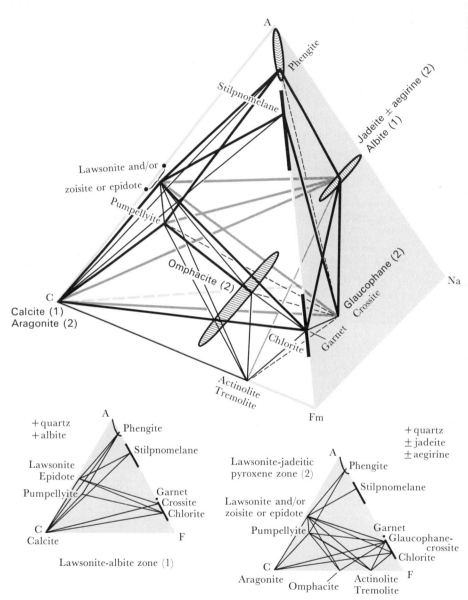

Fig. 8-14 *Blueschist facies ACFmNa and ACF diagrams. Minerals joined by heavy tie lines occur throughout the facies. Minerals in plain type and joined by thin tie lines occur in individual zones as indicated in parentheses. Restrictive-zone minerals are given in parentheses: (1) lawsonite-albite zone (albite, calcite); (2) lawsonite-jadeitic pyroxene zone (jadeite ±aegirine, omphacite, glaucophane, aragonite, actinolite-tremolite).*

Table 4-8). A few analyses showed slightly higher K_2O (but not when compared with continental tholeiites) and slightly lower MgO and CaO. Only one analysis showed slightly higher Na_2O (4.6%) and one slightly lower Na_2O. Although the average Na_2O of 3.0% is slightly higher than that for the oceanic tholeiites (2.7%) several of the metabasalts were described as spilitic. The difference in Na_2O is probably not significant.

Suggested Origins of the Blueschist Facies

The rather distinctive mineralogy of many rocks in the blueschist facies, along with their localization along the continental margins since the Paleozoic era, has given rise to considerable discussion of their mode of origin. Most writers agree that experimental work on high-density minerals, such as lawsonite, jadeitic pyroxene and metamorphic aragonite, indicates an origin involving high pressures and relatively low temperatures. However, the details are not clear. Suggestions include:

Rapid (low-temperature) burial to great depth (high pressure). Conditions could be:

5–10 kbars— > 15-km depth at 150–300°C (Essene et al., 1965)

6–10 kbars— at 200–300°C (Brown et al., 1962; Paterson, 1971)

> 20-km depth (Bailey et al., 1964)

5–8 kbars—20–30-km depth at 150–200–300°C (Ernst, 1965, 1971)

4–8 kbars—15–30-km depth at 150–400°C (Ernst et al., 1970)

(also Seki, 1960; McKee, 1962; Page, 1966).

Localized areas of abnormally high confining pressure, the apparent depth of burial (e.g., stratigraphic) being too shallow (that is, 9 to 12 km) for the required pressure (Turner, 1948, p. 100; Coleman, 1961; Page, 1966; Blake et al., 1967, 1969).

Metamorphic grade increases regionally to higher stratigraphic levels (Iwasaki, 1963; Kanehira, 1967; Blake et al., 1967) and high water pressures are considered to result from thrusting of the rocks under a regional serpentinite body (Blake et al., 1967, 1969; Bailey et al., 1970).

Sediments dragged to considerable depths (for example, 20 to 40 km) along Benioff zone at continental margin but squeezed upward before heating up notably (Gilluly, 1969; Hamilton, 1969; Ernst, 1971).

P_{H_2O} may be less than P_{load} if concurrent serpentinization used up H_2O during metamorphism (Coleman, 1961). Fyfe and Valpy (1959) show that a dry environment favors formation of jadeite. Brown and coworkers (1962) note that a dry environment would favor preservation of aragonite.

Metastable growth of blueschist minerals at relatively shallow depths (Gresens, 1969, 1970).

Development in metamorphic aureoles (especially narrow local ones) around serpentinites (Taliaferro, 1943; Wilson, 1943; Crittenden, 1951; Suzuki and Suzuki, 1959; Chesterman, 1960; Bloxam, 1960, 1966; Essene et al., 1965; Gresens, 1969).

Control by soda metasomatism (Taliaferro, 1943; Brothers, 1954; Suzuki and Suzuki, 1959; Essene et al., 1965).

A satisfactory origin for the blueschist facies must take into account the following common features:

Experimental confirmation of the high-pressure, relatively low temperature origin of minerals such as lawsonite, jadeitic pyroxene, and metamorphic aragonite.

Most rocks of the facies are chemically similar to equivalent unmetamorphosed basalts and sedimentary rocks in the same stratigraphic environment.

Admitting difficulties in estimation of the total stratigraphic depth of burial, the pressure requirements appear to be greater than that obtained by burial alone.

Cases are documented (e.g., California, Japan, and Venezuela) of a regional increase in metamorphic grade to higher stratigraphic levels, culminating in a regional thrust fault or series of serpentinite bodies.

Several areas show a local increase in grade to glaucophane schists adjacent to serpentinite bodies.

Perhaps the most promising model yet proposed for blueschist formation is that of Blake, Irwin, and Coleman (1967, 1969). They suggest development of blueschist metamorphism "during thrusting in a zone of anomalously high water pressure in the lower plate along the sole of the [regional] thrust fault" (1969, p. 237). The interstitial water could be "retained by an impermeable barrier such as serpentinite in the fault zone, [and] fluid pressure might exceed the lithostatic pressure, assuming . . . the rocks have sufficient strength to contain this overpressure" (1969, p. 243). In this connection, Heard and Carter (1968) have shown experimentally that quartz (normally considered to be one of the most easily deformed minerals) can sustain a differential stress of 1 kbar at a total pressure of 8 kbars P_{H_2O}. Somewhat related is the suggestion of rapid burial to great depth, probably accompanied by downbuckling and underthrusting in the vicinity of an active "subduction zone" (zone of downward movement of material such as in a convection cell and coincident with the Benioff zone of dipping earthquake foci). The necessity of rapid burial derives from the requirement of low temperature to accompany the high pressure. The low temperature, however, may result from an abnormally low thermal gradient due to the low content of radioactive elements (K, U, Th) in the "eugeosynclinal" environment. The graywackes, basalts, and associated rocks contain very little K-feldspar or other minerals that contain these elements.

THE ECLOGITE FACIES

The Eclogite Facies Associated with Blueschists and Serpentinites of the Northern Coast Ranges of California[1]

Eclogites are moderately common in the Franciscan formation of Sonoma and Marin Counties, immediately north of San Francisco, California. They occur as isolated, rounded blocks or boulders a few feet to a few hundred feet long, surrounded by prehnite-pumpellyite facies graywackes, blueschists, serpentinites, fault gauge, or weathered out on the surface in such an environment. Most are inferred to lie along major shear zones. Some blocks are encased in a rind of chlorite-actinolite schist or show scratched, slickensided surfaces. All workers consider them to be out of place in the environment of their presently enclosing rocks. Occurrences have been noted, however, of eclogites as small-scale interlayers in glaucophane schist blocks (Coleman et al., 1965).

Texturally, most of the eclogites consist of medium- to fine-grained, massive, green to dark green omphacite, studded with reddish-brown porphyroblasts of garnet up to about 1/2 cm across. In a few blocks, eclogites are interlayered on a small scale with glaucophane schists.

The omphacite is a jadeitic-diopside with minor iron solid solution,

[1]Bailey et al., 1964; Coleman and Lee, 1963; Coleman et al., 1965; Lanphere and Coleman, 1970; Dudley, 1969; Essene and Fyfe, 1967; Lee et al., 1963; Bloxam, 1959; Borg, 1956; Switzer, 1945.

and the garnet is a grossularitic-almandine with minor spessartite (Mn) and pyrope (Mg). Other minerals include minor amounts of muscovite, chlorite, epidote or clinozoisite, rutile, sphene, and glaucophane. All these texturally appear to be primary—i.e., crystallized at the same time as the omphacite and garnet. In other instances, a somewhat different chlorite and lawsonite appear to be altering from garnet, epidote from clinozoisite, sphene from rutile, and glaucophane from omphacite. In most of these eclogites, omphacite and garnet are the dominant minerals, amounting to about 40 to 90% and 6 to 30% respectively.

Chemically, these rocks are all equivalent to basalts. Because they occur as isolated blocks, it is difficult to correlate other mineral assemblages or other rock compositions. Other blocks occurring in the same environment (Coleman and Lee's [1963] Type IV blocks; but not necessarily of the same metamorphic grade) include muscovite-epidote-glaucophane schists, amphibole-garnet gneisses, and metachert.

On the basis of the presence of epidote instead of lawsonite and of the presence of rutile and the absence of pumpellyite, Coleman and Lee (1963) suggest that these eclogites crystallized not only at high pressures but at higher temperatures than blueschists. This and the greater K−Ar age of some of the contained minerals suggests that these blocks may have been derived from an older, deeper metamorphic terrane (Lanphere and Coleman, 1970). Essene and Fyfe (1967) came to a similar conclusion from a thermodynamic calculation of the stability of omphacite (in eclogite) compared with jadeite (in blueschists).

General Features of Eclogites

Eclogites are rare rocks, but because of their distinctive and nearly constant mineralogy and appearance and controversial petrology, they have been studied by many geologists. They occur in a wide range of rocks of all ages from Precambrian to Tertiary (Church, 1967), all of which indicate a high-pressure origin but a wide range of temperatures. For convenience of discussion, they can be divided into those that occur as blocks or layers in regional metamorphic rocks ranging from the blueschist, through the amphibolite, to the granulite facies and those that occur as xenoliths in basalts or peridotite pipes.

Blueschist facies occurrences of eclogites, in addition to those in northern California described above, include others in northern California (Brothers, 1954; Davis and Pabst, 1960; Bailey et al., 1964; Soliman, 1964), Panoche Pass and vicinity in west-central California (Coleman, 1961; McKee, 1962; Ernst, 1965; Bailey et al., 1964), Santa Catalina Island off southern California (Bailey et al., 1964; Dudley, 1969), Guatamala (McBirney et al., 1967), the Alps (Bearth, 1959, 1965), U.S.S.R. (Chesnokov,

1960), and New Caledonia (Coleman et al., 1965; Essene and Fyfe, 1967). Upper greenschist facies eclogite occurrences include those in Venezuela (Dengo, 1950; Morgan, 1970), Norway (Binns, 1967; Gjelsvik, 1952), Japan (Banno, 1964) and Tasmania (Spry, 1963a). Amphibolite facies occurrences include those in Scotland (Alderman, 1936; Davidson, 1943; Mercy and O'Hara, 1968), Norway (Eskola, 1921; Kolderup, 1960; Lappin, 1966; Bryhni, 1966; Bryhni et al., 1970), France (Briere, 1920; Velde et al., 1970), Spain (den Tex and Vogel, 1962; Vogel, 1969), Poland (e.g., Smulikowski, 1960a, b, 1968), and Japan (Shido, 1959; Banno, 1966; Ernst et al., 1970). Granulite facies occurrences include those in Norway (Lappin, 1966), Poland (Smulikowski, 1960a, b, 1968) and Uganda (Groves, 1935).

Eclogites also occur as inclusions (or "nodules") in basalt in Australia (e.g., Lovering, 1964; Lovering and White, 1964, 1970) and Kenya (e.g., Saggerson, 1968), and apparently on Nunivak Island off the west coast of Alaska (Hoare, in Forbes and Kuno, 1967), in peridotite or kimberlite (mica peridotite) in Arizona (e.g., Watson, 1960; O'Hara and Mercy, 1966; Watson and Morton, 1969; Essene and Ware, 1970), New Mexico (O'Hara and Mercy, 1966), South Africa (e.g., Williams, 1932; Gurney et al., 1966; Berg, 1968; Godovikov and and Kennedy, 1968; Kushiro and Aoki, 1968; Rickwood et al., 1968; Mathias et al., 1970), and the U.S.S.R. (e.g., V. S. Sobolev, 1960; N. V. Sobolev, 1968; Sobolev et al., 1968).

Eclogites in the blueschist facies environment are typified by those in California described above. They occur as isolated tectonic blocks along shear zones. Those in the greenschist facies appear as layers and lenses, in some instances tectonically emplaced (e.g., Binns, 1967), whereas in others the relationships are not clear. In the amphibolite and granulite facies, eclogites occur as layers up to 500 m thick (den Tex and Vogel, 1962), lenses, boudins (e.g., Morgan, 1970), and small massifs (Velde et al., 1970) that are in most instances apparently gradational and in equilibrium with the surrounding rocks. Those in basalt or peridotite are inclusions sharply bounded to the enclosing rock.

In all these occurrences, the eclogites are massive to somewhat layered, in some instances foliated parallel to the enclosing rocks. They are striking green (omphacite) and red (garnet) rocks containing a few accessory minerals. Although the color of these minerals varies between one locality and another, little obvious correlation with the type of occurrence is apparent. The omphacite is dark green in most blueschist areas and is variously described as gray-green, apple green, bright green, or dark green in other areas. The garnet ranges from pink to red to brown-red. Omphacite generally occupies somewhat more than half of the rock and the garnet less than half.

Coexisting mineral phases and apparent equilibrium assemblages for the eclogite facies are illustrated on an ACFmK diagram in Fig. 8-15.

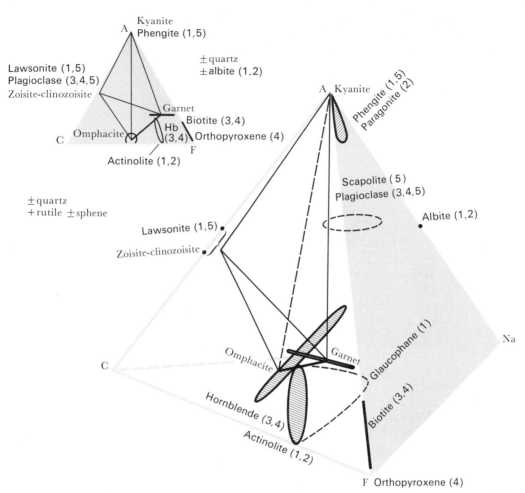

Fig. 8-15 *Eclogite facies ACFNa and ACF diagrams. Heavy line is tie line between minerals that coexist in eclogites associated with all facies. Minerals in plain type appear to coexist in eclogites associated with individual facies indicated by numbers in parentheses: (1) blueschist facies eclogite (lawsonite, albite, actinolitic glaucophane, muscovite or phengite, pyrite); (2) greenschist facies eclogite (green amphibole, Na–Al-rich amphibole, paragonite); (3) amphibolite facies eclogite (plagioclase, brown hornblende, biotite, pyrrhotite); (4) granulite facies eclogite (orthopyroxene, olive hornblende, biotite); (5) eclogite inclusions in kimberlite (lawsonite, plagioclase An_{18-24}, scapolite, phengite, ± diamond, pyrite).*

A variety of minor minerals are described as in apparent equilibrium or as later alteration products—in many areas both are present. Those texturally in equilibrium include omphacite, almandine garnet, zoisite-clinozoisite-epidote, quartz, kyanite, rutile, ± sphene, and apatite. In addition, certain minerals may occur in eclogites associated with country rocks of different facies (see Fig. 8-15). Many of these minerals are characteristic of rocks in the enclosing facies.

A similar group of minerals is texturally recognized as altering from the primary minerals: lawsonite, albite, glaucophane, actinolite, pumpellyite, epidote, chlorite, muscovite, paragonite, stilpnomelane, aragonite, quartz, sphene, ilmenite (blueschist or greenschist facies); plagioclase, scapolite, hornblende, diopside, anthophyllite, epidote-zoisite, chlorite, biotite, calcite, sphene, ilmenite, corundum (amphibolite or granulite facies). Again, many of the minerals are characteristic of the enclosing facies. Although individual minerals may occur in a single sample both as texturally "primary" and "alteration" products, their common occurrence in both forms throws some doubt on their "primary" nature. The ubiquitous nature of these alteration products suggests that most eclogites are not completely in equilibrium with their enclosing facies. Possible reactions transitional to the eclogite facies include (e.g., Yoder and Tilley, 1962; Green and Ringwood, 1967a; Church, 1967), beginning with basalt or gabbro:

Diopside + plagioclase + olivine ↔ omphacite + garnet + quartz
Albite ↔ jadeite + quartz
Albite + diopside ↔ omphacite + quartz
Anorthite + 2 olivine ↔ garnet

Beginning with amphibolite:

2 hornblende + 4 plagioclase ↔ 7 omphacite + 3 garnet + $2H_2O$

Chemistry of Eclogites

As with glaucophane schists, most eclogites are chemically equivalent to basalts—especially low-K oceanic tholeiites (see Table 8-2; also Forbes, 1965). A few of the higher-Na eclogites resemble spilites and a few of the higher-K eclogites resemble continental tholeiites. Plotting these analyses on a diagram of $Na_2O + K_2O : SiO_2$ (Fig. 4-19) suggests a wide range between tholeiitic and alkaline olivine basalts but the combination of high Na_2O and low K_2O for most of those in the "alkaline olivine" field suggests that these are instead spilitic.

Suggested Origins of Eclogites

The very high density of eclogites has long been recognized as indicating a high-pressure origin. Many early workers considered them to form by igneous crystallization under high temperatures and pressures (e.g., Eskola, 1921; Alderman, 1936). Only inclusions in kimberlites or other peridotites are now considered to be crystal cumulates in the upper mantle from melt

Table 8-2 *Average and range of chemical composition of eclogites and (for comparison) oceanic tholeiites*

	Eclogites		Oceanic Tholeiites	
	AVERAGE	RANGE	AVERAGE	RANGE
SiO_2	47.6	41.5 –51.01	49.8	43.49–52.24
TiO_2	1.3	0.26– 3.84	1.8	0.35– 3.65
Al_2O_3	14.6	7.49–20.68	15.0	10.28–17.07
Fe_2O_3	3.6	0.96– 7.05	2.8	0.95– 7.90
FeO	8.6	3.09–16.94	7.6	4.42–11.05
MnO	0.2	0.06– 0.45	0.16	0.11– 0.24
MgO	8.5	4.2 –16.23	8.2	5.44–17.30
CaO	10.6	7.3 –16.1	10.6	6.69–12.68
Na_2O	2.8	0.54– 6.24	2.7	1.29– 4.45
K_2O	0.5	0– 1.96	0.24	0.04– 0.70

SOURCE: Analyses of 36 eclogites from all occurrences: from Briere (1920); Eskola (1921); Williams (1932); Alderman (1936); Angel (1957); Yoder and Tilley (1962); Spry (1963); Coleman and Lee (1963); Coleman et al. (1965); O'Hara and Mercy (1966); McBirney et al. (1967); Mercy and O'Hara (1968); Church (1967); Kushiro and Aoki (1968); Velde et al. (1970); Morgan (1970).

formed by partial fusion of garnet peridotite (e.g., Williams, 1932; Ringwood, 1958; Kuno, 1967; Kushiro and Aoki, 1968; O'Hara, 1969; Mathias et al., 1970). Davidson (1967), however, indicates that at least some eclogite inclusions may be xenoliths from the lower crust. Eclogites occurring in metamorphic environments are now generally thought to form by metamorphic crystallization under various possible conditions. Suggestions include:

Very high pressures resulting from downwarping volcanic rocks to extreme depths in active orogenic belt or in mantle (Bailey et al., 1964; Bryhni, 1966; Lappin, 1966; Binns, 1967; Mercy and O'Hara, 1968).

High pressure due to tectonic overpressure, the erratic and limited distribution of blueschist occurrences indicating local rather than regional conditions (Coleman and Lee, 1962; Coleman et al., 1965).

Higher lithostatic pressure, lower water pressure, or higher temperature than blueschists (Coleman and Lee, 1963).

High pressure resulting from descent of oceanic basaltic crust into the mantle along inclined seismic (Benioff) zones (Green and Ringwood, 1969; Dickinson, 1970a; Gilluly, 1971).

Dry (low P_{H_2O}) metamorphism of basaltic rocks (e.g., Yoder, 1952; Borg, 1956; Bryhni, 1966) at 525°C and 7 kbars, 500°C and 10 to 14 kbars, 700°C and > 10 kbars, or 1000°C and > 13 to 18 kbars (Morgan, 1970; Lovering, 1958; Kennedy, 1959; Green and Ringwood, 1967a; Cohen et al., 1967; Fry and Fyfe, 1969) where hydrous metamorphism of the same rocks leads to greenschist, glaucophane schist, or amphibolite (Green and Ringwood, 1967a; Binns, 1967; Morgan, 1970). Decrease in P_{H_2O} may occur during partial melting of pre-orogenic basement rocks (Bryhni et al., 1970).

Wet metamorphism of amphibolite at 400 to 500°C and 9 to 14 kbars (Yoder and Tilley, 1962; Ernst, 1965) or 700°C and 21 kbars P_{H_2O} (Essene et al., 1969). If $P_{load} = P_{H_2O}$, eclogite is not stable in the crust (Fry and Fyfe, 1969).

Wet metamorphism at 650°C and 6.5 kbars P_{H_2O}, because of coexistence of sillimanite-zone gneisses (Velde et al., 1970).

Metasomatic granitization of ultrabasic intercalations in a schist paragneiss complex under normal P_{H_2O}-T conditions of the amphibolite facies (Smulikowski, 1960b, 1968).

Blueschist eclogites are of lower temperature than amphibolite eclogites and, in turn, granulite and kimberlite eclogites have lowest preference of Mg for garnet in the former (lower Mg/Fe of garnet to Mg/Fe of omphacite); thus each eclogite type then is being formed at about the same temperature as the enclosing rock (Coleman et al., 1965; Essene and Fyfe, 1967). Except for the Arizona kimberlite occurrences, therefore, they are not accidental inclusions (Essene and Fyfe, 1967).

Formed as a metastable step in the change of basic igneous rocks to true amphibolites (Dengo, 1950).

A satisfactory origin for eclogites must take into account the following:

Experimental confirmation of high pressure and/or relatively low temperature (compare Fig. 4-20 in the section on basalts)

Chemical similarity to basalts (or amphibolites)

Wide range of occurrences from blueschist, upper greenschist, amphibolite, and granulite facies and as inclusions in kimberlites and peridotites

Regular variation in distribution of Fe^{++} and Mg^{++} between garnet and omphacite in these occurrences

Common separation of the eclogite from the enclosing rock, but well-documented instances of interlayering with enclosing rock

Very common, in many instances thorough, alteration to metamorphic minerals similar to those in the enclosing rock

The origin that best seems to fit these characteristics involves high-pressure
"dry" metamorphism of basaltic rocks (dry) where concurrent "wet" metamorphism of the enclosing rocks gives rise to glaucophane schist or the
high-pressure parts of the greenschist, amphibolite, or granulite facies.
This may well be by increasing pressure on basaltic oceanic crust descending a subduction zone as suggested by Dickinson (1970a). Small slices or
"tectonic inclusions" could be incorporated in the lower continental crust
during deformation. Separation of the eclogite from the enclosing rock
could result from difference in competence between the two rocks during
deformation accompanying or following metamorphism. Alteration could
result from access of water during the later stages of metamorphism or
accompanying a later deformation.

DESCRIPTION AND DEVELOPMENT OF THE CONTACT METAMORPHIC ROCKS

CONTACT METAMORPHIC ROCKS OF THE ALBITE-EPIDOTE HORNFELS FACIES, HORNBLENDE HORNFELS FACIES, PYROXENE HORNFELS FACIES, AND SANIDINITE FACIES

Contact Metamorphic Rocks Bordering the Sierra Nevada Batholith[1]

The contact metamorphic effects bordering the Sierra Nevada batholith are
apparent in the country rocks where exposed for 200 miles along the
western margin adjacent to the northern half of the batholith and in roof
pendants (roof rock remnants) throughout the batholith (compare Fig.
4-6, p. 127). Elsewhere, the contact is faulted or not exposed. Shortly
before intrusion of the granitic rocks, predominantly low-grade greenschist facies metamorphism developed slates, phyllites, and low-grade
metavolcanic rocks. Granitic rocks at least as old as Upper Jurassic (K–Ar
"ages") intrude strata as young as Upper Jurassic. Increase in grade to the
amphibolite facies was accompanied by continued regional deformation,
development of minor folds, cleavages, and intrusion of early granitic
rocks. Following deformation, further intrusion of granitic rocks developed
hornfels and other massive contact metamorphic rocks for up to 2 miles
but most commonly less than about ½ mile from the contacts.

The contact metamorphic rocks range in grade from the low-grade
albite epidote hornfels (mineralogically about equivalent to the greenschist
facies) to the high-grade pyroxene hornfels facies (mineralogically about

[1]Bateman et al., 1963; Loomis, 1966; Rinehart and Ross, 1964; Kistler, 1966; Kerrick, 1970.

equivalent to the granulite facies)—see Fig. 7-9. The hornblende hornfels facies is most common, the pyroxene hornfels facies developing in overlapping aureoles from adjacent plutons (Loomis, 1966) and the albite-epidote hornfels being commonly masked by the earlier regional metamorphism. Parent rocks include shale, siltstone, graywacke, conglomerate, chert, quartzite, and limestone along with andesite and basalt tuffs and flows. These have developed pelitic and siliceous hornfelses, quartzite, calc-hornfels, and marble along with intermediate and mafic amphibole- and pyroxene-bearing hornfels. Parent-rock textures are preserved in many rocks—for example, relict bedding, cross-bedding, pebbles, concretions, pyroclastic fragments, and phenocrysts of pyroxene or plagioclase. The grain size is very fine in the low-grade rocks (for example, 0.01 mm), grading to fine in the higher-grade rocks (0.1 to 0.5 mm) near the batholith. Contact metamorphic textures are massive but may contain porphyroblasts or clusters of porphyroblasts to form the spots of "spotted hornfels" and "spotted slates."

The three contact metamorphic facies have developed a wide range of mineral assemblages in rocks of different compositions. A few assemblages from representative areas follow (Loomis, 1966; Rinehart and Ross, 1964; Best, 1963; Durrell, 1940; Chesterman, 1942; Best and Weiss, 1964; Kerrick, 1970):

Albite epidote hornfels facies:

   quartz-albite-actinolite-biotite

      albite-actinolite-biotite-epidote (±chlorite)

  ±quartz-relict plagioclase-actinolite-biotite-magnetite

  quartz   -tremolite   -calcite

Hornblende hornfels facies:

   quartz-plagioclase-tremolite or actinolite   (±K-feldspar)

   quartz-plagioclase-hornblende-diopside-biotite (±microcline)

   quartz-An_{30-50}-hornblende (±diopside)   (±almandine±magnetite)

   quartz   -diopside-wollastonite   -grossularite

   quartz-plagioclase-muscovite   -biotite

   quartz-plagioclase   -andalusite-biotite-cordierite

   quartz   -biotite-cordierite

   quartz   -calcite-tremolite-diopside

     bytownite -calcite   -wollastonite -idocrase-grossularite

       calcite -brucite (after periclase)

Pyroxene hornfels facies:

quartz-labradorite±diopsidic augite-hypersthene -biotite-K-feldspar

quartz-plagioclase	-sillimanite-biotite	-almandine
quartz-An_{24-37}	-andalusite-biotite	-garnet-cordierite
quartz	-sillimanite-biotite-orthoclase	-cordierite

Local patches of higher grade—local hot spots—in several areas (e.g., wollastonite in place of quartz + calcite) have been explained as resulting from a greater permeability for access of hot water-rich fluid such as from an adjacent pluton or from a greater thermal diffusivity (greater conduction of heat) through certain rocks (Compton, 1955; Parker, 1961; Rinehart and Ross, 1964; Loomis, 1966; Kerrick, 1970). An increase in P_{H_2O}/P_{CO_2} lowers the effective temperature of carbonate reactions as may be seen from Figs. 6-6, 6-7, and 7-12.

Restriction of pyroxene hornfels facies assemblages in the area of overlap of two granitic plutons has been interpreted as predrying of the country rocks by the first intrusion so that $P_{H_2O} < P_{load}$. The apparent reaction, biotite + quartz $\leftrightarrow$ hypersthene + orthoclase + water, has left biotite, quartz, and hypersthene in apparent equilibrium in the rock, so intermittent escape of water has been held to keep the assemblage from complete hydration (Loomis, 1966). Similarly, the occurrence of sillimanite-orthoclase ("higher-grade") assemblages farther from the igneous contact than sillimanite-muscovite ("lower-grade") assemblages has been attributed to $P_{H_2O} < P_{total}$ initially, P_{H_2O} increasing because of water being driven off during reactions. The "higher-grade" assemblage would occur in the more permeable rocks (Kerrick, 1970).

Temperatures of formation of the contact metamorphic rocks as estimated by comparison of mineralogy with experimental data are 400 to 450°C for the low-temperature limit of the hornblende hornfels and 600 to 700°C for the high-temperature limit (Bateman et al., 1963; Loomis, 1966; Kerrick, 1970). The first group of workers suggest different pressures of 5,000 to 6,000 bars versus 1,500 to 1,900 bars and 1,000 to 2,000 bars, for the other workers, respectively. Loomis used geomorphic considerations and structural projections of pluton contacts, and the others used mineralogical and experimental data from the contact and igneous rocks.

General Features of Contact Metamorphic Rocks

Contact metamorphic rocks are widespread but are not nearly so voluminous as the regional metamorphic rocks of comparable grade. They occur where magmas of all kinds intrude lower-temperature rocks. They are generally recognized where the intruded rocks consist of lower-grade minerals than those produced by the contact metamorphism. Where the country rocks are initially higher grade, the superposed metamorphism is separated with greater difficulty.

As may be inferred from Fig. 7-10 (p. 309), contact metamorphism adjacent to granitic plutons commonly reaches the hornblende horn-fels facies, the albite-epidote hornfels facies appearing farther away if the country rocks are not already in the greenschist facies of regional metamorphism (a very common situation). Inclusions of country rock completely enclosed in the magma may approach the temperature of the magma, reaching the pyroxene hornfels or even the sanidinite facies. Contact metamorphism adjacent to the higher temperatures of diabase or gabbro plutons commonly reaches the pyroxene hornfels facies or imme-diately adjacent the contact of near-surface intrusions, the sanidinite facies, the lower-grade facies appearing farther away. Although country rocks adjacent to diabase or gabbro contacts may be hornfelsed to as much as ½ mile from the contact, they in other instances show little or no effect, presumably because of the dryness of the basaltic magma and/or wetness of the intruded sediments. By contrast, aureoles around granitic plutons are generally ¼ to 2 miles in width.

Low- to high-grade contact metamorphic rocks have been described from many areas including those around the Sierra Nevada batholith described above, those in southern and eastern California (e.g., Burnham, 1959; Carpenter, 1967; Knopf, 1938), Nevada (e.g., Buseck, 1967), Mon-tana (e.g., Barrell, 1907; Taylor, 1935; Knopf, 1950; Melson, 1966), Colo-rado (e.g., Steiger and Hart, 1967; Wright, 1967), Arizona (e.g., Perry, 1969), Minnesota (e.g., Grout, 1933), Pennsylvania (e.g., Chapman, 1950), New Jersey (van Houten, 1971), New York (e.g., Ratcliffe, 1968), New Hampshire (e.g., Green, 1963), Maine (e.g., Woodard, 1957; Harwood and Larson, 1969), British Columbia (e.g., Rapson, 1963; Sangster, 1969), Quebec (e.g., deRosen-Spence, 1969), Nova Scotia (e.g., Taylor and Schiller, 1966), Mexico (e.g., Buseck, 1966), Haiti (e.g., Kesler, 1968), Great Britain (e.g., Tozer, 1955; Pitcher and Sinha, 1957; Pitcher and Read, 1960, 1963; Leake and Skirrow, 1960; Evans, 1964; Tilley, 1924, 1926; Stewart, 1942; Phemister, 1942; MacKenzie, 1949; Chinner, 1962; Floyd, 1964, 1965), Norway (e.g., Goldschmidt, 1911), Sweden (e.g., Lund-qvist, 1968, pp. 22–30), Finland (e.g., Eskola, 1914), Bavaria (Richarz, 1924), the U.S.S.R. (e.g., Reverdatto, 1964), New Zealand (e.g., Challis, 1965a; Brothers and Hopkins, 1967), and New South Wales (e.g., Osborne, 1931; Binns, 1965; Joyce, 1970).

Extremely high grade contact metamorphic rocks (sanidinite facies) have been described from far fewer areas. These include the Crestmore area of southern California (e.g., Burnham, 1959; Carpenter, 1967), Texas (Clabaugh, 1953), Maine (Woodard, 1968), Quebec (Philpotts et al., 1967), Mexico (Spurr and Garrey, 1908; Temple and Heinrich, 1964), Eire (Evans, 1964b), Scotland (Wyllie, 1961; Butler, 1961; Agrell, 1965; Smith, 1969), and South Africa (Ackermann and Walker, 1960).

Contact metamorphic rocks are characteristically massive because of lack of deformation and, except for skarns, are commonly finer grained than their regional metamorphic equivalents. Originally fine-grained rocks such as shales or basalts remain fine grained and dark colored in many instances into the pyroxene hornfels facies. In hand specimen, some closely resemble unmetamorphosed basalt. In thin section, many are seen to contain some medium to coarse grains (e.g., cordierite) but these are generally poikilitic and indistinct in hand specimen. Other minerals such as biotite, andalusite, and garnet not uncommonly form porphyroblasts. Original textures such as sand grains, pebbles, phenocrysts, amygdules, fossils, and bedding are commonly preserved because of lack of accompanying deformation and lesser increase in grain size. This restricted increase in grain size is presumably related to more rapid heating and therefore nucleation, less fluids, and less time to aid in grain growth. A striking exception is the development of skarns—very coarse-grained calc-silicate rocks accompanying metasomatism of limestone by granitic magma. In this instance, the coarse grains, up to several inches across, probably result from an abundance of magmatic fluids to inhibit nucleation and to supply constituents to the growing grains.

Although most contact metamorphic rocks are massive, many show a weak schistosity, in most cases relict from an earlier regional metamorphism. As described above for the Sierra Nevada, intrusion of granites and contact metamorphism closely followed regional metamorphism in many such areas. Other North American examples have been described in New England and the adjacent Maritimes (Woodard, 1957; Thompson and Norton, 1968; Taylor and Schiller, 1966; Taylor, 1969) and the northern Cordillera (Aitken, 1959; Hyndman, 1968).

A few contact metamorphic rocks may develop a schistosity during intrusion. A possible example is the Main Donegal Granite in Ireland (Pitcher and Read, 1960, 1963).

Minerals in contact metamorphic rocks for the most part are similar to those in regional metamorphic rocks of comparable grade. Minerals in the albite-epidote hornfels facies are much like those in the greenschist facies, in the hornblende hornfels facies like those in the amphibolite facies, in the pyroxene hornfels facies like those in the granulite facies (compare Fig. 7-9). Exceptions to this equivalence exist, of course, the most prominent being cordierite, andalusite, and wollastonite. However, even these probably should not be considered diagnostic of contact metamorphic facies. Contact metamorphic rocks are recognized by their location adjacent to igneous bodies and by evidence indicating a genetic including temporal relationship. In most instances, massive (no schistosity) rocks spatially associated with an intrusive body can be taken as contact metamorphic. As noted above, however, the converse may not necessarily always be true.

Mineral assemblages of the four contact metamorphic facies are illustrated on ACFmK and ACF phase diagrams in Figs. 8-16a, 8-17a, 8-18a, and 8-19a. The boundary of the albite-epidote hornfels and horn-

Fig. 8-16a *Albite-epidote hornfels facies ACFmK and ACF diagrams. Solid heavy lines are tie lines between coexisting minerals on ACF and AFmK facies. Dashed lines are tie lines within ACFmK tetrahedron. Minerals restricted to this (±one other) contact metamorphic facies are: albite, chlorite restricted to this facies; epidote, talc also in lower hornblende hornfels facies; calcite, tremolite-actinolite also in hornblende hornfels facies; (   ) andalusite may occur in the highest-temperature part of the facies. Pyrophyllite not yet found in hornfels but as noted by Winkler (1967, p. 69) experimental data suggests that it should occur in this facies.*

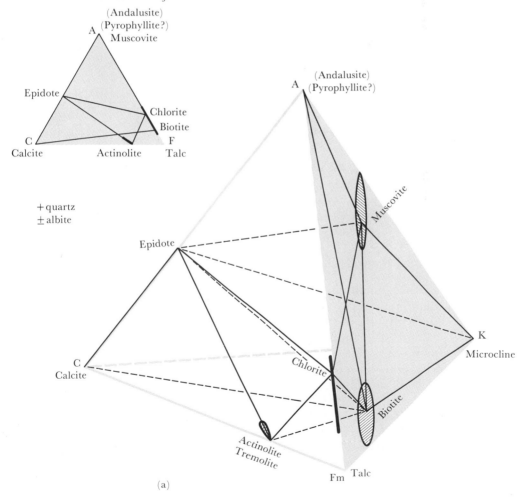

(a)

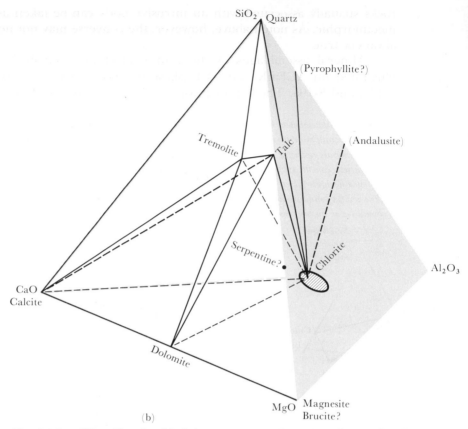

Fig. 8-16b *Albite-epidote hornfels facies* $CaO-MgO-Al_2O_3-SiO_2$ *diagram for silica-deficient, carbonate-rich assemblages. Calcite-tremolite join if high* P_{CO_2}/P_{H_2O}; *calcite-talc join if low* P_{CO_2}/P_{H_2O}.

blende hornfels facies is defined as the first appearance of hornblende (not the aluminum-free tremolite or actinolite) in rocks of appropriate composition. Epidote may persist into the hornblende hornfels facies. The beginning of the pyroxene hornfels facies is defined as the first appearance of orthopyroxene such as hypersthene in rocks of appropriate composition. Note that hornblende may persist into the pyroxene hornfels facies, and diopside may occur in the hornblende hornfels, pyroxene hornfels, and sanidinite facies. The beginning of the sanidinite facies is defined as the first appearance of sanidine in K-feldspar-bearing rocks. Restrictive minerals useful in recognition of individual facies or pairs of facies are listed in Figs. 8-16 to 8-19. The usual accessory minerals may occur in most of these contact metamorphic facies—magnetite, ilmenite, pyrite, sphene, zircon, apatite, and tourmaline.

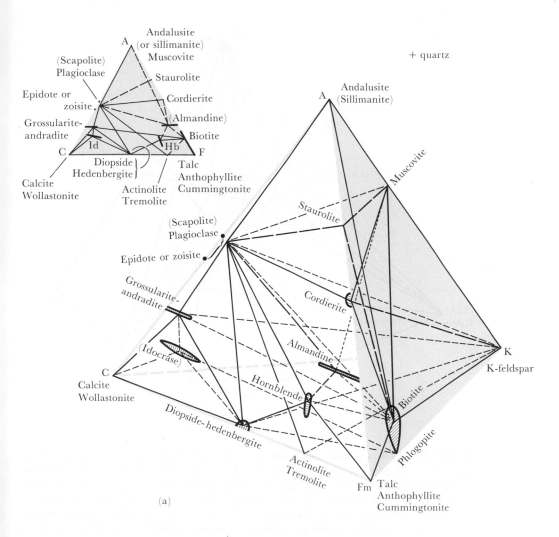

Fig. 8-17a *Hornblende hornfels facies ACFmK and ACF diagrams. Solid heavy lines are tie lines on ACFm and AFmK facies. Dashed lines are tie lines within ACFmK tetrahedron. Minerals restricted to this (±one other) contact metamorphic facies are: anthophyllite-cummingtonite restricted to this facies; tremolite-actinolite, talc, calcite also in albite-epidote hornfels facies; muscovite also in albite-epidote hornfels (± rarely in lowermost pyroxene hornfels facies); hornblende, sillimanite also in lower pyroxene hornfels facies; grossularite-andradite, idocrase (= vesuvianite), biotite, almandine also in pyroxene hornfels facies; ( ) sillimanite may occur in the highest-temperature part of the facies at higher pressure; staurolite may occur in higher pressure Fe-rich rocks. Almandine or spessartite may occur in Fe- or Mn-rich rocks since cordierite can accommodate only limited Fe^{++} or Mn^{++}. Wollastonite may occur in the highest-temperature part of the facies, K-feldspar does not occur with andalusite or sillimanite and probably not with cordierite except in the highest-temperature part of the facies. Calcite not with tremolite-actinolite or epidote or plagioclase.*

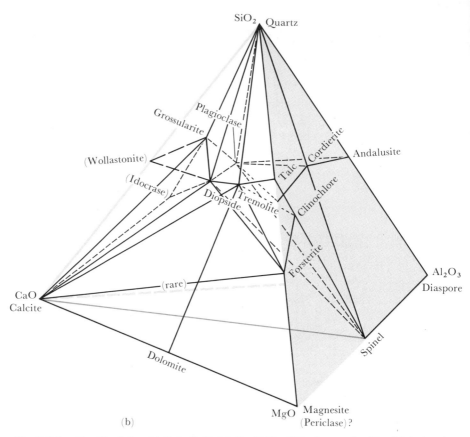

Fig. 8-17b *Hornblende hornfels facies* CaO−MgO−Al₂O₃−SiO₂ *diagram for silica-deficient, carbonate-rich assemblages. ( ) periclase, wollastonite may occur in the highest-temperature part of the facies;* ±*anthophyllite, cummingtonite;* ±*phlogopite:* K₂(Mg, Fe)₆Si₆Al₂O₂₀(OH)₄: *low Fe, Al biotite;* ±*diaspore:* AlO(OH); ±*clintonite:* Ca₂(Mg, Al)₆(Si, Al)₈O₂₀(OH)₄: *a "brittle mica";* ±*chondrodite:* Mg(OH, F)₂·2Mg₂SiO₄: *yellow or brown olivinelike mineral;* ±*ludwigite:* (Mg, Fe)₄Fe³⁺₂O₇B₂O₃: *black to green or dark brown;* ±*scheelite:* CaWO₄; ±*scapolite:* NaAlSi₃O₈·1/3NaCl−CaAl₂Si₂O₈·1/3 CaCO₃.

Possible reactions transitional to and within each of the contact metamorphic facies include (e.g., Tilley, 1923, 1951a; Bowen, 1940; Bowen and Tuttle, 1949; Chapman, 1950; Harker and Tuttle, 1955; Tuttle and England, 1955; Weeks, 1956; Winkler, 1957, 1965, 1967; Carr and Fyfe, 1960; Carr, 1963; Wyllie, 1962; Walter, 1963a, b; Metz and Winkler, 1963, 1964; Evans, 1965b; Fawcett and Yoder, 1966; Loomis, 1966; Melson, 1966; Newton, 1966a; Turner, 1967, 1968).

Below (lower grade than) the hornblende hornfels facies:

Kaolinite + 2 quartz ↪ pyrophyllite + H₂O
Pyrophyllite ↪ andalusite + 3 quartz + H₂O

Illite + calcite + quartz $\leftrightarrow$

tremolite + clinozoisite + K-feldspar + H_2O + CO_2

3 dolomite + 4 quartz + H_2O $\leftrightarrow$ talc + 3 calcite + $3CO_2$

5 dolomite + 8 quartz + H_2O $\leftrightarrow$ tremolite + 3 calcite + CO_2

Fig. 8-18a *Pyroxene hornfels facies ACFmK and ACF diagrams. Solid heavy lines are tie lines on ACFm and AFmK faces. Dashed lines are tie lines within ACFmK tetrahedron. Restrictive minerals are orthoclase with andalusite or sillimanite restricted to this facies; biotite, hornblende, idocrase, grossularite-andradite, almandine also in hornblende hornfels facies; sillimanite also in upper hornblende hornfels facies; hypersthene, glass also in sanidinite facies; ( ) ± muscovite may occur only in the lowest-temperature part of the facies. Although not characteristic, glass (resulting from partial fusion) has been reported in rocks containing orthoclase rather than sanidine (i.e., in pyroxene hornfels facies; Knopf, 1938; Wyllie, 1961). Likewise not characteristic, brown hornblende has been reported in rocks containing hypersthene (i.e., in pyroxene hornfels facies; Binns, 1965). Scapolite in place of plagioclase and idocrase in place of grossularite may form in the presence of high Cl^-, CO_3^{2-}, or high F^-, respectively.*

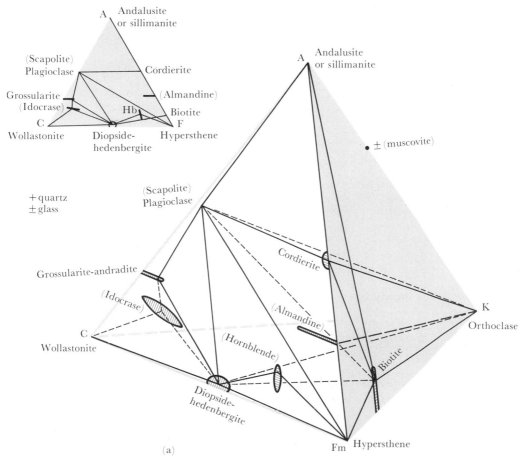

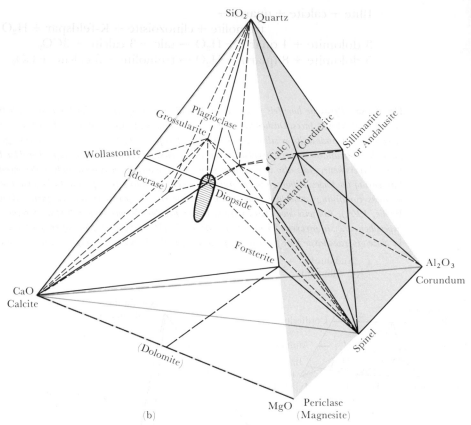

Fig. 8-18b *Pyroxene hornfels facies* $CaO-MgO-Al_2O_3-SiO_2$ *diagram for silica-deficient carbonate-rich assemblages. Note that grossularite and plagioclase are on the back* $SiO_2-CaO-Al_2O_3$ *face. ( ) dolomite, magnesite, talc may occur only in the lowest-temperature part of the facies or at high* P_{CO_2}. *Idocrase in calc-silicate rocks may form in the presence of high* F^-. *Periclase formed by contact metamorphism but normally completely altered to brucite by later addition of water. Hypersthene may form in place of enstatite if* Fe^{++} *present.* ± *biotite, phlogopite;* ± *orthoclase.*

Below the pyroxene hornfels facies:

$$6 \text{ chlorite} + 4 \text{ muscovite} + \text{quartz} \leftrightarrows 3 \text{ cordierite} + \text{biotite} + 24H_2O$$
$$\text{Chlorite} + \text{tremolite} + \text{epidote} + \text{quartz} \leftrightarrows \text{hornblende}$$
$$\text{Albite} + \text{actinolite} + \text{epidote} + \text{chlorite} \leftrightarrows$$
$$\text{hornblende} + \text{biotite} + \text{more-calcic plagioclase}$$
$$\text{Dolomite} + 2 \text{ quartz} \leftrightarrows \text{diopside} + 2CO_2$$
$$3 \text{ tremolite} + 5 \text{ calcite} \leftrightarrows 11 \text{ diopside} + 2 \text{ forsterite} + 5CO_2 + 3H_2O$$
$$\text{Tremolite} + 3 \text{ calcite} + 2 \text{ quartz} \leftrightarrows 5 \text{ diopside} + 3CO_2 + H_2O$$
$$\text{Calcite} + \text{quartz} \leftrightarrows \text{wollastonite} + CO_2$$
$$\text{Dolomite} \leftrightarrows \text{calcite} + \text{periclase} + CO_2$$

Below the sanidinite facies:

Muscovite + quartz $\leftrightarrow$ orthoclase + andalusite + H_2O $\pm$ biotite

(or sillimanite)   (if muscovite
contained Fe, Mg)

Biotite + 6 quartz $\leftrightarrow$ 2 orthoclase + 3 hypersthene + $2H_2O$

6 muscovite + 2 biotite + 15 quartz $\leftrightarrow$

8 orthoclase + 3 cordierite + $5H_2O$

Hornblende + quartz $\leftrightarrow$ 2 anorthite + 3 hypersthene + H_2O

Tremolite $\leftrightarrow$ 3 enstatite + 2 diopside + quartz + H_2O

Dolomite $\leftrightarrow$ periclase + calcite + CO_2

Fig. 8-19a *Sandinite facies ACFmK and ACF diagrams. Solid heavy lines are tie lines on ACFm and AFmK facies. Dashed lines are tie lines within ACFmK tetrahedron. Restrictive minerals are: sanidine, mullite, tridymite, pigeonite restricted to this facies; hypersthene, glass also in pyroxene hornfels facies; although not characteristic, quartz has been reported in sanidine-bearing rocks (i.e., in sanidinite facies; Butler, 1961; Woodard, 1968), presumably in higher-pressure parts of the facies (compare Fig. 7-11).*

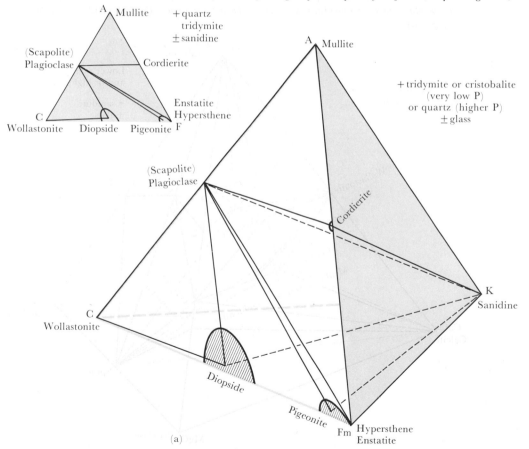

Transitional to and within the sanidinite facies:

Orthoclase ↔ sanidine

3 andalusite ↔ mullite + SiO_2
        ($Al_6Si_2O_{13}$)

Quartz ↔ tridymite ↔ cristobalite

Grossularite + quartz ↔ anorthite + wollastonite

Diopside + calcite ↔ akermanite   + CO_2
        (CaMg-melilite: $Ca_2MgSi_2O_7$)

Forsterite + calcite ↔ monticellite + periclase + CO_2

Fig. 8-19b *Sanidinite facies $CaO-MgO-Al_2O_3-SiO_2$ diagrams for silica-deficient, carbonate-rich assemblages. Wollastonite, grossularite, and plagioclase are on the back $SiO_2-CaO-Al_2O_3$ face. Lack of assemblages on the side $SiO_2-MgO-Al_2O_3$ face of the high-temperature zone reflects lack of data, not instability. ( ) idocrase, chondrodite may form in the presence of high F^-. Periclase normally completely altered to brucite by later addition of water. In lower-temperature zone: ±kalsilite = $KAlSiO_4$ rarely with melilite, diopside, spinel (Philpotts et al., 1970); ±pseudobrookite = Fe_2TiO_5; in higher-temperature zone: ±perovskite = $CaTiO_3$; ±pleonaste = spinel, (Mg, Fe)Al_2O_4; ±cuspidine = $Ca_4Si_2O_7F_2$.*

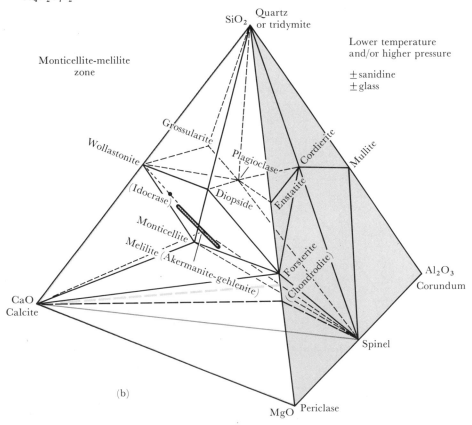

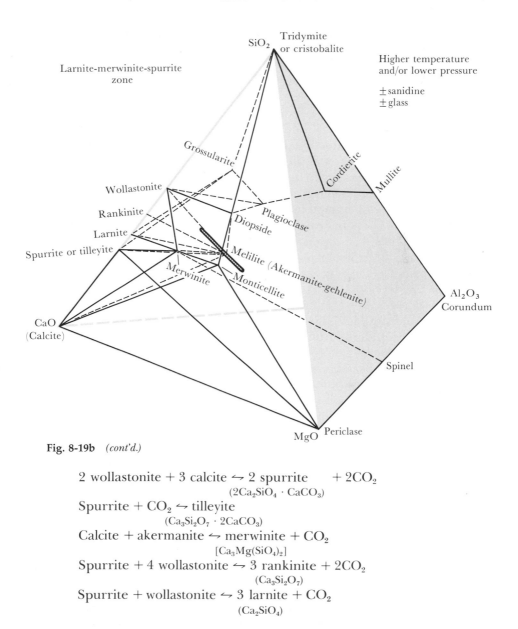

Fig. 8-19b *(cont'd.)*

$$2 \text{ wollastonite} + 3 \text{ calcite} \leftrightarrows 2 \text{ spurrite} + 2CO_2$$
$$(2Ca_2SiO_4 \cdot CaCO_3)$$
$$\text{Spurrite} + CO_2 \leftrightarrows \text{tilleyite}$$
$$(Ca_3Si_2O_7 \cdot 2CaCO_3)$$
$$\text{Calcite} + \text{akermanite} \leftrightarrows \text{merwinite} + CO_2$$
$$[Ca_3Mg(SiO_4)_2]$$
$$\text{Spurrite} + 4 \text{ wollastonite} \leftrightarrows 3 \text{ rankinite} + 2CO_2$$
$$(Ca_3Si_2O_7)$$
$$\text{Spurrite} + \text{wollastonite} \leftrightarrows 3 \text{ larnite} + CO_2$$
$$(Ca_2SiO_4)$$

Water Content and Attainment of Equilibrium in Contact Metamorphic Rocks

Many regional metamorphic rocks probably originate as water-saturated marine sediments and associated undersaturated volcanic rocks that become water saturated through deformation and addition of water in the early stages of metamorphism. Most contact metamorphic rocks probably originate as relatively dry sedimentary or volcanic rocks long after deposition or as dehydrated regional metamorphic rocks. Such parent rocks are undersaturated with water (e.g., Loomis, 1966; Kerrick, 1970), giving a situation analogous with granulite facies metamorphism of initially dry rocks such as plutonic igneous rocks (see the discussion of water in the granulite facies, p. 363).

Under such conditions, P_{H_2O} may be less than P_{load} during metamorphism unless abundant water is introduced from the intruding magma. Such an introduction of water and accompanying elements presumably occurs for the tens of feet involved in the formation of the coarse-grained skarns but probably not in most fine-grained hornfels. P_{H_2O} may approach P_{load} in the higher grades where water is liberated during the dehydration reactions of rising temperature and where less water is required for saturation of a less-hydrous assemblage.

The development of conditions of water undersaturation may have at least three important effects on contact metamorphic rocks. As in the granulite facies situation, the lower partial pressure of water would result in the development of some minerals at a lower temperature than expected where $P_{H_2O} = P_{load}$. Recall also that temperature limits inferred for a mineral from Fig. 7-11 are for P_{H_2O}, not P_{load}. A second effect of conditions of water undersaturation is to inhibit the attainment of equilibrium between the mineral grains during metamorphism. This effect, which is aggravated by rapid heating and lack of deformation, is especially apparent in the low-grade albite-epidote hornfels facies. In many instances, these rocks are spotted slates or spotted schists in which 1- to 6-mm aggregates of new minerals (e.g., biotite) have formed in a matrix consisting of relict minerals and relict textures. Such spots, if developed in the pelitic layers of graded beds, may give rise to an apparent reversal of the relative grain size of the alternate layers. The original finer clay-rich layers, being more susceptible to reaction, recrystallize to coarser-grained layers (commonly coarser mica), whereas the original coarser silty or sandy quartzofeldspathic layers remain relatively fine. A third effect of conditions of water undersaturation is to restrict grain growth by inhibiting diffusion of constituents while permitting nucleation during rapid heating. An analogy may be made between this situation and the contrast in grain size between the relatively dry crystallization of aplites and the wet crystallization of pegmatites discussed in Chap. 3.

Equilibrium is rarely reached also in the sanidinite facies. Under these extremely high temperatures and low pressures of metamorphism, the very rapid heating and rapid cooling (shallow conditions) presumably hinder recrystallization to any individual equilibrium assemblage.

The lack of apparent contact metamorphism in medium- to high-grade regional metamorphic rocks around many plutons (e.g., Eskola, 1914, p. 167) is generally attributed to the country rocks already being of a relatively high grade, so that little change in mineralogy is necessary, or to the composition of the country rocks (e.g., quartzofeldspathic) not being suitable for the development of distinctive minerals. In other instances, a subtle contact effect can be recognized in the recrystallization to polygonal arcs of micas tightly bent around crenulations (e.g., Hyndman, 1968b, p. 33). In such instances, the contact effect may not be strong enough to destroy the preferred orientation in a schist by nucleation of new grains, but the strain energy stored in a bent mica may be enough to permit further crystallization on segmented parts of a bent grain.

Partial Fusion of Contact Rocks

As noted above, glass resulting from partial fusion of country rocks may occur in the sanidinite or upper pyroxene hornfels facies. Although such fusion is caused by heat from an adjacent higher-temperature intrusion rather than buildup of radioactive heat and conduction from the mantle, the result is analogous to anatexis during regional metamorphism.

These partially fused rocks, called "buchites," have been described from a number of areas, but many features are common to all. In each instance, a granodiorite to diabase or basalt magma has invaded a quartz-orthoclase-plagioclase rock, most commonly an arkose or feldspathic sandstone but in at least two cases a granitic rock (Richarz, 1924; Knopf, 1938; Larsen and Switzer, 1939). The country rock has been converted partly into a colorless to yellow, brown, green, or black glass for several inches to 140 ft from the contact. Analyses of the glass and less-fused rock shows that the melts lie closer to and trend toward the minimum melting trough in the SiO_2-albite-orthoclase system (Larsen and Switzer, 1939; Ackermann and Walker, 1960; Butler, 1961; Wyllie, 1961) or toward a cotectic in the system $SiO_2 - Al_2O_3 - MgO$ (Spry and Solomon, 1964). Ackermann and Walker (1960) note that melting begins at the contacts between clastic quartz and orthoclase. Experimental melting of shales at between about 1 and 3 kbars shows generally similar trends (Winkler, 1960; Winkler and von Platen, 1957, 1958, 1960, 1961a, b; von Platen, 1965b; von Platen and Höller, 1966; Wyllie and Tuttle, 1961a).

Conditions of Formation of Contact Metamorphic Rocks

Correlation of the mineralogy of contact metamorphic rocks with experimental work on individual minerals as described above (see Fig. 7-11) suggests temperatures ranging from about 350 to more than 1000°C. The actual temperature depends on the temperature and size of intrusion, initial temperature, heat conductivity and permeability of the country rocks, fluid content of the intrusion and country rocks, and distance from the contacts. Temperatures at the contact are normally expected to reach to at least a few hundred degrees below magma temperature, as suggested by both heat-flow calculations and resultant mineralogy (see Fig. 7-10).

Determination of prevailing pressure during contact metamorphism is somewhat more difficult. A brief examination of Figs. 7-11 and 7-12 indicates why most minerals are better indicators of temperature than pressure. Nevertheless, a few minerals in pelitic hornfels—andalusite, cordierite, K-feldspar in place of muscovite, and hypersthene in unfused rocks in granitic aureoles—indicate formation below the intersection of the stability boundary of the mineral and the minimum melting curve for granite. This would be less than about 5 kbars or, in the case of hypersthene or sanidine, less than about 2 kbars. The low slopes on stability boundaries of many minerals in the sanidinite facies—e.g., tridymite, monticellite, spurrite—suggest fluid pressures of about 1 kbar or less. Geological estimates based, for example, on stratigraphic cover or structural projections at the time of intrusion give a similar answer.

The boundary between contact and regional metamorphism cannot be recognized on mineralogical grounds, the distinction being a genetic one. For this reason, such a boundary is not shown on Fig. 8-2. Pressures in contact metamorphism presumably overlap those of regional metamorphism, the overlap perhaps ranging from 1 or 2 to 5 kbars.

Very low pressure regional metamorphic rocks containing such minerals as andalusite and cordierite have been described in a number of areas: the Front Range of Colorado (Gable and Sims, 1969), Maine (Osberg, 1971), Georgia (Salotti and Fouts, 1967), the Scottish Highlands (Read, 1952; Johnson, 1963; Chinner, 1966), the Pyrenees and northern Portugal (Zwart, 1962; Brink, 1960 *in* Turner, 1968, pp. 221–222), Japan (Miyashiro, 1958, 1961), and southeastern Australia (Vallance, 1967). Such low-pressure regional metamorphism has been called Buchan type by Read and Abukuma type by Miyashiro. Mineral assemblages for these intermediate rocks have been described and compared by Winkler (1967, pp. 116–130). Because of the gradational nature of conditions and mineral assemblages, it seems best to treat these low-pressure regional metamorphic rocks as an intermediate group, as suggested by Turner (1968).

Metasomatism or introduction of chemical constituents such as Fe, Mg, Si, and Al from an adjacent pluton is very striking in the formation of the coarse calc-silicate minerals of skarns by intrusion of a limestone. These skarn zones, from a few to a few hundred feet thick, occur in place of the limestone immediately adjacent to the igneous contact. Many skarns form essentially monomineralic zones several feet thick. Also common but less prominent is the metasomatism of noncalcareous rocks. Both of these cases have been described above in the discussion of metasomatism (pp. 282–287).

Sulfide-ore mineralization associated with skarns is somewhat later than but may overlap the main silicate mineralization.

CHAPTER 9
REVIEW
OF PREFERRED ORIGINS
OF IGNEOUS AND
METAMORPHIC ROCK
ASSOCIATIONS

In order to develop a synthesis of the origin of the different igneous and metamorphic rock associations, the "preferred" origin of each association is brought together here for comparison.

>*Association and "Preferred" Origin*
>
>*Spilites*—contamination of normal basalt by eruption into saline water; low-grade metamorphism of solid basalt at shallow depths under ocean floor (either of these would be followed by leaching of pillow rims during cooling)
>
>*Alpine peridotite-serpentinite*—slices sheared off upper mantle along inclined earthquake zone or upper mantle exposed by buckling upward against continental margin
>
>*Basalt-andesite-rhyolite (mafic stratovolcano part)*—melting of water-bearing basaltic crust thrust under edge of continent; partial melting of peridotite in mantle, fractional crystallization and/or assimilation of crustal rocks, all probably under hydrous conditions
>
>*(Felsic ash-flow tuff part)*—anatexis (partial melting) of crustal rocks largely on continental side of the granitic batholiths derived by the same origin

Granitic batholiths and stocks—partial melting of deep "geosynclinal" rocks; smaller bodies over previous oceanic crust, formed by differentiation of basaltic magma

Anorthosites—moderately high pressure partial melting of base of crust or uppermost mantle to produce basaltic magma or differentiation of basaltic magma

Flood basalts (tholeiitic)—partial melting at high pressure in the mantle or extensive fractionation of olivine from basalt at high pressures, the dry basaltic magma slowly moving toward the surface

Alkaline olivine—partial melting at high pressure in the mantle, with fractionation of olivine and plagioclase under moderately low pressure conditions to restrict the magma to the silica-deficient field

Diabase dikes and sills—near-surface injection of tholeiitic ± alkaline olivine basalt magma and differentiation in place

Lamprophyre dikes—alkaline olivine basalt magma (formed by partial melting in the mantle) injected into fractures in or around an earlier granitic mass; contamination by gneissic, granitic, or possibly sedimentary rocks to contribute to the high alkali content and volatiles

Potassium-rich "basaltic" rocks—alkaline olivine basalt magma contaminated by K-rich granitic, gneissic, or sedimentary wall rocks. Melting of low Si, low Na, K-feldspar-rich granulite facies basement rocks of the lower crust, possibly with upper mantle rocks, to give low Si, low Na, K-rich melt

Nepheline syenites and related soda-rich rocks—differentiation, or possibly partial melting, of alkaline olivine basalt or partial melting of low Si, low K, plagioclase-rich granulite facies basement rocks of the dry, stable lower crust

Carbonatites and mafic alkaline rocks—differentiation, or partial melting of peridotite to form alkaline olivine basalt under very high pressure in the mantle, rise toward the surface, extreme differentiation under stable crustal conditions, and separation of carbonatite as an immiscible liquid

Gabbroic layered intrusions—injection of tholeiitic or high-alumina basalt magma at deeper level in crust than diabase sills, and differentiation in place; melting in mantle possibly caused by increase in temperature and decrease in pressure following impact of large meteorite or fracturing and decrease in pressure in the mantle

Zeolite facies—very low grade (very low-temperature, low-pressure) regional metamorphism; shallow, coastal part of "eugeosynclinal" environment, not much deformation, not much igneous activity

Prehnite-pumpellyite facies—low-grade (low-temperature, low-pressure) regional metamorphism; moderately shallow, coastal part of "eugeosynclinal" environment, not much deformation, not much igneous activity

Greenschist facies—moderately low grade (moderately low temperature, low- to moderate-pressure) regional metamorphism; moderate depth; mainly "eugeosynclinal" environment, moderate deformation during metamorphism; commonly associated with higher-level granitic plutons

Amphibolite facies—high-grade (moderate- to high-temperature, moderate-pressure) regional metamorphism; moderately great depth; mainly "eugeosynclinal" environment, strong deformation during metamorphism; commonly associated with moderate-level granitic plutons

Granulite facies—very high grade (very high-temperature, moderate- to high-pressure) regional metamorphism; moderately great depth; high-grade metamorphism of earlier dry-basement gneisses or igneous rocks ($P_{H_2O} < P_{load}$); possibly extreme metamorphism of deepest "geosynclinal" rocks

Blueschist facies—moderate- to high-pressure, low-temperature regional metamorphism; deep coastal part of "eugeosynclinal" environment, may or may not be much deformation; associated with serpentinites and perhaps developed with high water pressures during thrusting under a regional serpentinite body

Eclogite facies—high-pressure, low- to high-temperature, deep coastal part of "eugeosynclinal" environment, to deep basement where "dry" metamorphism of basaltic rocks (dry) accompanies "wet" metamorphism of the enclosing rocks and gives rise to blueschist or high-pressure parts of greenschist, amphibolite, or granulite facies

A generalized and simplified model of igneous and metamorphic activity of all stages of evolution of a "typical" continental margin is given in Fig. 9-1. This emphasizes the approximate spatial relationships among the various rock associations, taking into account the possible origins suggested above.

The evolution of a continental margin has been described classically in terms of two parallel belts of sedimentation, the outer volcanic-rich "eugeosyncline" and inner volcanic-poor "miogeosyncline," which gradually receive accumulations of sediment until their thickness and rise in temperature is sufficient for the onset of orogeny. Concepts embodied in the new global tectonics—ocean-floor spreading, continental drift, and plate tectonics—have recently led to the evolution of a new concept of the development of an orogenic belt as we now know it (e.g., Dewey, 1969; Hamilton, 1969; Dickinson, 1971; Dewey and Bird, 1970; see Chap. 1). A stable continental margin (e.g., the present Atlantic margin of North

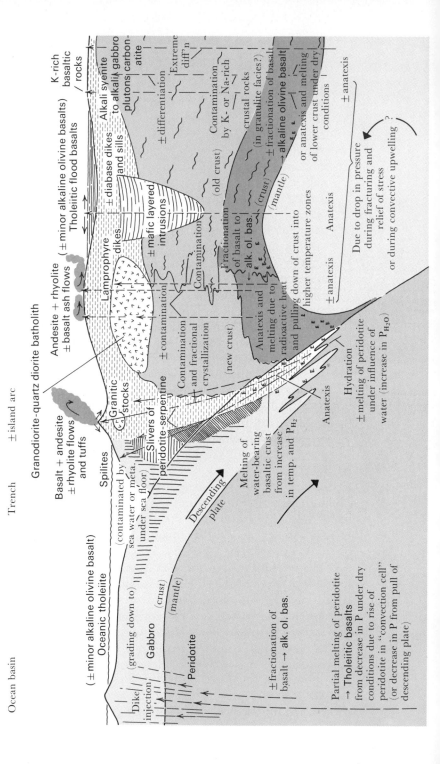

400

and South America), where the oceanic and continental crust is continuous with no break, is considered to be converted into an unstable orogenic margin (e.g., the present or recent-past Pacific margin of North, Central, and South America) when the boundary between the oceanic and continental crust is ruptured, the oceanic plate normally being thrust under the lower-density continental plate (an exception has been described by Davies, 1968; Davies and Milsom, 1969). Whether this rupture is coincidental, depending on migration of the continent or descending plate margin (trench) or descending limb of a one-sided convection cell, or is caused somehow by the accumulation of sediment at the continental margin is not yet clear. However, different orogenic belts may undergo a somewhat different sequence of events (Dickinson, 1970a). Thus, the development of particular igneous and metamorphic associations may come at different relative times or be more prolonged in different orogenic belts. More consistent is the orogenic environment in which the different rock associations develop (compare Table 2-1, p. 44, and Fig. 9-1).

The ophiolite suite—spilites and associated basalts, gabbros, peridotites, and serpentinites—develop along the continental margin in the early stages of development of plate rupture, with descent of the edge of the oceanic plate and accompanying formation of a trench. Melting of basaltic oceanic crust in the descending edge of the oceanic plate may result from increase in temperature and water pressure, the water accompanying thin sediments overlying basalts on the ocean floor and any continental-margin graywackes dragged down with them. Hydration of peridotite to form serpentinite in the upper mantle part of the descending plate would decrease the density and permit slivers of serpentinized peridotite to be squeezed upward into the graywackes of the continental margin. Or peridotite-serpentinite of the upper mantle could be buckled upward against the continental margin as suggested by Bailey, Blake, and Jones (1970). The mafic stratovolcano part of the basalt-andesite-rhyolite association could similarly be formed by melting or partial melting of basaltic crust or peridotite upper mantle in the edge of the descending plate. These basalts and andesites thereby could erupt more or less continuously throughout the life of the oceanic trench and descending plate margin (or convection), a situation that fits the present worldwide correlation of trench-Benioff earthquake zone and volcanism and the extensive period of basalt-andesite-rhyolite volcanism in many "orogenic cycles," such as that recorded for the North American cordillera.

Fig. 9-1 *A generalized model for the formation of the major igneous rock associations by reasonable mechanisms involving partial melting, contamination, and differentiation. Adjacent spatial relationships in some cases overlap or are reversed. Times of development are not all concurrent. Vertical scale is exaggerated greatly. Zeolite facies; prehnite-pumpellyite facies; blueschist facies; greenschist facies; amphibolite facies; granulite facies; eclogite facies; old metamorphic rocks.*

Granitic batholiths and the felsic ash-flow-tuff part of the basalt-andesite-rhyolite association develop inland from the continental margin and low-temperature–higher-pressure regional metamorphic rocks of the zeolite, prehnite-pumpellyite, blueschist facies. These granitic magmas appear most commonly in the "eugeosynclinal" belt adjacent to and overlapping "miogeosynclinal" rocks of the greenschist facies or amphibolite facies. They probably originate by partial melting of new lower-crustal rocks caused by the buildup of heat from radioactive decay of K, U, and Th and the thickening of the new sedimentary ± volcanic crust during crumpling of the continental margin against the descending oceanic plate. Heat could also be added by conduction from the mantle and rising melts of the mafic part of the basalt-andesite-rhyolite association. Presumably, heat buildup would be gradual, the melts finally developing later than the above associations as the conditions finally reached those of the uppermost amphibolite facies. The lighter, hydrous, more mobile granitic melts would rise through the overlying lower-grade, cooler amphibolite and greenschist facies zones until they approached the solidus in the granite melt system by decrease in pressure, assimilation of higher melting constituents, or cooling. There they would become more viscous, would spread out and would solidify as huge multi-intrusion granitic batholiths. Differentiation of certain high-level, very hydrous magmas or special tectonic conditions could permit continued slow rise to the surface where they would explosively erupt to form extensive ash-flow tuffs. Minor granitic plutons could presumably form by separation, followed by increase in viscosity, and emplacement of granitic magmas resulting from somewhat lower temperature partial melting, assimilation of more felsic constituents, and differentiation from the more mafic part of the basalt-andesite-rhyolite association.

Flood basalts, diabase dikes and sills, and gabbroic layered intrusions form by crystallization at the surface, near the surface, and well below the surface, respectively, of basalt magma formed by partial melting of peridotite in the "low-velocity" zone of the upper mantle. Partial melting could result from a decrease in pressure accompanying rise or pulling apart of peridotite in the upwelling limb of a convection cell to form oceanic basalt of the midoceanic ridges. Or it could result from a decrease in pressure accompanying fracturing and relief of tectonic overpressure in the orogenic belt near the continental margin. The dominant tholeiitic basalt in both environments would form or differentiate under different (perhaps higher) pressure conditions than the minor, commonly late alkaline olivine basalt.

Lamprophyre dikes, potassium-rich "basaltic" rocks, and sodium-rich mafic alkaline rocks that occur on the continental side of the main orogenic belt all show strong affinities to alkaline olivine basalt and probably result from contamination of alkaline olivine basalt by various high-grade gneissic, granitic, or possibly sedimentary rocks that may be low Na, K rich or low K, Na rich in different areas and thus control the direction of contamination. K-rich granulite facies rocks, for example, could remain as a K-feldspar-plus-mafic-rich residuum after anatexis and removal of quartz and sodic plagioclase in the resulting melt.

Nepheline syenites and related rocks develop in the same general environment as the above mafic alkaline rocks but appear to be much further evolved. They probably form either by differentiation of alkaline olivine basalt in this quiet, stable environment or possibly by partial melting of low-K, plagioclase-rich basement rocks of the dry stable lower crust. Carbonatites also appear strongly evolved in this environment, but they and many of their associated rocks appear silica deficient to the extreme. The carbonatites may separate as an immiscible liquid, possibly at very low pressure under very stable conditions, from a volatile- and carbonate-rich mafic or ultramafic alkaline magma. Melting in dry lower-crust or upper-mantle rocks could result from decrease in pressure accompanying rifting, possibly above convective upwelling in the mantle, to form any of these alkalic rock associations.

BIBLIOGRAPHY

Abdullah, M. I., and M. P. Atherton (1964): The Thermometric Significance of Magnetite in Low Grade Metamorphic Rocks, *Am. J. Sci.*, **262**, pp. 904–917.

Ackermann, P. B., and F. Walker (1960): Vitrification of Arkose by Karoo Dolerite near Heilbron, Orange Free State, *Geol. Soc. Lond. Quart. J.*, **116**, pp. 239–254.

Adams, F. D. (1929): Geology of Ceylon, *Canadian J. of Research.*

—— and A. E. Barlow (1910): Geology of the Haliburton and Bancroft Areas, Province of Ontario, *Geol. Surv. Can. Mem.* **6**, p. 419.

Afansyev, W. A. (1939): Alkaline Rocks of the Ozernaya Varaka of the Knabozero Region (Southwest Kola Peninsula), *Dokl. Akad. Nauk SSSR*, **25**, pp. 508–512.

Agard, J. (1960): Les carbonatites et les roches à silicates et carbonates associés du massif de roches alcalines du Tamazert (Haut Atlas de Mioelt. Maroc) et les problèmes de leur genèse, *Intern. Geol. Congr.* **21st,** Copenhagen, 1960, *Rept. Session,* Norden, pt. 13, pp. 293–303.

Agrell S. O. (1939): The Adinoles of Dinas Head, Cornwall, *Mineral. Mag.,* **25**, pp. 305–337.

—— (1941): Dravite-bearing Rocks from Dinas Head, Cornwall, *Mineral. Mag.,* **26**, pp. 81–93.

—— (1965): Polythermal Metamorphism of Limestones at Kilchoan, Ardnamurchan, *Mineral. Mag.*, **34**, pp. 1–15.

—— and J. M. Langley (1958): The Dolerite Plug at Tievebulliagh near Cushendall, Co. Antrim, *Proc. Roy. Irish Acad.,* sec. B, **59**, pp. 93–127.

Ahrens, L. H. (1952): The Use of Ionization Potentials—I. Ionic Radia of the Elements, *Geochim. Cosmochim. Acta,* **2**, pp. 155–169.

—— (1955a): The Convergent Lead Ages of the Oldest Monazites and Uraninites (Rhodesia, Manitoba, Madagascar, Transvaal), *Geochim. Cosmochim. Acta,* **7**, pp. 294–300.

—— (1955b): Implications of the Rhodesia Age Pattern, *Geochim. Cosmochim. Acta,* **8**, pp. 1–15.

Aitken, J. D. (1959): Atlin Map-area, British Columbia, *Geol. Surv. Can. Mem.* **307,** 89 pp.

Akella, J., and H. G. F. Winkler (1966): Orthorhombic Amphibole in some Metamorphic Reactions, *Contr. Mineral. Petrol.,* **12,** pp. 1–12.

Albee, A. L. (1965a): Distribution of Fe, Mg, and Mn between Garnet and Biotite in Natural Mineral Assemblages, *J. Geol.,* **73,** pp. 155–164.

——— (1965b): Phase Equilibria in Three Assemblages of Kyanite-zone Pelitic Schists, Lincoln Mountain Quadrangle, Central Vermont, *J. Petrol.,* **6,** pp. 246–301.

——— (1968): Metamorphic Zones in Northern Vermont, *in* "Studies of Appalachian Geology: Northern and Maritime," E-an Zen, W. S. White, J. B. Hadley, and J. B. Thompson, Jr., editors, Interscience, a division of Wiley, New York, chap. 25, pp. 329–341.

——— and A. A. Chodos (1969): Minor Element Content of Coexistent Al_2SiO_5 Polymorphs, *Am. J. Sci.,* **267,** pp. 310–316.

Alderman, A. R. (1936): Eclogites from the Neighborhood of Glenelg, Inverness-shire, *Geol. Soc. Lond. Quart. J.,* **92,** pp. 488–530.

Aldrich, L. T., and G. W. Wetherill (1958): Geochronology by Radioactive Decay, *Ann. Rev. Nucl. Sci.,* **8,** pp. 257–298.

———, ———, M. N. Bass, W. Compston, G. L. Davis, and G. R. Tilton (1959): Mineral Age Measurements, *Carnegie Inst. Wash. Yearbook,* **58,** pp. 237–250.

———, ———, G. L. Davis, and G. R. Tilton (1958): Radioactive Ages of Micas from Granitic Rocks by Rb-Sr and K-A Methods, *Trans. Am. Geophys. Union,* **39,** pp. 1124–1134.

Allen, J. C., Jr. (1966): Structure and Petrology of the Royal Stock, Flint Creek Range, Central-western Montana, *Geol. Soc. Am. Bull.,* **77,** pp. 291–302.

Althaus, E. (1966): Der stabilitatsbereich des Pyrophyllits unter dem Einfluss von Sauren. I. Mitteilung, Experimentelle Untersuchungen, *Contr. Mineral. Petrol.,* **13,** pp. 31–50.

——— (1967): The Triple Point Andalusite-Sillimanite-Kyanite, *Contr. Mineral. Petrol.,* **16,** pp. 29–44.

Amaral, G., J. Bushee, U. G. Cordani, K. Kawashita, and J. H. Reynolds (1967): Potassium-argon Ages of Alkaline Rocks from Southern Brazil, *Geochim. Cosmochim. Acta,* **31,** pp. 117–142.

Ambrose, J. W. (1936): Progressive Kinetic-metamorphism of the Missi Series near Flinflon, Manitoba, *Am. J. Sci.,* ser. 5, **32,** pp. 257–286.

——— and C. A. Burns (1956): Structures in the Clare River Syncline–A Demonstration of Granitization, *in* "The Grenville Problem," J. E. Thompson, editor, *Roy. Soc. Can. Spec. Publ.,* **no. 1,** pp. 42–53.

Anderson, A. T., Jr. (1968): Massif-type Anorthosite: A Widespread Precambrian Igneous Rock, *in* "Origin of Anorthosite and Related Rocks," Y. W. Isachson, editor, *N.Y. State Mus. Sci. Serv. Mem.* **18,** pp. 47–55.

——— and M. Morin (1968): Two Types of Massif Anorthosites and Their Implications Regarding the Thermal History of the Crust, *N.Y. State Mus. Sci. Serv. Mem.* **18,** pp. 57–69.

Anderson, D. L., C. Sammis, and T. Jordan (1971): Composition and Evolution of the Mantle and Core, *Science,* **171,** pp. 1103–1112.

Anderson, E. M. (1942): "The Dynamics of Faulting and Dyke Formation with Applications to Britain," Oliver & Boyd, London, 206 pp.

Anderson, G. M. (1970): Some Thermodynamics of Dehydration Equilibria, *Am. J. Sci.*, **269,** pp. 392–401.

Anderson, P. A. M., and O. J. Kleppa (1969): The Thermochemistry of the Kyanite-silli-manite Equilibrium, *Am. J. Sci.*, **267,** pp. 285–290.

Angel, F. (1957): Einige ausgewählte problems eklogitischer Gesteinsgruppen der österreich-ischen Ostalpen, *Neues Jahrb. Mineral.*, **91,** pp. 151–192.

Aniruddha De (1968): Anorthosites of the Eastern Ghats, India, *N.Y. State Mus. Sci. Serv. Mem.* **18,** pp. 425–434.

Aoki, K., and Y. Oji (1966): Calc-alkaline Volcanic Rock Series Derived from Alkali-olivine Basalt Magma, *J. Geophys. Res.*, **71,** pp. 6127–6135.

Appledorn, C. R., and H. E. Wright (1957): Volcanic Structures in the Chuska Mountains, Navajo Reservation, Arizona-New Mexico, *Geol. Soc. Am. Bull.*, **68,** pp. 445–468.

Armstrong, J. (1949): Fort St. James Map-area. Cassiar and Coast Districts, British Columbia, *Geol. Surv. Can. Mem.* **252,** 210 pp.

Armstrong, R. L. (1966): K-Ar Dating of Plutonic and Volcanic Rocks in Orogenic Belts, *in* "Potassium-argon Dating," Springer-Verlag, New York, pp. 117–133.

——— (1968a): Mantled Gneiss Domes in the Albion Range, Southern Idaho, *Geol. Soc. Am. Bull.*, **79,** pp. 1295–1314.

——— (1968b): A Model for the Evolution of Strontium and Lead Isotopes in a Dynamic Earth, *Rev. Geophys.*, **6,** pp. 175–199.

———, E. B. Ekren, E. H. McKee, and D. C. Noble (1969): Space-time Relations of Cenozoic Silicic Volcanism in the Great Basin of the Western United States, *Am. J. Sci.*, **267,** pp. 478–490.

——— and E. Hansen (1966): Cordilleran Infrastructure in the Eastern Great Basin, *Am. J. Sci.*, **264,** pp. 112–127.

Arriens, P. A., C. Brooks, V. M. Bofinger, and W. Compston (1966): The Discordance of Mineral Ages in Granitic Rocks Resulting from the Redistribution of Rubidium, *J. Geophys. Res.*, **71,** no. 20, pp. 4981–4994.

Atamanov, A. F., S. F. Lugov, and Ya. M. Feygin (1962): New Results on the Geology of the Lovozero Massif, *Intern. Geol. Rev.*, **4,** no. 5, pp. 570–577.

Atherton, M. P. (1964): The Garnet Isograd in Pelitic Rocks and its Relation to Metamorphic Facies, *Am. Minerologist*, **49,** pp. 1331–1349.

——— (1965): The Composition of Garnet in Regionally Metamorphosed Rocks, *in* "Controls of Metamorphism," W. S. Pitcher and G. W. Flinn, editors, Wiley, New York, pp. 281–290.

——— and W. M. Edmonds (1966): An Electron Microprobe Study of Some Zoned Garnets from Metamorphic Rocks, *Earth Planetary Sci. Letters,* **1,** pp. 185–193.

Atkins, F. B. (1968): Pyroxenes of the Bushveld Intrusion South Africa, *J. Petrol.*, **10,** pp. 222–249.

Aubouin, J. (1965): "Geosynclines," Elsevier, Amsterdam, 335 pp.

Aumento, F. (1967): Magmatic Evolution on the Mid-Atlantic Ridge, *Earth Planetary Sci. Letters*, **2**, pp. 225–230.

——— (1968): The Mid-Atlantic Ridge near 45°N. II. Basalts from the Area of Confederation Peak, *Can. J. Earth Sci.*, **5**, pp. 1–21.

——— (1969): Geological Investigations, Mid-Atlantic Ridge, *Geol. Surv. Can. Rept. Activ.*, *Paper* **69–1**, pt. A, pp. 253–257.

Ave'Lallemant, H. G., and N. L. Carter (1970): Syntectonic Recrystallization of Olivine and Modes of Flow in the Upper Mantle, *Geol. Soc. Am. Bull.*, **81**, pp. 2203–2220.

Avent, J. C. (1970): Correlation of the Steens-Columbia River Basalts: Some Tectonic and Petrogenetic Implications, *in* "Proceedings of the Second Columbia River Basalt Symposium," E. H. Gilmour and D. Stradling, editors, Eastern Wash. State Coll. Press, pp. 133–156.

Ayrton, S. N. (1963): A Contribution to the Geological Investigations in the Region of Ivigtut, S. W. Greenland, *Medd. Groenland*, **167**, no. 3, 139 pp.

Baadsgaard, H., G. L. Comming, R. E. Folinsbee, and J. D. Godfrey (1964): Limitations of Radiometric Dating, *Roy. Soc. Can., Spec. Publ.* **8**, pp. 20–38.

Bagdasarov, E. A. (1959): Alkaline Pegmatites of the Afrikanda Massif, *Intern. Geol. Rev.*, **3**, pp. 463–473.

Bailey, D. K. (1960): Carbonatites of the Rofunsa Valley, Feira District, *N. Rhodesia Geol. Surv. Bull.*, **5**.

——— (1964): Feldspar-liquid Equilibria in Peralkaline Liquids–The Orthoclase Effect, *Am. J. Sci.*, **262**, pp. 1198–1206.

——— (1966): Carbonatite Volcanoes and Shallow Intrusions in Zambia, *in* "Carbonatites," O. T. Tuttle and J. Gittins, editors, Interscience, a division of Wiley, New York, pp. 127–154.

——— and J. F. Schairer (1966): The System $Na_2O-Al_2O_3-Fe_2O_3-SiO_2$ at One Atmosphere and the Petrogenesis of Alkaline Rocks, *J. Petrol.*, **7**, pp. 114–170.

Bailey, E. B., and H. H. Thomas (1924): The Tertiary and Post-Tertiary Geology of Mull, Loch Aline and Oban, *Mem. Geol. Surv. Scotland*.

——— and W. J. McCallien (1953): Serpentine Lavas, the Ankara Mélange and the Anatolian Thrust, *Trans. Roy. Soc. Edinburgh*, **62**, pt. 2, pp. 403–442.

——— and ——— (1960): Some Aspects of the Steinmann Trinity, *Geol. Soc. Lond. Quart. J.*, **116**, pp. 365–395.

Bailey, E. H. (1962): Metamorphic Facies of the Franciscan Formation of California and Their Geologic Significance, *Geol. Soc. Am., Spec. Paper* **68**, pp. 4–5.

——— and M. C. Blake (1969): Tectonic Development of Western California in the Late Mesozoic, *Geotectonika*, pt. 3, pp. 17–30, pt. 4, pp. 24–34.

———, ———, and D. L. Jones (1970): On-land Mesozoic Oceanic Crust in California, *U. S. Geol. Surv. Prof. Paper* **700-C**, pp. C70–C81.

———, W. P. Irwin, and D. L. Jones (1964): Franciscan and Related Rocks and Their Significance in the Geology of Western California, *Calif. Div. Mines Geol. Bull.*, **183**, 177 pp.

Baird, A. K. (1962): Superposed Deformations in the Central Sierra Nevada Foothills of the Mother Lode, *Calif. Univ. Publ. Geol. Sci*, **42**, pp. 1–70.

Balk, R. (1931): Structural Geology of the Adirondack Anorthosite, *Mineral. Petrog. Mitt.*, **41**, pp. 341–347.

——— (1937): Structural Behavior of Igneous Rocks, *Geol. Soc. Am. Mem.* **5**, 177 pp.

Banno, S. (1958): Glaucophane Schist and Associated Rocks in the Ōmi District, Japan, *Japan. J. Geol. Geography*, **24**, no. 1-3, pp. 29–44.

——— (1964): Petrologic Studies on Sanbagawa Crystalline Schists in the Bessi-Ino District, Central Sikoku, Japan, *J. Fac. Sci., Univ. Tokyo*, sec. 11, **15**, pp. 203–319.

——— (1966): Eclogite and Eclogite Facies, *Japan. J. Geol. Geography*, **37**, pp. 105–122.

Baragar, W. R. A. (1967): Wakuach Lake Map-area, Quebec-Labrador (23 0), *Geol. Surv. Can. Mem.* **344**, 177 pp.

——— and L. E. Long (1969): Feldspathoidal Syenite in Quartz Diabase Sill, Brookville, New Jersey, *J. Petrol.*, **10**, pp. 202–221.

Bard, J. P. (1970): Composition of Hornblendes Formed during the Hercynian Progressive Metamorphism of the Aracena Metamorphic Belt (SW Spain), *Contr. Mineral. Petrol.*, **28**, pp. 117–134.

Barker, D. S. (1967): Texture, Composition, and Origin of Graphic Granite (abs.), *Can. Mineralogist*, **9**, pt. 2, p. 284.

——— (1970): Compositions of Granophyre, Myrmekite, and Graphic Granite, *Geol. Soc. Am. Bull.*, **81**, pp. 3339–3350.

Barker, F. (1961): Phase Relations in Cordierite-Garnet-bearing Kingsman Quartz Monzonite and Enclosing Schist, New Hampshire, *Am. Mineralogist*, **46**, pp. 1166–1176.

——— (1962): Cordierite-garnet Gneiss and Associated Microcline-rich Pegmatite at Sturbridge, Massachusetts and Union, Connecticut, *Am. Mineralogist*, **47**, pp. 907–918.

Barksdale, J. D. (1937): The Shonkin Sag Laccolith, *Am. J. Sci.*, ser. 5, **33**, pp. 321–359.

Barnes, I. (1970): Metamorphic Waters from the Pacific Tectonic Belt of the West Coast of the United States, *Science*, **168**, pp. 973–975.

———, V. C. LaMarche, Jr., and G. R. Himmelberg (1967): Geochemical Evidence of Present-day Serpentinization, *Science*, **156**, pp. 830–832.

——— and J. R. O'Neil (1969): The Relationship between Fluids in some Fresh Alpine-type Ultramafics and Possible Modern Serpentinization, Western United States, *Geol. Soc. Am. Bull.*, **80**, pp. 1947–1960.

Barrell, J. (1907): Geology of the Marysville Mining District, Montana, *U. S. Geol. Surv. Prof. Paper* **57**, 178 pp.

Barrow, G. (1893): On an Intrusion of Muscovite-biotite Gneiss in the Southeast Highlands of Scotland, and its Accompanying Metamorphism, *Geol. Soc. Lond. Quart. J.*, **49**, pp. 330–358.

——— (1912): On the Geology of Lower Dee-side and the Southern Highland Border, *Geol. Assoc. Proc.*, **23**, pp. 274–290.

Barth, T. F. W. (1930): Genesis der Pegmatite im Urgebirge, *Chem. Erde*, **4**, pp. 95–136.

—— (1938): Structural and Petrologic Studies in Dutchess County, New York, Part II, Petrology and Metamorphism of the Paleozoic Rocks, *Geol. Soc. Am. Bull.*, **47**, pp. 775–850.

—— (1945): Studies on the Igneous Rock Complex of the Oslo Region: Systematic Petrography of the Plutonic Rocks, *Norske Vid.-Akad. Skrift*, **no. 1**, *Mat.-Naturv. Kl.*, **9**, 104 pp.

—— (1948): Recent Contributions to the Granite Problem, *J. Geol.*, **56**, pp. 235–240.

—— (1954): The Igneous Rock Complex of the Oslo Region. Part XIV, Provenance of the Oslo Magmas, Norske Videnskaps-Akad. Oslo, I, *Mat.-Naturv. Kl.*, **no. 4**, 20 pp.

—— (1956a): Geology and Petrology of the Pribilof Islands, Alaska, *U.S. Geol. Surv. Bull.* **1028-F**, 160 pp.

—— (1956b): Studies in Gneiss and Granite I and II, *Norske Vid.-Akad. Skrift.*, **no. 1**, p. 35.

—— (1962): "Theoretical Petrology," 2d ed., Wiley, New York.

——, C. W. Correns, and P. Eskola (1939): Die Entstehung der Gesteine, Springer, Berlin, p. 422.

—— and J. A. Dons (1960): Geology of Norway, Precambrian of Southern Norway, *Norsk. Geol. Unders.*, **208**, pp. 6–67.

—— and I. B. Ramberg (1966): The Fen Circular Complex, *in* "Carbonatites," O. F. Tuttle and J. Gittins, editors, Interscience, a division of Wiley, New York, pp. 225–257.

Bass, M. N. (1970): North American Feldspathoidal Rocks in Space and Time: Discussion, *Geol. Soc. Am. Bull.*, **81**, pp. 3493–3500.

—— (1971): Variable Abyssal Basalt Populations and Their Relation to Sea-floor Spreading Rates, *Earth Planet. Sci. Letters*, **11**, pp. 18–22.

Bateman, P. C., L. D. Clark, N. K. Huber, J. G. Moore, and C. D. Rinehart (1963): The Sierra Nevada Batholith–A Synthesis of Recent Work across the Central Part, *U.S. Geol. Surv. Prof. Paper* **414-D**, pp. D1–D46.

—— and F. C. W. Dodge (1970): Variations of Major Chemical Constituents across the Central Sierra Nevada Batholith, *Geol. Soc. Am. Bull.* **81**, pp. 409–420.

—— and J. P. Eaton (1967): Sierra Nevada Batholith, *Science*, **158**, pp. 1409–1417.

—— and C. Wahrhaftig (1966): Geology of the Sierra Nevada, *in* "Geology of Northern California," E. H. Bailey, editor, *Calif. Div. Mines Geol. Bull.*, **190**, pp. 107–172.

Battey, M. H. (1956): The Petrogenesis of a Spilitic Rock Series from New Zealand, *Geol. Mag.*, **93**, pp. 89–110.

—— (1966): The "Two-magma" Theory and the Origin of Ignimbrites, *Bull. Volcanol.*, **29**, pp. 407–422.

Bearth, P. (1959): Über Ecklogite, Glaucophanschiefer und metamorphe Pillowlaven, *Schweiz. Mineral. Petrog. Mitt.*, **39**, pp. 267–286.

—— (1965): Zur Entstehung alpinotyper Eclogite, *Schweiz. Mineral. Petrog. Mitt.*, **45**, pp. 179–188.

—— (1966): Zur Mineral faziellen stellung der Glaucophangesteine der Westalpen, *Schweiz. Mineral. Petrog. Mitt.*, **46**, pp. 13–23.

Becke, F. (1913): Über Mineralbestand and Struktur der Krystallinischen Schiefer, *Denkschr. Akad. Wiss. Wien,* **75,** pp. 1–53.

Bederke, E. (1947a): Über den Wärmehaushalt der Regional metamorphose, *Geol. Rundschau,* **35,** pp. 26–32.

———— (1947b): Zum Problem der Lamprophyre, *Akad. Wiss. Göttingen Nachricht, Math.-Phys. Kl.,* pp. 53–57.

Beger, P. J. (1923): Der Chemismus der Lamprophyre, *Gesteins Mineral.,* **band 1,** pp. 571–574.

Bell, K., and J. L. Powell (1969): Strontium Isotopic Studies of Alkalic Rocks: The Potassium-rich Lavas of the Birunga and Toro-Ankole Regions, East and Central Equatorial Africa, *J. Petrology,* **10,** pp. 536–572.

———— and ———— (1970): Strontium Isotopic Studies of Alkalic Rocks: The Alkalic Complexes of Eastern Uganda, *Geol. Soc. Am. Bull.,* **81,** pp. 3481–3490.

Bell, P. M. (1963): Aluminum Silicate System: Experimental Determination of Triple Point, *Science,* **139,** no. 3559, pp. 1055–1056.

———— and J. Kalb (1969): Stability of Omphacite in the Absence of Excess Silica, *Ann. Rept., Dir. Geophys. Lab.,* 1967–1968, pp. 97–98.

Belousov, A. F. (1971): Basic Question in Origin of the Magmas of Basaltoid Associations, *Intern. Geol. Rev.,* **13,** pp. 118–122.

Belousov, V. V. (1966): Modern Concepts of the Structure and Development of the Earth's Crust and the Upper Mantle of Continents, *Geol. Soc. Lond. Quart. J.,* **122,** pp. 293–314.

———— (1969): Interrelations between the Earth's Crust and Upper Mantle, *in* "The Earth's Crust and Upper Mantle," P. J. Hart, editor, *Am. Geophys. Union, Geophys. Monogr.* **13,** pp. 698–712.

Benioff, H. (1954): Orogenesis and Deep Crustal Structure–Additional Evidence from Seismology, *Geol. Soc. Am. Bull.,* **65,** pp. 385–400.

Benson, N. N. (1918): The Origin of Serpentine, *Am. J. Sci.,* **46,** pp. 693–731.

———— (1926): The Tectonic Conditions Accompanying the Intrusion of Basic and Ultrabasic Igneous Rocks, *Mem. Nat. Acad. Sci.,* **19,** no. 1.

Benson, W. N. (1915): The Geology and Petrology of the Great Serpentine Belt of New South Wales, Part IV, The Dolerites, Spilites and Keratophyres of the Nundle District, *Proc. Linn. Soc.,* New South Wales, **40,** pp. 122–173.

Berg, G. W. (1968): Secondary Alteration in Eclogites from Kimberlite Pipes, *Am. Mineralogist,* **53,** pp. 1336–1346.

Berg, J. J. van den (1946): Petrofabric Analysis of the Bushveld Gabbro from Bon Accord, *Trans. Geol. Soc. S. Africa,* **49,** pp. 156–203.

Berg, R. B. (1968): Petrology of Anorthosites of the Bitterroot Range, Montana, *in* "Origin of Anorthosite and Related Rocks," Y. W. Isachsen, editor, *N. Y. State Mus. Sci. Serv. Mem.* **18,** pp. 387–398.

Bernal, J. D. (1960): The Structure of Liquids, *Sci. Am.,* Aug. 1960.

———— (1964): The Structure of Liquids, *Proc. Roy. Soc.* (London), **A280,** pp. 299–322.

Berrange, J. P. (1966): Some Critical Differences between Orogenic-plutonic and Gravity-stratified Anorthosites, *Geol. Rundsch.*, **55**, pp. 617–642.

Berthelsen, A. (1957): The Structural Evolution of an Ultra- and Polymetamorphic Gneiss Complex, West Greenland, *Geol. Rundsch.*, **46**, pp. 173–185.

—— (1960): An Example of the Structural Approach to the Migmatite Problem, *Intern. Geol. Congr.*, **21st**, Copenhagen, 1960, *Rept. Session*, Norden, pt. 14, pp. 149–157.

Best, M. G. (1963): Petrology and Structural Analysis of Metamorphic Rocks *in* "The Southwestern Sierra Nevada Foothills," *Calif. Univ. Publ. Geol. Sci.*, **42**, pp. 111–158.

—— (1969): Differentiation of Calc-alkaline Magmas, *in* "Proceedings of the Andesite Conference," A. R. McBirney, editor, *Oregon Dept. Geol. Mineral. Ind. Bull.*, **65**, pp. 65–75.

—— and L. E. Weiss (1964): Mineralogical Relations in Some Pelitic Hornfelses from the Southern Sierra Nevada, California, *Am. Mineralogist*, **49**, pp. 1240–1266.

Bhattacharji, S., and C. H. Smith (1963): Experimental Studies of Flowage Differentiation Applied to the "Feeder Dike" of the Muskox Intrusion, 13th *I.U.G.G. Meeting (Upper-Mantle Symp.)*, **1**, p. 42.

—— and —— (1964): Flowage Differentiation, *Science*, **145**, pp. 150–153.

Biehler, S., and W. E. Bonini (1969): A Regional Gravity Study of the Boulder Batholith, Montana, *in* "Igneous and Metamorphic Geology," L. H. Larsen, editor, *Geol. Soc. Am. Mem.* **115**, pp. 401–422.

Biggar, G. M. (1967a): Apatite Compositions and Liquidus Phase Relations on the Join $Ca(OH)_2$–CaF_2–$Ca_3(PO_4)_2$–H_2O from 250-4000 bars, *Mineral. Mag.*, **36**, pp. 539–564.

—— (1967b): Phase Relations in the Join $Ca(OH)_2$–$CaCO_3$–$Ca_3(PO_4)_2$–H_2O at 1000 bars, *Mineral. Mag.*, **37**, pp. 75–82.

Biljon, S. van (1949): The Transformation of the Pretoria Series in the Bushveld Complex, *Trans. Geol. Soc. S. Africa.*, **52**, pp. 1–198.

Billings, M. P. (1937): Regional Metamorphism of the Littleton-Moosilauke Area, New Hampshire, *Geol. Soc. Am. Bull.*, **48**, pp. 463–566.

—— (1938): Introduction of Potash during Regional Metamorphism in Western New Hampshire, *Geol. Soc. Am. Bull.*, **49**, pp. 289–302.

—— (1943): Ring-dikes and Their Origin, *N.Y. Acad. Sci. Trans.*, ser. 2, **5**, pp. 131–144.

—— (1956): The Geology of New Hampshire, Part II, Bedrock Geology, New Hampshire State Planning Comm., Concord, N.H., pp. 1–200.

—— and N. B. Keevil (1946): Petrography and Radioactivity of Four Paleozoic Magma Series in New Hampshire, *Geol. Soc. Am. Bull.*, **57**, pp. 797–828.

Bingham, J. W. (1970): Several Probable Source Vents for the Roza and Priest Rapids Type Basalts in Whitman and Adams Counties, Washington, *in* "Proceedings of the Second Columbia River Basalt Symposium," E. H. Gilmour and D. Stradling, editors, Eastern Wash. State Coll. Press, pp. 171–172.

Binns, R. A. (1962): Metamorphic Pyroxenes from the Broken Hill District, New South Wales, *Mineral. Mag.*, **33**, pp. 320–338.

——— (1964): Zones of Progressive Regional Metamorphism in the Willyama Complex, Broken Hill District, New South Wales, *J. Geol. Soc. Australia,* **11,** pp. 281–330.

——— (1965): Hornblendes from Some Basic Hornfelses in the New England Region, New South Wales, *Mineral. Mag.,* **Tilley vol.,** pp. 52–65.

——— (1967): Barroisite-bearing Eclogite from Naustdal, Sogn og Fjordane, Norway, *J. Petrol.,* **8,** pp. 349–371.

Birch, F. (1955): Physics of the Earth, *in* "Crust of the Earth," A. Poldervaart, editor, *Geol. Soc. Am. Spec. Paper* **62,** pp. 101–118.

Bishopp, D. W. (1952): Some New Features of the Geology of Cyprus, *Congr. Geol. Intern., Compt. Rend.,* **19,** Algiers, 1952, 15, p. 13.

Blake, M. C., Jr., W. P. Irwin, and R. G. Coleman (1967): Upside-down Metamorphic Zonation, Blueschist Facies, Along a Regional Thrust in California and Oregon, *U.S. Geol. Surv. Prof. Paper* **575-C,** pp. C1–C9.

———, ———, and ——— (1969): Blueschist Facies Metamorphism Related to Regional Thrust Faulting, *Tectonophysics,* **8,** pp. 237–246.

Bloxam, T. W. (1956): Jadeite Bearing Metagraywackes in California, *Am. Mineralogist,* **41,** pp. 488–496.

——— (1959): Glaucophane Schists and Associated Rocks near Valley Ford, California, *Am. J. Sci.,* **257,** pp. 95–112.

——— (1960): Jadeite-rocks and Glaucophane Schists from Angel Island, San Francisco Bay, California, *Am. J. Sci.,* **258,** pp. 555–573.

——— (1966): Jadeite-rocks and Blue Schists in California, *Geol. Soc. Am. Bull.,* **77,** pp. 781–786.

Boettcher, A. L. (1967): The Rainy Creek Alkaline-ultramafic Igneous Complex near Libby, Montana–Part 1, Ultramafic Rocks and Fenite, *J. Geol.,* **75,** pp. 526–553.

——— and P. J. Wyllie (1967): Revision of the Calcite-aragonite Transition, with Location of a Triple Point between Calcite I, Calcite II, and Aragonite, *Nature,* **213,** pp. 792–793.

——— and ——— (1968): Jadeite Stability Measured in the Presence of Silicate Liquids in the System $NaAlSiO_4$-SiO_2-H_2O, *Geochim. Cosmochim. Acta,* **32,** pp. 999–1012.

Bofinger, V. M., and W. Compston (1967): A Reassessment of the Age of the Hamilton Group, New York and Pennsylvania, and the Role of Inherited Sr^{87}, *Geochim. Cosmochim. Acta,* **31,** pp. 2353–2359.

Bonatti, E. (1968): Ultramafic Rocks from the Mid-Atlantic Ridge, *Nature,* **219,** pp. 363–364.

——— and D. E. Fisher (1971): Oceanic Basalts: Chemistry versus Distance from Oceanic Ridges, *Earth Planet. Sci. Letters,* **11,** pp. 307–311.

———, J. Honnorez, and G. Ferrara (1970): Equatorial Mid-Atlantic Ridge: Petrologic and Sr Isotopic Evidence for an Alpine-type Rock Assemblage, *Earth Planet. Sci. Letters,* **9,** pp. 247–256.

Boone, G. M. (1962): Potassic Feldspar Enrichment in Magma–Origin of Syenite, in Deboullie District, Northern Maine, *Geol. Soc. Am. Bull.,* **73,** pp. 1451–1476.

—— and W. D. Romey (1966): Oscillatory Zoning in Calcic Andesine-Sodic Labradorite Phenocrysts in the Anorthosite of North Creek Dome, Adirondack Mountains (abs.), *in* George H. Hudson symposium "Origin of Anorthosite," State Univ. Coll., Plattsburgh, N.Y., pp. 2–3.

Borg, I. Y. (1956): Glaucophane Schists and Eclogites near Healdsburg, California, *Geol. Soc. Am. Bull.*, **67**, pp. 1563–1584.

Bose, M. K. (1967): The Upper Mantle and Alkalic Magmas, *Norsk. Geologisk. Tidsskrift*, **47**, pp. 121–129.

Bottinga, Y., and D. F. Weill (1970): Viscosity of Anhydrous Silicate Melts, *Am. Geophys. Union Trans.*, **51**, no. 4, p. 439.

Bottino, M. L., and P. D. Fullagar (1968): The Effects of Weathering on Whole Rock Rb-Sr Ages of Granitic Rocks, *Am. J. Sci.*, **266**, pp. 661–670.

Boulanger, J. (1957): Les anorthosites de Madagascar, *Comm. Tech. Coop. in Africa South of the Sahara, Conf. de Tananarive, Second meeting*, pp. 55–60.

—— (1959): Les anorthosites de Madagascar, *Ann. Geol. Madagascar*, **fasc. XXVI,** 71 pp.

Bowen, N. L. (1913): The Melting Phenomena of the Plagioclase Feldspars, *Am. J. Sci.*, ser. 4, **35**, pp. 577–599.

—— (1914a): The Ternary System: Diopside-forsterite-silica, *Am. J. Sci.*, **38**, pp. 207–264.

—— (1914b): The Binary System $MgO-SiO_2$, *Am. J. Sci.*, ser. 4, **37**, pp. 487–500.

—— (1915): The Later Stages of Evolution of the Igneous Rocks, *J. Geol.*, **23**, no. 8, suppl., 91 pp.

—— (1917): The Problem of the Anorthosites, *J. Geol.*, **25**, pp. 209–243.

—— (1919): Crystallization-differentiation in Igneous Magmas, *J. Geol.*, **27**, pp. 393–430.

—— (1922): The Reaction Principle in Petrogenesis, *J. Geol.*, **30**, pp. 177–198.

—— (1924): The Fen Area in Telemark, Norway, *Am. J. Sci.*, ser. 5, **8**, pp. 1–11.

—— (1926): The Carbonate Rocks of the Fen Area in Norway, *Am. J. Sci.*, **12**, pp. 499–502.

—— (1927): The Origin of Ultrabasic and Related Rocks, *Am. J. Sci.*, **14**, pp. 89–108.

—— (1928): "The Evolution of Igneous Rocks," Princeton University Press, Princeton, N.J., 332 pp.

—— (1934): Viscosity Data for Silicate Melts, *Am. Geophys. Union Trans.*, 1934, pp. 249–255.

—— (1940): Progressive Metamorphism of Siliceous Limestone and Dolomite, *J. Geol.*, **48**, pp. 225–274.

—— and O. F. Tuttle (1949): The System $MgO-SiO_2-H_2O$, *Geol. Soc. Am. Bull.*, **60**, pp. 439–460.

—— and —— (1950): The System $NaAlSi_3O_8-KAlSi_3O_8-H_2O$, *J. Geol.*, **58**, pp. 498–511.

Bowes, D. R., and R. G. Park (1966): Metamorphic Segregation Banding in the Loch Kerry Basite Sheet from the Lewisian of Gairloch, Ross-shire, Scotland, *J. Petrol.*, **7**, pp. 306–330.

Bowin, C. O. (1966): Geology of Central Dominican Republic (A Case History of Part of an Island Arc) *in* "Caribbean Geological Investigations," H. H. Hess, editor, *Geol. Soc. Am. Mem.* **98,** pp. 11–84.

———, A. J. Nalwalk, and J. B. Hersey (1966): Serpentinized Peridotite from the North Wall of the Puerto Rico Trench, *Geol. Soc. Am. Bull.,* **77,** pp. 257–270.

Boyd, F. R. (1954): Amphiboles, *Ann. Rept. Dir. Geophys. Lab.,* **no. 53,** pp. 108–111.

——— (1959): Hydrothermal Investigation of Amphiboles, *in* "Researches in Geochemistry," P. H. Abelson, editor, Wiley, New York, pp. 377–396.

——— (1961): Welded Tuffs and Flows in the Rhyolite Plateau of Yellowstone Park, *Geol. Soc. Am. Bull.,* **72,** pp. 387–426.

——— and G. M. Brown (1969): Electron-probe Study of Pyroxene Exsolution, *Mineral. Soc. Am. Spec. Paper* **2,** pp. 211–216.

——— and J. L. England (1960): The Quartz-coesite Transition, *J. Geophys. Res.,* **65,** pp. 749–756.

———, ———, and B. T. C. Davis (1964): Effects of Pressure on the Melting and Polymorphism of Enstatite, $MgSiO_3$, *J. Geophys. Res.,* **69,** pp. 2101–2109.

——— and J. F. Schairer (1964): The System $MgSiO_3$–$CaMgSi_2O_6$, *J. Petrol.,* **5,** pp. 275–309.

Brace, W. F., and J. B. Walsh (1962): Some Direct Measurements of the Surface Energy of Quartz and Orthoclase, *Am. Mineralogist,* **47,** pp. 1111–1122.

Braddock, W. A. (1970): The Origin of Slaty Cleavage: Evidence from Precambrian Rocks in Colorado, *Geol. Soc. Am. Bull.,* **81,** pp. 589–600.

Bridgwater, D. (1967): Feldspathic Inclusions in the Gardar Igneous Rocks of South Greenland and Their Relevance to the Formation of the Major Anorthosites of the Canadian Shield, *Can. J. Earth Sci.,* **4,** no. 6, pp. 995–1014.

Briere, Y. (1920): Les eclogites francaises–leur composition mineralogique et chimique: leur origine, *Bull. Soc. Franc. Min.,* **42,** pp. 72–222.

Brill, R. H., R. L. Fleischer, P. B. Price, and R. M. Walker (1964): Fission Track Dating of Man-made Glasses: Preliminary Results, *J. Glass Studies,* **6,** pp. 151–155.

Brink, A. H. (1960): Petrology and Ore Geology of the Villa Real Region, Northern Portugal, *Comm. Serv. Geol. Portugal,* **no. 93.**

Brögger, W. C. (1921): Die Eruptivgesteine des Kristianiagebietes IV. Das Fengebiet in Telemark Norwegen, *Norske Vid.-Akad. Skrift,* **no. 1,** *Mat. Naturv. Kl.,* **no. 9.**

Brothers, R. N. (1954): Glaucophane Schists from the North Berkeley Hills, California, *Am. J. Sci.,* **252,** pp. 614–626.

——— (1956): The Structure and Petrography of Greywackes near Auckland, New Zealand, *Trans. Roy. Soc. New Zealand,* **83,** pp. 465–482.

——— (1964): Petrofabric Analyses of Rhum and Skaergaard Layered Rocks, *J. Petrol.,* **5,** pp. 255–274.

——— and J. C. Hopkins (1967): Igneous Rocks and Hornfelses from the Hen and Chickens Islands, *Trans. Roy. Soc. New Zealand, Geol.,* **5,** pp. 123–129.

Brouwer, H. A. (1945): The Association of Alkali Rocks and Metamorphic Limestone in a Block Ejected by the Volcano Merapi (Java), *Koninkl. Nederl. Akad. Wetens.*, **48,** pp. 166–189.

―――― and C. G. Egeler (1952): The Glaucophane Facies Metamorphism in the Schistes Lustres Nappe of Corsica, *Koninkl. Nederl. Akad. Wetens. Verh. Afd. Nat.* (Tweede Reeks) **Dl. XLVIII,** pp. 1–71.

Brown, E. H. (1967): The Greenschist Facies in Part of Eastern Otago, New Zealand, *Contr. Mineral. Petrol.*, **14,** pp. 259–292.

Brown, G. C. (1970): A Comment on the Role of Water in the Partial Fusion of Crustal Rocks, *Earth Planet. Sci. Letters*, **9,** pp. 355–358.

―――― and W. S. Fyfe (1970): The Production of Granitic Melts during Ultrametamorphism, *Contr. Mineral. Petrol.*, **28,** pp. 310–318.

Brown, G. M. (1956): The Layered Ultrabasic Rocks of Rhum, Inner Hebrides, *Phil. Trans. Soc. Lond.*, ser. B., **240,** pp. 1–53.

―――― (1957): Pyroxenes from the Early and Middle Stages of Fractionation of the Skaergaard Intrusion, East Greenland, *Mineral. Mag.*, **31,** pp. 511–543.

―――― and E. A. Vincent (1963): Pyroxenes from Late Stages of Fractionation of the Skaergaard Intrusion, East Greenland, *J. Petrol.*, **4,** pp. 175–197.

Brown, P. E. (1966): Major Element Composition of the Loch Coire Migmatite Complex, Sutherland, Scotland, *Contr. Mineral. Petrol.*, **14,** pp. 1–26.

Brown, R. E. (1970): Some Suggested Rates of Deformation of the Basalts in the Pasco Basin, and Their Implications, *in* "Proceedings of the Second Columbia River Basalt Symposium," E. H. Gilmour and D. Stradling, editors, Eastern Wash. State Coll. Press, pp. 179–187.

Brown, S. E., and T. P. Thayer (1963): Low-grade Mineral Facies in Upper Triassic and Lower Jurassic Rocks of the Aldrich Mountains, Oregon, *J. Sedim. Petrol.*, **33,** pp. 411–425.

Brown, W. H., W. S. Fyfe, and F. J. Turner (1962): Aragonite in California Glaucophane Schists, and the Kinetics of the Aragonite-Calcite Transformation, *J. Petrol.*, **3,** pp. 566–582.

Brown, W. L. (1960): The Crystallographic and Petrologic Significance of Peristerite Unmixing in the Acid Plagioclases, *Zeit. Kryst.*, **113,** pp. 330–344.

―――― (1962): Peristerite Unmixing in the Plagioclase and Metamorphic Facies Series, *Norsk. Geol. Tidssk.*, **42,** pt. 2, pp. 354–382.

Brunn, J. H. (1952): Les eruptions ophiolithiques dans le NW de la Grece; Leurs relations avec l'orogenese, *Congr. Geol. Intern., Compt. Rend.*, **19,** Algiers, 1952, 17, pp. 19–27.

Bryan, W. B. (1967): Geology and Petrology of Clarion Island, Mexico, *Geol. Soc. Am. Bull.*, **78,** pp. 1461–1476.

Bryant, B. (1967): The Occurrence of Green Iron-rich Muscovite and Oxidation during Regional Metamorphism in the Grandfather Mountain Window, Northwestern North Carolina, *U.S. Geol. Surv. Prof. Paper* **575-C,** pp. C10–C16.

――――, R. G. Schmidt, and W. T. Pecora (1960): Geology of the Maddux Quadrangle, Bearpaw Mountains, Blaine County, Montana, *U.S. Geol. Surv. Bull.*, **1081-C.**

Bryhni, I. (1966): Reconnaissance Studies of Gneisses, Ultrabasites, Eclogites and Anorthosites in Outer Nordfjord, Western Norway, *Norges Geol. Undersok.*, **241,** pp. 5–68.

——, D. H. Green, K. S. Heier, and W. S. Fyfe (1970): On the Occurrence of Eclogite in Western Norway, *Contr. Mineral. Petrol.*, **26,** pp. 12–19.

Bucher, W. H. (1953): Fossils in Metamorphic Rocks: A Review, *Geol. Soc. Am. Bull.*, **64,** pp. 275–300.

Buddington, A. F. (1939): Adirondack Igneous Rocks and Their Metamorphism, *Geol. Soc. Am. Mem.* **7,** 354 pp.

—— (1948): Origin of Granitic Rocks of the Northwest Adirondacks, *in* "Origin of Granite," J. Gilluly, editor, *Geol. Soc. Am. Mem.* **28,** pp. 21–43.

—— (1952): Chemical Petrology of Some Metamorphosed Adirondack Gabbroic, Syenitic and Quartz Syenitic Rocks, *Am. J. Sci.*, **250A,** Bowen vol., pp. 37–84.

—— (1957): Interrelated Precambrian Granitic Rocks, Northwest Adirondacks, *Geol. Soc. Am. Bull.*, **68,** pp. 291–305.

—— (1959): Granite Emplacement with Special Reference to North America, *Geol. Soc. Am. Bull.*, **70,** pp. 671–747.

—— (1963): Isograds and the Role of H_2O in Metamorphic Facies of Orthogneisses of the Northwest Adirondack Area, New York, *Geol. Soc. Am. Bull.*, **74,** pp. 1155–1181.

—— (1965): The Origin of Three Garnet Isograds in Adirondack Gneisses, *Mineral. Mag.*, Tilley vol., **34,** pp. 71–81.

—— (1966): The Occurrence of Garnet in the Granulite-facies Terrane of the Adirondack Highlands–A Discussion [of paper by D. DeWaard, 1965], *J. Petrol.*, **7,** pp. 331–335.

—— (1968): Adirondack Anorthositic Series, *N.Y. State Mus. Sci. Serv. Mem.* **18,** pp. 215–231.

Buerger, M. J. (1947): The Relative Importance of the Several Faces of a Crystal, *Am. Mineralogist*, **32,** pp. 593–606.

Bugge, J. A. W. (1940): Geological and Petrographical Investigations in the Arendal District, *Norsk. Geol. Tidsskr.*, **20.**

Buie, B. F. (1941): Igneous Rocks of the Highwood Mountains, Montana, Part III, Dikes and Related Intrusives, *Geol. Soc. Am. Bull.*, **52,** pp. 1753–1808.

Bullard, F. M. (1961): Volcanoes: In History, in Theory, in Eruption, University of Texas Press, Austin, Texas.

Bultitude, R. J., and D. H. Green (1968): Experimental Study at High Pressures on the Origin of Olivine Nephelinite and Olivine Melilite Nephelinite Magmas, *Earth Planet. Sci. Letters*, **3,** pp. 325–337.

Burbank, W. S. (1927): Additional Data on the Properties of Pumpellyite, and its Occurrence in the Republic of Haiti, West Indies, *Am. Mineralogist*, **12,** pp. 421–424.

Burnham, C. W. (1959): Contact Metamorphism of Magnesian Limestones at Crestmore, California, *Geol. Soc. Am. Bull.*, **70,** pp. 879–920.

—— and R. H. Jahns (1962): A Method for Determining the Solubility of Water in Silicate Melts, *Am. J. Sci.*, **260,** pp. 721–745.

Burns, R. G. (1970a): Mineralogical Applications of Crystal Field Theory, Cambridge University Press, London, 244 pp.

——— (1970b): Crystal Field Spectra and Evidence of Cation Ordering in Olivine Minerals, *Am. Mineralogist,* **55,** pp. 1608–1632.

——— and W. S. Fyfe (1964): Site Preference Energy and Selective Uptake of Transition-metal Ions from a Magma, *Science,* **144,** pp. 1001–1003.

——— and ——— (1966): Distribution of Elements in Geological Processes, *Chem. Geol.,* **1,** pp. 49–56.

Buryanova, E. Z. (1960): Analcite- and Zeolite-bearing Rocks of Tuva, *Akad. Nauk. SSSR, Izvestiya, Geol.,* **no. 6,** pp. 54–65.

Buseck, P. R. (1966): Contact Metasomatic and Ore Deposition: Concepcion del Oro, Mexico, *Econ. Geol.,* **61,** pp. 97–136.

——— (1967): Contact Metasomatism and Ore Deposition, Tem Piute, Nevada, *Econ. Geol.,* **62,** pp. 331–353.

Butler, B. C. M. (1961): Metamorphism and Metasomatism of Rocks of the Moine Series by a Dolerite Plug in Glenmore, Ardnamurchan, *Mineral. Mag.,* **32,** pp. 866–897.

——— (1967): Chemical Study of Minerals from the Moine Schists of the Ardnamurchan Area, Argyllshire, Scotland, *J. Petrol.,* **8,** pp. 233–267.

Butler, J. R. (1969): Origin of Precambrian Granitic Gneiss in the Beartooth Mountains, Montana and Wyoming, *in* "Igneous and Metamorphic Geology," L. H. Larsen, editor, *Geol. Soc. Am. Mem.* **115,** pp. 73–101.

Byers, F. M., Jr. (1959): Geology of Umnak and Bogoslof Islands, Aleutian Islands, Alaska, *U.S. Geol. Surv. Bull.,* **1028-L,** pp. 267–369.

——— (1961): Petrology of Three Volcanic Suites, Umnak and Bogoslof Islands, Aleutian Islands, Alaska, *Geol. Soc. Am. Bull.,* **72,** pp. 93–128.

———, P. P. Orkild, W. J. Carr, and W. D. Quinlivan (1968): Timber Mountain Tuff, Southern Nevada, and its Relation to Cauldron Subsidence, *in* "Nevada Test Site, Studies in Geology and Hydrology," *Geol. Soc. Am. Mem.* **110,** pp. 87–99.

Cady, W. M. (1969): Regional Tectonic Synthesis of Northwestern New England and Adjacent Quebec, *Geol. Soc. Am. Mem.* **120,** 181 pp.

———, A. L. Albee, and A. H. Chidester (1963): Petrology and Geochemistry of Selected Talc-bearing Ultramafic Rocks and Adjacent Rocks in North-central Vermont, *U.S. Geol. Surv. Bull.,* **1122-B,** pp. B1–B78.

Callisten, K. (1943): Igneous Rocks of the Ivigtut Region, Greenland, *Medd. Groenland,* **131,** no. 8, 74 pp.

Cameron, E. N. (1963): Symposium on Layered Intrusions: Structure and Rock Sequences of the Critical Zone of the Eastern Bushveld Complex, *Mineral. Soc. Am. Spec. Paper* **1,** pp. 93–107.

Campbell, A. S., and W. S. Fyfe (1965): Analcite-Albite Equilibria, *Am. J. Sci.,* **263,** pp. 807–816.

Campbell, F. A. (1961): Differentiation Trends in the Ice River Complex, British Columbia, *Am. J. Sci.,* **259,** pp. 173–180.

Campbell, K. V., and R. B. Campbell (1970): Quesnel Lake Map-area, British Columbia (93A), *in* "Report of Activities," Part A, *Geol. Surv. Can. Paper* **70-1,** pt. A, pp. 32–35.

Campbell, R. B. (1970): Structural and Metamorphic Transitions from Infrastructure to Suprastructure, Cariboo Mountains, British Columbia, *Geol. Assoc. Can. Spec. Paper* **6,** pp. 67–72.

Cann, J. R. (1968): Geological Processes at Mid-ocean Ridge Crests, *Geophys. J. R. Astr. Soc.,* **15,** pp. 331–341.

—— (1969): Spilites from the Carlsberg Ridge, Indian Ocean, *J. Petrol.,* **10,** pp. 1–19.

—— (1970) : Upward Movement of Granitic Magma, *Geol. Mag.,* **107,** pp. 335–340.

—— and F. J. Vine (1966): An Area on the Crest of the Carlsberg Ridge–Petrology and Magnetic Survey, *Roy. Soc. Lond. Phil. Trans.,* ser. A, **259,** no. 1099, pp. 198–217.

Cannon, R. T. (1966): Plagioclase Zoning and Twinning in Relation to the Metamorphic History of Some Amphibolites and Granulites, *Am. J. Sci.,* **264,** pp. 526–542.

Carey, S. W. (1961): Relation of Basic Intrusions to Thickness of Sediments, *in* "Dolerite, a Symposium," Univ. of Tasmania, Geology Dept., pp. 165–169.

Carlisle, D. (1963): Pillow Breccias and Their Aquagene Tuffs, Quadra Island, British Columbia, *J. Geol.,* **71,** pp. 48–71.

—— (1971): Low Grade Metamorphism in the Karmutsen Group, *in* "Metamorphism in the Canadian Cordillera," *Geol. Assoc. Can.,* Cordill. Sect. Prog. and Abst., p. 7.

Carmichael, D. M. (1969): On the Mechanism of Prograde Metamorphic Reactions in Quartz-bearing Pelitic Rocks, *Contr. Mineral. Petrol.,* **20,** pp. 244–267.

—— (1970): Intersecting Isograds in the Whetstone Lake Area, Ontario, *J. Petrol.,* **11,** pp. 147–181.

——, R. A. Price, and H. R. Wynne-Edwards (1971): Pressure-sensitive Isograds as a Measure of Post-metamorphic Uplift and Erosion in some North American Metamorphic Terranes, *in* "Metamorphism in the Canadian Cordillera," *Geol. Assoc. Can.,* Cordill. Sect. Prog. and Abst., p. 8.

Carmichael, I. S. E. (1967): The Mineralogy and Petrology of the Volcanic Rocks from the Leucite Hills, Wyoming, *Contr. Mineral. Petrol.,* **15,** pp. 24–66.

Carpenter, A. B. (1967): Mineralogy and Petrology of the System $CaO\text{-}MgO\text{-}CO_2\text{-}H_2O$ at Crestmore, California, *Am. Mineralogist,* **52,** pp. 1341–1363.

Carr, J. M. (1954): Zoned Plagioclases in Layered Gabbros of the Skaergaard Intrusion, East Greenland, *Mineral. Mag.,* **30,** pp. 367–375.

Carr, M. H., and K. K. Turekian (1961): The Geochemistry of Cobalt, *Geochim. Cosmochim. Acta,* **23,** pp. 9–60.

Carr, R. M. (1963): Synthesis Fields of Some Aluminum Silicates; Further Studies, *Geochim. Cosmochim. Acta,* **27,** pp. 133–135.

—— and W. S. Fyfe (1960): Synthesis Fields of Some Aluminum Silicates, *Geochim. Cosmochim. Acta*, **21**, pp. 99–109.

Carrigy, M. A., and G. B. Mellon (1964): Authigenic Clay Mineral Cements in Cretaceous and Tertiary Sandstones of Alberta, *J. Sedim. Petrol.*, **34**, pp. 461–472.

Carter, N. L., J. M. Christie, and D. T. Griggs (1964): Experimental Deformation and Recrystallization of Quartz, *J. Geol.*, **72**, pp. 687–733.

—— and M. Friedman (1965): Dynamic Analysis of Deformed Quartz and Calcite from the Dry Creek Ridge Anticline, Montana, *Am. J. Sci.*, **263**, pp. 747–785.

Catalogue of Active Volcanoes of the World, Parts 1-9 (1951–1959), Intern. Volcanol. Assoc.

Catanzaro, E. J., and J. L. Kulp (1964): Discordant Zircons from the Little Belt (Montana), Beartooth (Montana) and Santa Catalina (Arizona) Mountains, *Geochim. Cosmochim. Acta*, **28**, pp. 87–124.

Cater, F. W. (1969): The Cloudy Pass Epizonal Batholith and Associated Subvolcanic Rocks, *Geol. Soc. Am. Spec. Paper* **116**, 54 pp.

Challis, G. A. (1965a): High-temperature Contact Metamorphism at the Red Hills Ultramafic Intrusion–Wairau Valley–New Zealand, *J. Petrol.*, **6**, pp. 395–419.

—— (1965b): The Origin of New Zealand Ultramafic Intrusions, *J. Petrol.*, **6**, pp. 322–364.

—— and W. R. Lauder (1966): The Genetic Position of "Alpine" Type Ultramafic Rocks, *Bull. Volcanol.*, **29**, pp. 283–302.

Chamberlain, J. A. (1967): Sulfides in the Muskox Intrusion, *Can. J. Earth Sci.*, **4**, pp. 105–153.

Chapman, C. A. (1962): Bays-of-Maine Igneous Complex, *Geol. Soc. Am. Bull.*, **73**, pp. 883–888.

—— (1967): Magmatic Central Complexes and Tectonic Evolution of Certain Orogenic Belts, *in* "Etages Tectoniques–Colloque de Neuchâtel," Neuchâtel Univ. Inst. Geol., 1966.

Chapman, R. W. (1950): Contact-metamorphic Effects of Triassic Diabase at Safe Harbor, Pennsylvania, *Geol. Soc. Am. Bull.*, **61**, pp. 191–220.

—— and C. A. Chapman (1940): Cauldron Subsidence at Ascutney Mountain, Vermont, *Geol. Soc. Am. Bull.*, **51**, pp. 191–212.

—— and C. R. Williams (1935): Evolution of the White Mountain Magma Series, *Am. Mineralogist*, **20**, pp. 502–530.

Chappell, B. W. (1968): Volcanic Greywackes from the Upper Devonian Baldwin Formation, Tamworth-Barraba District, New South Wales, *Geol. Soc. Australia J.*, **15**, pp. 87–102.

Chase, R. B. (1968): "Petrology of the Northeastern Border of the Idaho Batholith, Bitterroot Range," Montana, Univ. of Montana, unpubl. dissert., 188 pp.

Chatterjee, N. D. (1961): The Alpine Metamorphism in the Simplon Area, Switzerland and Italy, *Geol. Rundsch.*, **51**, pp. 1–72.

—— (1971): Phase Equilibria in the Alpine Metamorphic Rocks of the Environs of the Dora-Maria-Massif, Western Italian Alps, *Neues Jahrb. Mineral., Abhandl.*, **114**, pp. 181–210.

Chatterjee, S. C. (1968): An Alkali-Olivine Basalt Sub-province in the Deccan Traps, *Intern. Geol. Congr.*, **22d,** India, 1964, pt. 7, pp. 35–41.

Chayes, F. (1942): Alkaline and Carbonate Intrusives near Bancroft, Ontario, *Geol. Soc. Am. Bull.*, **53,** pp. 440–512.

—— (1964a): A Petrographic Distinction between Cenozoic Volcanics in and around the Open Oceans, *J. Geophys. Res.*, **69,** pp. 1573–1588.

—— (1964b): Variance-covariance Relations in Harker Diagrams of Volcanic Suites, *J. Petrol.*, pp. 219–237.

—— (1966): Alkaline and Subalkaline Basalts, *Am. J. Sci.*, **264,** pp. 128–145.

—— (1969): The Chemical Composition of Cenozoic Andesite, *in* "Proceedings of the Andesite Conference," A. R. McBirney, editor, *Oregon Dept. Mineral. Ind. Bull.*, **65,** pp. 1–11.

—— (1970): On Estimating the Magnitude of the Hidden Zone and the Compositions of the Residual Liquids of the Skaergaard Layered Series, *J. Petrol.*, **11,** pp. 1–14.

Chesnokov, B. V. (1960): Rutile-bearing Eclogite from Shubino Village Deposit in the Southern Urals, *Intern. Geol. Rev.*, **2,** pp. 936–945.

Chesterman, C. W. (1942): Contact Metamorphism of the Twin Lakes Region, Fresno County, California, *Calif. J. Mines Geol.*, **38,** pp. 243–281.

—— (1960): Intrusive Ultrabasic Rocks and Their Metamorphic Relationships at Leech Lake Mountain, Mendocino County, California, *Intern. Geol. Cong.*, **21st,** Copenhagan, 1960, *Rept. Session*, Norden, pt. 13, pp. 208–215.

Chidester, A. H. (1968): Evolution of the Ultramafic Complexes of Northwestern New England, *in* "Studies of Appalachian Geology: Northern and Maritime," E-an Zen, W. S. White, J. J. Hadley, and J. B. Thompson, Jr., editors, Interscience, a division of Wiley, New York, pp. 343–354.

Chinner, G. A. (1960a): The Origin of Sillimanite in Glen Clova, Angus, *J. Petrol.*, **2,** pp. 312–323.

—— (1960b): Pelitic Gneisses with Varying Ferrous/Ferric Ratios from Glen Clova, Angus, Scotland, *J. Petrol.*, **1,** pp. 178–217.

—— (1962): Almandine in Thermal Aureoles, *J. Petrol.*, **3,** pp. 316–340.

—— (1966): The Distribution of Pressure and Temperature during Dalradian Metamorphism, *Geol. Soc. Lond. Quart. J.*, **122,** pp. 159–186.

Christensen, N. I. (1970): Composition and Evolution of the Oceanic Crust, *Marine Geol.*, **8,** pp. 139–154.

Christie, O. H. J. (1959): Note on the Equilibrium between Plagioclase and Epidote, *Norsk. Geol. Tidssk.*, **39,** pp. 268–271.

Church, W. R. (1967): Eclogites, *in* "Basalts," H. H. Hess, editor, vol. **2,** Interscience, a division of Wiley, New York, pp. 755–798.

Clabaugh, S. E. (1953): Contact Metamorphism in the Christmas Mountains, Brewster County, Texas, *Geol. Soc. Am. Bull.*, **64,** p. 1408.

Clapp, C. H. (1912): Southern Vancouver Island, *Geol. Surv. Can. Mem.* **13,** 208 pp.

———— (1927): Sooke and Duncan Map Areas, Vancouver Island, *Geol. Surv. Can. Mem.* **96,** p. 270.

Clark, B. R. (1970): Mechanical Formation of Preferred Orientation in Clays, *Am. J. Sci.,* **269,** pp. 250–266.

Clark, R. H. (1960): Petrology of the Volcanic Rocks of Tongariro Subdivision, Appendix 2, *in* "The Geology of Tongariro Subdivision," by D. R. Gregg, *New Zealand Geol. Surv. Bull.,* ns. **40,** pp. 107–123.

———— and W. S. Fyfe (1961): Ultrabasic Liquids, *Nature,* **191,** no. 4784, pp. 158–159.

Clarke, D. B., and B. G. J. Upton (1971): Tertiary Basalts of Baffin Island: Field Relation and Tectonic Setting, *Can. J. Earth Sci.,* **8,** pp. 248–258.

Clavan, W., W. M. McNabb, and E. H. Watson (1954): *Am. Mineralogist,* **39,** pp. 566–599.

Closs, H. (1969): Explosion Seismic Studies in Western Europe, *in* "The Earth's Crust and Upper Mantle," P. J. Hart, editor, *Am. Geophys. Union, Geophys. Monograph* **13,** pp. 178–188.

Coats, R. R. (1962): Magma Type and Crustal Structure in the Aleutian Arc, *in* "The Crust of the Pacific Basin," G. A. Macdonald and H. Kuno, editors, *Am. Geophys. Union, Geophys. Monograph* **6,** p. 92.

———— (1968): Basaltic Andesites, *in* "Basalts: The Poldervaart Treatise on Rocks of Basaltic Composition," Interscience, a division of Wiley, New York, pp. 689–736.

Cogne, J. (1962): La sizunite (cap Sizun, Finistere) et le probleme de l'origine des lampro-phyres, *Soc. Geol. Fr. Bull.,* ser. 7, **4,** no. 2, pp. 141–156.

Cohen, L. H., K. Ito, and G. C. Kennedy (1967): Melting and Phase Relations in an An-hydrous Basalt to 40 kilobars, *Am. J. Sci.,* **265,** pp. 475–518.

Coleman, R. G. (1961): Jadeite Deposits of the Clear Creek Area, New Idria District, San Benito County, California, *J. Petrol.,* **2,** pp. 209–247.

———— (1962): Metamorphic Aragonite as Evidence Relating Emplacement of Ultramafic Rocks to Thrust Faulting in New Zealand (abs.), *Trans. Am. Geophys. Union,* **43,** p. 447.

———— (1963): Serpentinites, Rodingites, and Tectonic Inclusions in Alpine-type Mountain Chains, *Geol. Soc. Am. Spec. Paper* **76,** p. 130.

———— (1966): New Zealand Serpentinites and Associated Metasomatic Rocks, *New Zeal. and Geol. Surv. Bull.,* ns. **76,** 102 pp.

———— (1967a): Glaucophane Schists from California and New Caledonia, *in* "Age and Nature of the Circum-Pacific Orogenesis," *Pacific Sci. Congr.* **11th,** Tokyo, 1966, Symposium, *Tectonophysics,* **4,** no. 4-6, pp. 479–498.

———— (1967b): Low-temperature Reaction Zones and Alpine Ultramafic Rocks of California, Oregon, and Washington, *U.S. Geol. Surv. Bull.,* **1247,** 49 pp.

———— and D. E. Lee (1962): Metamorphic Aragonite in the Glaucophane Schists of Cazadero, California, *Am. J. Sci.,* **260,** pp. 577–595.

———— and ———— (1963): Glaucophane-bearing Metamorphic Rock Types of the Cazadero Area, California, *J. Petrol.* **4,** pp. 260–301.

————, ————, L. B. Beatty, and W. W. Brannock (1965): Eclogites and Eclogites: Their Differences and Similarities, *Geol. Soc. Am. Bull.,* **76,** pp. 483–508.

Combe, A. D., and A. Holmes (1945): The Kalsilite-bearing Lavas of Kabirenge and Lyakauli, Southwest Uganda, *Trans. Roy. Soc. Edinburgh* **62,** pt. 2, pp. 359–379.

Compston, W., and P. M. Jeffrey (1959): Anómalous 'Common Strontium' in Granite, *Nature,* **184,** no. 4701, pp. 1792–1793.

Compton, R. R. (1955): Trondhjemite Batholith near Bidwell Bar, California, *Geol. Soc. Am. Bull.,* **66,** pp. 9–44.

———— (1958): Significance of Amphibole Paragenesis in the Bidwell Bar Region, California, *Am. Mineralogist,* **43,** pp. 890–907.

———— (1960): Charnockitic Rocks of Santa Lucia Range, California, *Am. J. Sci.,* **258,** pp. 609–636.

Condie, K. C., and R. A. Heimlich (1969): Interpretation of Precambrian K-Ar Biotite Dates in the Bighorn Mountains, Wyoming, *Earth Planet. Sci. Letters,* **6,** pp. 209–212.

———— and J. A. Madison (1969): Compositional and Volume Changes Accompanying Progressive Serpentinization of Dunites from the Webster-Addie Ultramafic Body, North Carolina, *Am. Mineralogist,* **54,** pp. 1173–1179.

Cook, K. L. (1968): Evidence for the East Pacific Rise and Mantle Convection Currents under Western North America (abs.), *Geol. Soc. Am. Spec. Paper* **101,** p. 43.

Cooke, H. C. (1919): Gabbros of East Sooke and Rocky Point, *Can. Geol. Surv. Mus. Bull.,* **30,** pp. 1–48.

Coombs, D. S. (1954): The Nature and Alteration of Some Triassic Sediments from Southland, New Zealand, *Trans. Roy. Soc. New Zealand* **82,** pp. 65–109.

———— (1960): Lower Grade Mineral Facies in New Zealand, *Intern. Geol. Congr.,* **21st,** Copenhagen, 1960, *Rept. Session,* Norden, 1960, **13,** pp. 339–351.

———— (1961): Some Recent Work on the Lower Grades of Metamorphism, *Australian J. Sci.,* **24,** pp. 203–215.

————, A. J. Ellis, W. S. Fyfe, and A. H. Taylor (1959): The Zeolite Facies with Comments on the Interpretation of Hydrosynthesis, *Geochim. Cosmochim. Acta,* **17,** pp. 53–107.

————, R. J. Horodyski, and R. S. Naylor (1970): Occurrence of Prehnite-Pumpellyite Facies Metamorphism in Northern Maine, *Am. J. Sci.,* **268,** pp. 142–156.

———— and J. F. G. Wilkinson (1969): Lineages and Fractionation Trends in Undersaturated Volcanic Rocks from the East Otago Volcanic Province (New Zealand) and Related Rocks, *J. Petrol.,* **10,** pp. 440–501.

Coombs, H. A. (1936): The Geology of Mount Rainier National Park, *Wash. Univ. Pub. in Geol.,* **3,** pp. 131–212.

Cooper, J. R. (1936): Geology of the Southern Half of the Bay of Islands Igneous Complex, *Nfld. Dept. Nat. Res. Geol. Soc. Bull.* **4,** pp. 1–62.

Cooray, P. G. (1962): Charnockites and Associated Gneisses in the Pre-Cambrian of Ceylon, *Geol. Soc. Lond. Quart. J.* **118,** pp. 239–273.

Cowan, D. S., and C. F. Mansfield (1970): Serpentinite Flows on Joaquin Ridge, Southern Coast Ranges, California, *Geol. Soc. Am. Bull.,* **81,** pp. 2615–2628.

Cox, A., and R. R. Doell (1964): Long Period Variations of the Geomagnetic Field, *Bull. Seis. Soc. Am.,* **54,** pp. 2243–2270.

Cox, D. P. (1967): Reconnaissance Geology of the Helena Quadrangle, Trinity County, California, *in* "Short Contributions to California Geology," *Calif. Div. Mines Geol. Spec. Rept.* **92,** pp. 43–55.

Crawford, M. L. (1966): Composition of Plagioclase and Associated Minerals in Some Schists from Vermont, U.S.A., and South Westland, New Zealand, with Inferences about the Peristerite Solvus, *Contr. Mineral. Petrol.,* **13,** pp. 269–294.

——— (1970): Phase Relations of Feldspars in Contact Metamorphic Rocks, *Geol. Soc. Am., Abst. with Prog.,* **2,** pp. 528–529.

Crawford, W. A., and W. S. Fyfe (1964): Calcite-Aragonite Equilibrium at 100°C, *Science,* **144,** pp. 1569–1570.

——— and ——— (1965): Lawsonite Equilibria, *Am. J. Sci.,* **263,** pp. 262–270.

Crittenden, M. D. (1951): Geology of the San Jose–Mount Hamilton Area, California, *Calif. Dept. Nat. Res., Div. Mines Bull.,* **157,** pp. 1–74.

Crook, K. A. W. (1961): Diagenesis in the Wahgi Valley Sequence, New Guinea, *Roy. Soc. Victoria Proc.,* **74,** pp. 77–81.

——— (1963): Burial Metamorphic Rocks from Fiji, *New Zealand J. Geol. Geophys.,* **6,** pp. 681–704.

Crosby, P. (1968): Petrogenetic and Statistical Implications of Modal Studies in Adirondack Anorthosite, *N. Y. State Mus. Sci. Serv. Mem.* **18,** pp. 289–303.

Cross, W. (1897): The Igneous Rocks of the Leucite Hills and Pilot Butte, Wyoming, *Am. J. Sci.,* **4,** pp. 115–141.

———, J. P. Iddings, L. V. Pirsson, and H. S. Washington (1902): A Quantitative Chemicomineralogical Classification and Nomenclature of Igneous Rocks, *J. Geol.,* **10,** pp. 555–690.

Crowder, D. F. (1959): Granitization, Migmatization, and Fusion in the Northern Entiat Mountains, Washington, *Geol. Soc. Am. Bull.,* **70,** pp. 827–878.

Crowell, J. C., and J. W. R. Walker (1962): Anorthosite and Related Rocks along the San Andreas Fault, Southern California, *Calif. Univ. Dept. Geol. Sci. Bull.,* **40,** pp. 219–288.

Crowley, M. S., and R. Roy (1964): Crystalline Solubility in the Muscovite and Phlogopite Groups, *Am. Mineralogist,* **49,** pp. 348–362.

de Cserna, Z. (1961): Tectonic Map of Mexico, Scale 1:2,500,000, Geol. Soc. Am.

Cuenod, Y., and J. Martini (1967): On the Presence of Zeolite Rocks in the Rio Chirique Basin (Western Panama), *Soc. Physique Histoire Nat. Geneve Compte. Rendu.,* new ser., **1,** no. 3 (*Archives Sci.,* **19,** 1966 supp.), pp. 152–158.

Currie, K. L. (1970): An Hypothesis on the Origin of Alkaline Rocks Suggested by the Tectonic Setting of the Monteregian Hills, *Can. Mineral.,* **10,** pp. 411–420.

—— and J. Ferguson (1970): Ice River Alkaline Complex, British Columbia (82N), *in* "Report of Activities," part A, *Can. Geol. Surv. Paper* **70-1,** pp. 112–113.

Curtis, C. D. (1964): Applications of the Crystal-field Theory to the Inclusion of Trace Transition Elements in Minerals during Magmatic Differentiation, *Geochim. Cosmochim. Acta,* **28,** pp. 389–403.

Dalrymple, G. B., and J. G. Moore (1968): Argon 40: Excess in Submarine Pillow Basalts from Kilauea Volcano, Hawaii, *Science,* **161,** pp. 1132–1135.

Daly, R. A. (1910): Origin of the Alkaline Rocks, *Geol. Soc. Am. Bull.,* **21,** pp. 87–118.

—— (1918): Genesis of the Alkaline Rocks, *J. Geol.,* **26,** pp. 97–134.

—— (1928): Bushveld Igneous Complexes of the Transvaal, *Geol. Soc. Am Bull.,* **39,** pp. 703–768.

—— (1933): "Igneous Rocks and the Depths of the Earth," McGraw-Hill Book Co., New York, 598 pp.

Damon, P. E. et al. (1969): Correlation and Chronology of Ore Deposits and Volcanic Rocks, *A.E.C. Prog. Rept.* C00-689-120, pp. 25–27, 65–68.

——, A. W. Laughlin, and J. K. Percious (1967): Problem of Excess Argon-40 in Volcanic Rocks, *in* "Radioactive Dating and Methods of Low-level Counting," Int. Atomic Energy Agency, pp. 463–482.

Davidson, A. (1963): "A Study of Okaite and Related Rocks near Oka, Quebec," Univ. of British Columbia, Canada, unpubl. M.S. thesis, 146 pp.

Davidson, C. F. (1943): The Archean Rocks of the Rodil District, South Harris, Outer Hebrides, *Roy. Soc. Edinburgh Trans.,* **61,** pp. 71–112.

—— (1967): The So-called "Cognate Xenoliths" of Kimberlite, *in* "Ultramafic and Related Rocks," P. J. Wyllie, editor, Wiley, New York, pp. 342–346.

Davies, G. L., S. R. Hart, and G. R. Tilton (1968): Some Effects of Contact Metamorphism on Zircon Ages, *Earth Planet. Sci. Letters,* **5,** pp. 27–34.

Davies, H. L. (1968): Papuan Ultramafic Belt, *Intern. Geol. Congr.,* **23d,** Czech., 1968, vol. 1, pp. 209–220.

—— and J. S. Milsom (1969): Eastern Papua Geology and Gravity, *Am. Geophys. Union Trans.,* **50,** p. 333.

Davies, K. A. (1952): The Building of Mount Elgon (East Africa), *Uganda Geol. Surv. Mem.* **7.**

Davies, R. G. (1956): The Pen-y-gader Dolerite and Its Metasomatic Effects on the Llyn-y-gader Sediments, *Geol. Mag.,* **93,** pp. 153–172.

Davis, B. T. C. (1968): Anorthositic and Quartz Syenitic Series of the St. Regis Quadrangle, New York, *N. Y. State Mus. Sci. Serv. Mem.* **18,** pp. 281–287.

Davis, G. A., (1968): Westward Thrusting in the South-central Klamath Mountains, California, *Geol. Soc. Am. Bull.,* **79,** pp. 911–934.

Davis, G. L., and A. Pabst (1960): Lawsonite and Pumpellyite in Glaucophane Schist, North Berkeley Hills, California, with Notes on the X-ray Crystallography of Lawsonite, *Am. J. Sci.,* **258,** pp. 689–704.

Dawson; J. B. (1962a): Sodium Carbonate Lavas from Oldoinyo Lengai, Tanganyika, *Nature*, **195,** pp. 1075–1076.

—— (1962b): The Geology of Oldoinyo Lengai, *Bull. Volcanol.*, **24,** pp. 349–387.

—— (1964a): Reactivity of the Cations in Carbonate Magmas, *Geol. Assoc. Can. Proc.*, **15,** pt. 2, pp. 103–113.

—— (1964b): Carbonate Tuff Cones in Northern Tanganyika, *Geol. Mag.*, **101,** pp. 129–137.

—— (1964c): Carbonatitic Volcanic Ashes in Northern Tanganyika, *Bull. Volcanol.*, **27,** pp. 1–11.

—— (1966): Oldoinyo Lengai—An Active Volcano with Sodium Carbonatite Lava Flows, *in* "Carbonatites," O. F. Tuttle and J. Gittins, editors, Interscience, a division of Wiley, New York, p. 155.

——, P. Bowden, and G. C. Clark (1968): Activity of the Carbonatite Volcano Oldoinyo Lengai, 1966, *Geol. Rundsch.*, **57,** pp. 865–879.

Deans, T. (1966): Economic Mineralogy of African Carbonatites, *in* "Carbonatites," O. F. Tuttle and J. Gittins, editors, Interscience, a division of Wiley, New York, pp. 385–413.

Dearnley, R. (1963): The Lewisian Complex of South Harris; With Some Observations on the Metamorphosed Basic Intrusions of the Outer Hebredes, *Geol. Soc. Lond. Quart. J.* **119,** pp. 243–312.

Deer, W. A., and D. Abbott (1965): Clinopyroxenes of the Gabbro Cumulates of the Kap Edvard Holm Complex, East Greenland, *Mineral. Mag.*, **34,** pp. 177–193.

——, R. A. Howie, and J. Zussman (1962): "Rock-forming Minerals," vol. 2: Chain Silicates, Wiley, New York.

—— and L. R. Wager (1939): Olivines from the Skaergaard Intrusion Kangerdlugssuag, East Greenland, *Am. Mineralogist*, **24,** pp. 18–25.

Degens, E. T. (1965): "Geochemistry of Sediments," Prentice-Hall, N.J., 342 pp.

—— and G. V. Chilingar (1967): Diagenesis of Subsurface Waters, *in* "Diagenesis in Sediments," G. Larsen and G. V. Chilingar, editors, Elsevier, Amsterdam, pp. 447–502.

Deines, P. (1970): The Carbon and Oxygen Isotopic Composition of Carbonates from the Oka Carbonatite Complex, Quebec, Canada, *Geochim. Cosmochim. Acta,* **34,** pp. 1199–1225.

—— and D. P. Gold (1969): The Change in Carbon and Oxygen Isotopic Composition during Contact Metamorphism of Trenton Limestones by the Mount Royal Pluton, *Geochim. Cosmochim. Acta,* **33,** no. 3, pp. 421–424.

Dengo, G. (1950): Eclogitic and Glaucophane Amphibolites in Venezuela, *Am. Geophys. Union Trans.,* **31,** pp. 873–878.

—— (1953): Geology of the Caracas Region, Venezuela, *Geol. Soc. Am. Bull.*, **64,** pp. 7–40.

deWaard, D. (1959): Anorthite Content of Plagioclase in Basic and Pelitic Crystalline Schists as Related to Metamorphic Zoning in the Usu Massif, Timor, *Am. J. Sci.*, **257,** pp. 553–562.

—— (1965a): The Occurrence of Garnet in the Granulite-facies Terrane of the Adirondack Highlands, *J. Petrol.*, **6,** pp. 165–191.

—— (1965b): A Proposed Subdivision of the Granulite Facies, *Am. J. Sci.*, **263**, pp. 455–461.

—— (1966a): The Biotite-Cordierite-Almandite Subfacies of the Hornblende Granulite Facies, *Can. Mineralogist*, **8**, pt. 4, pp. 481–492.

—— (1966b): On Water-vapor Pressure in Zones of Regional Metamorphism and the Nature of the Hornblende-Granulite Facies, *Koninkl. Nederlandse Akad. Wetensch. Proc.*, ser. B, **69**, pp. 453–458.

—— (1967): The Occurrence of Garnet in the Granulite-facies Terrane of the Adirondack Highlands and Elsewhere, an Amplification and a Reply [of 1965 Paper and to Discussion by A. F. Buddington, 1966], *J. Petrol.*, **8**, pp. 210–232.

—— (1968a): The Anorthosite Problem: The Problem of the Anorthosite-Charnockite Suite of Rocks, *N. Y. State Mus. Sci. Serv. Mem.* **18**, pp. 71–91.

—— (1968b): Metamorphism and Magmatism in the Charnockitic Terrane of the Adirondack Highlands, U.S.A., *Intern. Geol. Congr.*, **22d**, India, 1964, pt. 13, p. 185.

—— and W. D. Romey (1968): Petrogenetic Relationships in the Anorthosite-Charnockite Series of Snowy Mountain Dome, South Central Adirondacks, *N. Y. State Mus. Sci. Serv. Mem.* **18**, pp. 307–315.

—— and —— (1969): Chemical and Petrologic Trends in the Anorthosite-Charnockite Series of the Snowy Mountain Massif, Adirondack Highlands, *Am. Mineralogist*, **54**, pp. 529–538.

Dewey, J. F. (1969): Continental Margins: A Model for Conversion of Atlantic Type to Andean Type, *Earth Planet. Sci. Letters*, **6**, pp. 189–197.

—— and J. M. Bird (1970): Mountain Belts and the New Global Tectonics, *J. Geophys. Res.*, **75**, pp. 2625–2647.

—— and —— (1971): Origin and Emplacement of the Ophiolite Suite: Appalachian Ophiolites in Newfoundland, *J. Geophys. Res.*, **76**, pp. 3179–3206.

Dickinson, W. R. (1962): Petrogenetic Significance of Geosynclinal Andesitic Volcanism along the Pacific Margin of North America, *Geol. Soc. Am. Bull.*, **73**, pp. 1241–1256.

—— (1968): Circum-Pacific Andesite Types, *J. Geophys. Res.*, **73**, pp. 2261–2269.

—— (1969): Evolution of Calc-alkaline Rocks in the Geosynclinal System of California and Oregon, *in* "Proceedings of the Andesite Conference," A. R. McBirney, editor, *Oregon Dept. Geol. Mineral. Ind. Bull.*, **65**, pp. 151–156.

—— (1970a): Relations of Andesites, Granites, and Derivative Sandstones to Arc-trench Tectonics, *Rev. Geophys. Space Phys.*, **8**, pp. 813–860.

—— (1970b): Table Mountain Serpentinite Extrusion in California Coast Ranges, *Geol. Soc. Am. Bull.*, **77**, pp. 451–472.

—— (1971): Plate Tectonic Models of Geosynclines, *Earth Planet. Sci. Letters*, **10**, pp. 165–174.

—— and T. Hatherton (1967): Andesitic Volcanism and Seismicity around the Pacific, *Science*, **157**, no. 3780, pp. 801–803.

——, R. W. Ojakangas, and R. J. Stewart (1969): Burial Metamorphism of the Late Mesozoic Great Valley Sequence, Cache Creek, California, *Geol. Soc. Am. Bull.*, **80**, pp. 519–525.

Dickson, F. W. (1958): Zone Melting as a Mechanism of Intrusion–A Possible Solution of the Room and Superheat Problems (abs.), *Trans. Am. Geophys. Union,* **39,** p. 513.

Dietrich, R. V. (1954): Fish Creek Phacolith, Northwestern New York, *Am. J. Sci.,* **252,** pp. 513–531.

———— (1959): Development of Ptygmatic Features within a Passive Host during Partial Anatexis, *Beitr. Mineral. Petrog.,* B.d.b., pp. 357–365.

———— (1960): Banded Gneisses, *J. Petrol.,* **1,** pp. 99–120.

Dietrich, V. (1967): Geosynklinaler Vulkanismus in den oberen penninischen Decken Graubundens (Schweiz), *Geol. Rundschau,* **57,** pp. 246–264.

Dietz, R. S. (1963a): Alpine Serpentinites as Oceanic Rind Fragments, *Geol. Soc. Am. Bull.,* **74,** pp. 947–952.

———— (1963b): Collapsing Continental Rises: An Actualistic Concept of Geosynclines and Mountain Building, *J. Geol.,* **71,** pp. 314–333.

Dobretsov, N. L., V. V. Reverdatto, V. S. Sobolev, N. V. Sobolev, Ye. N. Ustakova, and V. V. Khlestov (1965): Distribution of Regional Metamorphic Facies in the USSR, *Intern. Geol. Rev.* **8,** no. 11, pp. 1335–1346.

Doe, B. R., and S. R. Hart (1963): The Effect of Contact Metamorphism on Lead in Potassium Feldspars near the Eldora Stock, Colorado, *J. Geophys. Res.,* **68,** pp. 3521–3530.

————, P. W. Lipman, and C. E. Hedge (1969): Radiogenic Tracers and the Source of Continental Andesites: A Beginning at the San Juan Volcanic Field, Colorado, *in* "Proceedings of the Andesite Conference," A. R. McBirney, editor, *Oregon Dept. Geol. Mineral. Ind. Bull.,* **65,** pp. 143–149.

————, ————, ————, and H. Kurasawa (1969): Primitive and Contaminated Basalts from the Southern Rocky Mountains, U.S.A., *Contr. Mineral. Petrol.,* **21,** pp. 142–156.

————, R. I. Tilling, C. E. Hedge, and M. R. Klepper (1968): Lead and Strontium Isotope Studies of the Boulder Batholith, Southwestern Montana, *Econ. Geol.,* **63,** pp. 884–906.

————, G. R. Tilton, and C. A. Hopson (1965): Lead Isotopes in Feldspars from Selected Granitic Rocks Associated with Regional Metamorphism, *J. Geophys. Res.,* **70,** pp. 1947–1968.

Donnelly, T. W. (1963): Genesis of Albite in Early Orogenic Volcanic Rocks, *Am. J. Sci.,* **261,** pp. 957–972.

———— (1966): Geology of St. Thomas and St. John, U.S. Virgin Islands, *in* "Caribbean Geological Investigations,' *Geol. Soc. Am. Mem.* **98,** pp. 85–176.

Dons, J. A. (1952): The Igneous Rock Complex of the Olso Region, XI, Compound Volcanic Necks, Igenous Dykes, and Fault Zones in the Ullern-Husebyasen Area, *Norske Vid.-Akad. Skrift,* no. 1, *Mat.-Naturv.* Kl.

Dott, R. H., Jr. (1965): Mesozoic-Cenozoic Tectonic History of the South-western Oregon Coast in Relation to Cordilleran Orogenesis, *J. Geophys. Res.,* **70,** pp. 4687–4708.

Drever, H. I. (1960): Immiscibility in the Picritic Intrusion at Igdlorssuit, West Greenland, *Intern. Geol. Congr.,* **21st,** Copenhagen, 1960, *Rept. Session,* Norden, pt. 13, pp. 47–58.

——— and R. Johnston (1967): The Ultrabasic Facies in Some Sills and Sheets, *in* "Ultramafic and Related Rocks," P. J. Wyllie, editor, Wiley, New York, pp. 51–63.

Drugova, G. M., and V. D. Bugrova (1965): Garnets of the Granulite Facies of the Aldan Shield under Conditions of Polymetamorphism, *Geochem. Intern.*, **1964,** pp. 365–371, trans. from Vsesoyuznoe Mineralogicheskoe Obshchestvo, Zapiski, **93,** pp. 37–45, 1964.

DuBois, C. G. B., J. Furst, N. J. Guest, and D. J. Jennings (1963): Fresh Natro Carbonatite Lava from Oldoinyo L'Engai, *Nature,* **197,** pp. 445–446.

Dudley, P. P. (1969): Electron Microprobe Analyses of Garnet in Glaucophane Schists and Associated Eclogites, *Am. Min.,* **54,** pp. 1139–1150.

Duke, M. B., and L. T. Silver (1967): Petrology of Eucrites, Howardites, and Mesosiderites, *Geochim. Cosmochim. Acta,* **31,** pp. 1637–1665.

Dunham, K. C. (1935): Geology of the Organ Mountains, *N. Mex. School Mines Bull.,* **11,** 272 pp.

Dunn, J. A. (1942): Granite and Magmation and Metamorphism, *Econ. Geol.,* **37,** pp. 231–238.

Durrell, C. (1940): Metamorphism in the Southern Sierra Nevada Northeast of Visalia, California, *Univ. Calif. Dept. Geol. Sci. Bull.,* **25,** pp. 1–118.

Eaton, J. P. (1963): Crustal Structure from San Francisco, Calif. to Eureka, Nev. from Seismic-refraction Measurements, *J. Geophys. Res.,* **68,** no. 20, pp. 5789–5806.

——— and K. J. Murata (1960): How Volcanoes Grow, *Science,* **132,** no. 3432, pp. 925–938.

Eckelmann, F. D., and A. Poldervaart (1957): Geologic Evolution of the Beartooth Mountains, Montana and Wyoming, Part 1: Archean History of the Quad Creek Area, *Geol. Soc. Am. Bull.,* **68,** pp. 1225–1262.

Eckermann, H. von (1948): The Alkaline District of Alnö Island, *Sveriges. Geol. Undersok.,* ser. ca. no. 36.

——— (1961): The Petrogenesis of the Alnö Alkaline Rocks, *Bull. Geol. Inst. Univ. Upsala,* **40,** pp. 25–36.

——— (1966): Progress of Research on the Alnö Carbonatite, *in* "Carbonatites," O. F. Tuttle and J. Gittins, editors, Interscience, a division of Wiley, New York, pp. 3–31.

——— (1967): A Comparison of Swedish, African, and Russian Kimberlites, *in* "Ultramafic and Related Rocks," P. J. Wyllie, editor, Wiley, New York, pp. 302–312.

Edwards, A. B. (1942): Differentiation of the Dolerites of Tasmania, *J. Geol.,* **50,** pp. 451–480; 579–610.

——— and G. Baker (1953): Scapolitization in the Cloncurry District of NW Queensland, *Geol. Soc. Australia,* **1,** pp. 1–33.

Egeler, C. G. (1956): The Alpine Metamorphism in Corsica, *Geol. Mijnb.,* **18,** pp. 115–118.

Egorov, L. S. (1970): Carbonatites and Ultrabasic-alkaline Rocks of the Maimecha-Kotui Region, N. Siberia, *Lithos,* **3,** pp. 341–359.

Ehinger, R. (1971): Magmatic Processes in the Philipsburg Batholith, Montana, *Northwest Geol.,* **1,** pp. 47–51.

Ellenberger, F. (1960): Sur un paragénèse ephémère à lawsonite et glaucophane dans le métamorphisme alpin en Haute-Maurienne (Savoie), *Bull. Soc. Géol. France,* **7,** pp. 190–194.

Ellert, R. (1959): Contribuição á geologia do maciço alcalino de Poços de Caldas, *Bd. F.F. C. L. Univ. São Paulo,* no. **237,** *Geologia,* no. 18, pp. 5–63.

Elliot, J. D. (1968): Low-grade Regional Metamorphism in the Waipapa Group of the Whangarei Coastal Region, Northland, *Roy. Soc. New Zealand Trans., Geology,* **6,** pp. 63–74.

Elliot, R. B. (1966): The Association of Amphibolite and Albitite, Kragerø, South Norway, *Geol. Mag.,* **103,** pp. 1–7.

Elston, W. E., and E. I. Smith (1970): Determination of Flow Direction of Rhyolitic Ash-flow Tuffs from Fluidal Textures, *Geol. Soc. Am. Bull.,* **81,** pp. 3393–3406.

Emeleus, C. H. (1964): The Grønnedal-Ika Alkaline Complex, South Greenland–The Structure and Geological History of the Complex, *Medd. Grønland,* **172,** no. 3, 75 pp.

Emmons, R. C. (1940): The Contribution of Differential Pressure to Magmatic Differentiation, *Am. J. Sci.,* **238,** pp. 1–21.

Emslie, R. F. (1965): The Michikamau Anorthositic Intrusion, Labrador, *Can. J. Earth Sci.,* **2,** pp. 385–399.

——— (1968): Crystallization and Differentiation of the Michikamau Intrusion, *N. Y. State Mus. Sci. Serv. Mem.* **18,** pp. 163–173.

——— (1970): The Geology of the Michikamau Intrusion, Labrador (13L, 23I), *Geol. Surv. Can. Paper* **68–57,** 85 pp.

——— and D. H. Lindsley (1969): Experiments Bearing on the Origin of Anorthositic Intrusions, *Ann. Rept., Dir. Geophys. Lab.,* 1967–1968, pp. 108–112.

Engel, A. E. J. and C. G. Engel (1953): Grenville Series in the Northwest Adirondack Mountains, New York, Part 1: General Features of the Grenville Series, *Geol. Soc. Am. Bull.,* **64,** pp. 1013–1097.

——— and ——— (1958): Progressive Metamorphism and Granitization of the Major Paragneiss, Northwest Adirondack Mountains, New York, Part 1: Total Rock, *Geol. Soc. Am. Bull.,* **69,** pp. 1369–1414.

——— and ——— (1962a): Progressive Metamorphism of Amphibolite, Northwest Adirondack Mountains, New York, *in* "Petrologic Studies," A. E. J. Engel, H. L. James, and B. F. Leonard, editors, *Geol. Soc. Am.,* **Buddington vol.,** pp. 37–82.

——— and ——— (1962b): Hornblendes Formed during Progressive Metamorphism of Amphibolites, Northwest Adirondack Mountains, New York, *Geol. Soc. Am. Bull.,* **73,** pp. 1499–1514.

——— and ——— (1963): Metasomatic Origin of Large Parts of the Adirondack Phacoliths, *Geol. Soc. Am. Bull.,* **74,** pp. 349–354.

——— and ——— (1964a): Composition of Basalts from the Mid-Atlantic Ridge, *Science,* **144,** pp. 1330–1333.

——— and ——— (1964b): Igneous Rocks of the East Pacific Rise, *Science,* **146,** pp. 477–485.

———, ———, and R. G. Havens, (1964): Mineralogy of Amphibolite Interlayers in the Gneiss Complex, Northwest Adirondack Mountains, New York, *J. Geol.,* **72,** pp. 131–156.

———, ———, and ——— (1965): Chemical Characteristics of Ocean Basalts and the Upper Mantle, *Geol. Soc. Am. Bull.,* **76,** pp. 719–734.

Engel, C. G., and A. E. J. Engel (1966): Volcanic Rocks Dredged Southwest of the Hawaiian Islands, *U.S. Geol. Surv. Prof. Paper* **550-D,** pp. D104–D108.

——— and R. L. Fisher (1969): Lherzolite, Anorthosite, Gabbro, and Basalt Dredged from the Mid-Indian Ocean Ridge, *Science,* **166,** pp. 1136–1141.

Engelhardt, W. von, and K. H. Gaida (1963): Concentration Changes of Pore Solutions during the Compaction of Clay Sediments, *J. Sedim. Petrol.,* **33,** pp. 919–930.

Erickson, R. L., and L. V. Blade (1963): Geochemistry and Petrology of the Alkalic Igneous Complex at Magnet Cove, Arkansas, *U.S. Geol. Surv. Prof. Paper* **425,** 95 pp.

Erikson, E. H., Jr. (1969): Petrology of the Composite Snoqualmie Batholith, Central Cascade Mountains, Washington, *Geol. Soc. Am. Bull.,* **80,** pp. 2213–2236.

Ernst, W. G. (1960): Diabase-granophyre Relations in the Endion Sill, Duluth, Minnesota, *J. Petrol.,* **1,** pp. 286–303.

——— (1963a): Significance of Phengitic Micas from Low-grade Schists, *Am. Mineralogist,* **48,** pp. 1357–1373.

——— (1963b): Petrogenesis of Glaucophane Schists, *J. Petrol.,* **4,** pp. 1–30.

——— (1964): Petrochemical Study of Coexisting Minerals from Low-grade Schists, Eastern Shikoku, Japan, *Geochim. Cosmochim. Acta,* **28,** pp. 1631–1668.

——— (1965): Mineral Paragenesis in Franciscan Metamorphic Rocks, Panoche Pass, California, *Geol. Soc. Am. Bull.,* **76,** pp. 879–914.

——— (1966): Synthesis and Stability Relation of Ferrotremolite, *Am. J. Sci.,* **264,** pp. 37–65.

——— (1971): Do Mineral Parageneses Reflect Unusually High-pressure Conditions of Franciscan Metamorphism?, *Am. J. Sci.,* **270,** pp. 81–108.

——— and Y. Seki (1967): Petrologic Comparison of the Franciscan and Sanbagawa Metamorphic Terranes, *in* "Age and Nature of the Circum-Pacific Orogenesis," Pacific Sci. Congr., **11th,** Tokyo, 1966, Symposium, *Tectonophysics,* **4,** no. 4–6, pp. 463–478.

———, ———, H. Onuki, and M. C. Gilbert (1970): Comparative Study of Low-grade Metamorphism in the California Coast Ranges and the Outer Metamorphic Belt of Japan, *Geol. Soc. Am. Mem.* **124,** 276 pp.

Eskola, P. (1914): On the Petrology of the Orijärvi Region in South-western Finland, *Bull. Comm. Géol. Finlande,* **40,** 277 pp.

——— (1915): On the Relation between Chemical and Mineralogical Composition in the Metamorphic Rocks of the Orijärvi Region, *Bull. Comm. Géol. Finlande,* **44.**

——— (1920): The Mineral Facies of Rocks, *Norsk Geol. Tidsskr.,* **6,** pp. 143–194.

——— (1921): On the Eclogites of Norway, *Skrift. Vidensk. Selsk. Christiania, Mat.-Naturv. Kl.* **1,** no. 8, pp. 1–118.

——— (1925): On the Petrology of Eastern Fennascandia, Part 1: The Mineral Development of Basic Rocks in the Karelian Formations, *Fennis,* **45,** no. 19, 93 pp.

—— (1932a): On the Origin of Granitic Magmas, *Tschermaks Mineral. Petrog. Mitt.*, **42,** pp. 445–481.

—— (1932b): On the Principles of Metamorphic Differentiation, *Comm. Géol. Finlande Bull.*, **97,** pp. 68–77.

—— (1933): On the Differential Anatexis of Rocks, *Comm. Géol. Finlande Bull.*, **103,** pp. 12–25.

—— (1937): An Experimental Illustration of the Spilite Reaction, *Comm. Géol. Finlande Bull.*, **119,** pp. 61–68.

—— (1939): Die metamorphen Gesteine, *in* "Die Entstenhung der Gesteine," T. F. W. Barth, C. W. Correns, and P. Eskola, Springer, Berlin, pp. 263–407.

—— (1949): The Problem of Mantled Gneiss Domes, *Geol. Soc. Lond. Quart. J.*, **104,** pp. 461–476.

—— (1952): On the Granulites of Lapland, *Am. J. Sci.*, Bowen vol., **250A,** pp. 133–171.

—— (1954): Ein Lamprophyrgang in Helsinki und die Lamprophyrprobleme, *Tschermaks Mineral. Petrog. Mitt.*, bd. **4,** pp. 329–337.

Espenshade, G. H., and E. L. Boudette (1964): Geology of the Greenville Quadrangle, Maine, *U.S. Geol. Surv., Geol. Quad Map* GQ-330.

—— and D. B. Potter (1960): Kyanite, Sillimanite, and Andalusite Deposits of the Southeastern States, *U.S. Geol. Surv. Prof. Paper* **336,** 121 pp.

Essene, E. J., and W. S. Fyfe (1967): Omphacite in Californian Metamorphic Rocks, *Contr. Mineral. Petrol.*, **15,** pp. 1–23.

——, ——, and F. J. Turner (1965): Petrogenesis of Franciscan Glaucophane Schists and Associated Metamorphic Rocks, California, *Beitr. Mineral. Petrog.*, **11,** pp. 695–704.

—— and N. G. Ware (1970): The Low Temperature Xenolithic Origin of Eclogites in Diatremes, N. E. Arizona, *Geol. Soc. Am. Abst.* with *Prog.*, **2,** pp. 547–548.

Eugster, H. P. (1956): Muscovite-Paragonite Join and Its Use as a Geologic Thermometer, *Geol. Soc. Am. Bull.*, **67,** p. 1693.

—— (1959): Reduction and Oxidation in Metamorphism, *in* "Researches in Geochemistry," Wiley, New York, pp. 397–426.

—— and D. R. Wones (1958): Phase Relations of Hydrous Silicates with Intermediate Mg/Fe Ratios, *Carnegie Inst. Wash., Ann. Rept. Dir. Geophys. Lab.*, 1957–1958, **57,** p. 193.

—— and —— (1962): Stability Relations of the Ferruginous Biotite, Annite, *J. Petrol.*, **3,** pp. 82–125.

Evans, A. M. (1966): The Development of Lit-par-lit Gneiss at the Bicroft Uranium Mine, Ontario, *Can. Mineralogist*, **8,** pp. 593–609.

Evans, B. W. (1964a): Coexisting Albite and Oligoclase in Some Schists from New Zealand, *Am. Mineralogist*, **49,** pp. 173–179.

—— (1964b): Fractionation of Elements in the Pelitic Hornfelses of the Cashell-Lough Wheelaun Intrusion, Connemara, Eire, *Geochim. Cosmochim. Acta*, **28,** pp. 127–156.

——— (1965a): Pyrope Garnet–Piezometer or Thermometer?, *Geol. Soc. Am. Bull.*, **76**, pp. 1295–1300.

——— (1965b): Application of a Reaction-rate Method to the Breakdown Equilibria of Muscovite and Muscovite Plus Quartz, *Am. J. Sci.*, **263**, pp. 647–667.

——— (1968): Mineralogy as a Function of Depth in the Prehistoric Makaopuhi Tholeiitic Lava Lake, Hawaii, *Contrib. Mineral. Petrol.*, **17**, pp. 85–115.

——— and C. V. Guidotti (1966): The Sillimanite-Potash Feldspar Isograd in Western Maine, U.S.A., *Contr. Mineral. Petrol.*, **12**, pp. 25–62.

——— and B. E. Leake (1960): The Composition and Origin of the Striped Amphibolites of Connemara, Ireland, *J. Petrol.*, **1**, pp. 337–363.

Evernden J. F., and G. H. Curtis (1965): The Potassium-argon Dating of Late Cenozoic Rocks in East Africa and Italy, *Current Anthro.*, **6**, pp. 343–385.

———, ———, R. W. Kistler, and J. Obradovich (1960): Argon Diffusion in Glauconite, Microcline, Sanidine, Leucite, and Phlogopite, *Am. J. Sci.*, **258**, pp. 583–604.

——— and R. W. Kistler (1970): Chronology of Emplacement of Mesozoic Batholithic Complex in California and Western Nevada, *U.S. Geol. Surv. Prof. Paper* **623**, 42 pp.

Ewart, A. (1965): Mineralogy and Petrogenesis of the Whakamaru Ignimbrite in the Maraetai Area of the Taupo Volcanic Zone, New Zealand, *New Zealand J. Geol. Geophys.*, **8**, pp. 611–677.

——— and J. J. Stipp (1968): Petrogenesis of the Volcanic Rocks of the Central North Island as Indicated by a Study of Sr^{87}/Sr^{86} Ratios, and Sr, Rb, K, U, and Th Abundances, *Geochim. Cosmochim. Acta*, **32**, pp. 699–736.

Ewing, J., and M. Ewing (1967): Sediment Distribution on the Mid-ocean Ridges with Respect to Spreading of the Sea Floor, *Science,* **156**, pp. 1590–1592.

Fahrig, W. F. (1962): Petrology and Geochemistry of the Griffis Lake Ultrabasic Sill of the Central Labrador Trough, Quebec, *Geol. Surv. Can. Bull.*, **77**, pp. 36–37.

Fairbairn, H. W. (1934): Spilite and the Average Metabasalt, *Am. J. Sci.*, ser. 5, **27**, pp. 92–97.

———, G. Faure, W. H. Pinson, P. M. Hurley, and J. L. Powell (1963): Initial Ratio of Strontium-87 to Strontium-86, Whole Rock Age, and Discordant Biotite in the Monteregian Igneous Province, Quebec, *J. Geophys. Res.*, **68**, pp. 6515–6522.

———, P. M. Hurley, and W. H. Pinson (1964): Initial Sr^{87}/Sr^{86} and Possible Sources of Granitic Rocks in Southern British Columbia, *J. Geophys. Res.*, **69**, pp. 4889–4893.

Faul, H., editor (1954): "Nuclear Geology: A Symposium on Nuclear Phenomena in the Earth Sciences," Wiley, New York, 414 pp.

——— (1963): Age and Extent of the Hercynian Complex, *Geol. Rundsch.*, **52**, pp. 767–781.

———, T. W. Stern, H. H. Thomas, and P. L. D. Elmore (1963): Ages of Intrusion and Metamorphism in the Northern Appalachians, *Am. J. Sci.*, **261**, pp. 1–19.

Faure, G., and P. M. Hurley (1963): Isotopic Composition of Sr in Oceanic and Continental Basalts, *J. Petrol.*, **4**, pp. 31–50.

Fawcett, J. J. (1964): The Muscovite-Chlorite-Quartz Assemblage, *Carnegie Inst. Wash. Year-book,* **63,** pp. 137–141.

—— and H. S. Yoder, Jr. (1966): Phase Relationships of Chlorites in the System MgO–Al_2O_3–SiO_2–H_2O, *Am. Mineralogist,* **51,** pp. 353–380.

Fawley, A. P., and T. C. James (1955): A Pyrochlore (Columbium) Carbonatite, Southern Tanganyika, *Econ. Geol.,* **50,** pp. 571–585.

Fenner, C. N. (1923): The Origin and Mode of Emplacement of the Great Tuff Deposit in the Valley of Ten Thousand Smokes, *Nat. Geog. Soc. Tech. Papers,* Katmai ser., **no. 1.**

—— (1926): The Katmai Magmatic Province, *J. Geol.,* **34,** pp. 673–772.

—— (1929): The Crystallization of Basalts, *Am. J. Sci.,* **18,** pp. 225–253.

Ferguson, J. (1970): The Differentiation of Agpaitic Magmas: The Ilimaussaq Intrusion, South Greenland, *Can. Mineral.,* **10,** pp. 335–349.

Fick, L. J., and C. van der Heyde (1959): Additional Data on the Geology of the Mbeya Carbonatite, *Econ. Geol.,* **54,** pp. 842–872.

Findlay, A. (1951): "The Phase Rule and its Applications," Dover, New York.

Findlay, D. C., and C. H. Smith (1965): The Muskox Drilling Project, *Geol. Surv. Can. Paper* **64–44.**

Fisher, D. J., A. J. Fruch, Jr., C. S. Hurlbut, Jr., and C. E. Tilley, editors (1963): Symposium on Layered Intrusions, *Mineral. Soc. Am. Spec. Paper* **1,** pp. 1–134.

Fisher, G. W. (1967): The Effect of Variable Oxygen Activity on Isograd Reactions in Pelitic Rocks, *Carnegie Inst. Wash. Yearbook,* **65,** 1965–1966, pp. 279–283.

—— (1970): The Application of Ionic Equilibria to Metamorphic Differentiation: An Example, *Contrib. Mineral. Petrol.,* **29,** pp. 91–103.

Fisher, J. R., and E-an Zen (1971): Thermochemical Calculations from Hydrothermal Phase Equilibria Data and the Free Energy of H_2O, *Am. J. Sci.,* **270,** pp. 297–314.

Fisher, R. L., and C. G. Engel (1969): Ultramafic and Basaltic Rocks Dredged from the Nearshore Flank of the Tonga Trench, *Geol. Soc. Am. Bull.,* **80,** pp. 1373–1378.

Fisher, R. V. (1966): Geology of a Miocene Ignimbrite Layer, John Day Formation, Eastern Oregon, *Calif. Univ. Publ. in Geol. Sci.,* **67,** 58 pp.

Fiske, R. S., C. A. Hopson, and A. C. Waters (1963): Geology of the Mount Rainier National Park, Washington, *U.S. Geol. Surv. Prof. Paper* **444,** 93 pp.

Flawn, P. T. and P. B. King (1953): Geology and Mineral Deposits of Pre-cambrian Rocks of the Van Horn Area, Texas, *Univ. Texas Publ.* **no. 5301,** 218 pp.

Fleischer, R. L., and P. B. Price (1964a): Glass Dating by Fission Fragment Tracks, *J. Geophys. Res.,* **69,** pp. 331–339.

—— and —— (1964b): Techniques for Geological Dating of Minerals by Chemical Etching of Fission Fragment Tracks, *Geochim. Cosmochim. Acta,* **28,** pp. 1705–1714.

——, ——, and R. M. Walker (1964): Fission Track Ages of Zircons, *J. Geophys. Res.,* **69,** pp. 4885–4888.

——, ——, and —— (1965): Ion Explosion Spike Mechanism for Formation of Charged Particle Tracks in Solids, *J. Appl. Phys.*, **36,** pp. 3645–3652.

Flett, J. S. (1946): Geology of the Lizard and Meneage, *Great Britain Geol. Surv. Mem.,* 208 pp.

Floyd, P. A. (1964): Progressive Desilication of Basic Hornfelses, *Nature,* **203,** no. 4944, pp. 510–511.

—— (1965): Metasomatic Hornfelses of the Land's End Aureole at Tater-du, Cornwall, *J. Petrol.,* **6,** pp. 223–245.

Fonteilles, M., and G. Guitard (1968): L'effet de socle dans les terrain metamorphiques autour des noyaux precambriens, *Intern. Geol. Congr.,* **23d,** Czech., 1968, sec. 4, pp. 9–25.

Forbes, R. B. (1965): The Comparative Chemical Composition of Eclogite and Basalt, *J. Geophys. Res.,* **70,** pp. 1515–1521.

——and H. Kuno (1967): Peridotite Inclusions and Basaltic Host Rocks, *in* "Ultramafic and Related Rocks," P. J. Wyllie, editor, Wiley, New York, pp. 328–337.

——, D. K. Ray, T. Katsura, H. Matsumoto, H. Haramura, and M. J. Furst (1969): The Comparative Chemical Composition of Continental Island Arc Andesites in Alaska, *in* "Proceedings of the Andesite Conference," A. R. McBirney, editor, *Oregon Dept. Geol. Mineral. Ind. Bull.,* **65,** pp. 111–120.

Fourmarier, P. (1953): Schistosite et phenomenes connexes dans le series plissees, *Congr. Geol. Intern., Compt. Rend.* **19** Algeria, 1952, sec. 3, f. 3, p. 117.

——(1959): Le granite et les deformations mineures des roches, *Mem. Acad. Roy. Belgique,* Cl. Sci., t. **31,** fasc. 3, 191 pp.

Fowler, K. S. (1930): The Anorthosite Area of the Laramie Mountains, Wyoming, *Am. J. Sci.,* **19,** pp. 373–403.

Fox, K. F., Jr., and C. D. Rinehart (1971): Okanogan Gneiss Dome, North-central Washington, *in* "Metamorphism in the Canadian Cordillera," *Geol. Assoc. Can.,* Cordill. Sect. Prog. and Abst., p. 10.

Francis, G. H. (1956a): Facies Boundaries in Pelites in the Middle Grades of Metamorphism, *Geol. Mag.,* **93,** pp. 353–368.

—— (1956b): The Serpentine Mass in Glen Urquhart, Inverness-shire, Scotland, *Am. J. Sci.,* **254,** pp. 201–226.

Franco, R. R., and J. F. Schairer (1951): Liquidus Temperatures in Mixtures of the Feldspars of Soda, Potash, and Lime, *J. Geol.,* **59,** pp. 259–267.

Franz, G. W., and P. J. Wyllie (1966): Melting Relationships in the System $CaO–MgO–SiO_2–H_2O$ at 1 Kilobar Pressure, *Geochim. Cosmochim. Acta,* **30,** pp. 9–22.

—— and —— (1967): Experimental Studies in the System $CaO–MgO–SiO_2–CO_2–H_2O$, *in* "Ultramafic and Related Rocks," P. J. Wyllie, editor, Wiley, New York, pp. 323–326.

Franzgrote, E. J., J. H. Patterson, A. L. Turkevich, T. E. Economou, and K. P. Sowinski (1970): Chemical Composition of the Lunar Surface in Sinus Medii, *Science,* **167,** pp. 376–379.

Freitas, R. O. de (1947): Geologia e Petrologia da Llha de São Sebastião, *Bol. F.F.C.L. Univ. São Paulo,* no. **85,** *Geologia,* no. 3.

French, B. M. (1971): Possible Relations between Meteorite Impact and Igneous Petrogenesis, as Indicated by the Sudbury Structure, Ontario, Canada, *Bull. Volcanol.,* **34,** pp. 466–517.

Friedman, I., W. Long, and R. L. Smith (1963): Viscosity and Water Content of Rhyolite Glass, *J. Geophys. Res.,* **68,** pp. 6523–6535.

Froese, E. (1970): Chemical Petrology of Some Pelitic Gneisses and Migmatites from the Thor-Odin Area, British Columbia, *Can. J. Earth Sci.,* **7,** pp. 164–175.

—— and H. G. F. Winkler (1966): The System $CaCO_3$–$SrCO_3$ at High Pressures and 500°C to 700°C, *Can. Mineralogist,* **8,** pp. 551–566.

Frondel, C. (1945): Secondary Dauphine Twinning in Quartz, *Am. Mineralogist,* **30,** pp. 447–460.

Fry, N., and W. S. Fyfe (1969): Eclogites and Water Pressure, *Contr. Mineral. Petrol.,* **24,** pp. 1–6.

Fryklund, V. C., Jr., R. S. Harner, and E. P. Kaiser (1954): Niobium (Columbium) and Titanium at Magnet Cove and Potash Sulphur Springs, Ark., *U.S. Geol. Surv. Bull.,* **1015-B,** pp. 23–56.

Fyfe, W. S. (1958): A Further Attempt to Determine the Vapor Pressure of Brucite, *Am. J. Sci.,* **256,** pp. 729–732.

—— (1960): Hydrothermal Synthesis and Determination of Equilibrium between Minerals in the Subliquidus Region, *J. Geol.,* **68,** pp. 553–566.

—— (1962): On the Relative Stability of Talc, Anthophyllite, and Enstatite, *Am. J. Sci.,* **260,** pp. 460–466.

—— (1964): "Geochemistry of Solids," McGraw-Hill Book Co., New York, 199 pp.

—— (1969a): Experimental Introduction of Excess Ar^{40} into a Granitic Melt, *Contr. Mineral. Petrol.,* **23,** pp. 189–193.

—— (1969b): Some Second Thoughts on Al_2O_3–SiO_2, *Am. J. Sci.,* **267,** pp. 291–296.

—— and L. H. Godwin (1962): Further Studies on the Approach to Equilibrium in the Simple Hydrate Systems MgO–H_2O and Al_2O_3–H_2O, *Am. J. Sci.,* **260,** pp. 289–293.

—— and F. J. Turner (1966): Reappraisal of the Concept of Metamorphic Facies, *Contr. Mineral. Petrol.,* **12,** pp. 354–364.

——, ——, and J. Verhoogen (1958): Metamorphic Reactions and Metamorphic Facies, *Geol. Soc. Am. Mem.* **73,** 259 pp.

——, ——, and —— (1961): Coupled Reactions in Metamorphism; A Correction, *Geol. Soc. Am. Bull.,* **72,** pp. 169–170.

—— and G. W. Valpy (1959): The Analcite-Jadeite Phase Boundary: Some Indirect Deductions, *Am. J. Sci.,* **257,** pp. 316–320.

Fyles, James T. (1970): Structure of the Shuswap Metamorphic Complex in the Jordan River Area, Northwest of Revelstoke, British Columbia, *Geol. Assoc. Can. Spec. Paper* **6,** pp. 87–98.

Fyson, W. K. (1970): Structural Relations in Metamorphic Rocks, Shuswap Lake Area, British Columbia, *Geol. Assoc. Can. Spec. Paper* **6,** pp. 107–122.

Gable, D. J., and P. K. Sims (1969): Geology and Regional Metamorphism of Some High-grade Cordierite Gneisses, Front Range, Colorado, *Geol. Soc. Amer. Spec. Paper* **128,** 87 pp.

Gabrielse, H., and J. E. Reesor (1964): Geochronology of Plutonic Rocks in Two Areas of the Canadian Cordillera, *Roy. Soc. Can. Spec. Publ.* **8,** pp. 96–128.

Gandhi, S. S. (1970): Petrology of the Monteregian Intrusions of Mount Yamaska, Quebec, *Can. Mineralogist,* **10,** pp. 452–484.

Ganguly, J. (1968): Analysis of the Stabilities of Chloritoid and Staurolite and Some Equilibria in the System $FeO-Al_2O_3-SiO_2-H_2O-O_2$, *Am. J. Sci.,* **266,** pp. 277–298.

——— (1969): Chloritoid Stability and Related Parageneses: Theory, Experiments, and Applications, *Am. J. Sci.,* **267,** pp. 910–944.

Garson, M. S. (1961): The Geology of the Namangali Vent, Mlanje District, *Nyasaland Geol. Surv. Rec. 1959,* **1,** pp. 51–62.

——— (1966): Carbonatites of Malawi, *in* "Carbonatites," O. F. Tuttle and J. Gittins, editors, Interscience a division of Wiley, New York, pp. 33–71.

Gaskell, T. F., editor (1967): "The Earth's Mantle," Academic Press, New York, 509 pp.

Gass, I. G. (1958): Ultrabasic Pillow Lavas from Cyprus, *Geol. Mag.,* **95,** pp. 241–251.

——— (1967): The Ultrabasic Volcanic Assemblage of the Troodos Massif, Cyprus, *in* "Ultramafic and Related Rocks," P. J. Wyllie, editor, Wiley, New York, pp. 121–134.

——— (1968): Is the Troodos Massif of Cyprus a Fragment of Mesozoic Ocean Floor?, *Nature,* **220,** pp. 39–42.

Gast, P. W. (1967): Isotopic Geochemistry of Volcanic Rocks, *in* "Basalts," H. H. Hess and A. Poldervaart, editors, Interscience, a division of Wiley, New York, vol. 1, pp. 325–358.

——— (1968): Trace Element Fractionation and the Origin of Tholeiitic and Alkaline Magma Types, *Geochim. Cosmochim. Acta,* **32,** pp. 1057–1086.

———, G. R. Tilton, and C. Hedge (1964): Isotopic Composition of Lead and Strontium from Ascension and Gough Islands, *Science,* **145,** pp. 1181–1185.

Gastil, R. G., M. DeLisle, and J. R. Morgan (1967): Some Effects of Progressive Metamorphism on Zircons, *Geol. Soc. Am. Bull.,* **78,** pp. 879–905.

Gates, R. M. (1967): Amphibolites–Syntectonic Intrusives?, *Am. J. Sci.,* **265,** pp. 118–131.

Gavrilova, S. P., and V. G. Khryukin (1964): Relationship between Alkalic and Granitic Rocks of Southeast Tuva, *Dokl. Akad. Nauk SSSR,* **154,** no. 1–6, pp. 66–68.

Gay, P., and J. V. Smith (1955): Phase Relations in the Plagioclase Feldspars: Composition Range An_0 to An_{70}, *Acta Cryst.,* **8,** pp. 64–65.

——— and I. D. Muir (1962): Investigation of the Feldspars of the Skaergaard Intrusion, Eastern Greenland, *J. Geol.,* **70,** pp. 565–581.

Gees, R. A. (1956): Ein Beitrag zum ophiolith-Problem behandelt an einigen Beispielen aus dem Gebiet von Klosters-Davos (Graubünden), *Schweiz. Mineral. Petrog. Mitt.,* **36,** pp. 454–488.

Geological Survey of Greenland (1970): Tectonic/Geological Map of Greenland, Scale 1:

2,500,000, *Geol. Surv. Greenland,* Østervoldgade 10, DK 1350 Copenhagen, Denmark.

Gerling, E. K., I. M. Morozova, and V. V. Kurabatov (1961): Retention of Radiogenic Argon in Powdered Potassium-bearing Minerals, *Geokhimiya,* no. 1, pp. 45–56.

Geze, B. (1962): Relations entre volcans et plutons dans la Montagne Noire, les Causses et le Bas-Languedoc (Sud de la France), *Bull. Volcanol.,* ser. 2, **24,** pp. 87–91.

Ghaly, T. S. (1969): Metamorphic Differentiation in Some Lewisian Rocks of North West Scotland, *Contr. Mineral. Petrol.,* **22,** pp. 276–289.

Ghent, E. D. (1965): Glaucophane-schist Facies Metamorphism in the Black Butte Area, Northern Coast Ranges, California, *Am. J. Sci.,* **263,** pp. 385–400.

Gibson, I. L. (1970): A Pantelleritic Welded Ash-flow Tuff from the Ethiopian Rift Valley, *Contr. Mineral. Petrol.,* **28,** pp. 89–111.

———and G. P. L. Walker (1964): Some Composite Rhyolite/Basalt Lavas and Related Composite Dykes in Eastern Iceland, *Proc. Geol. Assoc.,* **74,** pp. 301–318.

Gilbert, C. M. (1938): Welded Tuff in Eastern California, *Geol. Soc. Am. Bull.,* **49,** pp. 1829–1862.

Gilbert, G. K. (1880): Report on the Geology of the Henry Mountains, 2d ed., *U.S. Geog. Geol. Surv. Rocky Mount. Region,* 179 pp.

Gilbert, M. C., P. M. Bell, and S. W. Richardson (1969): The Andalusite-Sillimanite Transition and the Aluminum Silicate Triple Point, *Ann. Rept., Dir. Geophys. Lab.,* 1967–1968, pp. 135–137.

Giletti, B. J. (1971): Discordant Isotopic Ages and Excess Argon in Biotites, *Earth Planet. Sci. Letters,* **10,** pp. 157–164.

———and P. E. Damon (1961): Rubidium-Strontium Ages of Some Basement Rocks from Arizona and Northwestern Mexico, *Geol. Soc. Am. Bull.,* **72,** pp. 639–644.

———and J. L. Kulp (1955): Radon Leakage from Radioactive Minerals, *Am. Mineral.,* **40,** pp. 481–496.

Gilluly, J. (1935): Keratophyres of Eastern Oregon and the Spillite Problem, *Am. J. Sci.,* ser. 5, **29,** pp. 225–252.

———, editor (1948): Origin of Granite, *Geol. Soc. Am. Mem.* **28,** 139 pp.

———(1965): Volcanism, Tectonism, and Plutonism in the Western United States, *Geol. Soc. Am. Spec. Paper* **80,** 69 pp.

———(1969): Oceanic Sediment Volumes and Continental Drift, *Science,* **166,** pp. 992–994.

———(1971): Plate Tectonics and Magmatic Evolution, *Geol. Soc. Am. Bull.,* **82,** pp. 2383–2396.

Giraud, P. (1968): Les roches a caractere charnockitique de laserie d'In Quzzal en Ahaggar (Sahara central), *Intern. Geol. Congr.,* **22d.,** India, 1964, pt. 13, pp. 1–20.

Girifalco, L. A. (1964): "Atomic Migration in Crystals," Blaisdell, a division of Ginn, Waltham, Mass., 162 pp.

Gittins, J. (1961): Nephelinization in the Haliburton-Bancroft District, Ontario, *J. Geol.,* **69,** pp. 291–308.

———(1966a): Summaries and Bibliographies of Carbonatite Complexes, *in* "Carbonatites," O. F. Tuttle and J. Gittins, editors, Interscience, a division of Wiley, New York, pp. 417–541.

———(1966b): Russian Views on the Origin of Carbonatite Complexes, *in* "Carbonatites," O. F. Tuttle and J. Gittins, editors, Interscience, a division of Wiley, New York, pp. 379–382.

Gjelsvik, T. (1952): Metamorphosed Dolerites in the Gneiss Area of Sunnmore on the West Coast of Southern Norway, *Norsk. Geol. Tidsskrift,* **30,** pp. 33–134.

Goddard, E. N., chairman, North American Geologic Map Committee (1965): Geologic Map of North America, *U.S. Geol. Surv.,* Map scale 1:5,000,000.

Godovikov, A. A., and G. C. Kennedy (1968): Kyanite Eclogites, *Contrib. Mineral. Petrol.,* **19,** pp. 169–176.

Gold, D. P. (1966a): The Minerals of the Oka Carbonatite and Alkaline Complex, Oka., Quebec, I. M. A. Volume, Int. Mineralog. Assoc., 4th Gen. Mtg., New Delhi, 1964, Papers and Proc.: New Delhi, India, *Mineral. Soc. India,* pp. 109–125.

———(1966b): The Average and Typical Chemical Composition of Carbonatites, I. M. A. Volume, Int. Mineralog. Assoc., 4th Gen. Mtg., New Delhi, 1964, Papers and Proc.: New Delhi, India, *Mineral. Soc. India,* pp. 83–91.

———(1967a): Alkaline Ultrabasic Rocks in the Montreal Area, Quebec, *in* "Ultramafic and Related Rocks," P. J. Wyllie, editor, Wiley, New York, pp. 288–302.

———(1967b): Economic Geology and Geophysics of the Oka Alkaline Complex, Quebec, *Can. Mining Metal. Bull.,* **60,** no. 666, pp. 1131–1144.

———(1969): The Oka Carbonatite and Alkaline Complex, *Can. Mineralogist,* **10,** pp. 134–135.

Goldich, S. S. (1971): Lunar and Terrestrial Ilmenite Basalt, *Science,* **171,** pp. 1245–1246.

———, H. Baadsgaard, and A. O. Nier (1957): Investigations in Ar^{40}/K^{40} Dating, *Trans. Am. Geophys. Union,* **38,** pp. 547–551.

Goldschmidt, V. M. (1911): Die Kontaktmetamorphose im Kristianiagebiet, Oslo *Vidensk. Skr.,* I, *Math.-Nat. Kl.,* no. 11.

———(1937): The Principles of Distribution of Chemical Elements in Minerals and Rocks, *J. Chem. Soc.,* 1937, pp. 655–673.

———and C. L. Peters (1932): Zur Geochemie der Bors, *Gesell. Wiss. Gottingen Nachr.,* Mat-Phys. Kl., pp. 403–407, 528–545.

Goldsmith, J. R., and D. L. Graf (1958): Relation between the Lattice Constants and the Composition of the Ca-Mg Carbonates, *Am. Mineralogist,* **43,** pp. 84–101.

———and H. C. Heard (1961): Subsolidus Phase Relations in the System $CaCO_3$-$MgCO_3$, *J. Geol.,* **69,** pp. 45–74.

———and R. C. Newton (1969): P-T-X Relations in the System $CaCO_3$-$MgCO_3$ at High Temperatures and Pressures, *Am. J. Sci.,* **Schairer vol.,** pp. 160–190.

Goldsmith, R. (1959): Granofels, A New Metamorphic Rock Name, *J. Geol.*, **67,** pp. 109–110.

—— (1962): Geologic Map of New England, *U.S. Geol. Surv.*, open-file map.

Gomes, C. B. de, P. Santini, and C. V. Dutra (1964): Petrochemistry of a Precambrian Amphibolite from the Jaragua Area, São Paulo, Brazil, *J. Geol.*, **72,** pp. 664–680.

Goode, A. D. T., and G. W. Krieg (1967): The Geology of Ewarara Intrusion, Giles Complex, Central Australia, *J. Geol. Soc. Australia,* **14,** pp. 185–194.

Gorshkov, G. S. (1965): On the Relations of Volcanism and the Upper Mantle, *Bull. Volcanol.,* **23,** pp. 159–168.

—— (1969): Geophysics and Petrochemistry of Andesite Volcanism of the Circum-Pacific Belt, *in* "Proceedings of the Andesite Conference," A. R. McBirney, editor, *Oregon Dept. Geol. Mineral. Ind. Bull.,* **65,** pp. 91–98.

—— (1970): "Volcanism and the Upper Mantle," Plenum, New York, 385 pp.

Graf, D. F., and J. R. Goldsmith (1955): Dolomite-Magnesian Calcite Relations at Elevated Temperatures and CO_2 Pressures, *Geochim. Cosmochim. Acta,* **7,** pp. 109–128.

Grantham, D. R. (1928): Petrology of the Shap Granite, *Proc. Geol. Assoc.,* **39,** pp. 299–331.

Green, D. H. (1964): The Petrogenesis of the High-temperature Peridotite Intrusion in the Lizard Area, Cornwall, *J. Petrol.,* **5,** pp. 131–188.

—— (1970): Peridotite-Gabbro Complexes as Keys to Petrology of Midoceanic Ridges: Discussion, *Geol. Soc. Am. Bull.,* **81,** pp. 2161–2166.

—— and A. E. Ringwood (1967a): An Experimental Investigation of the Gabbro to Eclogite Transformation and its Petrological Applications, *Geochim. Cosmochim. Acta,* **31,** pp. 767–834.

—— and —— (1967b): The Genesis of Basaltic Magmas, *Contrib. Mineral. Petrol.,* **15,** pp. 103–190.

Green, J. C. (1963a): Alkali Metasomatism in a Thermal Gradient: Two Possible Examples, *J. Geol.,* **71,** pp. 653–657.

—— (1963b): High-level Metamorphism of Pelitic Rocks in Northern New Hampshire, *Am. Mineralogist,* **48,** pp. 991–1023.

Green, T. H. (1968): Experimental Fractional Crystallization of Quartz Diorite and its Application to the Problem of Anorthosite Origin, *N. Y. State Mus. Sci. Serv. Mem.* **18,** pp. 23–29.

—— (1969): High-pressure Experimental Studies on the Origin of Anorthosite, *Can. J. Earth Sci.,* **6,** pp. 427–440.

—— and A. E. Ringwood (1967): Crystallization of Basalt and Andesite under High Pressure Hydrous Conditions, *Earth Planet. Sci. Letters,* **3,** pp. 481–489.

—— and —— (1968): Genesis of the Calc-alkaline Igneous Rock Suite, *Contr. Mineral. Petrol.,* **18,** pp. 105–162.

—— and —— (1969): High Pressure Experimental Studies on the Origin of Andesites, *in* "Proceedings of the Andesite Conference," A. R. McBirney, editor, *Oregon Dept. Geol. Mineral. Ind. Bull.,* **65,** pp. 21–32.

Greenly, E. (1923): Further Researches on the Succession and Metamorphism in the Mona Complex of Anglesey, *Geol. Soc. Lond. Quart. J.,* **79,** pp. 334–351.

Greenwood, H. J. (1961): The System $NaAlSi_2O_6-H_2O$-argon: Total Pressure and Water Pressure in Metamorphism, *J. Geophys. Res.,* **66,** pp. 3923–3946.

—— (1962): Metamorphic Reactions Involving Two Volatile Components, *Ann. Rept. Geophys. Lab.,* **61,** pp. 82–85.

—— (1963): The Synthesis and Stability of Anthophyllite, *J. Petrol.,* **4,** pp. 317–351.

—— (1967): Wollastonite: Stability in H_2O-CO_2 Mixtures and Occurrence in a Contact-metamorphic Aureole near Salmo, British Columbia, Canada, *Am. Mineralogist,* **52,** pp. 1669–1680.

—— (1971): Mass Transport of Heat in Metamorphism, *in* "Metamorphism in the Canadian Cordillera," *Geol. Assoc. Can.,* Cordill. Sect. Prog. and Abst., p. 12.

—— and K. C. McTaggart (1957): Correlation of Zones in Plagioclase, *Am. J. Sci.,* **255,** pp. 656–666.

Gresens, R. L. (1966): The Effect of Structurally Produced Pressure Gradients on Diffusion in Rocks, *J. Geol.,* **74,** pp. 307–321.

—— (1967): Tectonic-hydrothermal Pegmatites, Part I: The Model, *Contr. Mineral. Petrol.,* **15,** pp. 345–355.

—— (1969): Blueschist Alteration during Serpentinization, *Contr. Mineral. Petrol.,* **24,** pp. 93–113.

—— (1970): Serpentinites, Blueschists, and Tectonic Continental Margins, *Geol. Soc. Am. Bull.,* **81,** pp. 307–310.

Griggs, D. (1939): A Theory of Mountain Building, *Am. J. Sci.,* **237,** pp. 611–650.

Grindley, G. W. (1958): The Geology of the Eglinton Valley, Southland, *Bull. Geol. Surv. New Zealand,* **58,** 68 pp.

Gross, E. B., and E. W. Heinrich (1966): Petrology and Mineralogy of the Mount Rosa Area, El Paso and Teller Counties, Colorado–Part 3, Lamprophyres and Mineral Deposits, *Am. Mineralogist,* **51,** pp. 1433–1442.

Grout, F. F. (1918a): The Lopolith; An Igneous Form Exemplified by the Duluth Gabbro, *Am. J. Sci.,* ser. 4, **46,** pp. 516–522.

—— (1918b): A Type of Igneous Differentiation, *J. Geol.,* **26,** pp. 626–658.

—— (1918c): Internal Structures of Igneous Rocks: Their Significance and Origin with Special Reference to the Duluth Gabbro, *J. Geol.,* **26,** pp. 439–458.

—— (1933): Contact Metamorphism of the Slates of Minnesota by Granite and Gabbro Magmas, *Geol. Soc. Am. Bull.,* **44,** pp. 989–1040.

Groves, A. W. (1935): The Charnockite Series of Uganda, British East Africa, *Geol. Soc. Lond. Quart. J.,* **91,** pp. 150–207.

—— (1951): Discussion of Paper by J. Sutton and J. Watson: The Pre-Torridonian Meta-morphic History of the Loch Torridon and Scourie Areas in the North-west Highlands and its Bearing on the Chronological Classification of the Lewisian, *Geol. Soc. Lond. Quart. J.,* **106,** p. 299.

Grummer, W. K., and S. V. Burr (1943): The Nephelinized Paragneisses of the Bancroft · Region, Ontario, *Science,* **97,** pp. 286–287.

—— and —— (1946): Nephelinized Paragneisses in the Bancroft Area, Ontario, *J. Geol.,* **54,** pp. 137–168.

Guest, N. J. (1963): Description of Exhibit of Fresh "Natro-carbonatite" from Oldoinyo Lengai, Tanganyika, *Geol. Soc. Lond. Proc.,* no. 1606, pp. 54–57.

Guidotti, C. V. (1963): Metamorphism of Pelitic Schists in the Bryant Pond Quadrangle, Maine, *Am. Mineralogist,* **48,** pp. 772–791.

—— (1969): A Comment on "Chemical Study of Minerals From the Moine Schists of the Ardnamurchan Area, Argyllshire, Scotland," by B. C. M. Butler, and Its Implications for the Phengite Problem, *J. Petrol.,* **10,** pp. 164–170.

Gunn, B. M. (1962): Differentiation in Ferrar Dolerites, Antarctica, *New Zealand J. Geol. Geophys.,* **5,** pp. 820–863.

Gurney, J. J., G. W. Berg, and L. H. Ahrens (1966): Observations on Caesium Enrichment and the Potassium/Rubidium/Caesium Relationship in Eclogites from the Roberts Victor Mine, South Africa, *Nature,* **210,** pp. 1025–1027.

Hagner, A. F., and L. G. Collins (1967): Magnetite Ore Formed during Regional Metamorphism, Ausable Magnetite District, New York, *Econ. Geol.,* **62,** pp. 1034–1071.

Hall, A. L. (1932): The Bushveld Igneous Complex of the Central Transvaal, *Geol. Surv., S. Africa Mem.* **28.**

Haller, J. (1956a): Die Strukturelemente Ostgrønlands zwischen 74° und 78°N, *Medd. Grœnland,* bd. **154,** no. 2, with English summary, pp. 22–24.

—— (1956b): Probleme der Tiefentektonik: Bauformen im Migmatit Stockwerk der ostgronlandischen Kaledoniden, *Geol. Rundsch.,* **45,** pp. 159–167.

—— (1958): Der "Zentrale Metamorphe Komplex" von NE-Grönland, Feil II, Die geologische Karte der staunings Aoper und des Forsblads Fjordes, *Medd. Grœnland,* bd. **154,** no. 3, with English summary, pp. 138–153.

—— (1962): Structural Control of Regional Metamorphism in the East Greenland Caledonides, *Geol. Soc. Lond. Proc.,* no. 1962 (1594) pp. 21–25, 30–33.

Hamilton, D. L., C. W. Burnham, and E. F. Osborn (1964): The Solubility of Water and Effect of Oxygen Fugacity and Water Content on Crystallization in Mafic Magmas, *J. Petrol.,* **5,** pp. 21–39.

Hamilton, E. I. (1963): The Isotopic Composition of Strontium in the Skaergaard Intrusion, *J. Petrol.,* **4,** pp. 383–391.

—— (1965): Distribution of Some Trace Elements and the Isotopic Composition of Strontium in Hawaiian Lavas, *Nature,* **206,** pp. 251–253.

—— (1966): The Isotopic Composition of Lead in Igneous Rocks, *Earth Planet. Sci. Letters,* **1,** pp. 30–37.

—— (1968): The Isotopic Composition of Strontium Applied to Problems of the Origin of the Alkaline Rocks, *in* "Radiometric Dating for Geologists," E. I. Hamilton and R. M. Farquhar, editors, Interscience, a division of Wiley, New York, pp. 437–463.

—— and L. H. Ahrens (1965): "Applied Geochronology," Academic Press, New York, 267 pp.

—— and T. Deans (1963): Isotopic Composition of Strontium in Some African Carbonatites and Limestones and in Strontium Minerals, *Nature,* **198,** pp. 776–777.

Hamilton, W. B. (1964): Origin of High-alumina Basalt, Andesite, and Dacite Magmas, *Science,* **146,** pp. 635–637.

——— (1965a): Geology and Petrogenesis of the Island Park Caldera of Rhyolite and Basalt, Eastern Idaho, *U. S. Geol. Surv. Prof. Paper* **504-C,** pp. C1–C37.

——— (1965b): Diabase Sheets of the Taylor Glacier Region, Antarctica, *U. S. Geol. Surv. Prof. Paper* **456-B,** 71 pp.

——— (1966): Origin of the Volcanic Rocks of Eugeosynclines and Island Arcs, *Geol. Surv. Can. Paper* **66-15,** pp. 348–356.

——— (1969): Mesozoic California and the Underflow of Pacific Mantle, *Geol. Soc. Am. Bull.,* **80,** pp. 2409–2430.

——— and W. B. Meyers (1966): Cenozoic Tectonics of the Western United States, *Rev. Geophys.,* **4,** pp. 509–549.

——— and ——— (1967): The Nature of Batholiths, *U. S. Geol. Surv. Prof. Paper* **554-C,** pp. C1–C30.

——— and W. Mountjoy (1965): Alkali Content of Alpine Ultramafic Rocks, *Geochim. Cosmochim. Acta,* **29,** pp. 661–671.

Hanson, G. N., and P. W. Gast (1967): Kinetic Studies in Contact Metamorphic Zones, *Geochim. Cosmochim. Acta,* **31,** pp. 1119–1153.

Hapuarachchi, D. J. A. C. (1967): Hornblende-Granulite Subfacies Mineral Assemblages from Areas in Ceylon, *Geol. Mag.,* **104,** pp. 29–34.

——— (1968): Cordierite and Wollastonite-bearing Rocks of Southwestern Ceylon, *Geol. Mag.,* **105,** pp. 317–324.

Hargraves, R. B. (1962): Petrology of the Allard Lake Anorthosite Suite, Quebec, *Geol. Soc. Am.,* **Buddington vol.,** pp. 163–189.

Harker, A. (1904): The Tertiary Igneous Rocks of Skye, *Mem. Geol. Surv. Scotland.*

——— (1908): The Geology of the Small Isles of Inverness-shire (Sheet 60), *Mem. Geol. Surv. Scotland.*

——— (1909): "The Natural History of Igneous Rocks," Macmillan, New York.

Harker, R. I., and O. E. Tuttle (1953): Studies in the System CaO-MgO-CO$_2$, Part 1, *Am. J. Sci.,* **253,** pp. 209–224.

——— and ——— (1955): Studies in the System CaO-MgO-CO$_2$, Part 2: Limits of Solid Solution along the Binary Join, CaCO$_3$-MgCO$_3$, *Am. J. Sci.,* **253,** pp. 274–282.

——— and ——— (1956): Experimental Data on the P$_{CO_2}$-T Curve for the Reaction Calcite + Quartz = Wollastonite + CO$_2$, *Am. J. Sci.,* **254,** pp. 239–256.

Harper, C. T. (1964): Potassium-Argon Ages of Slates and Their Geological Significance, *Nature,* **203,** no. 4944, pp. 468–470.

——— (1967): On the Interpretation of Potassium-Argon Ages from Precambrian Shields and Phanerozoic Orogens, *Earth Planet. Sci. Letters,* **3,** pp. 128–132.

Harris, P. G. (1957): Zone Refining and the Origin of Potassic Basalts, *Geochim. Cosmochim. Acta,* **12,** pp. 195–208.

Harry, W. T. (1954): The Composite Granite Gneiss of Western Ardgour, *Geol. Soc. Lond. Quart. J.*, **109,** pp. 285–308.

—— (1961): Gneisses of the Kipawa District, Western Quebec, Grenville Sub-Province of Canadian Shield, *Geol. Surv. Can. Bull.*, **64,** 26 pp.

—— and T. C. R. Pulvertaft (1963): The Nunarssuit Intrusive Complex, South Greenland, *Medd. Grønland,* **169,** pp. 1–136.

Hart, S. R. (1963): Excess Argon in Pyroxenes, U. S. Natl. Acad. Sci., Natl. Res. Council, Publ. 1075, *Nucl. Sci. Ser. Rept.* **no. 38,** pp. 68–69.

—— (1964): The Petrology and Isotopic-mineral Age Relations of a Contact Zone in the Front Range, Colorado, *J. Geol.*, **72,** pp. 493–525.

—— and G. L. Davis (1969): Zircon U-Pb and Whole-rock Rb-Sr Ages and Early Crustal Development near Rainy Lake, Ontario, *Geol. Soc. Am. Bull.*, **80,** pp. 595–616.

——, ——, R. H. Steiger, and G. R. Tilton (1968): A Comparison of the Isotopic Mineral Age Variations and Petrologic Changes Induced by Contact Metamorphism, *in* "Radiometric Dating for Geologists," E. Hamilton and R. Farquhar, editors, Wiley, New York, pp. 73–110.

—— and R. T. Dodd (1962): Excess Radiogenic Argon in Pyroxenes, *J. Geophys. Res.*, **67,** no. 7, pp. 2998–2999.

——, B. M. Gunn, and N. D. Watkins (1971): Intra-lava Variation of Alkali Elements in Icelandic Basalt, *Am. J. Sci.*, **270,** pp. 315–318.

Harwood, D. S., and R. R. Larson (1969): Variations in the Delta Index of Cordierite around the Cupsuptic Pluton, West-central Maine, *Am. Mineralogist,* **54,** pp. 896–908.

Hashimoto, M. (1966): On the Prehnite-Pumpellyite Metagraywacke Facies [in Japanese with English abs.], *J. Geol. Soc. Japan,* **72,** no. 5, pp. 253–265.

Hatch, F. H., A. K. Wells, and M. K. Wells (1949): "Textbook of Petrology," vol. I: Igneous Rocks, Murby, London, 469 pp.

Hawkins, J. W., Jr. (1967): Prehnite-Pumpellyite Facies Metamorphism of a Graywacke-shale Series, Mount Olympus, Washington, *Am. J. Sci.,* **265,** pp. 798–818.

—— (1968): Regional Metamorphism, Metasomatism, and Partial Fusion in the Northwestern Part of the Okanogan Range, Washington, *Geol. Soc. Am. Bull.*, **79,** pp. 1785–1820.

—— (1970): Petrology and Possible Tectonic Significance of Late Cenozoic Volcanic Rocks, Southern California and Baja California, *Geol. Soc. Am. Bull.*, **81,** pp. 3323–3338.

Hay, R. L. (1966): Zeolites and Zeolitic Reactions in Sedimentary Rocks, *Geol. Soc. Am. Spec. Paper* **85,** 130 pp.

Hayama, Y. (1959): Some Considerations on the Color of Biotite and its Relation to Metamorphism, *J. Geol. Soc. Japan,* **65,** pp. 21–30.

Heald, M. T. (1950): Structure and Petrology of the Lovewell Mountain Quadrangle, New Hampshire, *Geol. Soc. Am. Bull.*, **61,** pp. 43–89.

—— (1956): Cementation of Triassic Arkoses in Connecticut and Massachusetts, *Geol. Soc. Am. Bull.*, **67,** 1133–1154.

Healy, J. H., and D. H. Warren (1969): Explosion Seismic Studies in North America, *in* "The

Earth's Crust and Upper Mantle," P. J. Hart, editor, *Am. Geophys. Union, Geophys. Monograph* **13,** pp. 208–220.

Heard, H. C. (1963): Effect of Large Changes of Strain Rate in the Experimental Deformation of Yule Marble, *J. Geol.,* **71,** pp. 162–195.

——— and N. L. Carter (1968): Experimentally Induced "Natural" Intergranular Flow in Quartz and Quartzite, *Am. J. Sci.,* **266,** pp. 1–42.

Heath, S. A., and H. W. Fairbairn (1968): Sr^{87}/Sr^{86} Ratios in Anorthosites and Some Associated Rocks, *N. Y. State Mus. Sci. Serv. Mem.* **18,** pp. 99–110.

Hedge, C. (1966): Variations in Radiogenic Strontium Found in Volcanic Rocks, *J. Geophys. Res.,* **71,** no. 24, pp. 6119–6126.

———, R. A. Hildreth, and W. T. Henderson (1970): Strontium Isotopes in Some Cenozoic Lavas from Oregon and Washington, *Earth Planet. Sci. Letters,* **8,** pp. 434–438.

Hedge, C. E., and F. Walthall (1963): Radiogenic Sr^{87} as an Index of Geologic Processes, *Science,* **140,** pp. 1214–1217.

Heezen, B. C., and M. Ewing (1963): The Mid-oceanic Ridge, *in* "The Sea," **3,** Interscience, a division of Wiley, New York, pp. 388–410.

Heier, K. S. (1957): Phase Relations of Potash Feldspar in Metamorphism, *J. Geol.,* **65,** pp. 468–479.

——— (1960): Petrology and Geochemistry of High-grade Metamorphic and Igneous Rocks on Langøy, Northern Norway, *Norges Geol. Undersøkelse,* nr. **207,** 246 pp.

——— (1961): Layered Gabbro, Hornblendite, Carbonatite and Nepheline Syenite on Stjernøy, North Norway, *Norsk. Geol. Tidsskr.,* **41,** pp. 109–155.

——— (1962): The Possible Origins of Amphibolites in an Area of High Metamorphic Grade, *Norsk. Geol. Tidsskr.,* **42,** pp. 157–165.

——— (1964a): Geochemistry of the Nepheline Syenite on Stjernøy, North Norway, *Norsk. Geol. Tidsskr.,* **44,** pp. 204–215.

——— (1964b): Rubidium/Strontium and Strontium-87/Strontium-86 Ratios in Deep Crustal Material, *Nature,* **202,** pp. 477–478.

———, W. Compston, and I. McDougall (1965): Thorium and Uranium Concentrations, and the Isotopic Composition of Strontium in the Differentiated Tasmanian Dolerites, *Geochim. Cosmochim. Acta,* **29,** pp. 643–659.

Heimlich, R. A. (1965): Petrology of the Flora Lake Stock, Lake of the Woods Region, Ontario, Canada, *Geol. Soc. Am. Bull.,* **76,** pp. 1–26.

——— (1969): Reconnaissance Petrology of Precambrian Rocks in the Bighorn Mountains, Wyoming, *Contrib. Geol.,* **8,** pp. 47–61.

Heinrich, E. W. (1960): Pre-Beltian Geology of the Cherry Creek and Ruby Mountains Areas, Southwestern Montana, Part 2: Geology of the Ruby Mountains, *Montana Bur. Mines Geol. Mem.* **38,** pp. 15–40.

——— (1967): Carbonatites–Nil-silicate Igneous Rocks, *Earth Sci. Rev.,* **3,** pp. 203–210.

——— (1970): The Palabora Carbonatitic Complex–A Unique Copper Deposit, *Can. Mineralogist,* **10,** pp. 585–598.

—— and D. G. Moore, Jr. (1970): Metasomatic Potash Feldspar Rocks Associated with Igneous Alkalic Complexes, *Can. Mineralogist,* **10,** pp. 571–584.

Heirtzler, J. R. (1968): Sea-floor Spreading, *Sci. Am.,* **219,** pp. 60–70.

—— (1969): Geomagnetic Studies in the Atlantic Ocean, *in* "The Earth's Crust and Upper Mantle," P. J. Hart, editor, *Am. Geophys. Union, Geophys. Monograph* **13,** pp. 431–436.

Hekinian, R. (1968): Rocks from the Mid-oceanic Ridge in the Indian Ocean, *Deep-Sea Res.,* **15,** pp. 195–213.

Henson, F. R. S., R. V. Browne, and J. McGinty (1949): A Synopsis of the Stratigraphy and Geological History of Cyprus, *Geol. Soc. Lond. Quart. J.,* **105,** pp. 1–25.

Hepworth, J. V. (1968): The Charnockites of Southern West Nile, Uganda, and Their Parageneses, *Intern. Geol. Congr.,* **22d,** India, 1964, pt. 13, pp. 168–184.

Hermes, J. J. (1968): The Papuan Geosyncline and the Concept of Geosynclines, *Geol. Mijnbouw,* **47,** pp. 81–97.

Herz, N. (1968): The Roseland Alkalic Anorthosite Massif, Virginia, *N. Y. State Mus. Sci. Serv. Mem.* **18,** pp. 357–367.

—— (1969): Anorthosite Belts, Continental Drift, and the Anorthosite Event, *Science,* **164,** pp. 944–947.

Herzen, R. P. von, and W. H. K. Lee (1969): Heat Flow in Oceanic Regions, *in* "The Earth's Crust and Upper Mantle," P. J. Hart, editor, *Am. Geophys. Union, Geophys. Monograph* **13,** pp. 88–95.

—— and S. Uyeda (1963): Heat Flow through the Eastern Pacific Ocean Floor, *J. Geophys. Res.,* **68,** pp. 4219–4250.

Hess, H. H. (1933): The Problem of Serpentinization and the Origin of Certain Chrysotile Asbestos, Talc and Soapstone Deposits, *Econ. Geol.,* **28,** pp. 634–657.

—— (1938a): Gravity Anomalies and Island Arc Structure with Particular Reference to the West Indies, *Am. Phil. Soc. Proc.,* **79,** pp. 71–96.

—— (1938b): A Primary Peridotite Magma, *Am. J. Sci.,* **35,** pp. 321–344.

—— (1941): Pyroxenes of Common Mafic Magmas, Part II, *Am. Mineralogist,* **26,** pp. 573–594.

—— (1950): Vertical Mineral Variation in the Great Dyke of Southern Rhodesia, *Trans. Geol. Soc. S. Africa,* **53,** pp. 159–166.

—— (1954): Geological Hypotheses and the Earth's Crust Under the Oceans, *Proc. Roy. Soc. Lond.,* ser. A, **222,** p. 341.

—— (1955): Serpentines, Orogeny and Epeirogeny, *Geol. Soc. Am. Spec. Paper* **62,** pp. 391–408.

—— (1956): Rock Magnetism and the Differentiation of Dolerite Sills: A Discussion of J. C. Jaeger and G. Joplin (1955) and F. Walker (1956), *Am. J. Sci.,* **254,** pp. 446–451.

—— (1960a): Stillwater Igneous Complex, Montana, A Quantitative Mineralogical Study, *Geol. Soc. Am. Mem.* **80,** 225 pp.

—— (1960b): Caribbean Research Project: Progress Report, *Geol. Soc. Am. Bull.,* **71,** pp. 235–240.

—— (1962): History of Ocean Basins, *in* "Petrologic Studies," A volume to honor A. F. Buddington; A. E. J. Engel, H. L. James, and B. F. Leonard, editors, *Geol. Soc. Am.,* **Buddington vol.,** pp. 599–620.

—— (1964): The Oceanic Crust, The Upper Mantle and the Mayaguez Serpentinized Peridotite, *in* "A Study of Serpentinite," C. A. Burk, editor, *Nat. Acad. Sci.–Nat. Res. Coun. Publ.* **1188,** pp. 169–175.

—— (1965): Mid-oceanic Ridges and Tectonics of the Sea Floor, *in* "Submarine Geology and Geophysics"–Colston Papers no. 17, W. F. Whittard and R. Bradshaw, editors, Butterworth, London, pp. 317–332.

—— (1966a): Comments on the Pacific Basin, *in* "Continental Margins and Island Arcs," W. H. Poole, editor, *Geol. Surv. Can. Paper* **66–15,** pp. 311–312.

—— (1966b): Caribbean Research Project, 1965, and Bathymetric Chart, *in* "Caribbean Geological Investigations," *Geol. Soc. Am. Mem.* **98,** pp. 1–10.

—— and G. Otalora (1964): Mineralogical and Chemical Composition of the Mayaguez Serpentinite Cores, *in* "A Study of Serpentinite," C. A. Burk, editor, *Nat. Acad. Sci.–Nat. Res. Coun. Publ.* **1188,** pp. 152–168.

Hewitt, D. F. (1957): Geology of Cardiff and Faraday Townships, *Ontario Dept. Mines Ann. Rept.,* **66,** pt. 3.

—— (1960): Nepheline Syenite Deposits of Southern Ontario, *Ontario Dept. Mines Ann. Rept.,* **69,** pt. 8.

Hibbard, M. J. (1965): Origin of Some Alkali Feldspar Phenocrysts and Their Bearing on Petrogenesis, *Am. J. Sci.,* **263,** pp. 245–261.

Hietanen, A. (1947): Archean Geology of the Turku District in Southwestern Finland, *Geol. Soc. Am. Bull.,* **58,** pp. 1019–1084.

—— (1951): Metamorphic and Igneous Rocks of the Merrimac Area, Plumas National Forest, California, *Geol. Soc. Am. Bull.,* **62,** pp. 565–608.

—— (1956): Kyanite, Andalusite, and Sillimanite in the Schist in Boehls Butte Quadrangle, Idaho, *Am. Mineralogist,* **41,** pp. 1–27.

—— (1961): Metamorphic Facies and Style of Folding in the Belt Series Northwest of the Idaho Batholith, *Bull. Comm. Géol. Finlande,* **196,** pp. 73–103.

—— (1962): Metasomatic Metamorphism in Western Clearwater County, Idaho, *U.S. Geol. Surv. Prof. Paper* **344-A,** 116 pp.

—— (1963a): Anorthosite and Assoc. Rocks in the Boehls Butte Quadrangle and Vicinity, Idaho, *U. S. Geol. Surv. Prof. Paper* **344-B,** 78 pp.

—— (1963b): Idaho Batholith near Pierce and Bungalow, Clearwater County, Idaho, *U. S. Geol. Surv. Prof. Paper* **344-D,** 42 pp.

—— (1963c): Metamorphism of the Belt Series in the Elk River-Clarkia Area, Idaho, *U.S. Geol. Surv. Prof. Paper* **344-C,** 49 pp.

—— (1967): Scapolite in the Belt Series in the St. Joe-Clearwater Region, Idaho, *Geol. Soc. Am. Spec. Paper* **86,** 54 pp.

—— (1968): Metamorphic Environment of Anorthosite in the Boehls Butte Area, Idaho, *N. Y. State Mus. Sci. Serv. Mem.* **18,** pp. 371–386.

—— (1969): Distribution of Fe and Mg between Garnet, Staurolite, and Biotite in Aluminum-rich Schist in Various Metamorphic Zones North of the Idaho Batholith, *Am. J. Sci.,* **267,** pp. 422–456.

Higazy, R. A. (1954): Trace Elements of Volcanic Ultrabasic Potassic Rocks of Southwestern Uganda and Adjoining Part of the Belgian Congo, *Geol. Soc. Am. Bull.,* **65,** pp. 39–70.

Higgs, D. W. (1954): Anorthosite and Related Rocks of the San Gabriel Mountains, Southern California, *Calif. Univ. Publ. Geol. Sci.,* **30,** no. 3, pp. 171–222.

Himmelberg, G. R., and W. C. Phinney (1967): Granulite-facies Metamorphism, Granite Falls-Montevideo Area, Minnesota, *J. Petrol.,* **8,** pp. 325–348.

Hoare, J. M., W. H. Condon, and W. W. Patton (1964): Occurrence and Origin of Laumontite in Cretaceous Sedimentary Rocks in Western Alaska, *U. S. Geol. Surv. Prof. Paper* **501-C,** pp. C74–C78.

Hoffer, A. (1966): Seismic Control of Layering in Intrusives, *Bull. Volcanol.,* **29,** pp. 817–821.

Hoffer, J. M. (1967): The Rock Creek Flow of the Columbia River Basalt, *Northwest Sci.,* **41,** pp. 23–31.

Hoffman, C. (1970): Die Glaukophangesteine, ihre stofflichen Aquivalente und Umwandlungsprodukte in Nordcalabrien (Suditalien), *Contr. Mineral. Petrol.,* **27,** pp. 283–320.

Holdaway, M. J. (1965): Basic Regional Metamorphic Rocks in Part of the Klamath Mountains, Northern California, *Am. Mineralogist,* **50,** p. 953.

—— (1966): Hydrothermal Stability of Clinozoisite Plus Quartz, *Am. J. Sci.,* **264,** pp. 643–667.

—— (1971): Stability of Andalusite and the Aluminum Silicate Phase Diagram, *Am. J. Sci.,* **271,** pp. 97–131.

Holgate, N. (1954): The Role of Liquid Immiscibility in Igneous Petrogenesis. *J. Geol.,* **62,** pp. 439–480.

Holland, T. H. (1893): The Petrology of Job Charnock's Tombstone, *J. Asiatic Soc. Bengal,* **62,** pt. 2, no. 3, pp. 162–164.

—— (1900): The Charnockite Series, A Group of Archean Hypersthenic Rocks in Peninsular India, *Geol. Surv. India Mem.* **28,** pt. 2, pp. 119–249.

Hollister, L. S. (1966): Garnet Zoning: An Interpretation Based on the Raleigh Fractionation Model, *Science,* **154,** pp. 1647–1651.

—— (1969a): Contact Metamorphism in the Kwoiek Area of British Columbia–An End Member of the Metamorphic Process, *Geol. Soc. Am. Bull.,* **80,** pp. 2465–2493.

—— (1969b): Metastable Paragenetic Sequence of Andalusite, Kyanite, and Sillimanite, Kwoiek Area, British Columbia, *Am. J. Sci.,* **267,** pp. 352–370.

Holm, J. L., and O. J. Kleppa (1966): The Thermodynamic Properties of the Aluminum Silicates, *Am. Mineralogist,* **51,** pp. 1608–1622.

Holmes, A. (1931): Radioactivity and Earth Movements, *Trans. Geol. Soc. Glasgow,* **18,** pp. 599–606.

—— (1932): The Origin of Igneous Rocks, *Geol. Mag.,* **69,** pp. 543–558.

―――― (1945): Leucitized Granite Xenoliths from the Potash-rich Lavas of Bunyaruguru, Southwest Uganda, *Am. J. Sci.*, **243A,** pp. 313–332.

―――― (1950): Petrogenesis of Katungite and its Associates, *Am. Mineralogist,* **35,** pp. 772–792.

―――― (1956): The Ejectamenta of Katwe Crater, Southwest Uganda, Verh. konink. Nederland Geol. Mijnbouw, Genootschap, *Geol. Serv.,* **16,** Brouwer vol., pp. 139–166.

―――― (1958): Spitskop Carbonatite, Eastern Transvaal, *Geol. Soc. Am. Bull.,* **69,** pp. 1525–1526.

―――― (1965): "Principles of Physical Geology," 2d ed., Ronald Press, New York.

―――― and H. F. Harwood (1932): Petrology of the Volcanic Fields East and Southeast of Ruwenzori, Uganda, *Geol. Soc. Lond. Quart. J.,* **88,** pp. 370–442.

―――― and ―――― (1937): The Petrology of the Volcanic Area of Bufumbira, *Uganda Geol. Surv. Mem.* **3,** pt. 2.

Holmquist, P. J. (1910): The Archean Geology of the Coast-regions of Stockholm, *Geol. Foren. Stockh. Forh.,* **34,** pp. 789–908.

―――― (1916): Swedish Archean Structures and Their Meaning, *Bull. Geol. Inst. Upsala,* **XV.**

―――― (1921): Typen und Nomenklatur der Adergesteine, *Geol. Foren. Stockh. Forh.,* **43,** pp. 612–631.

Holser, W. T. (1947): Metasomatic Processes, *Econ. Geol.,* **42,** pp. 384–395.

―――― (1950): Metamorphism and Associated Mineralization in the Philipsburg Region, Montana, *Geol. Soc. Am. Bull.,* **61,** pp. 1053–1090.

Hopgood, A. M. (1962): Radial Distribution of Soda in a Pillow of Spilitic Lava from the Franciscan, California, *Am. J. Sci.,* **260,** pp. 383–396.

Hopson, C. A. (1964): The Crystalline Rocks of Howard and Montgomery Counties, Maryland, *Maryland Geol. Surv.,* pp. 27–215.

Hori, F. (1964): On the Role of Water in Heat Transfer from a Cooling Magma, *Tokyo Univ., Coll. Gen. Educ. Sci. Paper* **14,** pp. 121–127.

Hoschek, G. (1967): Untersuchungen zum Stabilitatsbereich von chloritoid und Staurolith, *Contr. Mineral. Petrol.,* **14,** pp. 123–162.

Hostetler, P. B., R. G. Coleman, F. A. Mumpton, and B. Evans (1966): Brucite in Alpine Serpentines, *Am. Mineralogist,* **51,** pp. 75–98.

Hotz, P. E. (1953): Petrology of Granophyre in Diabase Near Dillsburg, Pennsylvania, *Geol. Soc. Am. Bull.,* **64,** pp. 675–704.

Houten, F. B. van (1971): Contact Metamorphic Mineral Assemblages, Late Triassic Newark Group, New Jersey, *Contr. Mineral. Petrol.,* **30,** pp. 1–14.

Howard, L. E., and J. H. Sass (1964): Terrestrial Heat Flow in Australia, *J. Geophys. Res.,* **69,** pp. 1617–1626.

Howie, R. A. (1955): The Geochemistry of the Charnockite Series of Madras, India, *Trans. Roy. Soc. Edinburgh,* **62,** pt. 3, pp. 725–768.

—— (1964): Some Orthopyroxenes from Scottish Metamorphic Rocks, *Mineral. Mag.*, **33,** pp. 903–911.

—— (1965): The Pyroxenes of Metamorphic Rocks, *in* "Controls of Metamorphism," W. S. Pitcher and G. W. Flinn, editors, Wiley, New York, pp. 319–326.

—— and A. P. Subramaniam (1957): The Paragenesis of Garnet in Charnockite, Enderbite, and Related Granulites, *Mineral. Mag.*, **31,** pp. 565–586.

Howland, A. L., J. W. Peoples, and E. Sampson (1936): The Stillwater Igneous Complex, *Montana Bur. Mines Geol. Misc. Contr.*, **7,** 15 pp.

Hriskevich, M. E. (1968): Petrology of the Nipissing Diabase Sill of the Cobalt Area, Ontario, Canada, *Geol. Soc. Am. Bull.*, **79,** pp. 1387–1404.

Hsu, K. J. (1955): Granulites and Mylonites of the Region about Cucamonga and San Antonio Canyons, San Gabriel Mountains, California, *Calif. Univ. Publ. Geol. Sci.*, **30,** no. 4, pp. 223–352.

Hsu, L. C. (1968): Selected Phase Relationships in the System Al-Mn-Fe-Si-O-H–A Model for Garnet Equilibria, *J. Petrol.*, **9,** pp. 40–83.

Huber, N. K., and C. D. Rinehart (1967): Cenozoic Volcanic Rocks of the Devils Postpile Quadrangle, Eastern Sierra Nevada, California, *U. S. Geol. Surv. Prof. Paper* **554-D,** pp. D2–D21.

Hughes, C. J. (1960): The Southern Mountains Igneous Complex, Isle of Rhum, *Geol. Soc. Lond. Quart. J.*, **116,** pp. 111–138.

—— (1970a): Lateral Cryptic Variation in the Great Dyke of Rhodesia, *Geol. Mag.*, **107,** pp. 319–325.

—— (1970b): The Significance of Biotite Selvedges in Migmatites (1970), *Geol. Mag.*, **107,** pp. 21–24.

Hunahashi, M., C. W. Kim, Y. Ohta, and T. Tsuchiya (1968): Co-existence of Plagioclases of Different Compositions in Some Plutonic and Metamorphic Rocks, *Lithos*, **1,** pp. 356–373.

Hurlbut, C. S., Jr., and D. T. Griggs (1939): Igneous Rocks of the Highwood Mountains, Montana, *Geol. Soc. Am. Bull.*, **50,** pp. 1043–1112.

Hurley, P. M. (1967): Rb[87] - Sr[87] Relationships in the Differentiation of the Mantle, *in* "Ultramafic and Related Rocks," Wiley, New York, pp. 372–375.

——, P. C. Bateman, H. W. Fairbairn, and W. H. Pinson, Jr. (1965): Investigation of Initial Sr[87]/Sr[86] Ratios in the Sierra Nevada Plutonic Province, *Geol. Soc. Am. Bull.*, **76,** pp. 165–174.

——, H. Hughes, G. Fauro, H. W. Fairbairn, and W. H. Pinson, Jr. (1962): Radiogenic Strontium-87 Model of Continent Formation, *J. Geophys. Res.*, **67,** pp. 5315–5334.

Hutchison, W. W. (1967): Prince Rupert and Skeena Map-area, British Columbia, Canada, *Geol. Surv. Paper* **66-33,** 27 pp.

—— (1970): Metamorphic Framework and Plutonic Styles in the Prince Rupert Region of the Central Coast Mountains, British Columbia, *Can. J. Earth Sci.*, **7,** pp. 376–405.

—— and J. A. Roddick (1971): Major Aspects of Metamorphism in the Coast Mountains, British Columbia, *in* "Metamorphism in the Canadian Cordillera," *Geol. Assoc. Can.,* Cordill. Sect. Prog. and Abst., p. 14.

Hyndman, D. W. (1968a): Mid-Mesozoic Multiphase Folding along the Border of the Shuswap Metamorphic Complex, *Geol. Soc. Am. Bull.,* **79,** pp. 575–588.

—— (1968b): Petrology and Structure of Nakusp Map Area, British Columbia, *Geol. Surv. Can. Bull.,* **161,** 95 pp.

—— (1969): The Development of Granitic Plutons through Anatexis in the Northern Cordillera, British Columbia, *Geol. Soc. Am. Spec. Paper* **121,** p. 146.

—— (1971): Environment and Conditions of Shuswap Metamorphism as Determined in the Nakusp Area of the Eastern Border Zone of the Complex, *in* "Metamorphism in the Canadian Cordillera," *Geol. Assoc. Can.,* Cordill. Sect. Prog. and Abst., pp. 15–20.

—— and D. Alt (1971): The Kyanite-Andalusite-Sillimanite Problem on Goat Mountain, Idaho, *Northwest Geol.,* **1,** pp. 42–46.

——, ——, and J. L. Nold (1967): Some Limits to Diffusion during Regional Metamorphism North of the Idaho Batholith, *Geol. Soc. Am. Spec. Paper* **151,** p. 106.

——, J. D. Obradovich, and R. Ehinger (1972): Potassium-argon Age Determinations of the Philipsburg Batholith, *Geol. Soc. Am. Bull.,* **83,** pp. 473–474.

Hyndman, R. D., and D. W. Hyndman (1968): Water Saturation and High Electrical Conductivity in the Lower Continental Crust, *Earth Planet. Sci. Letters,* **4,** pp. 427–432.

——, I. B. Lambert, H. S. Heier, J. C. Jaeger and A. E. Ringwood (1968): Heat Flow and Surface Radioactivity Measurements in the Precambrian Shield of Western Australia, *Phys. Earth Planet. Interiors,* **1,** pp. 129–135.

Iannello, P. (1971): The Bushveld Granites around Rooiberg, Transvaal, South Africa, *Geol. Rundsch.,* **60,** pp. 630–655.

Iddings, J. P. (1914): Some Examples of Magmatic Differentiation and Their Bearing on the Problem of Petrographical Provinces, *Intern. Geol. Congr.,* **12,** 1913, *Compt. Rend.,* pp. 209–228.

Iiyama, J. T. (1961): Étude préliminaire de la solubilité du basalte dans l'eau à haute température, *Soc. Franc. Min. Bull.,* **84,** pp. 128–130.

Ingerson, E. (1935): Layered Peridotitic Laccoliths of the Trout River Area, Newfoundland, *Am. J. Sci.,* **29,** pp. 422–440.

—— (1937): Layered Peridotitic Laccoliths of the Trout River Area, Newfoundland, a Reply, *Am. J. Sci.,* **33,** pp. 389–392.

International Association of Volcanology (1951–1965): Catalogue of the Active Volcanoes of the World Including Solfatara Fields, *Intern. Assoc. Volcanol.,* Instituto di Geologia Applicata, Roma (Italy), parts I-XVIII.

Irvine, T. N. (1967a): Chromian Spinel as a Petrogenetic Indicator, *Can. J. Earth Sci.,* **4,** pp. 71–103.

—— (1967b): The Ultramafic Rocks of the Muskox Intrusion, Northwest Territories,

Canada, *in* "Ultramafic and Related Rocks," P. J. Wyllie, editor, Wiley, New York, pp. 38–49.

——— (1970a): Geologic Age and Structural Relations of the Muskox Intrusion, *in* "Report of Activities April to October, 1969," *Geol. Surv. Can. Paper* **70-1,** pt. A, pp. 149–153.

——— (1970b): Heat Transfer during Solidification of Layered Intrusions, Part 1: Sheets and Sills, *Can. J. Earth Sci.,* **7,** pp. 1031–1061.

——— and C. H. Smith (1969): Primary Oxide Minerals in the Layered Series of the Muskox Intrusion, *in* "Magmatic Ore Deposits, a Symposium," H. D. B. Wilson, editor, *Econ. Geol. Monograph* **4,** pp. 76–94.

Isachsen, Y. W., and R. L. Moxham (1968): Chemical Variation in Plagioclase Megacrysts from Two Vertical Sections in the Main Adirondack Metanorthosite Massif, *N. Y. State Mus. Sci. Surv. Mem.* **18,** pp. 255–265.

Isacks, B., J. Oliver, and L. R. Sykes (1968): Seismology and the New Global Tectonics, *J. Geophys. Res.,* **73,** pp. 5855–5900.

———, ———, and J. Oliver (1969): Focal Mechanisms of Deep and Shallow Earthquakes in the Tanga-Kermadec Region and the Tectonics of Island Arcs, *Geol. Soc. Am. Bull.,* **80,** pp. 1143–1470.

Ito, K., and G. C. Kennedy (1968): Melting and Phase Relations in the Plane Tholeiite-Lherzolite-Nepheline Basanite to 40 Kilobars with Geological Implications, *Contr. Mineral. Petrol.,* **19,** pp. 177–211.

Iwasaki, M. (1963): Metamorphic Rocks of the Kotu-Bizan Area, Eastern Sikoku, *J. Fac. Sci., Univ. Tokyo,* sec. 11, **XV,** pp. 1–90.

Jackson, E. D. (1961): Primary Textures and Mineral Associations in the Ultramafic Zone of the Stillwater Complex, Montana, *U. S. Geol. Surv. Prof. Paper* **358,** 106 pp.

——— (1963): Symposium on Layered Intrusions: Stratigraphic and Lateral Variation of Chromite Composition in the Stillwater Complex, *Min. Soc. Am. Spec. Paper* **1,** pp. 46–54.

——— (1967): Ultramafic Cumulates in the Stillwater, Great Dyke, and Bushveld Intrusions, *in* "Ultramafic and Related Rocks," P. J. Wyllie, editor, Wiley, New York, pp. 20–38.

Jaeger, J. C. (1957): The Temperature in the Neighborhood of a Cooling Intrusive Sheet, *Am. J. Sci.,* **255,** pp. 306–318.

——— (1959): Temperatures Outside a Cooling Intrusive Sheet, *Am. J. Sci.,* **257,** pp. 44–54.

——— (1961): The Temperature in the Neighborhood of a Cooling Intrusive Sheet, *Am. J. Sci.,* **255,** pp. 306–318.

——— and G. A. Joplin (1955): Rock Magnetism and the Differentiation of a Dolerite Sill, *J. Geol. Soc. Australia,* **2,** pp. 1–19.

Jager, E. (1962): Rb-Sr Age Determinations on Micas and Total Rocks from the Alps, *J. Geophys. Res.,* **67,** pp. 5293–5306.

Jahns, R. H., and C. W. Burnham (1957): Preliminary Results from Experimental Melting and Crystallization of Harding Pegmatite (abs.), *Geol. Soc. Am. Bull.,* **68,** pp. 1751–1752.

———— and ———— (1969): Experimental Studies of Pegmatite Genesis: I. A Model for the Derivation and Crystallization of Granitic Pegmatites, *Econ. Geol.*, **64**, pp. 843–864.

James, O. B., and E. D. Jackson (1970): Petrology of the Apollo 11 Ilmenite Basalts, *J. Geophys. Res.*, **75**, pp. 5793–5824.

James, R. S., and D. L. Hamilton (1969): Phase Relations in the System $NaAlSi_3O_8 - KAlSi_3O_8 - CaAl_2Si_2O_8 - SiO_2$ at 1 Kilobar Water Vapour Pressure, *Contr. Mineral. Petrol.*, **21**, pp. 111–141.

Janse, A. J. A. (1969): Gross Brukkaros, a Probable Carbonatite Volcano in the Nama Plateau of Southwest Africa, *Geol. Soc. Am. Bull.*, **80**, pp. 573–586.

Jenness, S. E. (1966): The Anorthosite of Northern Cape Breton Island, Nova Scotia, a Petrological Enigma, *Geol. Surv. Can. Paper* **66-21**, 25 pp.

Johannes, W., and P. Metz (1968): Experimentelle bistimmungen von gleichgewichts-beziehungen im system $MgO - CO_2 - H_2O$, *Neues Jahrb. Mineral. Monatsh.*, **1**, pp. 15–26.

Johannsen, A. (1931): "A Descriptive Petrography of the Igneous Rocks," vol. 1, The University of Chicago Press, Chicago, 318 pp.

Johnson, M. R. W. (1962): Relations of Movement and Metamorphism in the Dalradians of Banffshire, *Geol. Soc. Edinburgh Trans.*, **19**, pp. 29–64.

———— (1963): Some Time Relations of Movement and Metamorphism in the Scottish Highlands, *Geol. Mijnbouw*, **42**, pp. 121–142.

Johnson, R. B. (1961): Patterns and Origin of Radial Dike Swarms Associated with West Spanish Peak and Dike Mountain, South-central Colorado, *Geol. Soc. Am. Bull.*, **72**, pp. 579–590.

———— (1964): Walsen Composite Dike near Walsenburg, Colorado, *U. S. Geol. Surv. Prof. Paper* **501-B**, pp. B69–B73.

———— (1968): Geology of the Igneous Rocks of the Spanish Peaks Region, Colorado, *U. S. Geol. Surv. Prof. Paper* **594-G**, pp. G1–G47.

Johnson, R. L. (1966): "The Shawa and Dorowa Carbonatite Complexes, Rhodesia," Interscience, a division of Wiley, New York, pp. 205–224.

Jolly, W. T. (1971): Potassium-rich Igneous Rocks from Puerto Rico, *Geol. Soc. Am. Bull.*, **82**, pp. 399–408.

———— and R. E. Smith (1970): Zeolite and Prehnite-Pumpellyite Facies in the Keewanawan Flood Basalts of Northern Michigan, l. Metamorphic Differentiation, *Geol. Soc. Am. Abst. with Prog.*, **2**, pp. 589–590.

Jones, A. G. (1959): Vernon Map Area, British Columbia, *Geol. Surv. Can. Mem.* **296**, 186 pp.

Jones, J. G. (1969): Pillow Lavas as Depth Indicators, *Am. J. Sci.*, **267**, pp. 181–195.

Jones, R. W. (1970): Comparison of Columbia River Basalts and Snake Plains Basalts, *in* "Proceedings of the Second Columbia River Basalt Symposium," E. H. Gilmour and D. Stradling, editors, Eastern Wash. State Coll. Press, pp. 209–221.

Jones, W. R., J. W. Peoples, and A. L. Howland (1960): Igneous and Tectonic Structures of the Stillwater Complex, Montana, *U. S. Geol. Surv. Bull.*, **1071-H**, pp. 281–340.

Joplin, G. A. (1937): An Interesting Occurrence of Lawsonite in Glaucophane-bearing Rocks from New Caledonia, *Mineral. Mag.*, **24,** pp. 534–537.

—— (1961): The Problem of the Quartz Dolerites, *in* "Dolerite, a Symposium," S. W. Carrey, convener, *Univ. Tasmania, Geol. Dept. Symp.*, July, 1957, pp. 38–51.

—— (1966): On Lamprophyres, *Roy. Soc. N. S. W., J. Proc.*, **99,** pp. 37–42.

Joyce, A. S. (1970): Chemical Variation in a Pelitic Hornfels, *Chem. Geol.*, **6,** pp. 51–58.

Jung, J., and M. Roques (1952): Introduction a l'etude zoneographique des formations cristal-lophylliennes, *Serv. Carte Geol. France Bull.*, no. 235, t.l., p. 152.

—— and R. Brousse (1959): "Classification modale des roches éruptives," Masson, Paris.

Kaitaro, S. (1956): On Central Complexes with Radial Lamprophyre Dikes, *Soc. Géol. Finlande Compt. Rend.*, no. **29,** pp. 55–65.

Kalsbeek, F., and H. J. Zwart (1967): Zircons from Some Gneisses and Granites in the Central and Eastern Pyrenees, *Geol. Mijnbouw*, **46,** no. 1, pp. 457–466.

Kamp, P. C. van de (1969): Origin of Amphibolites in the Beartooth Mountains, Wyoming and Montana: New Data and Interpretation, *Geol. Soc. Am. Bull.*, **80,** pp. 1127–1136.

—— (1970): The Green Beds of the Scottish Dalradian Series: Geochemistry, Origin, and Metamorphism of Mafic Sediments, *J. Geol.*, **78,** pp. 281–303.

Kania, J. E. A. (1929): Precipitation of Limestone by Submarine Vents, Fumaroles, and Lava Flows, *Am. J. Sci.*, ser. 5, **18,** pp. 347–359.

Katsui, Y. (1961): Petrochemistry of the Quaternary Volcanic Rocks of Hokkaido and Surrounding Areas, *J. Fac. Sci. Hokkaido Univ.*, ser. 4, *Geol. Mineral.*, **XI,** pp. 1–58.

Katz, M. B. (1969): Retrograde Contact Metamorphism in the Granulite Facies Terrain of Mont Tremblant Park, Quebec, Canada, *Geol. Mag.*, **105,** pp. 487–492.

—— (1969): The Nature and Origin of the Granulites of Mont Tremblant Park, Quebec, *Geol. Soc. Am. Bull.*, **80,** pp. 2019–2038.

—— (1970): Notes on the Mineralogy and Coexisting Pyroxenes from the Granulites of Mont Tremblant Park, Quebec, *Can. Mineralogist*, **10,** pp. 247–251.

Kelley, K. K. (1960): High-temperature Heat Content, Heat Capacity and Entropy Data for the Elements and Inorganic Compounds, *U.S. Bur. Mines Bull.*, **584.**

—— (1962): Heats and Free Energies of Formation of Anhydrous Silicates, *U.S. Bur. Mines Rept. Invest.*, **5901.**

Kennedy, G. C. (1955): Some Aspects of the Role of Water in Rock Melts, *Geol. Soc. Am. Spec. Paper* **62,** pp. 489–504.

—— (1956): The Brucite-Periclase Equilibrium, *Am. J. Sci.*, **254,** pp. 567–573.

—— (1959): The Origin of Continents, Mountain Ranges, and Ocean Basins, *Am. Sci.*, **47,** pp. 491–504.

Kennedy, W. Q. (1933): Trends of Differentiation in Basaltic Magmas, *Am. J. Sci.*, ser. 5, **25,** pp. 239–256.

Kern, R., and A. Weisbrod (1967): "Thermodynamics for Geologists," Freeman, Cooper, 304 pp.

Kerrick, D. M. (1968): Experiments on the Upper Stability Limit of Pyrophyllite at 1.8 Kilobars and 3.9 Kilobars Water Pressure, *Am. J. Sci.*, **266,** pp. 204–214.

―――― (1970): Contact Metamorphism in Some Areas of the Sierra Nevada, California, *Geol. Soc. Am. Bull.*, **81,** pp. 2913–2938.

Kesler, S. E., and S. A. Heath (1968): The Effect of Dissolved Volatiles on Magmatic Heat Sources at Intrusive Contacts, *Am. J. Sci.*, **266,** pp. 824–839.

Khalfin, S. L. (1961): Petrologiya differentsirovannogo massiva g. Kogtakh [Petrology of the Kogtakh Differentiated Massif], *Geol. Geofiz.*, no. 7.

Khitarov, N. I., V. A. Pugin, P. Chao, and A. B. Slutshii (1963): Relations between Andalusite, Kyanite and Sillimanite at Moderate Temperatures and Pressures, *Geochem.*, **no. 3,** pp. 235–244.

King, B. C. (1937): The Minor Intrusives of Kirkcudbrightshire, *Geol. Assoc. Lond. Proc.*, **48,** pp. 282–306.

―――― (1948): The Napak Area of Southern Karamoja, Uganda, *Mem. Geol. Surv. Uganda,* **5.**

―――― (1965): The Nature and Origin of Migmatites: Metasomatism or Anatexis, *in* "Controls of Metamorphism," W. S. Pitcher and G. W. Flinn, editors, Wiley, New York, pp. 218–234.

―――― (1965): Petrogenesis of the Alkaline Igneous Rock Suites of the Volcanic and Intrusive Centres of Eastern Uganda, *J. Petrol.*, **6, pp. 67**–100.

―――― and D. S. Sutherland (1960): Alkaline Rocks of Eastern and Southern Africa, Parts I-III, *Sci. Progress,* **48,** pp. 298–321, 504–524, 709–720.

―――― and ―――― (1966): The Carbonatite Complexes of Eastern Uganda, *in* "Carbonatites," O. F. Tuttle and J. Gittins, editors, Interscience, a division of Wiley, New York, pp. 73–126.

Kirsten, T. (1968): Incorporation of Rare Gasses in Solidifying Enstatite Melts, *J. Geophys. Res.*, **73,** pp. 2807–2810.

Kistler, R. W. (1966): Structure and Metamorphism of the Mono Craters Quadrangle, Sierra Nevada, California, *U.S. Geol. Surv. Bull.* **1221-E,** pp. E1–E53.

――――, P. C. Bateman, and W. W. Brannock (1965): Isotopic Ages of Minerals from Granitic Rocks of the Central Sierra Nevada and Inyo Mountains, California, *Geol. Soc. Am. Bull.*, **76,** pp. 155–164.

――――, J. F. Evernden, and H. R. Shaw (1971): Sierra Nevada Plutonic Cycle; Part 1: Origin of Composite Granitic Batholiths, *Geol. Soc. Am. Bull.*, **82,** pp. 853–868.

Kleeman, A. W. (1965): The Origin of Granitic Magmas, *J. Geol. Soc. Australia,* **12,** pp. 35–52.

Klepper, M. R., G. D. Robinson, and H. W. Smedes (1971): On the Nature of the Boulder Batholith of Montana, *Geol. Soc. Am. Bull.*, **82,** pp. 1563–1580.

――――, R. A. Weeks, and E. T. Ruppel (1957): Geology of the Southern Elkhorn Mountains, Jefferson and Broadwater Counties, Montana, *U. S. Geol. Surv. Prof. Paper* **292,** 82 pp.

Klimov, L. V., M. G. Ravich, and D. S. Solovjev (1968): East Antarctic Charnockites, *Intern. Geol. Congr.*, **22d,** India, 1964, pt. 13, pp. 79–87.

Klugman, M. A. (1965): Resume of the Geology of the Laramie Anorthosite Mass, *Geol. Soc. Am.*, Rocky Mtn. Sect., Ann. Mtg., Road Log of Field Trip #1, 21 pp.

—— (1968): The Geology and Origin of the Laramie Anorthosite Mass, Albany County, Wyoming (abs), *N. Y. State Mus. Sci. Serv. Mem.* **18**, p. 369.

Knopf, A. (1912): The Eagle River Region, Southeastern Alaska, *U. S. Geol. Surv. Bull.*, **502**, p. 36.

—— (1936): Igneous Geology of the Spanish Peaks Region, Colorado, *Geol. Soc. Am. Bull.*, **47**, pp. 1727–1784.

—— (1938): Partial Fusion of Granodiorite by Intrusive Basalt, *Am. J. Sci.*, **236**, pp. 373–376.

—— (1950): The Marysville Granodiorite Stock, Montana, *Am. Mineralogist*, **35**, pp. 834–844.

—— (1964): Time Required to Emplace the Boulder Bathylith, Montana: A First Approximation, *Am. J. Sci.*, **262**, pp. 1207–1211.

Knorring, O. von (1962): Geochemical Characteristics of Carbonatites, *Nature*, **194**, pp. 860–861.

—— and C. G. B. Dubois (1961): Carbonatitic Lava from Fort Portal Area in Western Uganda, *Nature*, **192**, pp. 1064–1065.

—— and W. Q. Kennedy (1958): The Mineral Paragenesis and Metamorphic Status of Garnet-Hornblende-Pyroxene-Scapolite Gneiss from Ghana (Gold Coast), *Mineral. Mag.*, **31**, pp. 846–859.

Kolderup, C. F. (1936): The Anorthosites of Western Norway, *Intern. Geol. Congr.*, **16th**, U.S.A., 1933, 1, pp. 289–296.

Kolderup, N. H. (1960): Origin of Norwegian Eclogites in Gneiss, *Norsk. Geol. Tidsskr.*, **40**, pp. 73–76.

Koporulin, V. I. (1961): The Origin of Zeolite Cement in the Sandstones and Gravel-stones of the Coal-bearing Stratum in the Southeastern Part of the Irkutsk Basin, *Dokl. Akad. Nauk SSSR*, **137**, pp. 467–470.

—— (1964): Types of Secondary Alterations in Sands and Gravels of Irkutsk Basin Coal Measures and Their Possible Relationship with Underground Water, *Intern. Geol. Rev.*, **6**, pp. 531–540.

Korn, H., and H. Martin (1954): The Messum Igneous Complex in Southwest Africa, *Trans. Geol. Soc. S. Africa*, **57**, pp. 83–124.

Kornprobst, J. (1969): Le massif ultrabasique des Beni Bouchera (Rif. Interne, Maroc), *Contr. Mineral. Petrol.*, **23**, pp. 283–322.

Korzhinskii, D. S. (1950): Phase Rule and Geochemical Mobility of Elements, *Intern. Geol. Congr.*, **18th**, 1948, pt. 2, sec. A, pp. 50–57.

—— (1959): "Physiochemical Basis of the Analysis of the Paragenesis of Minerals (English translation)," Consultant's Bureau, Inc., New York, 142 pp.

Koschmann, A. H. (1960): Mineral Paragenesis of Pre-Cambrian Rocks in the Tenmile Range, Colorado, *Geol. Soc. Am. Bull.*, **71**, pp. 1357–1370.

Kossovskaya, A. G., and V. D. Shutov (1961): The Correlation of Zones of Regional Epigenesis and Metagenesis in Terrigenous and Volcanic Rocks, *Dokl. Akad. Sci. USSR,* **139,** pp. 732–736.

Koster van Groos, A. F., and P. J. Wyllie (1963): Experimental Data Bearing on the Role of Liquid Immiscibility in the Genesis of Carbonatites, *Nature,* **199,** pp. 801–802.

——— and ——— (1966): Liquid Immiscibility in the System $Na_2O–Al_2O_3–SiO_2–CO_2$ at Pressures to 1 Kilobar, *Am. J. Sci.,* **264,** pp. 234–255.

Kouvo, O., and G. R. Tilton (1966): Mineral Ages from the Finnish Precambrian, *J. Geol.,* **74,** pp. 421–442.

Kramer, J. R. (1969): Subsurface Brines and Mineral Equilibria, *Chem. Geol.,* **4,** pp. 37–50.

Kranck, E. H. (1935): On the Crystalline Complex of Liverpool Land, *Medd. Grønland,* **95,** no. 7.

——— (1961): The Tectonic Position of the Anorthosites of Eastern Canada, *Bull. Comm. Géol. Finlande,* **196,** pp. 299–320.

——— (1968): Anorthosites and Rapakivi, Magmas from the Lower Crust, *N. Y. State Mus. Sci. Serv. Mem.* **18,** pp. 93–97.

Kranck, S. H. (1961): A Study of Phase Equilibria in a Metamorphic Iron Formation, *J. Petrol.,* **2,** pp. 137–184.

Krauskopf, K. B. (1967): "Introduction to Geochemistry," McGraw-Hill Book Co., New York, 721 pp.

Kretz, R. (1961): Some Applications of Thermodynamics to Coexisting Minerals of Variable Composition, Examples Orthopyroxene-Clinopyroxene and Orthopyroxene-Garnet, *J. Geol.,* **69,** pp. 361–387.

——— (1964): Analysis of Equilibrium in Garnet-Biotite-Sillimanite Gneisses from Quebec, *J. Petrol.,* **5,** pp. 1–20.

——— (1966a): Interpretation of the Shape of Mineral Grains in Metamorphic Rocks, *J. Petrol.,* **7,** pp. 68–94.

——— (1966b): Metamorphic Differentiation at Einasleigh, Northern Queensland, *Geol. Soc. Australia J.,* **13,** pp. 561–581.

——— (1967): Granite and Pegmatite Studies at Northern Indian Lake, Manitoba, *Geol. Surv. Can. Bull.* **148,** 42 pp.

Kudrin, V. S. (1963): Alkalic Intrusions in the Northeast of Tuva, *Intern. Geol. Rev.,* **5,** pp. 1280–1289, trans. from *Sovetskaya Geologiya,* 1962, no. 4, pp. 40–52.

Kudryatsev, V. A., V. G. Melamed, and V. N. Sharapov (1971): Dynamics of Temperature Field in the Halo of Emplaced Plutons, during Several Metamorphic Reactions, *Intern. Geol. Rev.,* **13,** pp. 294–300.

Kukharenko, A. A., and E. I. Dontsova (1964): A Contribution to the Problem of the Genesis of Carbonatites, *Econ. Geol. USSR,* **1,** no. 3-4, pp. 47–68.

Kulp, J. L. (1963): Potassium-Argon Dating of Volcanic Rocks, *Bull. Volcanol.,* **26,** pp. 247–258.

—— and J. Engels (1963): Discordances in K-Ar and Rb-Sr Isotopic Ages, *in* "Symposium on Radioactive Dating," *Intern. Atomic Energy Agency Proc.*, Vienna, pp. 219–238.

Kumarapeli, P. S. (1970): Monteregian Alkalic Magmatism and the St. Lawrence Rift System in Space and Time, *Can. Mineralogist,* **10,** pp. 421–431.

Kuno, H. (1950): Petrology of Hakone Volcano and the Adjacent Areas, Japan, *Geol. Soc. Am. Bull.,* **61,** pp. 957–1020.

—— (1954): Geology and Petrology of Omuroyama Volcano Group, North Izu, *Tokyo Univ. Fac. Sci. J.,* sect. 2, **9,** pt. 2, pp. 241–265.

—— (1957): Differentiation of Hawaiian Magmas, *Jap. J. Geol. Geog.,* **28,** pp. 179–218.

—— (1959): Origin of Cenozoic Petrographic Provinces of Japan and Surrounding Areas, *Bull. Volcanol.* ser. 11, **20,** pp. 37–76.

—— (1960): High-alumina Basalt, *J. Petrol.,* **1,** pp. 121–145.

—— (1966): Lateral Variation of Basalt Magma across Continental Margins and Island Arcs, *in* International Upper Mantle Project Symposium "Continental Margins and Island Arcs," W. H. Poole, editor, *Geol. Surv. Can. Paper* **66–15,** pp. 317–335.

—— (1968): Differentiation of Basalt Magmas, *in* "Basalts: The Poldervaart Treatise on Rocks of Basaltic Composition," Interscience, a division of Wiley, New York, pp. 623–688.

—— (1969a): Andesite in Time and Space, *in* "Proceedings of the Andesite Conference," A. R. McBirney, editor, *Oregon Dept. Geol. Mineral. Ind. Bull.,* **65,** pp. 13–20.

—— (1969b): Plateau Basalts, *in* "The Earth's Crust and Upper Mantle," P. J. Hart, editor, *Am. Geophys. Union, Geophys. Monograph* **13,** pp. 495–501.

Kunzler, R. H., and H. G. Goodell (1970): The Aragonite-Calcite Transformation: A Problem in the Kinetics of a Solid-solid Reaction, *Am. J. Sci.,* **269,** pp. 360–391.

Kushiro, I., and K. Aoki (1968): Origin of Some Eclogite Inclusions in Kimberlite, *Am. Mineralogist,* **53,** pp. 1347–1367.

—— and H. Haramura (1971): Major Element Variation and Possible Source Materials of Apollo 12 Crystalline Rocks, *Science,* **171,** pp. 1235–1237.

—— and H. Kuno (1963): Origin of Primary Basaltic Magmas and Classification of Basaltic Rocks, *J. Petrol.,* **4,** pp. 75–89.

Lachenbruch, A. H. (1968): Preliminary Geothermal Model of the Sierra Nevada, *J. Geophys. Res.,* **73,** pp. 6977–6989.

Lacroix, A. (1900): Le Granite des Pyrenees et ses phenomenes de contact, *Bull. Serv. Carte Geol. France,* **XI,** no. 71, p. 1.

—— (1930): La jadeite de Birmanie: les roches qu'elle constitue on qui l'accompagnent. Composition et origine, *Soc. Franc. Mineral. Bull.,* **53,** pp. 216–254.

Laitakari, A. (1947): The Scapolite Occurrence of Pusunsaari, *Comm. Géol. Finlande Bull.,* **140,** pp. 115–119.

Lambert, I. B., and P. J. Wyllie (1968): Stability of Hornblende and a Model for the Low Velocity Zone, *Nature,* **219,** pp. 1240–1241.

Lambert, R. St. J. (1959): The Mineralogy and Metamorphism of the Moine Schist of the Morar and Knoydart Districts of Inverness-shire, *Trans. Roy. Soc. Edinburgh*, **63,** pp. 553–588.

—— (1965): The Metamorphic Facies Concept, *Mineral. Mag.*, **34,** pp. 283–291.

Landes, K. K. (1931): A Paragenetic Classification of the Magnet Cove Minerals, *Am. Mineralogist*, **16,** pp. 313–326.

Landis, C. A., and D. S. Coombs (1967): Metamorphic Belts and Orogenesis in Southern New Zealand, *Tectonophysics*, **4,** no. 4-6, pp. 501–518.

Langseth, M. G., X. LePichon, and M. Ewing (1966): Crustal Structure of the Mid-ocean Ridges, Part 5: Heat Flow through the Atlantic Ocean Floor and Convection Currents, *J. Geophys. Res.*, **71,** pp. 5321–5355.

Lanphere, M. A., and R. G. Coleman (1970): Age and Geologic Implications of High-grade Blueschists and Associated Amphibolites from Oregon and California, *Geol. Soc. Am. Abst.* with *Prog.*, **2,** pp. 602–603.

——, G. J. Wasserburg, A. L. Albee, and G. R. Tilton (1964): "Distribution of Strontium and Rubidium Isotopes during Metamorphism, World Beater Complex, Panamint Range, California," North-Holland Publ. Co., Amsterdam, pp. 269–320.

Lapadu-Hargues P. (1947): Les directions de la tectonique hercynienne dans le nord de l'Aveyron, *Acad. Sci. Paris*, C.R. t.225, pp. 753–754.

Lappin, M. A. (1966): The Field Relationships of Basic and Ultrabasic Masses in the Basal Gneiss Complex of Stadtlandet and Almklovdalen, Nordfjord, Southwestern Norway, *Norsk. Geol. Tidsskr.*, **46,** pp. 439–495.

Larsen, E. S. (1938): Some New Variation Diagrams for Groups of Igneous Rocks, *J. Geol.*, **46,** pp. 505–520.

—— (1940): The Petrographic Province of Central Montana, *Geol. Soc. Am. Bull.*, **51,** pp. 887–948.

—— (1942): Alkalic Rocks of Iron Hill, Gunnison County, Colorado, *U. S. Geol. Surv. Prof. Paper* **197-A,** 64 pp.

Larsen, E. S., Jr. (1945): Time Required for the Crystallization of the Great Batholith of Southern and Lower California, *Am. J. Sci.*, **243A,** pp. 399–416.

——, C. S. Hurlbut, C. H. Burgess, and B. F. Buie (1941): Igneous Rocks of the Highwood Mountains, Montana; Part II: The Extrusive Rocks; Part III: Dikes and Related Intrusives; Part IV: The Stocks; Part V: Contact Metamorphism; Part VI: Mineralogy; Part VII: Petrology, *Geol. Soc. Am. Bull.*, **52,** pp. 1733–1868.

—— and G. Switzer (1939): An Obsidian-like Rock Formed from the Melting of a Granodiorite, *Am. J. Sci.*, **237,** pp. 562–568.

Latimer, W. M. (1952):."The Oxidation States of the Elements and Their Potentials in Aqueous Solutions," Prentice-Hall, Englewood Cliffs, N.J., 392 pp.

Lauder, W. R. (1965): The Petrology of Dun Mountain, *New Zealand J. Geol. Geophys.*, **8.**

Laves, F. (1954): The Coexistence of Two Plagioclases in the Oligoclase Composition Range, *J. Geol.*, **62,** pp. 409–411.

Leake, B. E. (1963): Origin of Amphibolites from Northwest Adirondacks, New York, *Geol. Soc. Am. Bull.*, **74,** pp. 1193–1202.

——— (1964): The Chemical Distinction between Ortho- and Para-amphibolites, *J. Petrol.*, **5,** pp. 238–254.

——— and G. Skirrow (1960): The Pelitic Hornfelses of the Cashel-Lough Wheelaon Intrusion, County Galway, Eire, *J. Geol.*, **68,** pp. 23–40.

Lebedev, A. P., and O. A. Bogatikov (1965): Plutonic Analogs of the Trachybasalt Suite Exemplified by the Kizir Massif (East Sayan), *Intern. Geol. Rev.*, **7,** no. 11, pp. 1917–1927; trans. from Akademiya Nauk SSSR, Izvestiya, *Seriya Geologicheskaya,* 1963, no. 10, pp. 15–29.

Lebedev, P. I. (1939): The Podolian Charnockite Formation, *Intern. Geol. Congr.*, **2,** U.S.S.R., p. 71.

Lee, D. E., R. G. Coleman, and R. C. Erd (1963): Garnet Types from the Cazadero Area, California, *J. Petrol.*, **4,** pp. 460–492.

———, H. H. Thomas, R. F. Marvin, and R. G. Coleman (1964): Isotopic Ages of Glaucophane Schists from the Area of Cazadero, California, *U.S. Geol. Surv. Prof. Paper* **475-D,** pp. 105–107.

Lee, W. H. K., and S. Uyeda (1965): Review of Heat Flow Data, *in* "Terrestrial Heat Flow," W. H. K., Lee, editor, *Am. Geophys. Union, Geophys. Monograph* **no. 8,** pp. 87–190.

———, ———, and P. T. Taylor (1966): Geothermal Studies of Continental Margins and Island Arcs, *in* "Continental Margins and Island Arcs," a symposium, W. H. Poole, editor, *Geol. Surv. Can. Paper* **66-15,** pp. 398–414.

Leelanandam, C. (1968): Zoned Plagioclase from the Charnockites of Kondapalli, Krishna District, Andhra Pradesh, India, *Mineral. Mag.*, **36,** pp. 805–813.

Leeman, W. P. (1970): The Isotopic Composition of Strontium in Late-Cenozoic Basalts from the Basin-Range Province, Western United States, *Geochim. Cosmochim. Acta*, **34,** pp. 857–872.

——— and J. J. W. Rogers (1970): Late Cenozoic Alkali-Olivine Basalts of the Basin-Range Province, U.S.A., *Contrib. Mineral. Petrol.*, **25,** pp. 1–24.

Lefebvre, R. H. (1970): Columbia River Basalts of the Grand Coulee Area, *in* "Proceedings of the Second Columbia River Basalt Symposium," E. H. Gilmour and D. Stradling, editors, Eastern Wash. State Coll. Press, pp. 1–38.

Leighton, M. W. (1954): Petrogenesis of a Gabbro-Granophyre Complex in Northern Wisconsin, *Geol. Soc. Am. Bull.*, **65,** pp. 401–442.

Lemasurier, W. E. (1968): Crystallization Behavior of Basalt Magma, Santa Rosa Range, Nevada, *Geol. Soc. Am. Bull.*, **79,** pp. 949–972.

Leonardos, G. H., and W. S. Fyfe (1967): Serpentinites and Associated Albites, Mocassin Quadrangle, California, *Am. J. Sci.*, **265,** pp. 609–618.

Leonardos, O. H. (1956): Carbonatitos com apatita e pirocloro, *Dep. Nac. Prod. Min., Minis. Agricultura,* Rio de Janeiro, Avulso, **8,** pp. 7–30.

LePichon, X. (1968): Sea-floor Spreading and Continental Drift, *J. Geophys. Res.,* **73,** pp. 3661–3697.

Letteney, C. D. (1968): The Anorthosite-Norite-Charnockite Series of the Thirteenth Lake Dome, South-central Adirondacks, *N. Y. State Mus. Sci. Serv. Mem.* **18,** pp. 329–342.

Leveson, D. J., and C. K. Seyfert (1969): The Role of Metasomatism in the Formation of Layering in Amphibolites of Twin Island, Pelham Bay Park, The Bronx, New York, *in* "Igneous and Metamorphic Geology," L. H. Larsen, editor, *Geol. Soc. Am. Mem.* **115,** pp. 379–399.

Levi, B. (1970): Burial Metamorphic Episodes in the Andean Geosyncline, Central Chile, *Geol. Rundsch.,* **59,** pp. 994–1013.

Lewis, J. V. (1908): The Palisade Diabase of New Jersey, *Am. J. Sci.,* ser. 4, **26,** pp. 155–162.

Lidiak, E. G. (1965): Petrology of Andesitic, Spilitic, and Keratophyric Flow Rock, North-central Puerto Rico, *Geol. Soc. Am. Bull.,* **76,** pp. 57–88.

Liese, H. C. (1964): A Correlative Geothermometric Mineral Study, *Am. J. Sci.,* **262,** pp. 223–230.

Lindgren, W. (1933): Differentiation and Ore Deposition, Cordilleran Region of the United States, *in* "Ore Deposits of the Western States," Lindgren volume, *Am. Inst. Mining Metall. Engineers,* pp. 152–180.

Lindsley, D. H. (1968): Melting Relations of Plagioclase at High Pressures, *N. Y. State Mus. Sci. Serv. Mem.* **18,** pp. 39–46.

Liou, J. G. (1969): P-T Stabilites of Laumontite, Wairakite, Lawsonite (abs), *Am. Geophys. Union. Trans.,* **50,** p. 352.

—— (1971): Synthesis and Stability Relations of Prehnite, $Ca_2Al_2Si_3O_{10}(OH)_2$, *Am. Mineralogist,* **56,** pp. 507–531.

Lipman, P. W. (1966): Water Pressures during Differentiation and Crystallization of Some Ash-flow Magmas From Southern Nevada, *Am. J. Sci.,* **264,** pp. 810–826.

—— (1967): Mineral and Chemical Variations within an Ash-flow Sheet from Aso Caldera, Southwestern Japan, *Contr. Mineral. Petrol.,* **16,** pp. 300–327.

——, R. L. Christiansen, and J. T. O'Connor (1966): A Compositionally Zoned Ash-flow Sheet in Southern Nevada, *U.S. Geol. Surv. Prof. Paper* **524-F,** pp. F1–F47.

——, T. A. Steven, and H. M. Mehnert (1970): Volcanic History of the San Juan Mountains, Colorado, as Indicated by Potassium-Argon Dating, *Geol. Soc. Am. Bull.,* **81,** pp. 2329–2352.

Little, H. W. (1960): Nelson Map Area, West Half, British Columbia, *Geol. Surv. Can. Mem.* **308,** 205 pp.

Livingston, D. E., P. E. Damon, R. L. Manger, R. Bennett, and A. W. Laughlin (1967): Argon 40 in Cogenetic Feldspar-Mica Mineral Assemblages, *J. Geophys. Res.,* **72,** pp. 1361–1375.

Loberg, B. (1963): The Formation of a Flecky Gneiss and Similar Phenomena in Relation to the Migmatite and Vein Gneiss Problem, *Geol. Foren. Stockh., Forh.,* **85,** pp. 3–109.

Lockwood, J. P. (1971): Sedimentary and Gravity-slide Emplacement of Serpentinite, *Geol. Soc. Am. Bull.*, **82,** pp. 919–936.

Lombard, A. F. (1934): On the Differentiation and Relationships of the Rocks of the Bushveld Complex, *Geol. Soc. S. Africa Trans.*, **37,** pp. 5–52.

Loomis, A. A. (1966): Contact Metamorphic Reactions and Processes in the Mt. Tallac Roof Remnant, Sierra Nevada, California, *J. Petrol.*, **7,** pp. 221–245.

Lovering, J. F. (1958): The Nature of the Mohorovicic Discontinuity, *Am. Geophys. Union Trans.*, **39,** pp. 947–955.

―――― (1964): The Eclogite-bearing Basic Igneous Pipe at Ruby Hill near Bingera, N.S.W., *Roy. Soc. N. S. W. J. Proc.*, **97,** pp. 73–79.

―――― and A. J. R. White (1964): The Significance of Primary Scapolite in Granulitic Inclusions from Deep-seated Pipes, *J. Petrol.*, **5,** pp. 195–218.

―――― and ―――― (1969): Granulitic and Eclogitic Inclusions from Basic Pipes at Delegate, Australia, *Contrib. Mineral. Petrol.*, **21,** pp. 9–52.

Lovering, T. S., and E. N. Goddard (1950): Geology and Ore Deposits of the Front Range, Colorado, *U. S. Geol. Surv. Prof. Paper* **223,** 319 pp.

Lowder, G. G., and I. S. E. Carmichael (1970): The Volcanoes and Caldera of Talasea, NE Britain, Geology and Petrology, *Geol. Soc. Am. Bull.*, **81,** pp. 17–38.

Loyendyk, B. P. (1970): Dips of Downgoing Lithospheric Plates beneath Island Arcs, *Geol. Soc. Am. Bull.*, **81,** pp. 3411–3416.

LSPET (Lunar Sample Preliminary Examination Team) (1969): Preliminary Examination of Lunar Samples from Apollo 11, *Science*, **165,** no. 3899, pp. 1211–1227.

―――― (1970): Preliminary Examination of Samples from Apollo 12, *Science*, **167,** pp. 1325–1339.

Lundgren, L. W. (1966): Muscovite Reactions and Partial Melting in Southeastern Connecticut, *J. Petrol.*, **7,** pp. 421–454.

Lundqvist, T. (1968): Precambrian Geology of the Los-Hamra Region, Central Sweden, *Sveriges Geol. Undersok.*, ser. Ba, no. 23, 255 pp.

Luth, W. C. (1967): $KAlSiO_4$-Mg_2SiO_4-SiO_2-H_2O–Part 1: Inferred Phase Relations and Petrologic Applications, *J. Petrol.*, **8,** pp. 372–416.

――――, R. H. Jahns, and O. F. Tuttle (1964): The Granite System at Pressures of 4 to 7 Kilobars, *J. Geophys. Res.*, **69,** no. 4, pp. 759–773.

―――― and G. Simmons (1968): Melting Relations in Natural Anorthosite, *N. Y. State Mus. Sci. Serv. Mem.* **18,** pp. 31–37.

―――― and O. F. Tuttle (1969): The Hydrous Vapor Phase in Equilibrium with Granite and Granite Magmas, *in* "Igneous and Metamorphic Geology," L. H. Larsen, editor, *Geol. Soc. Am. Mem.* **115,** pp. 513–548.

Lutz, B. G. (1968): The Charnockite Series of Anabar Massif in Siberia, *Intern. Geol. Congr.*, **22d,** India, 1964, pt. 13, pp. 142–150.

Lyons, J. B. (1944): Igneous Rocks of the Northern Big Belt Range, Montana, *Geol. Soc. Am. Bull.*, **55**, pp. 445–472.

——— (1955): Geology of the Hanover Quadrangle, New Hampshire-Vermont, *Geol. Soc. Am. Bull.*, **66**, pp. 105–146.

Macdonald, G. A. (1941): Geology of the Western Sierra Nevada between the Kings and San Joaquin Rivers, California, *Calif. Univ. Publ. Bull. Dept. Geol. Sci.*, **26**, no. 2, pp. 215–286.

——— (1949): Hawaiian Petrographic Province, *Geol. Soc. Am. Bull.*, **60**, pp. 1541–1596.

——— (1953): Pahoehoe, Aa and Block Lava, *Am. J. Sci.*, **251**, pp. 169–191.

——— (1955): Catalogue of the Active Volcanoes of the World, Part III: Hawaiian Islands, *Intern. Volcanol. Assoc.*, Napoli, Italia, 37 pp.

——— (1963): Physical Properties of Erupting Hawaiian Magmas, *Geol. Soc. Am. Bull.*, **74**, pp. 1071–1077.

——— (1969): Composition and Origin of Hawaiian Lavas, *Geol. Soc. Am. Mem.* **116**, pp. 477–522.

——— and T. Katsura (1964): Chemical Composition of Hawaiian Lavas, *J. Petrol.*, **5**, pp. 82–133.

——— and ——— (1965): Eruption of Lassen Peak, Cascade Range, California, in 1915, *Geol. Soc. Am. Bull.*, **76**, pp. 475–482.

MacDonald, G. J. F. (1955): Gibbs Free Energy of Water at Elevated Temperatures and Pressures with Applications to the Brucite-Periclase Equilibrium, *J. Geol.*, **63**, pp. 244–252.

Macdonald, R. (1968): "Charnockites" in the West Nile District of Uganda: A Systematic Study in the Groves' Type Area, *Intern. Geol. Congr.*, **22d**, India, 1964, pt. 13, pp. 227–249.

MacGregor, I. D. (1968): Mafic and Ultramafic Inclusions as Indicators of the Depth of Origin of Basaltic Magmas, *J. Geophys. Res.*, **73**, pp. 3737–3745.

MacKenzie, D. B. (1960): High-temperature Alpine-type Peridotite from Venezuela, *Geol. Soc. Am. Bull.*, **71**, pp. 303–318.

Mackenzie, D. H. (1957): On the Relation between Migmatization and Structure in Mid-Stratspey, *Geol. Mag.*, **94**, pp. 177–186.

MacKenzie, W. S. (1949): Kyanite Gneiss within a Thermal Aureole, *Geol. Mag.*, **86**, pp. 251–254.

——— (1965): Some Comments on the Application of Experimental Results to the Study of Metamorphism, *in* "Controls of Metamorphism," W. S. Pitcher and G. W. Flinn, editors, Wiley, New York, pp. 268–280.

——— and J. V. Smith (1961): Experimental and Geological Evidence for the Stability of Alkali Feldspars, *Instituto Lucas Mallada*, Cursillas y Conferencias, fasc. **8**, pp. 53–69.

Mackin, J. H. (1960): Structural Significance of Tertiary Volcanic Rock in Southwestern Utah, *Am. J. Sci.*, **258**, pp. 81–131.

——— (1961): A Stratigraphic Section of the Yakima Basalt and the Ellensburg Formation in South-central Washington, *Wash. Div. Mines Geol. Rept. Inv.*, **19**, 45 pp.

Malde, H. E., and H. A. Powers (1962): Upper Cenozoic Stratigraphy of Western Snake River Plain, *Geol. Soc. Am. Bull.,* **73,** pp. 1197–1220.

Marmo, V. (1958): Serpentinites of Central Sierra Leone, *Bull. Comm. Géol. Finlande,* **no. 180,** pp. 1–30.

—— (1967): On the Granite Problem, *Earth Sci. Rev.,* **3,** pp. 7–29.

Martin, H., M. Mathias, and E. S. W. Simpson (1960): The Damaraland Sub-volcanic Ring Complexes in South West Africa, *Intern. Geol. Congr.,* **21st,** Copenhagen, 1960, *Rept. Session, Norden,* pt. 13, pp. 156–174.

Martin, R. C. (1965): Lithology and Eruptive History of the Whakamaru Ignimbrites in the Maraetai Area of the Taupo Volcanic Zone, New Zealand, *New Zealand J. Geol. Geophys.,* **8,** pp. 680–701.

Martini, J. (1968): Etude petrographique des Gres de Taveyanne entre Arve et Giffre, *Schweizer. Mineral. Petrog. Mitt.,* **48,** pp. 539–654.

—— and M. Vuagnat (1965): Presence du facies a zeolites dans la formation des "gres" de Taveyanne (Alpes Franco-Suisses), *Schweiz. Mineral. Petrog. Mitt.,* **45,** pp. 281–293.

Mason, B. (1962): "Meteorites," Wiley, New York.

—— (1966): "Principles of Geochemistry," 2d and 3d eds., Wiley, New York, 329 pp.

—— (1967): Meteorites, *Am. Sci.,* **55,** pp. 429–455.

—— and W. G. Melson (1970): Comparison of Lunar Rocks with Basalts and Stony Meteorites, *in* "Proceedings of the Apollo 11 Lunar Science Conference," *Geochim. Cosmochim. Acta,* **34,** suppl. 1, pp. 661–671.

Mathews, D. H., F. J. Vine, and F. J. Cann (1965): Geology of an Area of the Carlsberg Ridge, Indian Ocean, *Geol. Soc. Am. Bull.,* **76,** pp. 675–682.

Mathias, M., J. C. Siebert, and P. C. Rickwood (1970): Some Aspects of the Mineralogy and Petrology of Ultramafic Xenoliths in Kimberlite, *Contr. Mineral. Petrol.,* **26,** pp. 75–123.

Mathison, C. I. (1967): The Somerset Dam Layered Basic Intrusion, Southeastern Queensland, *J. Geol. Soc. Australia,* **14,** pp. 57–86.

Mattson, P. H. (1960): Geology of the Mayaguez Area, Puerto Rico, *Geol. Soc. Am. Bull.,* **71,** pp. 319–362.

—— (1966): Geological Characteristics of Puerto Rico, *in* "Continental Margins and Island Arcs"–Internat. Upper Mantle Comm., Symposium, Ottawa, 1965, W. H. Poole, editor, *Geol. Surv. Can. Paper* **66-15,** pp. 124–138.

Mauger, R. L., P. E. Damon, and D. E. Livingston (1968): Cenozoic Argon Ages on Metamorphic Rocks from the Basin and Range Province, *Am. J. Sci.,* **266,** pp. 579–589.

Maxwell, D. T., and J. Hower (1967): High-grade Diagenesis and Low-grade Metamorphism of Illite in the Precambrian Belt Series, *Am. Mineralogist,* **52,** pp. 843–857.

Maxwell, J. C. (1962): Origin of Slaty and Fracture Cleavage in the Delaware Water Gap Area, New Jersey and Pennsylvania, *Geol. Soc. Am.,* **Buddington vol.,** pp. 281–311.

—— (1969): "Alpine" Mafic and Ultramafic Rocks: The Ophiolite Suite, *Tectonophysics,* **7,** pp. 489–494.

McBirney, A. R. (1969a): Compositional Variations in Cenozoic Calc-alkaline Suites of Central America, *in* "Proceedings of the Andesite Conference," A. R. McBirney, editor, *Oregon Dept. Geol. Mineral. Ind. Bull.,* **65,** pp. 185–189.

——— (1969b): Andesitic and Rhyolitic Volcanism of Orogenic Belts, *in* "The Earth's Crust and Upper Mantle," P. J. Hart, editor, *Am. Geophys. Union, Geophys. Monograph* **13.**

———, K. Aoki, and M. N. Bass (1967): Eclogites and Jadeite from the Motagua Fault Zone, Guatamala, *Am. Mineralogist,* **52,** pp. 908–918.

——— and D. F. Weill (1966): Rhyolite Magmas of Central America, *Bull. Volcanol.,* **29,** pp. 435–446.

——— and H. Williams (1965): Volcanic History of Nicaragua, *Calif. Univ. Publ. Geol. Sci.,* **55,** 65 pp.

——— and ——— (1969): Geology and Petrology of the Galapagos Islands, *Geol. Soc. Am. Mem.* **118,** 197 pp.

McCall, G. J. H. (1959): Alkaline and Carbonatite Ring Complexes in the Kauirondo Rift Valley, Kenya, *Intern. Geol. Congr.,* **20th,** Assoc. de Serv. Geol. Africains, pp. 327–334.

——— (1963): A Reconsideration of Certain Aspects of the Rangwa and Ruri Carbonatite Complexes in Western Uganda, *Geol. Mag.,* **100,** pp. 181–185.

McConnell, D., and J. W. Gruner (1940): The Problem of the Carbonate-Apatites; Part III: Carbonate-Apatite from Magnet Cove, Arkansas, *Am. Mineralogist,* **25,** pp. 157–167.

McDonald, J. A. (1967): Evolution of Part of the Lower Critical Zone, Farm Ruighoek, Western Bushveld, *J. Petrol.,* **8,** pp. 165–209.

McDougall, I. (1961): A Note on the Petrography of the Great Lake Dolerite Sill, *Univ. Tasmania, Geol. Dept. Symp.,* July, 1957, pp. 52–60.

——— (1962): Differentiation of the Tasmanian Dolerites: Red Hill Dolerite-Granophyre Association, *Geol. Soc. Am. Bull.,* **73,** pp. 279–316.

——— (1964): Differentiation of the Great Lake Dolerite Sheet, Tasmania, *J. Geol. Soc. Australia,* **11,** pp. 107–132.

——— (1966): Precision Methods of Potassium-Argon Isotopic Age Determination on Young Rocks, *in* "Methods and Techniques in Geophysics," vol. 2, Interscience, a division of Wiley, New York, pp. 279–304.

——— and W. Compston (1965): Strontium Isotope Composition and Potassium-Rubidium Ratios in Some Rocks from Reunion and Rodriguez, Indian Ocean, *Nature,* **207,** pp. 252–253.

McIntire, W. L. (1963): Trace Element Partition Coefficients–A Review of Theory and Applications to Geology, *Geochim. Cosmochim. Acta,* **27,** pp. 1209–1264.

McIver, J. R., and T. W. Gevers (1968): Charnockites and Associated Hypersthene-bearing Rocks in Southern Natal, South Africa, *Intern. Geol. Congr.,* **22d,** India, 1964, pt. 13, pp. 151–168.

McKee, B. (1962): Widespread Occurrence of Jadeite, Lawsonite, and Glaucophane in Central California, *Am. J. Sci.,* **260,** pp. 596–610.

McKenzie, D. P. (1969): Speculations on the Consequences and Causes of Plate Motions, *Geophys. J. R. Astr. Soc.*, **18**, pp. 1–32.

—— and D. L. Parker (1967): The North Pacific: An Example of Tectonics on a Sphere, *Nature*, **216**, pp. 1267–1276.

McKie, D. (1966): Fenitization, *in* "Carbonatites," O. F. Tuttle and J. Gittins, editors, Interscience, a division of Wiley, New York, pp. 261–294.

McLean, D. (1965): The Science of Metamorphism in Metals, *in* "Controls of Metamorphism," W. S. Pitcher and G. W. Flinn, editors, Wiley, New York, pp. 103–118.

McMillan, W. J. (1970): West Flank, Frenchman's Cap Gneiss Dome, Shuswap Terrane, British Columbia, *Geol. Assoc. Can. Spec. Paper* **6**, pp. 99–106.

McNamara, M. J. (1965): The Lower Greenschist Facies in the Scottish Highlands, *Geol. Foren. Stockh. Forh.*, **87**, pp. 347–389.

McTaggart, K. C. (1960): The Mobility of Nuées Ardentes, *Am. J. Sci.*, **258**, pp. 369–382.

—— (1962): Nuées Ardentes and Fluidization–A Reply, *Am. J. Sci.*, **260**, pp. 470–476.

—— (1971): On the Origin of Ultramafic Rocks, *Geol. Soc. Am. Bull.*, **82**, pp. 23–42.

Medaris, L. G., Jr., and R. H. Dott, Jr. (1970): Mantle-derived Peridotites in Southwestern Oregon: Relation to Plate Tectonics, *Science*, **169**, pp. 971–974.

Mehnert, K. R. (1951): Zur Frage des Stoffhaushalts anatektischer Gesteine, *Neues Jahrb. Mineral. Abhandl.*, **82**, pp. 155–198.

—— (1960): Zur Geochemie der Alkalien im tiefen Grundgebirge, *Beitr. Mineral. Petrol.*, **7**, pp. 318–339.

—— (1968): "Migmatites and the Origin of Granitic Rocks," Elsevier, Amsterdam, 403 pp.

Melcher, G. C. (1966): The Carbonatites of Jacupiranga, São Paulo, Brazil, *in* "Carbonatites," O. F. Tuttle and J. Gittins, editors, Interscience, a division of Wiley, New York, pp. 169–181.

Melson, W. G. (1966): Phase Equilibria in Calc-silicate Hornfels, Lewis and Clark County, Montana, *Am. Mineralogist*, **51**, pp. 402–421.

——, E. Jarosewitch, V. T. Bowen, and G. Thompson (1967): St. Peter and St. Paul Rocks: A High-temperature Mantle-derived Intrusion, *Science*, **155**, pp. 1532–1535.

—— and T. H. van Andel (1966): Metamorphism in the Mid-Atlantic Ridge, 22°N. Latitude, *Marine Geol.*, **4**, pp. 165–186.

Menard, W. H. (1958): Development of Median Elevations in Ocean Basins, *Geol. Soc. Am. Bull.*, **69**, pp. 1179–1186.

—— (1960): The East Pacific Rise, *Science*, **132**, pp. 1737–1746.

—— (1964): "Marine Geology of the Pacific," McGraw-Hill Book Co., New York, 271 pp.

Menzer, F. J., Jr. (1970): Geochronologic Study of Granitic Rocks from the Okanogan Range, North-central Washington, *Geol. Soc. Am. Bull.*, **81**, pp. 573–578.

Mercy, E. L. P., and M. J. O'Hara (1968): Nepheline Normative Eclogite from Loch Duich, Ross-shire, *Scot. J. Geol.*, **4**, pp. 1–9.

Merrill, R. B., J. K. Robertson, and O. J. Wyllie (1970): Melting Reactions in the System $NaAlSi_3O_8$-$KAlSi_3O_8$-SiO_2-H_2O to 20 Kilobars Compared with Results for Other Feldspar-Quartz-H_2O and Rock-H_2O Systems, *J. Geol.*, **78,** pp. 558–569.

Metz, P. W. (1966): Untersuchung eine heterogenen bivarianten Gleichgewichts mit CO_2 und H_2O als fluider phase bei hohen Drucken, *Ber. Bunsen Ges. Physik. Chem.*, **70,** pp. 1043–1045.

———— (1967): Experimentelle Bildung von Forsterit und Calcit aus Tremolit und Dolomit, *Geochim. Cosmochim. Acta,* **31,** pp. 1517–1532.

———— and V. Trommsdorff (1968): On Phase Equilibria in Metamorphosed Siliceous Dolomites, *Contr. Mineral. Petrol.,* **18,** pp. 305–309.

———— and H. G. F. Winkler (1963): Experimentelle Gesteinsmetamorphose–VII, Die Bildung von Talk aus kieseligem Dolomit, *Geochim. Cosmochim. Acta,* **27,** pp. 431–457.

———— and ———— (1964): Experimentelle Untersuchung der Diopsidbildung aus Tremolit, Calcit und Quarz, *Naturwiss.,* **51,** p. 460.

Michel-Levy, A. (1888): Sur l'origine des terrains cristallins primitifs, *Soc. Geol. France Bull.,* ser. 3^e, **XIV,** pp. 102–113.

Michot, P. (1955): Anorthosites et anorthosites, *Acad. Roy. Belgique Bull., classical Sci.,* ser. 5, **41,** pp. 275–294.

———— (1960): La Geologie de la Catazone: Le probleme des Anorthosites, la palingenese basique et la tectonique catazonal dans le Rogaland Meridional, *Intern. Geol. Congr.,* **21st,** Copenhagen, 1960, *Rept. Session, Norden,* Guidebook G, Guide de l'excursion A9, 54 pp.

———— (1968): Geological Environments of the Anorthosites of South Rogaland, Norway, *N. Y. State Mus. Sci. Serv. Mem.* **18,** pp. 411–423.

———— and P. Michot (1968): The Problem of Anorthosites: The South-Rogaland Igneous Complex, Southwestern Norway, *N. Y. State Mus. Sci. Serv. Mem.* **18,** pp. 399–410.

Middlemost, E. (1968): Petrogenetic Model for the Evolution of the Anorthosite Kindrid, *Lithos,* **1,** no. 2, pp. 193–198.

Mikhaleva, L. A., and V. A. Skuridin (1971): Granite Formation of the Anatectic Type in Zones of Compression in Gornyy Altay, *Intern. Geol. Rev.,* **13,** pp. 30–39.

Miller, W. J. (1914): Magmatic Differentiation and Assimilation in the Adirondack Regions, *Geol. Soc. Am. Bull.,* **25,** pp. 243–264.

———— (1931): Anorthosite in Los Angeles County, California, *J. Geol.,* **39,** pp. 331–344.

Milovanovic, B., and S. Karamata (1960): Uber den diapirismus serpentinischer Massen, *Intern. Geol. Congr.,* **21st,** Copenhagen, 1960, *Rept. Session, Norden,* pt. 18, pp. 409–417.

Minato, M., M. Gorai, and M. Hunahashi, editors (1965): "Geological Development of the Japanese Islands," Tsukiji Shokan, Tokyo, 442 pp.

Misch, P. (1949a): Metasomatic Granitization of Batholithic Dimensions, Part I, *Am. J. Sci.,* **247,** pp. 209–245.

———— (1949b): Metasomatic Granitization of Batholithic Dimensions, Part II: Static Granitization in Sheku Area, Northwest Yunnan (China), *Am. J. Sci.,* **247,** pp. 372–406.

—— (1952): Geology of the Northern Cascades, Washington, *The Mountaineers*, **55,** pp. 4–22.

—— (1966): Tectonic Evolution of the Northern Cascades of Washington State, *in* "Tectonic History and Mineral Deposits of the Western Cordillera," H. C. Gunning, editor, *Can. Inst. Mining Met.*, spec. vol. **8,** pp. 101–148.

—— (1968): Plagioclase Compositions and Non-anatectic Origin of Migmatitic Gneisses in Northern Cascade Mountains of Washington State, *Contr. Mineral. Petrol.*, **17,** pp. 1–70.

—— (1971): Metamorphic Facies Types in North Cascades, *in* "Metamorphism in the Canadian Cordillera," *Geol. Assoc. Can.*, Cordill. Sect. Prog. and Abst., pp. 22–23.

—— and J. C. Hazzard (1962): Stratigraphy and Metamorphism of Late Precambrian Rocks in Central Northeastern Nevada and Adjacent Utah, *Am. Assoc. Petrol. Geol. Bull.*, **46,** pp. 289–343.

Miyashiro, A. (1953): Calcium-poor Garnets in Relation to Metamorphism, *Geochim. Cosmochim. Acta,* **IV,** pp. 179–208.

—— (1956): Data on Garnet-Biotite Equilibria in some Metamorphic Rocks of the Ryoke Zone, *J. Geol. Soc. Japan,* **62,** pp. 700–702.

—— (1957a): Chlorite of Crystalline Schists, *J. Geol. Soc. Japan,* **63,** pp. 1–8.

—— (1957b): Cordierite-Indialite Relations, *Am. J. Sci.*, **255,** pp. 43–62.

—— (1958): Regional Metamorphism of the Gosaisyo-Takanuki District in the Central Abukuma Plateau, *J. Fac. Sci. Univ. Tokyo,* sec. 2, **11,** pp. 219–272.

—— (1960): Thermodynamics of Reactions of Rock-forming Minerals with Silica, *Japan. J. Geol. Geog.*, **31,** pp. 71–84, 107–111.

—— (1961): Evolution of Metamorphic Belts, *J. Petrol.*, **2,** pp. 277–311.

—— (1966): Some Aspects of Peridotite and Serpentinite in Orogenic Belts, *Japan. J. Geol. Geog.*, **37,** no. 1, pp. 45–61.

—— and F. Shido (1970): Progressive Metamorphism in Zeolite Assemblages, *Lithos,* **3,** pp. 251–260.

——, ——, and M. Ewing (1969a): Composition and Origin of Serpentinites from the Mid-Atlantic Ridge Near 24° and 30° North Latitude, *Contr. Mineral. Petrol.*, **23,** pp. 117–127.

——, ——, and —— (1969b): Diversity and Origin of Abyssal Tholeiite from the Mid-Atlantic Ridge Near 24° and 30° North Latitude, *Contr. Mineral Petrol.*, **23,** pp. 38–52.

Moench, R. H. (1966): Relation of S_2 Schistosity to Metamorphosed Clastic Dikes, Rangeley-Phillips Area, Maine, *Geol. Soc. Am. Bull.*, **77,** pp. 1449–1462.

Monger, J. W. H. (1970): Hope Map-area, West Half, British Columbia, *Geol. Surv. Can. Paper* **69–47.**

—— and W. W. Hutchison (1971): Metamorphic Map of the Canadian Cordillera, *Geol. Surv. Can. Paper* **70–33.**

Moor, G. (1940): The Charnockite Series of the Anabar Pre-Cambrian and Similar Metamorphic Formations (English summary), *Acad. Sci. S.S.S.R.*, ser. geol. **6.**

Moorbath, S. (1965): Isotopic Dating of Metamorphic Rocks, *in* "Controls of Metamorphism," W. S. Pitcher and G. W. Flinn, editors, Wiley, New York, pp. 235–267.

—— and J. D. Bell (1965): Strontium Isotope Abundance Studies and Rubidium-Strontium Age Determinations on Tertiary Igneous Rocks from the Isle of Skye North-west Scotland, *J. Petrol.*, **6,** pp. 37–66.

——, P. N. Hurley, H. W. Fairbairn, (1967): Evidence for the Origin and Age of some Mineralized Laramide Intrusives in the Southwestern U.S. from Strontium Isotopes and Rubidium-Strontium Measurements, *Econ. Geol.*, **62,** no. 2, pp. 228–236.

—— and H. Welke (1968): Lead Isotope Studies on Igneous Rocks from the Isle of Skye, Northwest Scotland, *Earth Planet. Sci. Letters*, **5,** pp. 217–230.

——, ——, and N. H. Gale (1969): The Significance of Lead Isotope Studies in Ancient, High-grade Metamorphic Basement Complexes, as Exemplified by the Lewisian Rocks of Northwest Scotland, *Earth Planet. Sci. Letters*, **6,** pp. 245–256.

Moore, J. G. (1959): The Quartz Diorite Boundary Line in the Western United States, *J. Geol.*, **67,** pp. 198–210.

—— (1963): Geology of the Mount Pinchot Quadrangle, Southern Sierra Nevada, California, *U. S. Geol. Surv. Bull.* **1130,** 152 pp.

—— (1965): Petrology of Deep-sea Basalt near Hawaii, *Am. J. Sci.*, **263,** pp. 40–52.

—— (1970): Pillow Lava in a Historic Lava Flow from Hualalai Volcano, Hawaii, *J. Geol.*, **78,** pp. 239–243.

—— and L. Calk (1971): Sulfide Spherules in Vesicles of Dredged Pillow Basalt, *Am. Mineralogist*, **56,** pp. 476–488.

—— and F. C. Dodge (1962): Mesozoic Age of Metamorphic Rocks in the Kings River Area, Southern Sierra Nevada, California, *in* "Short Papers in Geology, Hydrology and Typography," *U.S. Geol. Surv. Prof. Paper* **450-B,** pp. B19–B21.

——, A. Grantz, and M. C. Blake, Jr. (1963): The Quartz Diorite Line in North-western North America, *U. S. Geol. Surv. Prof. Paper* **450-E,** pp. 89–93.

—— and C. A. Hopson (1961): The Independence Dike Swarm in Eastern California, *Am. J. Sci.*, **259,** pp. 241–259.

Moore, J. M., Jr., and E. Froese (1971): Metamorphic Zoning in Pinacles and Thor-Odin Domes, Shuswap Complex, B.C., *in* "Metamorphism in the Canadian Cordillera," *Geol. Assoc. Can.*, Cordill. Sect. Prog. and Abst., pp. 25–26.

Moores, E. M. (1969): Petrology and Structure of the Vourinos Ophiolite Complex of Northern Greece, *Geol. Soc. Am. Spec. Paper* **118,** 74 pp.

—— and I. D. MacGregor (1968): Depth Classification of Alpine Peridotites (abst.), *Geol. Soc. Am. Spec. Paper* **115,** p. 155.

—— and F. J. Vine (1971): The Troodos Massif, Cyprus, and Other Ophiolites as Oceanic Crust, Evaluation and Implications, *Phil. Trans. Roy. Soc. Lond.*, ser. **268,** pp. 443–466.

Morgan, B. A. (1970): Petrology and Mineralogy of Eclogite and Garnet Amphibolite from Puerto Cabello, Venezuela, *J. Petrol.*, **11,** pp. 101–145.

Morgan, W. J. (1968): Rises, Trenches, Great Faults, and Crustal Blocks, *J. Geophys. Res.*, **73**, pp. 1961–1982.

Morse, S. A. (1968): Layered Intrusions and Anorthosite Genesis, *N. Y. State Mus. Sci. Serv. Mem.* **18**, pp. 175–187.

—— (1969): The Kiglapait Layered Intrusion, Labrador, *Geol. Soc. Am. Mem.* **111**, 204 pp.

—— (1970): Alkali Feldspars with Water at 5kb Pressure, *J. Petrol.*, **11**, pp. 221–251.

—— and G. L. Davis (1969): Fractionation of Potassium and Rubidium in a Layered Intrusion, *Carnegie Inst. Wash., Ann. Rept. Dir. Geophys. Lab.* **1967–1968**, pp. 231–233.

Moxham, R. L. (1965): Distribution of Minor Elements in Coexisting Hornblendes and Biotites, *Can. Mineralogist,* **8,** pp. 204–240.

Moyd, L. (1945): Petrology of the Corundum-bearing Rocks of Southeastern Ontario, *Am. Mineralogist,* **34,** pp. 736–751.

Mueller, R. F. (1961): Oxydation and High Temperature Petrogenesis, *Am. J. Sci.,* **259,** pp. 460–480.

—— (1966): Stability Relations of the Pyroxenes and Olivine in Certain High Grade Metamorphic Rocks, *J. Petrol.,* **7,** pp. 363–374.

—— (1967): Mobility of the Elements in Metamorphism, *J. Geol.,* **75,** pp. 565–582.

—— and K. C. Condie (1964): Stability Relations of Carbon Mineral Assemblages in the Southern California Batholith, *J. Geol.,* **72,** pp. 400–411.

—— and E. J. Olsen (1969): Mineral Assemblages and the Chemical History of Chondritic Meteorites, *Fieldiana Geol.,* **16,** pp. 377–410.

Muffler, L. J. P., and D. E. White (1969): Active Metamorphism of Upper Cenozoic Sediments in the Salton Sea Geothermal Field and the Salton Trough, Southeastern California, *Geol. Soc. Am. Bull.,* **80,** pp. 157–181.

Muir, I. D. (1951); The Clinopyroxenes of the Skaergaard Intrusion, Eastern Greenland, *Mineral. Mag.,* **29,** pp. 690–714.

—— and C. E. Tilley (1958): The Compositions of Coexisting Pyroxenes in Metamorphic Assemblages, *Geol. Mag.,* **95,** pp. 403–408.

Muir, J. D., and C. E. Tilley (1966): Basalts from the Northern Part of the Mid-Atlantic Ridge, Part 2: The Atlantis Collections near 30° N., *J. Petrol.,* **7,** pp. 193–201.

Mukherjee, A. (1967): Role of Fractional Crystallization in the Descent Basalt-Trachyte, *Contr. Mineral. Petrol.,* **16,** pp. 139–148.

Murata, K. J. (1960): A New Method of Plotting Chemical Analyses of Basaltic Rocks, *Am. J. Sci.,* **258-A,** pp. 247–252.

—— (1966): An Acid Fumarolic Gas From Kilauea Iki, Hawaii, *U.S. Geol. Surv. Prof Paper* **537-C,** 6 pp.

—— and D. H. Richter (1966): Chemistry of the Lavas of the 1959–60 Eruption of Kilauea Volcano, Hawaii, *U.S. Geol. Surv. Prof. Paper* **537-A,** 26 pp.

Mutch, T. A., and G. E. McGill (1962): Deformation in Host Rocks Adjacent to an Epizonal Pluton (the Royal Stock, Montana), *Geol. Soc. Am. Bull.,* **73,** pp. 1541–1543.

Naeser, C. W. (1967): The Use of Apatite and Sphene for Fission Track Age Determinations, *Geol. Soc. Am. Bull.,* **78,** pp. 1523–1526.

—— and F. C. W. Dodge (1969): Fission-track Ages of Accessory Minerals from Granitic Rocks of the Central Sierra Nevada Batholith, California, *Geol. Soc. Am. Bull.,* **80,** pp. 2201–2212.

—— and E. H. McKee (1970): Fission-track and K-Ar Ages of Tertiary Ash-flow Tuffs, North-central Nevada, *Geol. Soc. Am. Bull.,* **81,** pp. 3375–3384.

Naha, K. (1959): Time of Formation and Kinematic Significance of Deformation Lamellae in Quartz, *J. Geol.,* **67,** pp. 120–124.

Nalivkin, D. V., et al., editors (1957): The Geological Map of the U.S.S.R., *in* "The Geology of the U.S.S.R.," International Series of Monographs on Earth Sciences, vol. 8, Pergamon Press, New York (1960). Scale 1:7,500,000.

Nandi, K. (1967): Garnets as Indices of Progressive Regional Metamorphism, *Mineral. Mag.,* **36,** pp. 89–93.

Nash, W. P., and J. F. G. Wilkinson (1970): Shonkin Sag Laccolith, Montana, Part I: Mafic Minerals and Estimates of Temperature, Pressure, Oxygen Fugacity and Silica Activity, *Contr. Mineral. Petrol.,* **25,** pp. 241–269.

Naylor, R. S. (1969): Age and Origin of the Oliverian Domes, Central-western New Hampshire, *Geol. Soc. Am. Bull.,* **80,** pp. 405–428.

——, R. H. Steiger, and G. J. Wasserburg (1970): U-Th-Pb and Rb-Sr Systematics in 2700 × 10⁶-year-old Plutons from the Southern Wind River Range, Wyoming, *Geochim. Cosmochim. Acta,* **34,** pp. 1133–1159.

Němec, D. (1966): Plagioclase Albitization in the Lamprophyric and Lamproid Dykes at the Eastern Border of the Bohemian Mass, *Contrib. Mineral. Petrol.,* **12,** pp. 340–353.

—— (1971): On the Relationship of Lamprophyres to Hydrothermal Ore Veins in the Bohemian-Moravian Heights (Czechoslovakia), *Geol. Rundsch.,* **60,** pp. 718–726.

Nesbitt, R. W., and A. W. Kleeman (1964): Layered Intrusions of the Giles Complex, Central Australia, *Nature,* **203,** pp. 391–392.

—— and J. L. Talbot (1966): The Layered Basic and Ultrabasic Intrusives of the Giles Complex, Central Australia, *Contr. Mineral. Petrol.,* **13,** pp. 1–11.

Neumann van Padang, M. (1951): "Catalogue of the Active Volcanoes of the World Including Solfatara Fields, Part 1: Indonesia," Intern. Volcanol. Assoc., Napoli, Italy, 271 pp.

Newton, R. C. (1966a): Some Calc-Silicate Equilibrium Relations, *Am. J. Sci.,* **264,** pp. 204–222.

—— (1966b): Kyanite-Andalusite Equilibrium from 700°C to 800°C, *Science,* **153,** pp. 170–172.

——, J. R. Goldsmith, and J. V. Smith (1969): Aragonite Crystallization from Strained

Calcite at Reduced Pressures and Its Bearing on Aragonite in Low-grade Metamorphism, *Contr. Mineral. Petrol.,* **22,** pp. 335–348.

—— and G. C. Kennedy (1963): Some Equilibrium Reactions in the Join $CaAl_2Si_2O_8$-H_2O, *J. Geophys. Res.,* **68,** pp. 2967–2983.

Nichols, G. D. (1965): Basalts from the Deep Ocean Floor, *Mineral. Mag.,* **34,** pp. 373–388.

Nicholson, R. (1966): Metamorphic Differentiation in Crenulated Schists, *Nature,* **209,** pp. 68–69.

Niggli, P. (1952): The Chemistry of the Keweenawan Lavas, *Am. J. Sci.,* **Bowen vol.,** pp. 381–412.

Noble, C. S., and J. J. Naughton (1968): Deep Ocean Basalts: Inert Gas Content and Uncertainties in Age Dating, *Science,* **162,** pp. 265–267.

Noble, D. C. (1962): Plagioclase Unmixing and the Lower Boundary of the Amphibolite Facies, *J. Geol.,* **70,** pp. 234–239.

——, J. C. Drake, and M. K. Whallon (1969): Some Preliminary Observations on Compositional Variations within the Pumice- and Scoria-flow Deposits of Mount Mazama, *in* "Proceedings of the Andesite Conference," A. R. McBirney, editor, *Oregon Dept. Geol. Mineral. Ind. Bull.,* **65,** pp. 157–164.

—— and C. E. Hedge (1969): Sr^{87}/Sr^{86} Variations within Individual Ash-flow Sheets, *U.S. Geol. Surv. Prof. Paper* **650-C,** pp. C133–C139.

Nockolds, S. R. (1950): On the Occurrence of Neptunite and Eudialyte in Quartz-bearing Syenites from Barnauave, Carlingford, Ireland, *Mineral. Mag.,* **29,** pp. 27–33.

—— and R. Allen (1953): The Geochemistry of Some Igneous Rock Series, *Geochim. Cosmochim. Acta,* **4,** pp. 105–142.

—— and —— (1954): Average Chemical Composition of Some Igneous Rocks, *Geol. Soc. Am. Bull.,* **65,** pp. 1007–1032.

Nolan, J. (1966): Melting-relations in the System $NaAlSi_3O_8$-$NaAlSiO_4$-$NaFeSi_2O_6$-$CaMgSi_2O_6$-H_2O, and Their Bearing on the Genesis of Alkaline Undersaturated Rocks, *Geol. Soc. Lond. Quart. J.,* **122,** pp. 119–157.

Nold, J. L. (1968): "Geology of the Northeastern Border Zone of the Idaho Batholith, Montana and Idaho," Unpubl. dissert., Univ. of Montana, 159 pp.

Oftedahl, C. (1946): Studies on the Igneous Rock Complex of the Oslo Region, Part VI: Akerites, Felsites, and Rhomb Porphyries, *Norske Vidensk.-Akad. Oslo,* I, *Mat.-Naturv. Kl.,* **no. 1.**

—— (1948): Studies on the Igneous Rock Complex of the Oslo Region, Part IX: The Feldspars, *Norske Vidensk.-Akad. Oslo,* I, *Mat.-Naturv. Kl.,* **no. 3.**

—— (1952): Studies on the Igneous Rock Complex of the Oslo Region, Part XII: The Lavas, *Norske Vidensk.-Akad. Oslo,* I, *Mat.-Naturv. Kl.,* **no. 3.**

—— (1957): Studies on the Igneous Rock Complex of the Oslo Region, Part XV: *Norske Vidensk.-Akad. Oslo Skr., Mat.-Naturv. Kl.,* **no. 2.**

O'Hara, M. J. (1960): Co-existing Pyroxenes in Metamorphic Rocks, *Geol. Mag.,* **97,** pp. 498–503.

────── (1961): Zoned Ultrabasic and Basic Gneiss Masses in the Early Lewisian Metamorphic Complex at Scourie, Sutherland, *J. Petrol.*, **2**, pp. 248–276.

────── (1963): Melting of Garnet-Peridotite at 30 Kilobars; Melting of Bimineralic Eclogite at 30 Kilobars, *Carnegie Inst. Wash. Yearbook*, **62**, pp. 71–77.

────── (1965): Primary Magmas and the Origin of Basalts, *Scot. J. Geol.*, **1**, no. 1, pp. 19–40.

────── (1967): Mineral Parageneses in Ultrabasic Rocks, *in* "Ultramafic and Related Rocks," P. J. Wyllie, editor, Wiley, New York, pp. 393–403.

────── (1968): The Bearing of Phase Equilibria Studies in Synthetic and Natural Systems on the Origin and Evolution of Basic and Ultrabasic Rocks, *Earth Sci. Rev.*, **4**, pp. 69–133.

────── (1969): The Origin of Eclogite and Ariegite Nodules in Basalt, *Geol. Mag.*, **106**, pp. 322–330.

────── and E. L. P. Mercy (1966): Peridotite, Eclogite and Pyrope from the Navajo Reservation, *Am. Mineralogist*, **51**, pp. 336–352.

────── and H. S. Yoder (1967): Formation and Fractionation of Basic Magmas at High Pressures, *Scot. J. Geol.*, **3**, pp. 67–113.

Ojakangas, R. W. (1968): Cretaceous Sedimentation, Sacramento Valley, California, *Geol. Soc. Am. Bull.*, **79**, pp. 973–1008.

Oki, Y. (1961a): Metamorphism in the Northern Kiso Range, Nagano Prefecture, Japan, *Japan, J. Geol. Geog.*, **XXXII**, no. 3-4, pp. 479–496.

────── (1961b): Biotites in Metamorphic Rocks, *Japan. J. Geol. Geog.*, **XXXII**, no. 3-4, pp. 479–496.

Oliver, R. L. (1968): Note on Some Garnetiferous Rocks from Ceylon, *Intern. Geol. Congr.*, **22d**, India, 1964, pt. 13, pp. 59–78.

Olmsted, J. F. (1968): Petrology of the Mineral Lake Intrusion, Northwestern Wisconsin, *N. Y. State Mus. Sci. Serv. Mem.* **18**, pp. 149–161.

Olson, J. C., D. R. Shawe, L. C. Pray, and W. N. Sharp (1954): Rare-earth Mineral Deposits of the Mountain Pass District, San Bernardino County, California, *U. S. Geol. Surv. Prof. Paper* **261**, 75 pp.

O'nions, R. K., D. G. W. Smith, H. Baadsgaard, and R. D. Morton (1969): Influence of Chemical Composition on Argon Retentivity in Metamorphic Calcic Amphiboles from South Norway, *Earth Planet. Sci. Letters*, **5**, pp. 339–345.

Orville, P. M. (1963): Alkali Ion Exchange between Vapor and Feldspar Phases, *Am. J. Sci.*, **261**, pp. 201–237.

────── and H. J. Greenwood (1965): Determination of ΔH of Reaction from Experimental Pressure-Temperature Curves, *Am. J. Sci.*, **263**, pp. 678–683.

Osberg, P. H. (1971): An Equilibrium Model for Buchan-Type Metamorphic Rocks, South-central Maine, *Am. Mineralogist*, **56**, pp. 570–586.

Osborn, E. F. (1959): Role of Oxygen Pressure in the Crystallization and Differentiation of Basaltic Magma, *Am. J. Sci.*, **257**, pp. 609–647.

—— (1962): Reaction Series for Subalkaline Igneous Rocks Based on Different Oxygen Pressure Conditions, *Am. Mineralogist,* **47,** pp. 211–226.

—— (1969a): Experimental Aspects of Calc-Alkaline Differentiation, *in* "Proceedings of the Andesite Conference," A. R. McBirney, editor, *Oregon Dept. Geol. Mineral. Ind. Bull.,* **65,** pp. 33–42.

—— (1969b): The Complementariness of Orogenic Andesite and Alpine Peridotite, *Geochim. Cosmochim. Acta,* **33,** pp. 307–334.

—— and D. B. Tait (1952): The System Diopside-Forsterite-Anorthite, *Am. J. Sci.,* Bowen vol., **250A,** pp. 413–433.

Osborne, F. F. (1936): Petrology of the Shawinigan Falls District, *Geol. Soc. Am. Bull.,* **47,** pp. 197–227.

—— (1949): Coronite, Labradorite, Anorthosite, and Dikes of Andesine Anorthosite, New Glasgow, P. Q., *Roy. Soc. Can. Trans.,* ser. 3, **43,** pp. 85–112.

—— and E. J. Roberts (1931): Differentiation in the Shonkin Sag Laccolith, Montana, *Am. J. Sci.,* ser. 5, **22,** pp. 331–353.

Osborne, G. D. (1931): The Contact Metamorphism and Related Phenomena in the Neighborhood of Marulan, New South Wales, *Geol. Mag.,* **68,** pp. 289–314.

Osterwald, F. W. (1955): Petrology of Pre-Cambrian Granites in the Northern Bighorn Mountains, Wyoming, *J. Geol.,* **63,** pp. 310–327.

Otalora, G. (1964): Zeolites and Related Minerals in Cretaceous Rocks of East-central Puerto Rico, *Am. J. Sci.,* **262,** pp. 726–734.

Overstreet, W. C., and H. Bell III (1965): The Crystalline Rocks of South Carolina, *U. S. Geol. Surv. Bull.,* **1183,** 126 pp.

Oxburgh, E. R., and D. L. Turcotte (1968a): Mid-ocean Ridges and Geotherm Distribution during Mantle Convection, *J. Geophys. Res.,* **73,** pp. 2643–2668.

—— and —— (1968b): Problems of High Heat Flow and Volcanism Associated with Zones of Descending Mantle Convective Flow, *Nature,* **218,** pp. 1041–1043.

—— and —— (1970): Thermal Structure of Island Arcs, *Geol. Soc. Am. Bull.,* **81,** pp. 1665–1688.

Oyawoye, M. O. (1962): The Petrology of the District around Bauchi, Northern Nigeria, *J. Geol.,* **70,** pp. 604–615.

Pabst, A. (1955): Manganese Content of Garnets from the Franciscan Schists, *Am. Mineralogist,* **40,** pp. 919–923.

Packham, G. H., and K. A. W. Crook (1960): The Principle of Diagenetic Facies and Some of its Implications, *J. Geol.,* **68,** pp. 392–407.

Page, B. M. (1966): Geology of the Coast Ranges of California, *in* "Geology of Northern California," E. H. Bailey, editor, *Calif. Div. Mines Geol. Bull.,* **190,** pp. 255–276.

—— (1970): Oceanic Crust Remnant on Land near San Luis Obispo, California, *Geol. Soc. Am.,* Abst. with Prog., **2,** pp. 643–644.

Page, L. R. (1968): Devonian Plutonic Rocks in New England, *in* "Studies of Appalachian Geology: Northern and Maritime," E-an Zen, W. S. White, J. B. Hadley, and J. B. Thompson, Jr., editors, Interscience, a division of Wiley, New York, chap. 28, pp. 371–383.

Page, N. J. (1967): Serpentinization at Burro Mountain, California, *Contr. Mineral. Petrol.,* **14,** pp. 321–342.

——— (1968): Serpentinization in a Sheared Serpentinite Lens, Tiburon Peninsula, California, *U. S. Geol. Surv. Prof. Paper* **600-B,** pp. B21–B28.

Pakiser, L. C. (1963): Structure of the Crust and Upper Mantle in the Western United States, *J. Geophys. Res.,* **68,** pp. 5747–5756.

——— and R. Robinson (1966): Composition and Evolution of the Continental Crust as Suggested by Seismic Observations, *Tectonophysics,* **3,** pp. 547–557.

Papezik, V. S. (1965): Geochemistry of Some Canadian Anorthosites, *Geochim. Cosmochim. Acta,* **29,** pp. 673–709.

Pardee, J. T. (1927): Manganese-bearing Deposits near Lake Crescent and Humptulips, Washington, *U. S. Geol. Surv. Bull.,* **795A,** pp. 15–20.

Park, C. F. (1944): The Spillite and Manganese Problems of the Olympic Peninsula, *Am. J. Sci.,* **244,** pp. 305–323.

Parker, R. B. (1961): Petrology and Structural Geometry of Pregranitic Rocks in the Sierra Nevada, Alpine County, California, *Geol. Soc. Am. Bull.,* **72,** pp. 1789–1806.

Parker, R. L., and W. N. Sharp (1970): Mafic-ultramafic Igneous Rocks and Associated Carbonatites of the Gem Park Complex, Custer and Fremont Counties, Colo., *U. S. Geol. Surv. Prof. Paper* **649,** 24 pp.

Parras, K. (1958): On the Charnockites in the Light of a Highly Metamorphic Rock Complex in Southwestern Finland, Part I: The Charnockites–an Independent Problem or Part of a Greater Whole? Part 2: The West Uusimaa Complex, *Bull. Comm. Géol. Finlande,* no. **181,** 137 pp.

Parsons, G. E. (1961): Niobium-bearing Complexes East of Lake Superior, *Ontario Dept. Mines Geol. Rept.,* **3.**

Parsons, W. H. (1969): Criteria for the Recognition of Volcanic Breccias: Review, *in* "Igneous and Metamorphic Geology," L. H. Larsen, editor, *Geol. Soc. Am. Mem.* **115,** pp. 263–304.

Paterson, I. (1971): The Significance of "Blueschist" Facies Rocks Adjacent to the Pinchi Fault Zone, *in* "Metamorphism in the Canadian Cordillera," *Geol. Assoc. Can.,* Cordill. Sect. Prog. and Abst., p. 26.

Paterson, M. S. (1958): The Melting of Calcite in the Presence of Water and Carbon Dioxide, *Am. Mineralogist,* **43,** pp. 603–606.

Peacey, J. S. (1961): Rolled Garnets from Morar, Inverness-shire, *Geol. Mag.,* **98,** pp. 77–80.

Peacock, M. A. (1931): Classification of Igneous Rock Series, *J. Geol.,* **39,** pp. 54–67.

Pearce, T. H. (1970): Chemical Variations in the Palisade Sill, *J. Petrol.,* **11,** pp. 15–32.

Peck, D. L., and T. Minakami (1968): The Formation of Columnar Joints in the Upper Part of Kilauean Lava Lakes, Hawaii, *Geol. Soc. Am. Bull.,* **79,** pp. 1151–1166.

——, J. G. Moore, and G. Kojima (1964): Temperatures in the Crust and Melt of Alae Lava Lake Hawaii, after the August, 1963 Eruption of Kilauea Volcano–A Preliminary Report, *U. S. Geol. Surv. Prof. Paper* **501-D.**

Pecora, W. T. (1956): Carbonatites, A Review, *Geol. Soc. Am. Bull.,* **67,** pp. 1537–1556.

—— (1962): Carbonatite Problem in the Bearpaw Mountains, Montana, *in* "Petrologic Studies," a volume in honor of A. F. Buddington; A. E. J. Engel, H. L. James, and B. F. Leonard, editors, *Geol. Soc. Am.,* pp. 83–104.

Peng, C. J. (1970): Intergranular Albite in Some Granite and Syenites of Hong Kong, *Am. Mineralogist,* **55,** pp. 270–282.

Perchuk, L. L. (1966): Temperature Dependence of the Coefficient of Distribution of Calcium between Coexisting Amphibole and Plagioclase, *Dokl. Akad. Nauk. SSSR,* **169,** pp. 203–205.

Perret, F. A. (1935): Eruption of Mt. Pelée 1929–32, *Carnegie Inst. Wash. Publ.* **458.**

—— (1950): Volcanological Observations, *Carnegie Inst. Wash. Publ.* **549.**

Perrin, R. (1954): Granitization, Metamorphism and Volcanism, *Am. J. Sci.,* **252,** pp. 449–465.

—— (1956): Granite Again, *Am. J. Sci.,* **254,** pp. 1–18.

—— and M. Roubault (1939): Le granite et les reactions a l'état solide, *Bull. Serv. Carte Geol. l'Algerie,* ser. 5, Petrographie, no. 4.

—— and —— (1949): On the Granite Problem, *J. Geol.,* **57,** pp. 357–379.

Perry, D. V. (1969): Skarn Genesis at the Christmas Mine, Gila County, Arizona, *Econ. Geol.,* **64,** pp. 255–270.

Peterman, Z. E. (1966): Rb-Sr Dating of Middle Precambrian Metasedimentary Rocks of Minnesota, *Geol. Soc. Am. Bull.,* **77,** pp. 1031–1043.

——, I. S. E. Carmichael, and A. L. Smith (1970): Sr^{87}/Sr^{86} Ratios of Quaternary Lavas of the Cascade Range, Northern California, *Geol. Soc. Am. Bull.,* **81,** pp. 311–318.

——, B. R. Doe, and H. J. Prostka (1969): Lead and Strontium Isotopes in Rocks of the Absaroka Volcanic Field, Wyoming, *Geol. Soc. Am.* Abst. with Prog., pt. 3, Cordilleran Sec., p. 52.

—— and C. E. Hedge (1971): Related Strontium Isotopic and Chemical Variations in Oceanic Basalts, *Geol. Soc. Am. Bull.,* **82,** pp. 493–500.

——, ——, R. G. Coleman, and P. D. Snavely, Jr. (1967): Sr^{87}/Sr^{86} Ratios in Some Eugeosynclinal Sedimentary Rocks and Their Bearing on the Origin of Granitic Magma in Orogenic Belts, *Earth Planet. Sci. Letters,* **2,** pp. 433–439.

Petersil'ye, I. A., and R. M. Yashina (1971): Investigation of the Gas Phase as an Aid in Determining the Origin of Tuva Alkalic Rocks, *Intern. Geol. Rev.,* **13,** pp. 149–160.

Phemister, J. (1942): Note on Datolite and Other Minerals at a Contact-altered Limestone at Chapel Quarry near Kirkaldy, Fife, *Mineral. Mag.,* **26,** pp. 275–282.

Phillips, F. C. (1938): Mineral Orientation in Some Olivine-rich Rocks from Rum and Skye, *Geol. Mag.,* **75,** pp. 130–135.

Phillips, W. J. (1956): The Minor Intrusive Suite Associated with the Criffel-Dalbeattie Grano-diorite Complex, *Proc. Geol. Assoc.,* **67,** pp. 103–121.

Philpotts, A. R. (1966): Origin of the Anorthosite-Mangerite Rocks in Southern Quebec, *J. Petrol.,* **7,** pp. 1–64.

———— (1963): Contact Metamorphism in Relation to Manner of Emplacement of the Granites of Donegal, Ireland, *J. Geol.,* **71,** pp. 261–296.

———— (1968a): Parental Magma of the Anorthosite-Mangerite Suite, *N. Y. State Mus. Sci. Serv. Mem.* **18,** pp. 207–212.

———— (1968b): Igneous Structures and Mechanism of Emplacement of Mount Johnson, a Monteregian Intrusion, Quebec, *Can. J. Earth Sci.,* **5,** pp. 1131–1137.

———— (1970a): Igneous Structures and Mechanism of Emplacement of Mount Johnson, a Monteregian Intrusion, Quebec: Reply, *Can. J. Earth Sci.,* **7,** pp. 195–197.

———— (1970b): Mechanism of Emplacement of Monteregian Intrusions, *Can. Mineralogist,* **10,** pp. 395–410.

———— and C. J. Hodgson (1968): Role of Liquid Immiscibility in Alkaline Rock Genesis, *Intern. Geol. Congr.,* **23d,** Czech., *Rept. Session,* sec. 2, pp. 175–188.

————, E. F. Pattison, and J. S. Fox (1967): Kalsilite, Diopside and Melilite in a Sedimentary Xenolith from Brome Mountain, Quebec, *Nature,* **214,** no. 5095, pp. 1322–1323.

Philpotts, J. A., and C. C. Schnetzler (1970a): Phenocryst-matrix Partition Coefficients for K, Rb, Sr and Ba, with Applications to Anorthosite and Basalt Genesis, *Geochim. Cosmochim. Acta,* **34,** pp. 307–322.

———— and ———— (1970b): Speculations on the Genesis of Alkaline and Sub-alkaline Basalts Following Exodus of the Continental Crust, *Can. Mineralogist,* **10,** pp. 374–379.

Phinney, W. C. (1968): Anorthosite Occurrences in Keweenawan Rocks of Northeastern Minnesota, *N.Y. State Mus. Sci. Serv. Mem.* **18,** pp. 135–147.

Pichler, H., and W. Zeil (1969): Andesites of the Chilean Andes, *in* "Proceedings of the Andesite Conference," A. R. McBirney, editor, *Oregon Dept. Geol. Mineral. Ind. Bull.,* **65,** pp. 165–174.

Pirsson, L. V. (1905): Petrography and Geology of the Igneous Rocks of the Highwood Mountains, Montana, *U.S. Geol. Surv. Bull.,* **237.**

Pitcher, N. S., and G. W. Flinn, editors (1965): "The Controls of Metamorphism," Wiley, New York, 368 pp.

Pitcher, W. S., and H. H. Read (1960): The Aureole of the Main Donegal Granite, *Geol. Soc. Lond. Quart. J.,* **116,** pp. 1–36.

———— and R. C. Sinha (1957): The Petrochemistry of the Ardara Aureole, *Geol. Soc. Lond. Quart. J.,* **113,** pp. 393–408.

Piwinskii, A. J. (1968): Experimental Studies of Igneous Rock Series: Central Sierra Nevada Batholith, California, *J. Geol.,* **76,** pp. 548–570.

Plas, L. van der (1959): Petrology of the Northern Adula Region, Switzerland, *Leid. Geol. Meded.,* **24,** pp. 418–598.

Platen, H. von (1965a): Kristallisation granitischer Schmelzen, *Beitr. Mineral. Petrog.*, **11**, pp. 334–381.

—— (1965b): Experimental Anatexis and Genesis of Migmatites, *in* "Controls of Metamorphism," W. S. Pitcher and G. W. Flynn, editors, Wiley, New York, pp. 203–218.

—— and H. Höller (1966): Experimentelle Anatexis des Stainzer Plattengneisses von der Koralpa, Steiermark, bei 2, 4, 7 and 10 kb H_2O-Druck, *Neues Jahrb. Mineral. Abhandl.*, **106**, pp. 106–130.

Poldervaart, A. (1944): The Petrology of the Elephant's Head Dike and the New Amalfi Sheet (Matatiele), *Roy. Soc. S. Africa Trans.*, **30**, pp. 85–119.

—— (1953): Metamorphism of Basaltic Rocks: A Review, *Geol. Soc. Am. Bull.*, **64**, pp. 259–273.

—— (1955): Zircon in Rocks, Part 1: Sedimentary Rocks, *Am. J. Sci.*, **253**, pp. 433–464.

—— (1956): Zircon in Rocks, Part 2: Igneous Rocks, *Am. J. Sci.*, **254**, pp. 521–554.

—— and H. H. Hess (1951): Pyroxenes in the Crystallization of Basaltic Magma, *J. Geol.*, **59**, pp. 472–489.

—— and A. B. Parker (1964): The Crystallization Index as a Parameter of Igneous Differentiation in Binary Variation Diagrams, *Am. J. Sci.*, **262**, pp. 281–289.

—— and K. R. Walker (1962): The Palisade Sill, *Intern. Mineral. Assoc. 3d Gen. Congr.*, Washington, D. C., Northern Field Excurs. Guidebook, pp. 5–7.

Polkanov, A. A. (1937): On the Genesis of the Labradorites of Volhynia, *J. Geol. (Acad. Sci. Ukranian SSR)*, **3**, pp. 51–82.

Pouliot, G. (1967): Paramorphisme du quartz dans le granophyre de l'intrusion Muskox (abs), *Assoc. Can.-Franc. Avan. Sci. Ann.*, 1965-1966, **33**, pp. 90–91.

Powell, J. L. (1965): Isotopic Composition of Strontium in Carbonate Rocks from Keshya and Mkwisi, Zambia, *Nature*, **206**, pp. 288–289.

—— (1966): Isotopic Composition of Strontium in Carbonatites and Kimberlites, *Intern. Mineral. Assoc. Proc.*, **I.M.A. vol.**, 1964, New Delhi, India.

—— and S. E. DeLong (1966): Isotopic Composition of Strontium in Volcanic Rocks from Oahu, *Science*, **153**, no. 3741, pp. 1239–1242.

——, G. Faure, and P. M. Hurley (1965): Strontium-87 Abundance in a Suite of Hawaiian Volcanic Rocks of Varying Silica Content, *J. Geophys. Res.*, **70**, pp. 1509–1513.

——, P. M. Hurley, and H. W. Fairbairn (1962): Isotopic Composition of Strontium in Carbonatites, *Nature*, **196**, p. 1085.

——, ——, and —— (1966): The Strontium Isotopic Composition and Origin of Carbonatites, *in* "Carbonatites," O. F. Tuttle and J. Gittins, editors, Interscience, a division of Wiley, New York, pp. 365–378.

Powers, H. A. (1955): Composition and Origin of Basaltic Magma of the Hawaiian Islands, *Geochim. Cosmochim. Acta*, **7**, pp. 77–107.

Press, F. (1969): The Suboceanic Mantle, *Science*, **165**, pp. 174–176.

Price, P. B., and R. M. Walker (1962a): Observations of Charged Particle Tracks in Solids, *J. Appl. Phys.*, **33,** pp. 3400–3406.

——— and ——— (1962b): Chemical Etching of Charged Particle Tracks in Solids, *J. Appl. Phys.*, **33,** pp. 3407–3412.

——— and ——— (1962c): Observation of Fossil Particle Tracks in Natural Micas, *Nature,* **196,** pp. 732–734.

——— and ——— (1963): Fossil Tracks of Charged Particles in Mica and the Age of Minerals, *J. Geophys. Res.,* **68,** pp. 4847–4862.

Prider, R. T. (1945): Charnockitic and Related Cordierite-bearing Rocks from Dangin, Western Australia, *Geol. Mag.,* **82,** pp. 145–172.

——— (1960): The Leucite Lamproites of the Fitzroy Basin, Western Australia, *J. Geol. Soc. Australia,* **6,** pt. 2, pp. 71–118.

Prodehl, C. (1970): Seismic Refraction Study of Crustal Structure in the Western United States, *Geol. Soc. Am. Bull.,* **81,** pp. 2629–2646.

Pulfrey, W. (1946): A Suite of Hypersthene-bearing Plutonic Rocks in the Meru District, Kenya, *Geol. Mag.,* **83,** pp. 67–88.

——— (1954): Alkaline Syenites at Ruri, Kenya, *Geol. Mag.,* **91,** pp. 209–219.

Pulvertaft, T. C. R. (1965): The Eqalogarfia Layered Dike, Nunarssuit, South Greenland, Greenland, *Geol. Undersøgelse Bull.,* no. **55,** 39 pp.

Pushkar, P. (1968): Strontium Isotope Ratios in Volcanic Rocks of Three Island Arc Areas, *J. Geophys. Res.,* **73,** no. 8, pp. 2701–2714.

Putman, G. W., and J. T. Alfors (1969): Geochemistry and Petrology of the Rocky Hill Stock, Tulare County, California, *Geol. Soc. Am. Spec. Paper* **120,** 109 pp.

Quensel, P. (1951): The Charnockite Series of Varberg District on the Southwestern Coast of Sweden, *Arkiv Mineral. Geol.,* **1,** pp. 227–332.

Quon, S. H. (1966): Geochemistry and Paragenesis of Carbonatitic Calcites and Dolomites (abs.), *Dissert.-Abs.,* sec. B, *Sci. Eng.,* **27,** p. 245B.

——— and E. G. Ehlers (1963): Rocks of Northern Part of Mid-Atlantic Ridge, *Geol. Soc. Am. Bull.,* **74,** pp. 1–8.

Radulescu, D. P. (1966): Rhyolites and Secondary Ultrapotassic Rocks in the Subsequent Neogene Volcanism from the East Carpathians, *Bull. Volcanol.,* **29,** pp. 425–433.

Raff, A. D., and R. G. Mason (1961): Magnetic Survey Off the West Coast of North America, 40° N. Latitude to 52° N. Latitude, *Geol. Soc. Am. Bull.,* **72,** pp. 1267–1270.

Raguin, E. (1965): "Geology of Granite," transl. from the second French ed. (1957) by E. H. Kranck, P. R. Eakins, and J. M. Eakins, Interscience, a division of Wiley, New York, 314 pp.

Raleigh, C. B. (1965): Crystallization and Recrystallization of Quartz in a Simple Piston-cylinder Device, *J. Geol.,* **73,** pp. 369–377.

Rama Rao, B. (1945): The Charnockite Rocks of Mysore (Southern India), *Bull. Mysore Geol. Dept.,* **18.**

Ramberg, H. (1944): The Thermodynamics of the Earth's Crust, Part 1: Preliminary Survey of Principal Forces and Reactions in the Earth's Crust, *Norsk. Geol. Tidsskr.*, **24**, p. 98.

—— (1948a): Titanic Iron Ore Formed by Dissociation of Silicates in Granulite Facies, *Econ. Geol.*, **43**, pp. 553–570.

—— (1948b): On Sapphirine-bearing Rocks in the Vicinity of Sukkertoppen, West Greenland, *Medd. Grønland*, **142**, p. 5.

—— (1949): The Facies Classification of Rocks, A Clue to the Origin of Quartzofeldspathic Massifs and Veins, *J. Geol.*, **57**, pp. 18–54.

—— (1952): Intergranular Precipitation of Albite Formed by Unmixing of Alkali Feldspar, *Neues Jahrb. Mineral. Abhandl.*, **98**, pp. 14–34.

—— (1952): "The Origin of Metamorphic and Metasomatic Rocks," The University of Chicago Press, Chicago, Ill., 317 pp.

—— (1964): Chemical Thermodynamics in Mineral Studies, *in* "Physics and Chemistry of the Earth," Pergamon, New York, pp. 226–252.

Ramsay, J. (1967): "Folding and Fracturing of Rocks," McGraw-Hill Book Co., New York, 568 pp.

Rapson, J. E. (1963): Age and Aspects of Metamorphism Associated with the Ice River Complex, British Columbia, *Bull. Can. Petrol. Geol.*, **11**, pp. 116–124.

—— (1966): Carbonatite in the Alkaline Complex of the Ice River Area, Southern Canadian Rocky Mountains, *Intern. Mineral. Assoc. Proc.*, 1964 meeting, India.

Rast, N. (1965): Nucleation and Growth of Metamorphic Minerals, *in* "Controls of Metamorphism," W. S. Pitcher and G. W. Flinn, editors, Wiley, New York, pp. 73–102.

Ratcliffe, N. M. (1968): Contact Relations of the Cortlandt Complex at Stony Point, New York, and Their Regional Implications, *Geol. Soc. Am. Bull.*, **79**, pp. 777–786.

Ratte, J. C., and T. A. Steven (1967): Ash Flows and Related Volcanic Rocks Associated with the Creede Caldera, San Juan Mountains, Colorado, *U.S. Geol. Surv. Prof. Paper* **524-H**, 58 pp.

Read, H. H. (1919): The Two Magmas of Strathbogie and Lower Banffshire, *Geol. Mag.*, **56**, pp. 364–371.

—— (1926): The Mica Lamprophyres of Wigtownshire, *Geol. Mag.*, **63**, pp. 422–429.

—— (1931): The Geology of Central Sutherland, *Scot. Geol. Surv. Mem.*, 238 pp.

—— (1933): On Quartz-Kyanite Rocks in Unst, Shetland Islands, and Their Bearing on Metamorphic Differentiation, *Mineral. Mag.*, **23**, pp. 317–328.

—— (1948): Granites and Granites, *in* "Origin of Granite," J. Gilluly, editor, *Geol. Soc. Am. Mem.* **28**, pp. 1–19.

—— (1949): A Contemplation of Time in Plutonism, *Geol. Soc. Lond. Quart. J.*, **105**, pp. 324–337.

—— (1951): Metamorphism and Granitization. *Geol. Soc. S. Africa Annex.*, **54**.

—— (1952): Metamorphism and Migmatism in the Ythan Valley, Aberdeenshire, *Trans. Geol. Soc. Edinburgh*, **15**, pp. 265–279.

——— (1955a): The Banff Nappe: An Interpretation of the Structure of the Dalradian Rocks of North-east Scotland, *Proc. Geol. Assoc.*, **66**, pp. 1–29.

——— (1955b): Granite Series in Mobile Belts, *Geol. Soc. Am. Spec. Paper* **62**, pp. 409–429.

——— (1957): "The Granite Controversy," Murby, London.

Reed, J. J. (1958): Regional Metamorphism in South-east Nelson, *New Zealand Geol. Surv. Bull.*, no. **60**.

Reesor, J. E. (1958): Dewar Creek Map Area with Special Emphasis on the White Creek Batholith, British Columbia, *Geol. Surv. Can. Mem.* **292**, pp. 58–60.

——— (1961a): White Creek Batholith, *in* "Age Determinations by the Geological Survey of Canada, Report 2, Isotopic Ages," compiled by J. A. Lowden, *Geol. Surv. Can. Paper* **61-17**, pp. 87–91.

——— (1961b): Valhalla Complex, *in* "Age Determinations by the Geological Survey of Canada, Report 2, Isotopic Ages," compiled by J. A. Lowden, *Geol. Surv. Can. Paper* **61-17**, pp. 92–97.

——— (1965): Structural Evolution and Plutonism in Valhalla Gneiss Complex, British Columbia, *Geol. Surv. Can. Bull.*, **129**, 128 pp.

——— (1970): Some Aspects of Structural Evolution and Regional Setting in Part of the Shuswap Metamorphic Compex, *in* "Structure of the Southern Canadian Cordillera," edited by J. O. Wheeler, *Geol. Assoc. Can. Spec. Paper* no. **6**, pp. 73–86.

Reiche, P. (1937): Geology of the Lucia Quadrangle, California, *Univ. Calif. Publ. Geol. Sci. Bull.*, **24**, pp. 115–168.

Reid, R. R. (1957): Bedrock Geology of the North End of the Tobacco Root Mountains, Madison County, Montana, *Montana Bur. Mines Geol. Mem.* **36**, 25 pp.

——— (1959): Reconnaissance Geology of the Elk City Region, Idaho, *Idaho Bur. Mines Geol.*, pamphlet 120, 74 pp.

——— (1968): Multiple Deformation and Associated Progressive Polymetamorphism in the Beltian Rocks North of the Idaho Batholith, Idaho, U.S.A., *Intern. Geol. Congr.*, **23d**, Czeck., 1968, *Rept. Session*, sec. 4, pp. 75–87.

Reinhardt, E. W. (1968): Phase Relations in Cordierite-bearing Gneisses from the Gananoque Area, Ontario, *Can. J. Earth Sci.*, **5**, pp. 455–482.

Reitan, P. (1956): Pegmatite Veins and the Surrounding Rocks, Part I: Petrography and Structure, *Norsk. Geol. Tidsskr.*, **36**, pp. 213–239.

——— (1968): Frictional Heat during Metamorphism –Quantitative Evolution of Concentration of Heat Generation in Time, *Lithos*, **1**, no. 2, pp. 151–163.

Rengarten, N. V. (1950): Laumontite and Analcime from Lower Jurassic Deposits in Northern Caucasus, *Dokl. Akad. Nauk. SSSR*, **70**, pp. 485–488.

Reverdatto, V. V. (1964): Metamorphic Tridymite from Contact-altered Limestones of the Larnite-Merwinite Facies, *Dokl. Akad. Nauk. SSSR*, **146**, pp. 160–163.

Reynolds, D. L. (1931): The Dikes of the Ards Peninsula, Co. Down., *Geol. Mag.*, **68**, p. 145–165.

—— (1946): The Sequence of Geochemical Changes Leading to Granitization, *Geol. Soc. Lond. Quart. J.*, **102**, pp. 389–446.

—— (1947): The Association of Basic "Fronts" with Granitization, *Sci. Prog.*, **35**, pp. 205–219.

Reynolds, R. C., Jr., P. R. Whitney, and Y. W. Isachsen (1968): K/Rb Ratios in Anorthositic and Associated Charnockitic Rocks of the Adirondacks, and Their Petrogenetic Implications, *N. Y. State Mus. Sci. Serv. Mem.* **18**, pp. 267–280.

Ribbe, P. H. (1960): X-ray and Optical Investigation of the Peristerite Plagioclases, *Am. Mineralogist*, **45**, pp. 626–644.

Ribeiro, F. E. (1966): The Itatiaia Alkaline Massif, transl. by N. Herz, *Am. Geol. Inst., Intern. Field Inst.*, Brazil, **1966,** guidebook, pp. VIII-1-VIII-4.

Richards, J. R., and R. T. Pidgeon (1963): Some Age Measurements on Micas from Broken Hill, Australia, *J. Geol. Soc. Australia*, **10,** pp. 243–260.

Richardson, S. W. (1968a): Staurolite Stability in a Part of the System Fe-Al-Si-O-H, *J. Petrol.*, **9,** pp. 467–488.

—— (1968b): The Stability of Fe-staurolite + Quartz, *Carnegie Inst. Wash., Ann. Rept. Geophys. Lab.*, 1966–1967, pp. 398–402.

——, M. C. Gilbert, and P. M. Bell (1969): Experimental Determination of Kyanite-Andalusite and Andalusite-Sillimanite Equilibria; the Aluminum Silicate Triple Point, *Am. J. Sci.*, **267**, pp. 259–272.

Richarz, S. (1924): Some Inclusions in Basalt, *J. Geol.*, **32,** pp. 685–689.

Richter, D. H., and J. G. Moore (1966): Petrology of the Kilauea Iki Lava Lake, Hawaii, *U. S. Geol. Surv. Prof. Paper* **537-B,** 26 pp.

—— and K. J. Murata (1966): Petrography of the Lavas of the 1959–60 Eruption of Kilauea Volcano, Hawaii, *U. S. Geol. Surv. Prof. Paper* **537-D,** 12 pp.

Rickard, M. J. (1964a): Contact Metamorphism in Relation to Manner of Emplacement of the Granites of Donegal, Ireland, A Discussion, *J. Geol.*, **72,** pp. 682–684.

—— (1964b): Reply by W. S. Pitcher, *J. Geol.*, **72**, p. 685.

Rickwood, P. C., M. Mathias, and J. C. Siebert (1968): A Study of Garnets from Eclogite and Peridotite Xenoliths Found in Kimberlite, *Contr. Mineral. Petrol.*, **19,** pp. 271–301.

Rinehart, C. D., and D. C. Ross (1964): Geology and Mineral Deposits of the Mount Morrison Quadrangle, Sierra Nevada, California, *U. S. Geol. Surv. Prof. Paper* **385,** 106 pp.

Ringwood, A. E. (1955): The Principles Governing Trace Element Distribution during Magmatic Differentiation, Part I: The Influence of Electronegativity, Part II: The Role of Complex Formation, *Geochim. Cosmochim. Acta*, **7,** pp. 189–202.

—— (1958): The Constitution of the Mantle, Part III: Consequences of the Olivine-Spinel Transition, *Geochim. Cosmochim. Acta*, **15,** pp. 195–212.

—— (1966): Chemical Evolution of the Terrestrial Planets, *Geochim. Cosmochim. Acta*, **30,** pp. 41–104.

—— (1969): Composition and Evolution of the Upper Mantle, *in* "The Earth's Crust and Upper Mantle," P. J. Hart, editor, *Am. Geophys. Union Monograph* **13,** pp. 1–17.

———— and E. Essene (1970): Petrogenesis of Apollo 11 Basalts, Internal Constitution and Origin of the Moon, *in* "Proceedings of the Apollo 11 Lunar Science Conference," *Geochim. Cosmochim. Acta*, **34,** suppl. 1, pp. 769–799.

Rittmann, A. (1933): Die geologische bedingte evolution und differentiation des Somma-Vessuvius magma, *Zeitschr. Vulkanol.*, **15,** pp. 8–94.

———— (1962): "Volcanoes and Their Activity," Wiley, New York.

———— (1971): The Probable Origin of High-alumina Basalts, *Bull. Volcanol.*, **34,** pp. 414–420.

Roach, R. A., and S. Duffell (1968): The Pyroxene Granulites of the Mount Wright Map-area, Quebec-Newfoundland, *Geol. Surv. Can. Bull.*, **162.**

Robertson, E. C., F. Birch, and G. J. F. Macdonald (1957): Experimental Determination of Jadeite Stability Relations to 25,000 Bars, *Am. J. Sci.*, **255,** pp. 115–137.

Robie, R. A. (1966): Thermodynamic Properties of Minerals, *in* "Handbook of Physical Constants," S. P. Clark, editor, *Geol. Soc. Am. Mem.* **97,** pp. 459–482.

Robinson, P. T. (1969): High-titania Alkali-Olivine Basalts of North-central Oregon, U.S.A., *Contrib. Mineral. Petrol.*, **22,** pp. 349–360.

Rocci, G., and T. Juteau (1968): Spilite-keratophyres et ophiolites influence de la traversee d'un socle sialique sur le magmatisme initial, *Geol. Mijnbouw*, **47,** pp. 330–339.

Roddick, J. (1965): Vancouver North, Coquitlam, and Pitt Lake Map Areas, British Columbia, with Special Emphasis on the Evolution of the Plutonic Rocks, *Geol. Surv. Can. Mem.* **335,** 276 pp.

Rodgers, J. (1967): Chronology of Tectonic Movements in the Appalachian Region of Eastern North America, *Am. J. Sci.*, **265,** pp. 407–427.

Roedder, E., and P. W. Weiblen (1970a): Silicate Liquid Immiscibility in Lunar Magmas, Evidenced by Melt Inclusions in Lunar Rocks, *Science*, **167,** pp. 641–644.

———— and ———— (1970b): Lunar Petrology of Silicate Melt Inclusions, Apollo 11 Rocks, *in* "Proceedings of the Apollo 11 Lunar Science Conference," *Geochim. Cosmochim. Acta*, **34,** suppl. 1, pp. 801–837.

Roeder, P. L., and E. F. Osborn (1966): Experimental Data for the System $MgO\text{-}FeO\text{-}Fe_2O_3$-$CaAl_2Si_2O_8\text{-}SiO_2$ and Their Petrologic Implications, *Am. J. Sci.*, **264,** pp. 428–480.

Roever, W. P. de (1947): Igneous and Metamorphic Rocks in Eastern Central Celebes, *in* "Geological Explorations in the Island of Celebes Under the Leadership of H. A. Brouwer," North-Holland Publishing Company, Amsterdam, pp. 65–173.

———— (1953): Tectonic Conclusions from the Distribution of the Metamorphic Facies in the Island of Kabaena, near Celebes, *Proc. 7th Pacific Sci. Congr.*, New Zealand, 1949, **2,** pp. 71–81.

———— (1955): Genesis of Jadeite by Low-temperature Metamorphism, *Am. J. Sci.*, **253,** pp. 283–298.

———— (1956): Some Differences between Post-Paleozoic and Older Regional Metamorphism, *Geol. Mijnbouw*, **18,** pp. 123–127.

——— and H. J. Nijhuis (1964): Plurifacial Alpine Metamorphism in the Eastern Betic Cordilleras (SE Spain) with Special Reference to the Genesis of the Glaucophane, *Geol. Rundsch.,* **35,** pp. 324–335.

Roots, E. F. (1954): Geology and Mineral Deposits of the Aiken Lake Map Area, British Columbia, *Geol. Surv. Can. Mem.* **274,** 246 pp.

Roques, M. (1941): Les schistes cristallins de la partie sud-ouest du Massif Central Français, *Mem. Serv. Carte Geol. Franc.*

Rose, E. R. (1960): Iron and Titanium in the Morin Anorthosite, Quebec, *Geol. Surv. Can. Paper* **60-11.**

Rose, R. L. (1958): Metamorphic Rocks of the May Lake Area, Yosemite Park, and a Metamorphic Facies Problem (abs.), *Geol. Soc. Am. Bull.,* **69,** p. 1703.

Rosenbusch, H. (1877): Die Steiger Schiefer und ihre Contactzone an den Graniten von Barr-Andlau und Hohwald, *Abh. Geol. Spezialkarte Elsass-Lothringen,* **1,** pp. 80–393.

Rosenfeld, J. L. (1961): The Contamination Reaction Rules, *Am. J. Sci.,* **259,** pp. 1–23.

——— (1970): Rotated Garnets in Metamorphic Rocks, *Geol. Soc. Am. Spec. Paper* **129,** 105 pp.

Rosen-Spence, A. de (1969): Genese des roches a cordierite-anthophyllite des gisements cupro-zinciferes de la region de Rouyn-Noranda, Quebec, Canada, *Can. J. Earth Sci.,* **6,** pp. 1339–1345.

Ross, C. S. (1958): Welded Tuff from Deep-well Cores from Clinch County, Georgia, *Am. Mineralogist,* **43,** pp. 537–543.

——— and R. L. Smith (1960): Ash Flow Tuffs, Their Origin, Geologic Relations and Identification, *U.S. Geol. Surv. Prof. Paper* **366,** 81 pp.

Rowe, R. B. (1954): Notes on Geology and Mineralogy of the Newman Columbian-Uranium Deposit, Lake Nipissing, Ontario, *Geol. Surv. Can. Paper* **54–55.**

Roy, R., and O. F. Tuttle (1956): Investigations under Hydrothermal Conditions, *in* "Physics and Chemistry of the Earth," 1, L. H. Ahrens, K. Rankama, and S. K. Runcorn, editors, Pergamon, New York, pp. 138–180.

Roy, R. F., D. D. Blackwell, and F. Birch (1968): Heat Generation of Plutonic Rocks and Continental Heat Flow Provinces, *Earth Planet. Sci. Letters,* **5,** pp. 1–12.

Runcorn, S. K., editor (1967): "International Dictionary of Geophysics," 2 vols., Pergamon, New York, 1728 pp.

Russell, H. D., S. A. Hiemstra, and D. Groenveld (1954): The Mineralogy and Petrology of the Carbonatite at Loolekop, Eastern Transvaal, *Geol. Soc. S. Africa Trans.,* **57,** pp. 197–208.

Rutland, R. W. R. (1961): The Control of Anorthite Content of Plagioclase in Metamorphic Crystallization, *Am. J. Sci.,* **259,** pp. 76–79.

——— (1962): Feldspar Structure and the Equilibrium between Plagioclase and Epidote: A Reply, *Am. J. Sci.,* **260,** pp. 153–157.

——— (1965): Tectonic Overpressures, *in* "Controls of Metamorphism," W. S. Pitcher and G. W. Flinn, editors, Wiley, New York, pp. 119–139.

Rutten, M. G. (1964): Formation of a Plateaubasalt, *Bull. Volcanol.*, **27**, pp. 93–112.

Saether, E. (1947): The Igneous Rock Complex of the Oslo Region, Part VIII: The Dykes in the Cambro-Silurian Lowland of Baerum, *Norske Vidensk.-Akad. Skr., Mat.-Naturv. Kl., Oslo,* **no. 7.**

———— (1948): On the Genesis of Peralkaline Rock Provinces, *Intern. Geol. Congr.*, **18th,** *Rept. Session,* pt. 2, sec. A, pp. 123–130.

———— (1957): The Alkaline Rock Province of the Fen Area in Southern Norway, *Det. Kgl., Norske Viedersk.-Akad. Selsk. Skr.,* **no. 1.**

Saggerson, E. P. (1968): Eclogite Nodules Associated with Alkaline Olivine Basalts, Kenya, *Geol. Rundsch.,* **57,** pp. 890–903.

Sahama, T. G. (1960): Kalsilite in the Lavas of Mt. Nyiragongo, Belgian Congo, *J. Petrol.,* **1,** pp. 146–170.

Salotti, C. A., and J. A. Fouts (1967): An Occurrence of Cordierite-Garnet Gneiss in Georgia, *Am. Mineralogist,* **52,** pp. 1240–1243.

Sandberg, C. G. S. (1926): On the Probable Origin of the Members of the Bushveld Igneous Complex (Transvaal), *Geol. Mag.,* **43,** pp. 210–219.

Sangster, D. F. (1969): The Contact Metasomatic Magnetite Deposits of Southwestern British Columbia, *Geol. Surv. Can. Bull.,* **172,** 85 pp.

Saxena, S. K. (1969a): Silicate Solid Solutions and Geothermometry, Part 3: Distribution of Fe and Mg between Coexisting Garnet and Biotite, *Contr. Mineral. Petrol.,* **22,** pp. 259–267.

———— (1969b): Distribution of Elements in Coexisting Minerals and the Problem of Chemical Disequilibrium in Metamorphosed Basic Rocks, *Contr. Mineral. Petrol.,* **20,** pp. 177–197.

———— and N. B. Hollander (1969): Distribution of Iron and Magnesium in Coexisting Biotite, Garnet, and Cordierite, *Am. J. Sci.,* **267,** pp. 210–216.

Scarfe, C. M., and P. J. Wyllie (1967): Experimental Redetermination of the Upper Stability Limit of Serpentine up to 3-kbar Pressure, *Trans. Am. Geophys. Union,* **48,** p. 225.

Schairer, J. F. (1950): The Alkali Feldspar Join in the System $NaAlSiO_4$-$KAlSiO_4$-SiO_2, *J. Geol.,* **58,** pp. 512–517.

———— and N. L. Bowen (1947): The System Anorthite-Leucite-Silica, *Soc. Géol. Finlande Bull.,* **20,** pp. 72–75.

Schmidt, D. L. (1964): Reconnaissance Petrographic Cross Section of the Idaho Batholith in Adams and Valley Counties, Idaho, *U.S. Geol. Surv. Bull.,* **1181-G,** 49 pp.

Schmidt, R. G., W. T. Pecora, B. Bryant, and W. G. Ernst (1961): Geology of the Lloyd Quadrangle, Bearpaw Mts., Blaine County, Montana, *U. S. Geol. Surv. Bull.,* **1081-E,** pp. 159–188.

Schmus, W. R. van (1969): The Mineralogy and Petrology of Chondritic Meteorites, *Earth Sci. Rev.,* **5,** pp. 145–184.

———— and J. A. Wood (1967): A Chemical-Petrologic Classification for the Chondritic Meteorites, *Geochim. Cosmochim. Acta,* **31,** pp. 747–765.

Schreyer, W. (1965): Zur stabilitat des Ferrocordierits, *Beitr. Mineral. Petrog.,* **11,** pp. 297–322.

—— and H. S. Yoder, Jr. (1964): The System Mg-Cordierite-H_2O and Related Rocks, *Neues Jahrb. Mineral. Abhandl.*, **101,** pp. 271–342.

Schuiling, R. D. (1958): Kyanite-Sillimanite Equilibrium at High Temperatures and Pressures, *Am. J. Sci.*, **256,** pp. 680–682.

—— (1964): The Limestone Assimilation Hypothesis, *Nature*, **204,** no. 4963, pp. 1054–1055.

Schultz, A. R., and W. Cross (1912): Potash-bearing Rocks of the Leucite Hills, *U. S. Geol. Surv. Bull.*, **512.**

Schurmann, H. M. E. (1956): The Geology of the Glaucophane Rocks in Turkey and Japan, *Geol. Mijnbouw*, **18,** pp. 119–122.

Schwarcz, H. P. (1967): The Effect of Crystal Field Stabilization on the Distribution of Transition Metals between Metamorphic Minerals, *Geochim. Cosmochim. Acta*, **31,** pp. 503–517.

Schwartz, G. M. (1939): Hydrothermal Alteration of Igneous Rocks, *Geol. Soc. Am. Bull.*, **50,** pp. 181–238.

—— and A. E. Sandberg (1940): Rock Series in Diabase Sills at Duluth, Minnesota, *Geol. Soc. Am. Bull.*, **51,** pp. 1135–1171.

Scott, B. (1951): A Note on the Occurrence of Intergrowth between Diopsidic Augite and Albite and of Hydrogrossular from King Island, Tasmania, *Geol. Mag.*, **88,** pp. 429–431.

Scott, W. H., E. C. Hansen, and R. J. Twiss (1965): Stress Analysis of Quartz Deformation Lamellae in a Minor Fold, *Am. J. Sci.*, **263,** pp. 729–746.

Searle, D. L. (1968): The Metamorphic History of Ceylon and the Origin of the Charnockite Series, *Intern. Geol. Congr.*, **22d,** India, 1964, pt. 13, pp. 45–58.

Sederholm, J. J. (1891): Studien uber archaische Eruptivgesteine aus dem sudwestlichen Finnland, *Tschech. Mineral. Petrog. Mitt.*, **12,** pp. 97–142.

—— (1907): Om granit och gneis, deras uppkomst, upptradande och utbredning inom urberget i Finnoskandia, English summary, *Bull. Comm. Géol. Finlande*, **no. 23,** 110 pp.

—— (1916): On Synantetic Minerals and Related Phenomena, *Bull. Comm. Géol. Finlande* **no. 48,** p. 134.

—— (1923): On Migmatites and Associated Pre-Cambrian Rocks of South-western Finland, Part I: The Pellinge Region, *Bull. Comm. Géol. Finlande*, **no. 58,** 153 pp.

—— (1926): On Migmatites and Associated Pre-Cambrian Rocks of South-western Finland, Part II: The Region around the Barosundsfjard W. of Helsingfors and Neighbouring Areas, *Bull. Comm. Géol. Finlande*, **no. 77,** 143 pp.

Seki, Y. (1958): Glaucophanic Regional Metamorphism in the Kanto Mountains, Central Japan, *Japan. J. Geol. Geog.*, **29,** pp. 233–258.

—— (1960): Jadeite in Sanbagawa Crystalline Schists of Central Japan, *Am. J. Sci.*, **258,** pp. 705–715.

—— (1965): Prehnite in Low-grade Metamorphism, *Saitama Univ. Sci. Rept. B.*, **5,** pp. 29–43.

——, M. Aiba, and C. Kato (1960): Jadeite and Associated Minerals of Metagabbroic Rocks in the Sibukawa District, Central Japan, *Am. Mineralogist*, **45,** pp. 668–679.

Shagam, R. (1960): The Geology of Central Aragua, Venezuela, *Geol. Soc. Am. Bull.*, **71,** pp. 249–302.

Shand, S. J. (1931): The Granite-Syenite-Limestone Complex of Palabora, Eastern Transvaal and the Associated Apatite Deposits, *Geol. Soc. S. Africa Trans.*, **34,** pp. 81–105.

———— (1945): The Present Status of Daly's Hypothesis of the Alkaline Rocks, *Am. J. Sci.*, **243A,** pp. 495–507.

———— (1949): Rocks of the Mid-Atlantic Ridge, *J. Geol.*, **57,** pp. 89–92.

———— (1951): "Eruptive Rocks," Murby, London.

Shannon, E. V. (1924): The Mineralogy and Petrology of the Intrusive Diabase at Goose Creek, Loudoun County, Virginia, *U. S. Nat. Mus. Proc.*, **66,** art. 2.

Shannon, R. D., and C. T. Prewitt (1969): Effective Crystal Radii in Oxides and Fluorides, *Acta Crystallogr.*, **B25,** pp. 925–946.

Sharp, W. N., and L. C. Pray (1952): Geologic Map of the Bastnaesite Deposits of the Birthday Claims, San Bernardino County, California, *U. S. Geol. Surv. Min. Inv. Field Studies,* **MF-4.**

Shaw, D. M. (1953): The "Camouflage" Principle and Trace Element Distribution in Magmatic Minerals, *J. Geol.*, **61,** pp. 142–151.

———— (1960a): The Geochemistry of Scapolite, Part I: Previous Work and General Mineralogy, *J. Petrol.*, **1,** pp. 218–260.

———— (1960b): The Geochemistry of Scapolite, Part II: Trace Elements, Petrology, and General Geochemistry, *J. Petrol.*, **1,** pp. 261–285.

Shaw, H. R. (1963): Obsidian-H_2O Viscosities at 1000 and 2000 Bars in the Temperature Range 700° to 900° C., *J. Geophys. Res.*, **68,** pp. 6337–6343.

———— (1965): Comments on Viscosity, Crystal Settling and Convection in Granitic Magmas, *Am. J. Sci.*, **263,** pp. 120–152.

———— (1969): Rheology of Basalt in the Melting Range, *J. Petrol.*, **10,** pp. 510–535.

———— and D. A. Swanson (1970): Eruption and Flow Rates of Flood Basalts, *in* "Proceedings of the Second Columbia River Basalt Symposium," E. H. Gilmour and D. Stradling, editors, Eastern Wash. State Coll. Press, pp. 271–299.

Shawe, D. R., and R. L. Parker (1967): Mafic-ultramafic Layered Intrusion at Iron Mountain, Fremont County, Colorado, *U. S. Geol. Surv. Bull.*, **1251-A,** pp. A1–A29.

Sherlock, D. G., and W. B. Hamilton (1958): Geology of the North Half of the Mount Abbot Quadrangle, Sierra Nevada, California, *Geol. Soc. Am. Bull.*, **69,** pp. 1245–1268.

Shido, F. (1959): Notes on Rock-forming Minerals, Part 9: Hornblende-bearing Eclogite from Gongenyama of Higasi-Akaishi in the Bessi District, Sikoku, *J. Geol. Soc. Japan*, **65,** pp. 701–703.

Short, N. M. (1970): Evidence and Implications of Shock Metamorphism in Lunar Samples, *Science,* **167,** pp. 673–675.

Simmons, G. (1964): Gravity Survey and Geological Interpretation, Northern New York, *Geol. Soc. Am. Bull.*, **75,** pp. 81–98.

Simms, F. E., Jr. (1966): "The Igneous Petrology, Geochemistry, and Structural Geology of Part of the Northern Crazy Mountains, Montana," *Dissert.-Abs.*, sec. B, *Sci. Eng.*, **27,** no. 6, p. 1991B. Unpubl. Ph.D. dissert., Univ. of Cincinnati, 339 pp.: (1970): manuscript.

Simonen, A. (1948): On the Petrochemistry of the Intracrustal Rocks in the Svecofennidic Territory of Southwestern Finland, *Bull. Comm. Géol. Finlande,* **no. 141.**

────── (1960): Plutonic Rocks of the Svecofennides in Finland, *Bull. Comm. Géol. Finlande,* **no. 189,** 101 pp.

Simpson, E. S. W. (1954): The Okonjeje Igneous Complex, Southwest Africa, *Trans. Geol. Soc. S. Africa,* **57,** pp. 125–172.

Singh, S. (1966): Orthopyroxene-bearing Rocks of Charnockitic Affinities in the South Savonna-Kanuko Complex of British Guiana, *J. Petrol.,* **7,** pp. 171–191.

Skiba, W., and J. R. Butler (1963): The Use of Sr-An Relationships in Plagioclases to Distinguish between Somalian Metagabbros and Country Rock Amphibolites, *J. Petrol.,* **4,** pp. 352–366.

Skinner, B. J., D. E. White, H. J. Rose, and R. E. Mays (1967): Sulfides Associated with the Salton Sea Geothermal Brine, *Econ. Geol.,* **62,** pp. 316–330.

Smedes, H. W. (1966): Geology and Igneous Petrology of the Northern Elkhorn Mountains, Jefferson and Broadwater Counties, Montana, *U. S. Geol. Surv. Prof. Paper* **510,** 116 pp.

Smith, A. L., and I. S. E. Carmichael (1968): Quaternary Lavas from the Southern Cascades, Western U.S.A., *Contr. Mineral. Petrol.,* **19,** pp. 212–238.

Smith, C. H. (1958): Bay of Islands Igneous Complex, Western Newfoundland, *Geol. Surv. Can. Mem.* **290,** pp. 1–132.

────── (1962): Notes on the Muskox Intrusion, *Geol. Surv. Can. Paper* **61–25.**

──────, T. N. Irvine, and D. C. Findlay (1967a): Geology, Muskox Intrusion, South Sheet, *Geol. Surv. Can. Map* **1214A,** scale 1:63,360.

──────, ──────, and ────── (1967b): Geology, Muskox Intrusion, North Sheet, *Geol. Surv. Can. Map* **1213A.**

────── and H. E. Kapp (1963): The Muskox Intrusion, a Recently Discovered Layered Intrusion in the Coppermine River Area, Northwest Territories, Canada, *in* "Symposium on Layered Intrusions," *Mineral. Soc. Am. Spec. Paper* **1,** pp. 30–35.

Smith, D. G. W. (1969): Pyrometamorphism of Phyllites by a Dolerite Plug, *J. Petrol.,* **10,** pp. 20–53.

Smith, H. G. (1946): The Lamprophyre Problem, *Geol. Mag.,* **83,** pp. 165–171.

Smith, I. F. (1922): Genesis of Anorthosites of Piedmont, Pennsylvania, *Pan-Am. Geologist,* **38,** pp. 29–50.

Smith, J. V., P. H. Ribbe, and D. B. Stewart (1966): Short Course on Feldspars, *Am. Geol. Inst.,* unpubl. notes, Univ. of Kansas.

Smith, R. E. (1968): Redistribution of Major Elements in the Alteration of Some Basic Lavas during Burial Metamorphism, *J. Petrol.,* **9,** pp. 191–219.

—— (1969): Zones of Progressive Regional Burial Metamorphism in Part of the Tasman Geosyncline, Eastern Australia, *J. Petrol.,* **10,** pp. 144–163.

Smith, R. L. (1960): Ash Flows, *Geol. Soc. Am. Bull.,* **71,** pp. 795–842.

—— and R. A. Bailey (1966): The Bandelier Tuff–A Study of Ash-flow Eruption Cycles from Zoned Magma Chambers, *Bull. Volcanol.,* **29,** pp. 83–104.

Smith, W. C. (1953): Carbonatites of the Chilwa Series of Southern Nyasaland, *Bull. Brit. Mus.* (Nat. Hist.) *Mineral.,* **1,** no. 4, pp. 95–120.

—— (1956): A Review of Some Problems of African Carbonatites, *Geol. Soc. Lond. Quart. J.,* **112,** pp. 189–219.

Smulikowski, K. (1960a): Evolution of the Granite-gneisses in the Snieznik Mountains, East Sudetes, *Intern. Geol. Congr.,* **21st,** Copenhagen, 1960, *Rept. Session, Norden,* pt. 14, pp. 120–130.

—— (1960b): Petrographic Notes on Some Eclogites of the Eastern Sudetes, *Bull. Acad. Polon. Sci., Ser. Sci. Geol. Geogr.* **8,** pp. 11–19.

—— (1960c): Comments on Eclogite Facies in Regional Metamorphism, *Intern. Geol. Congr.,* **21st,** Copenhagen, 1960, *Rept. Session, Norden,* pt. 13, pp. 372–382.

—— (1968): On So-Called "Dry Metamorphism," *Intern. Geol. Cong.,* **22d,** India, 1964, pt. 13, pp. 128–141.

Smyth, F. H., and L. H. Adams (1923): The System Calcium Oxide-Carbon Dioxide, *J. Am. Chem. Soc.,* **45,** pp. 1169–1184.

Snelling, N. J. (1957): Notes on the Petrology and Mineralogy of the Barrovian Metamorphic Zones, *Geol. Mag.,* **94,** pp. 297–304.

Snyder, G. L., and G. D. Fraser (1963): Pillowed Lavas, Part I: Intrusive Layered Lava Pods and Pillowed Lavas, Unalaska Island, Alaska; Part II: A Review of Selected Recent Literature, *U. S. Geol. Surv. Prof. Paper* **454-B, C,** pp. B1–B23, C1–C7.

Sobolev, N. V., Jr. (1968): The Xenoliths of Eclogites from the Kimberlite Pipes of Yakutia as Fragments of the Upper Mantle Substance, *Intern. Geol. Congr.,* **23d,** Prague, sec. 1, pp. 155–163.

——, I. K. Kuznetsova, and N. I. Zyuzin (1968): The Petrology of Grospydite Xenoliths from the Zagadochnaya Kimberlite Pipe in Yakutia, *J. Petrol.,* **9,** pp. 253–280.

Sobolev, V. S. (1960): Role of High Pressure in Metamorphism, *Intern. Geol. Congr.,* **21st,** Copenhagen, 1960, *Rept. Session, Norden,* pt. 14, pp. 72–82.

Soliman, S. M. (1964): Geology of the East Half of the Mount Hamilton Quadrangle, California, *Calif. Div. Mines Geol. Bull.,* **185,** 32 pp.

Sood, M. K., R. G. Platt, and A. D. Edgar (1970): Phase Relations in Portions of the System Diopside-Nepheline-Kalsilite-Silica and Their Importance in the Genesis of Alkaline Rocks, *Can. Mineralogist,* **10,** pp. 380–394.

Sørensen, H. (1955): Anorthosite from Buksefjorden, West Greenland, *Medd. Dansk Geol. Foren.,* bd. **13,** pp. 31–41.

—— (1958): The Ikmaussaq Batholith, *Medd. Grønland,* **162,** no. 3, 48 pp.

—— (1960): On the Agpaitic Rocks, *Intern. Geol. Congr.*, **21st,** Copenhagen, 1960, *Rept. Session, Norden,* pt. 13, pp. 319–327.

—— (1970): Internal Structures and Geological Setting of the Three Agpaitic Intrusions – Khibina and Lovozero of the Kola Peninsula and Ilimaussaq, South Greenland, *Can. Mineralogist,* **10,** pp. 299–334.

Souther, J. G. (1967): Acid Volcanism and Its Relationship to the Tectonic History of the Cordillera of British Columbia, Canada, *Bull. Volcanol.,* **30,** pp. 161–176.

—— (1970): Volcanism and Its Relationship to Recent Crustal Movements in the Canadian Cordillera, *Can. J. Earth Sci.,* **7,** pp. 553–568.

Spencer, A. B. (1969): Alkalic Igneous Rocks of the Balcones Province, Texas, *J. Petrol.,* **10,** pp. 272–306.

Spry, A. (1961): Some Observations of the Jurassic Dolerite of the Eureka Cone Sheet near Zeehan, Tasmania, *Univ. Tasmania, Geol. Dept.,* Symp., July, **1957,** pp. 52–60.

—— (1962): The Origin of Columnar Jointing, Particularly in Basalt Flows, *J. Geol. Soc. Australia,* **8,** pt. 2, pp. 191–216.

—— (1963a): The Occurrence of Eclogite on the Lyell Highway, Tasmania, *Mineral. Mag.,* **33,** pp. 589–593.

—— (1963b): The Chronological Analysis of Crystallization and Deformation of Some Tasmanian Precambrian Rocks, *J. Geol. Soc. Australia,* **10,** pp. 193–208.

—— and M. Solomon (1964): Columnar Buchites at Apsley, Tasmania, *Geol. Soc. Lond. Quart. J.,* **120,** pt. 4, pp. 519–545.

Spurr, J. E., and G. H. Garrey (1908): Ore-deposits of the Velardena District, Mexico, *Econ. Geol.,* **3,** pp. 688–725.

Sriramadas, A. (1966): Geology of the Manchester Quadrangle, New Hampshire, *New Hampshire Dept. Resources Econ. Devel. Bull.,* **2,** 78 pp.

Stearns, H. T. (1966): "Geology of the State of Hawaii," Pacific Books, Palo Alto, Calif.

——, L. Crandall, and W. G. Steward (1938): Geology and Ground Water Resources of the Snake River Plain in South-eastern Idaho, *U. S. Geol. Surv. Water Supply Paper* **774,** p. 1.

—— and G. A. MacDonald (1946): Geology and Ground-water Resources of the Island of Hawaii, *Hawaii Div. Hydrog. Bull.,* **9,** 363 pp.

Steiger, R. H. (1964): Dating of Orogenic Phases in the Central Alps by K-Ar Ages of Hornblende, *J. Geophys. Res.,* **69,** no. 24, pp. 5407–5421.

—— and S. R. Hart (1967): The Microcline-Orthoclase Transition within a Contact Aureole, *Am. Mineralogist,* **52,** pp. 87–116.

Steiner, A. (1958): Origin of Ignimbrites of the North Island, New Zealand: A New Petrogenetic Concept, *New Zealand Geol. Surv. Bull.,* ns. **68,** 42 pp.

Steinmann, G. (1927): Die ophiolitischen zonen in dem Mediterranean Kettengebirgen, *Intern. Geol. Congr.,* **14ᵉ** Madrid, 1926, *Compt. Rend.,* pt. 2, pp. 636–667.

Steven, T. A. (1957): Metamorphism and the Origin of Granitic Rocks, Northgate District, Colorado, *U.S. Geol. Surv. Prof. Paper* **274-M.**

Stewart, F. H. (1942): Chemical Data on a Silica-poor Hornfels and Its Constituent Minerals, *Mineral. Mag.,* **26,** pp. 260–266.

————— (1946): The Gabbroic Complex of Belhelvie in Aberdeenshire, *Geol. Soc. Lond. Quart. J.,* **102,** pp. 465–498.

————— and M. R. W. Johnson (1960): The Structural Problem of the Younger Gabbros of Northeast Scotland, *Trans. Geol. Soc. Edinburgh,* **18,** pp. 104–112.

Stewart, R. J., and R. J. Page (1971): Incipient Metamorphism of Sandstones in the Upper Cretaceous Nanaimo Group, Vancouver Island and Gulf Islands, British Columbia, *in* "Metamorphism in the Canadian Cordillera," *Geol. Assoc. Can.,* Cordill. Sect. Prog. and Abst., pp. 29–30.

Stieff, L. R., and T. W. Stern (1956): Interpretation of the Discordant Age Sequence of Uranium Ores, *U. S. Geol. Surv. Prof. Paper* **300,** pp. 549–555.

Stille, H. (1940): "Einfuhrung in den Bau Nordamerikas," Borntraeger, Berlin, 717 pp.

Stockwell, C. H. (1961): Structural Provinces, Orogenies, and Time Classification of Rocks of the Canadian Precambrian Shield, *Geol. Surv. Can. Paper* **61-17,** pp. 108–118.

————— (1964): Fourth Report on Structural Provinces, Orogenies, and Time Classification of Rocks of the Canadian Precambrian Shield, *Geol. Surv. Can. Paper* **64-17,** pp. 1–21.

————— (1965): Tectonic Map of the Canadian Shield, *Geol. Surv. Can. Map* **4-1965,** scale 1:5,000,000.

Stoiber, R. E., and M. J. Carr (1971): Lithospheric Plates, Benioff Zones, and Volcanoes, *Geol. Soc. Am. Bull.,* **82,** pp. 515–522.

Stose, G. W., and J. V. Lewis (1916): Triassic Igneous Rocks in the Vicinity of Gettysburg, Pennsylvania, *Geol. Soc. Am. Bull.,* **27,** pp. 623–644.

Strauss, E. A., and F. C. Truter (1951): The Alkali Complex at Spitskop, Sekukuniland, Eastern Transvaal, *Trans. Geol. Soc. S. Africa,* **53,** pp. 81–130.

Streckeisen, A. (1960): On the Structure and Origin of the Nepheline-Syenite Complex of Ditro (Transylvania, Roumania), *Intern. Geol. Congr.,* **21st,** Copenhagen, 1960, *Rept. Session, Norden,* pt. 13, pp. 228–238.

————— (1967): Classification and Nomenclature of Igneous Rocks, *Neues Jahrb. Mineral. Abhandl.,* **107,** pp. 144–240.

Strens, R. G. J. (1968): Reconnaissance of the Prehnite Stability Field, *Mineral. Mag.,* **36,** pp. 864–867.

Stuart, C. (1966): "Metamorphism in the Central Flint Creek Range, Montana," unpubl. M. Sci. thesis, Univ. of Montana, 103 pp.

Stueber, A. M., and V. R. Murthy (1966): Strontium Isotopes and Alkali Element Abundances in Ultramafic Rocks, *Geochim. Cosmochim. Acta,* **30,** no. 12, pp. 1243–1259.

Sturt, B. A. (1962): The Composition of Garnets from Pelitic Schists in Relation to the Grade of Metamorphism, *J. Petrol.,* **3,** pp. 181–191.

————— and D. M. Ramsay (1965): The Alkaline Complex of the Breivikbotn Area, Sørøy, Northern Norway, *Norges Geol. Undersokn.,* **no. 231,** 142 pp.

Subramanian, A. P. (1956): Petrology of the Anorthosite-Gabbro Mass at Kadavur, Madras, India, *Geol. Mag.*, **93**, pp. 287–300.

—— (1959): Charnockites of the Type Area near Madras–A Reinterpretation, *Am. J. Sci.*, **257**, pp. 321–353.

—— (1960): Petrology of the Charnockite Suite of Rocks from the Type Area Around St. Thomas Mount and Pallavaram near Madras City, India, *Intern. Geol. Congr.*, **21st,** Copenhagen, 1960, *Rept. Session, Norden*, pt. 13, pp. 394–403.

Sundius, N. (1915): Beitrage zur Geologie des sudlichen Teils des Kirunagabiets, *Vitensk. Prakt. Unders Lappl.*, pp. 1–237.

—— (1930): On the Spilitic Rocks, *Geol. Mag.*, **67**, pp. 1–17.

—— (1946): The Classification of the Hornblendes, *Sver. Geol. Undersok. Arsbok*, **40**, no. 4, 35 pp.

Suppe, J. (1969): Times of Metamorphism in the Franciscan Terrain of the Northern Coast Ranges, California, *Geol. Soc. Am. Bull.*, **80**, pp. 135–142.

Surdam, R. C. (1966): Analcime-Wairakite Mineral Series, *Geol. Soc. Am. Spec. Paper* **87**, pp. 169–170.

—— (1969): Electron Microprobe Study of Prehnite and Pumpellyite from the Karmutsen Group, Vancouver Island, British Columbia, *Am. Mineralogist*, **54**, pp. 256–266.

Sutton, J. (1965): Some Recent Advances in Our Understanding of the Controls of Metamorphism, *in* "Controls of Metamorphism," W. S. Pitcher and G. W. Flinn, editors, Wiley, New York, pp. 22–45.

—— and J. Watson (1950): The Pre-Torridonian Metamorphic History of the Loch Torridon and Scourie Area in the North-west Highlands, and Its Bearing on the Chronological Classification of the Lewisian, *Geol. Soc. Lond. Quart. J.*, **106**, pp. 241–296.

Suwa, K. (1968): Petrologic Studies on the Metamorphic Rocks from the Lutzow-Holrnbu Area, East Antarctica, *Intern. Geol. Congr.*, **23d,** Czech., 1968, *Rept. Session*, sect. 4, pp. 171–187.

Suzuki, J. (1930): Petrological Study of the Crystalline Schist System of Shikoku, Japan, *J. Fac. Sci. Hokkaido Univ.*, ser. IV, **1**, pp. 27–111.

—— (1954): On the Rodingitic Rocks in the Serpentinite Masses of Hokkaido, *J. Fac. Sci. Hokkaido Univ.*, ser. IV, **VIII**, no. 4, pp. 419–430.

—— and Y. Suzuki (1959): Petrological Study of the Kamuikotan Metamorphic Complex in Hokkaido, Japan, *J. Fac. Sci. Hokkaido Univ.*, ser. IV, **X**, no. 2, pp. 349–446.

Swanson, D. A. (1967): Yakima Basalt of the Tieton River Area, South-central Washington, *Geol. Soc. Am. Bull.*, **78**, pp. 1077–1110.

Swift, C. M., Jr. (1966): Geology of the Southeast Portion of the Averill Quadrangle, New Hampshire, *New Hampshire Dept. Resources Econ. Devel.*, Concord, N. H., 19 pp.

Switzer, G. (1945): Eclogite from the California Glaucophane Schists, *Am. J. Sci.*, **243**, pp. 1–8.

—— (1951): Mineralogy of the California Glaucophane Schists, *Calif. Div. Mines Geol. Bull.*, **161**, pp. 51–70.

Sykes, L. R. (1966): The Seismicity and Deep Structure of Island Arcs, *J. Geophys. Res.,* **71,** pp. 2981–3006.

——— (1967): Mechanism of Earthquakes and Nature of Faulting on the Mid-oceanic Ridges, *J. Geophys. Res.,* **72,** pp. 2131–2153.

Tabor, R. W., and D. F. Crowder (1969): On Batholiths and Volcanoes; Intrusion and Eruption of Late Cenozoic Magmas in the Glacier Peak Area, North Cascades, Washington, *U.S. Geol. Surv. Prof. Paper* **604,** 67 pp.

Takeuchi, H., and S. Uyeda (1965): A Possibility of Present-day Regional Metamorphism, *Tectonophysics,* **2,** pp. 59–68.

Talbot, J. L., and B. E. Hobbs (1968): The Relationship of Metamorphic Differentiation to Other Structural Features at Three Localities, *J. Geol.,* **76,** pp. 581–587.

———, ———, H. G. Wilshire, and T. R. Sweatman (1963): Xenoliths and Xenocrysts From Lavas of the Kergulen Archipelago, *Am. Mineralogist,* **48,** pp. 159–179.

Taliaferro, N. L. (1943): Franciscan-Knoxville Problem, *Am. Assoc. Petrol. Geol. Bull.,* **27,** pp. 109–219.

Tanguy, J. C. (1967): Les laves récentes de l'Etna, *Soc. Geol. Franc. Bull.,* ser. 7, **8** (1966), no. 2, pp. 201–217.

Tatsumoto, M., C. E. Hedge, and A. E. J. Engel (1965): Potassium, Rubidium, Strontium, Thorium, Uranium, and the Ratio of Strontium-87 to Strontium-86 in Oceanic Tholeiitic Basalt, *Science,* **150,** pp. 886–888.

Taubeneck, W. H. (1970): Dikes of Columbia River Basalt in Northeastern Oregon, Western Idaho, and Southeastern Washington, *in* "Proceedings of the Second Columbia River Basalt Symposium," E. H. Gilmour and D. Stradling, editors, Eastern Wash. State Coll. Press, pp. 73–96.

——— (1971): Idaho Batholith and Its Southern Extension, *Geol. Soc. Am. Bull.,* **82,** pp. 1899–1928.

Taylor, F. C. (1969): Geology of the Annapolis-St. Marys Bay Map-area, Nova Scotia, *Geol. Surv. Can. Mem.* **358,** 63 pp.

——— and E. A. Schiller (1966): Metamorphism of the Meguma Group of Nova Scotia, *Can. J. Earth Sci.,* **3,** pp. 959–974.

Taylor, H. P., Jr. (1963): O^{18}/O^{16} Ratios of Coexisting Minerals in Three Assemblages of Kyanite-zone Pelitic Schist, *J. Geol.,* **71,** pp. 513–522.

——— and S. Epstein (1963): O^{18}/O^{16} Ratios in Rocks and Coexisting Minerals of the Skaergaard Intrusion, East Greenland, *J. Petrol.,* **4,** pp. 51–74.

Taylor, J. H. (1935): A Contact Metamorphic Zone from the Little Belt Mountains, Montana, *Am. Mineralogist,* **20,** pp. 120–128.

Taylor, R. B. (1964): Geology of the Duluth Gabbro Complex near Duluth, Minnesota, *Minn. Geol. Surv. Bull.,* **44,** 63 pp.

Taylor, S. R. (1969): Trace Element Chemistry of Andesites and Associated Calc-Alkaline

Rocks, *in* "Proceedings of the Andesite Conference," A. R. McBirney, editor, *Oregon Dept. Geol. Mineral. Ind. Bull.,* **65,** pp. 43–63.

———— and A. J. R. White (1965): Geochemistry of Andesites and the Growth of Continents, *Nature,* **208,** no. 5007, pp. 271–273.

Temple, A. K., and R. M. Grogan (1965): Carbonatite and Related Alkalic Rocks at Powderhorn, Colorado, *Econ. Geol.,* **60,** pp. 672–692.

———— and E. W. Heinrich (1964): Spurrite from Northern Coahuila, Mexico, *Mineral. Mag.,* **33,** pp. 841–852.

Tex, E. den, and D. E. Vogel (1962): A "Granulitgebirge" at Cabo Ortegal (NW Spain), *Geol. Rundschau,* **52,** pp. 95–112.

Thayer, T. P. (1960): Some Critical Differences between Alpine-type and Stratiform Peridotite-Gabbro Complexes, *Intern. Geol. Congr.,* **21st,** Copenhagen, 1960, *Rept. Session, Norden,* pt. 13, pp. 247–259.

———— (1966): Serpentinization Considered as a Constant Volume Process, *Am. Mineralogist,* **51,** pp. 685–710.

———— (1967): Chemical and Structural Relations of Ultramafic and Feldspathic Rocks in Alpine Intrusive Complexes, *in* "Ultramafic and Related Rocks," P. J. Wyllie, editor, Wiley, New York, pp. 222–239.

———— (1969): Peridotite-Gabbro Complexes as Keys to Petrology of Mid-oceanic Ridges, *Geol. Soc. Am. Bull.,* **80,** pp. 1515–1522.

———— and C. E. Brown (1960): Upper Triassic Graywackes and Associated Rocks in the Aldrich Mountains, Oregon, *U. S. Geol. Surv. Prof. Paper* **400-B,** pp. 300–302.

———— and ———— (1961): Is the Tinaquillo, Venezuela, "Pseudogabbro" Metamorphic or Igneous?, *Geol. Soc. Am. Bull.,* **72,** pp. 1565–1570.

———— and G. R. Himmelberg (1968): Rock Succession in the Alpine-type Mafic Complex at Canyon Mountain, Oregon, *Intern. Geol. Congr.,* **23d,** Czech., 1968, pp. 175–186.

Thompson, A. B. (1970): Laumontite Equilibria and the Zeolite Facies, *Am. J. Sci.,* **269,** pp. 267–275.

Thompson, G. A., and M. Talwani (1964): Crustal Structure from Pacific Basin to Central Nevada, *J. Geophys. Res.,* **69,** no. 22, pp. 4813–4837.

Thompson, J. B., Jr. (1957): The Graphical Analysis of Mineral Assemblages in Pelitic Schists, *Am. Mineralogist,* **42,** pp. 842–858.

————(1959): Local Equilibrium in Metasomatic Processes, *in* "Researches in Geochemistry," P. H. Abelson, editor, Wiley, New York, pp. 427–457.

———— and S. A. Norton (1968): Paleozoic Regional Metamorphism in New England and Adjacent Areas, *in* "Studies of Appalachian Geology: Northern and Maritime," E-an Zen, W. S. White, J. B. Hadley, and J. B. Thompson, Jr., editors, Interscience, a division of Wiley, New York, chap. 24, pp. 319–327.

————, P. Robinson, T. N. Clifford, and N. J. Trask, Jr. (1968): Nappes and Gneiss Domes in West central New England, *in* "Studies of Appalachian Geology: Northern and Maritime,"

E-an Zen, W. S. White, J. B. Hadley, and J. B. Thompson, Jr., editors, Interscience, a division of Wiley, New York, pp. 203–218.

Thornton, C. P. (1964): Flowage of Fragmental Volcanic Material, *Mineral Ind.*, **34,** no. 3, pp. 1–8.

—— and O. F. Tuttle (1960): Chemistry of Igneous Rocks, Part 1: Differentiation Index, *Am. J. Sci.,* **258,** pp. 664–684.

Tilley, C. E. (1923): Contact Metamorphism in the Comrie Area of the Perthshire Highlands, *Geol. Soc. Lond. Quart. J.,* **80,** pp. 22–71.

—— (1924): The Facies Classification of Metamorphic Rocks, *Geol. Mag.,***61,** pp. 167–171.

—— (1925): A Preliminary Survey of Metamorphic Zones in the Southern Highlands of Scotland, *Geol. Soc. Lond. Quart. J.,* **81,** pp. 100–110.

—— (1926a): On Garnet in Pelitic Contact Zones, *Mineral. Mag.,* **21,** pp. 47–50.

—— (1926b): On Some Mineralogical Transformations in Crystalline Schist, *Mineral. Mag.,* **21,** pp. 34–46.

—— (1950): Some Aspects of Magmatic Evolution (presidential address), *Geol. Soc. Lond. Quart. J.,* **106,** pt. 1, no. 421, pp. 37–61.

—— (1951a): A Note on the Progressive Metamorphism of Siliceous Limestones and Dolomites, *Geol. Mag.,* **88,** pp. 175–178.

—— (1951b): The Zoned Contact Skarns of the Broadford Area, Skye, *Mineral. Mag.,* **29,** pp. 621–666.

—— (1958): Problems of Alkali Rock Genesis, *Geol. Soc. Lond. Quart. J.,* **113,** pp. 323–360.

—— (1960): Differentiation of Hawaiian Basalts. Some Variants in Lava Suites of Dated Kilauean Eruptions, *J. Petrol.,* **1,** pp. 47–55.

—— (1961): Igneous Nepheline-bearing Rocks of the Haliburton-Bancroft Province of Ontario, *J. Petrol.,* **2,** p. 38.

—— and R. N. Thompson (1970): Melting and Crystallization Relations of the Snake River Basalts of Southern Idaho, USA, *Earth Planet. Sci. Letters.,* **8,** pp. 79–92.

——, H. S. Yoder, Jr., and J. F. Schairer (1963): Melting Relations of Basalts, *Carnegie Inst. Wash. Yearbook,* pp. 77–84.

Tilling, R. I. (1964): Variation in Modes and Norms of a "Homogeneous" Pluton of the Boulder Batholith, Montana, *U.S. Geol. Surv. Prof. Paper* **501-D,** pp. D8–D13.

——, M. R. Klepper, and J. D. Obradovich (1968): K-Ar Ages and Time Span of Emplacement of the Boulder Batholith, Montana, *Am. J. Sci.,* **266,** pp. 671–689.

Tilton, G. R. (1960): Volume Diffusion as a Mechanism for Discordant Lead Ages, *J. Geophys. Res.,* **65,** pp. 2933–2945.

——, G. W. Wetherill, G. L. Davis, and C. A. Hobson (1958): Ages of Minerals from the Baltimore Gneiss near Baltimore, Maryland, *Geol. Soc. Am. Bull.,* **69,** pp. 1469–1474.

Tobisch, O. T. (1968): Gneissic Amphibolite at Las Palmas, Puerto Rico and Its Significance in the Early History of the Greater Antilles Island Arc, *Geol. Soc. Am. Bull.,* **79,** pp. 557–574.

Tomkeieff, S. I. (1937): Petrochemistry of the Scottish Carboniferous-Permian Igneous Rocks, *Bull. Volcanol.*, ser. 2, **1**, pp. 59–87.

—— (1949): The Volcanoes of Kamchatka, *Bull. Volcanol.*, **8**, pp. 87–113.

Touret, J. (1968): The Precambrian Metamorphic Rocks around the Lake Vegar (Aust-Adger, Southern Norway), *Norges Geol. Undersol.*, no. 257, 45 pp.

—— (1971): Le facies granulite en Norvege meridionale, I. les associations mineralogiques, *Lithos*, **4**, pp. 239–249.

Tozer, C. F. (1955): The Mode of Occurrence of Sillimanite in the Glen District, Co. Donegal, *Geol. Mag.*, **92**, pp. 310–320.

Turekian, K. K., and K. H. Wedepohl (1961): Distribution of the Elements in Some Major Units of the Earth's Crust, *Geol. Soc. Am. Bull.*, **72**, pp. 175–192.

Turkevich, A. L., E. J. Franzgrote, and J. H. Patterson (1967): Surveyor V Mission Report, Part 2: Science Results, *Science*, **158**, p. 635.

——, ——, and —— (1968a): Surveyor VI Mission Report, Part 2: Science Results, *Science*, **160**, p. 1108.

——, ——, and —— (1968b): Chemical Analysis of the Moon at the Surveyor 7 Landing Site: Preliminary Results, *Science*, **162**, pp. 117–118.

Turner, F. J. (1930): The Metamorphic and Ultrabasic Rocks of the Lower Cascade Valley, South Westland, *New Zealand Inst., Trans.*, **61**, p. 170.

—— (1933): The Metamorphic and Intrusive Rocks of Southern Westland, *New Zealand Inst. Trans.*, **63**, pp. 269–276.

—— (1941): The Development of Pseudostratification by Metamorphic Differentiation in the Schist of Otago, New Zealand, *Am. J. Sci.*, **239**, pp. 1–16.

—— (1942): Preferred Orientation of Olivine Crystals in Peridotites, with Special Reference to New Zealand Examples, *Roy. Soc. New Zealand Trans.*, **72**, pp. 280–300.

—— (1948): Mineralogical and Structural Evolution of the Metamorphic Rocks, *Geol. Soc. Am. Mem.* **30**, 342 pp.

—— (1967): Thermodynamic Appraisal of Steps in Progressive Metamorphism of Siliceous Dolomitic Limestone, *Neues Jahrb. Mineral. Monatsh.*, pp. 1–22.

—— (1968): "Metamorphic Petrology–Mineralogical and Field Aspects," McGraw-Hill Book Co., New York, 403 pp.

—— and J. Verhoogen (1960): "Igneous and Metamorphic Petrology," McGraw-Hill Book Co., New York, 694 pp.

—— and L. E. Weiss (1963): "Structural Analysis of Metamorphic Tectonites," McGraw-Hill Book Co., New York, 545 pp.

Tuttle, O. F. (1948): A New Hydrothermal Quenching Apparatus, *Am. J. Sci.*, **246**, pp. 628–635.

—— (1949): Two Pressure Vessels for Silicate-water Studies, *Geol. Soc. Am. Bull.*, **60**, pp. 1727–1729.

——— and N. L. Bowen (1958): Origin of Granite in the Light of Experimental Studies in the System $NaAlSi_3O_8$–$KAlSi_3O_8$–SiO_2–H_2O, *Geol. Soc. Am. Mem.* **74,** 153 pp.

——— and J. C. England (1955): Preliminary Report on the System SiO_2–H_2O, *Geol. Soc. Am. Bull.,* **66,** pp. 149–152.

——— and J. Gittins, editors (1968): "Carbonatites," Interscience, a division of Wiley, New York, 591 pp.

——— and R. I. Harker (1957): Synthesis of Spurrite and the Reaction Wollastonite + Calcite ⇆ Spurrite + Carbon Dioxide, *Am. J. Sci.,* **255,** pp. 226–234.

——— and J. V. Smith (1958): The Nepheline-Kalsilite System, Part II: Phase Relations, *Am. J. Sci.,* **256,** pp. 571–589.

Tuve, M. A., H. E. Tatel, and P. J. Hart (1954): Crustal Structure from Seismic Exploration, *J. Geophys. Res.,* **59,** pp. 415–422.

Tyrrell, G. W. (1928): Dolerite Sills Containing Analcite Syenite in Central Ayrshire, *Geol. Soc. Lond. Quart. J.,* **84,** pp. 540–569.

——— (1929): "The Principles of Petrology," 2d ed., Methuen, London, 349 pp.

——— (1932): The Basalts of Patagonia, *J. Geol.,* **40,** pp. 374–383.

——— (1955): Distribution of Igneous Rocks in Space and Time, *Geol. Soc. Am. Bull.,* **66,** pp. 405–425.

Uffen, R. J. (1959): On the Origin of Rock Magma, *J. Geophys. Res.,* **64,** pp. 117–122.

Ulbrich, M. C. (1970): Chemical Individuality of Lunar, Meteoritic, and Terrestrial Silicate Rocks, *Science,* **168,** pp. 1375–1376.

Upton, B. G. I. (1960): The Alkaline Igneous Complex of Küngnat Field, South Greenland, *Medd. Grønland,* bd 123, no. 4, 145 pp.

——— (1964): The Geology of Tugtutôq and Neighboring Islands, South Greenland, *Medd Grønland,* **169,** no. 2, 62 pp. Part 2, The Nordmarkitic Syenites and Related Alkaline Rocks, no. 3, pp. 50–80. Part 4, The Nepheline Syenites of the Hviddal Composite Dike.

——— (1965): The Petrology of a Camptonite Sill in South Greenland, *Medd. Grønland,* **169,** no. 11, 21 pp.

——— (1967): Alkaline Pyroxenites, *in* "Ultramafic and Related Rocks," P. J. Wyllie, editor, Wiley, New York, pp. 281–288.

Ustiev, E. K. (1965): Problems of Volcanism and Plutonism. Volcano-Plutonic Formations, *Intern. Geol. Rev.,* **7,** no. 11, pp. 1994–2016; transl. from *Akad. Nauk SSSR, Izvestiya, Seriya Geolog.,* **1963,** no. 12, pp. 3–30.

Vacquier, V. (1969): Magnetic Intensity Field in the Pacific, *in* "The Earth's Crust and Upper Mantle," P. J. Hart, editor, *Am. Geophys. Union. Geophys. Monograph* **13,** pp. 422–430.

Valiquette, G., and G. Archambault (1970): Les gabbros et les syenites du complexe de Brome, *Can. Mineralogist,* **10,** pp. 485–510.

Vallance, T. G. (1960): Concerning Spilites, *Proc. Linn. Soc. New South Wales,* **85,** pp. 8–52.

—— (1965): On the Chemistry of Pillow Lavas and the Origin of Spilites, *Mineral. Mag.,* **34,** pp. 471–481.

—— (1967): Palaeozoic Low-pressure Regional Metamorphism in Southeastern Australia, *Medd. fra Dansk. Geol. Foren.,* bd. **17,** pp. 494–503.

Vance, J. A. (1961): Zoned Granitic Intrusions–An Alternative Hypothesis of Origin, *Geol. Soc. Am. Bull.,* **72,** pp. 1723–1727.

—— (1966): Prehnite-Pumpellyite Facies of Metamorphism on Orcas Island, San Juan Islands, Northwestern Washington, *Geol. Soc. Am. Spec. Paper* **101,** p. 342.

—— (1968): Metamorphic Aragonite in the Prehnite-Pumpellyite Facies, Northwest Washington, *Am. J. Sci.,* **266,** pp. 299–315.

Van Hise, C. R. (1904): "A Treatise on Metamorphism," *U.S. Geol. Surv. Monograph* **47.**

Varne, R. (1968): The Petrology of Moroto Mountain, Eastern Uganda, and the Origin of Nephelinites, *J. Petrol.,* **9,** pp. 169–190.

——, R. D. Gee, and P. G. J. Quilty (1969): Macquarie Island and the Cause of Oceanic Linear Magnetic Anomalies, *Science,* **166,** pp. 230–233.

Velde, B. (1964): Upper Stability of Muscovite, and Low-grade Metamorphism of Micas in Pelitic Rocks, *Carnegie Inst. Wash., Ann. Rept. Dir. Geophys. Lab.,* 1963–1964, pp. 142–147.

—— (1966): Upper Stability of Muscovite, *Am. Mineralogist,* **51,** pp. 924–929.

—— (1969): The Compositional Join Muscovite-Pyrophyllite at Moderate Pressures and Temperatures, *Soc. Franc. Minéral. Crist. Bull.,* **92,** pp. 360–368.

——, F. Herve, and J. Kornprobst (1970): The Eclogite-Amphibolite Transition at 650°C and 6.5 Kbar Pressure, as Exemplified by Basic Rocks of the Uzerche Area, Central France, *Am. Mineralogist,* **55,** pp. 953–974.

Velde, D. (1967): Sur un lamprophyre hyperalcalin potassique; la minette de Sisco (ile de Corse), *Soc. Franc. Minéral. Crist. Bull.,* **90,** no. 2, pp. 214–223.

Vening Meinesz, F. A. (1948): Gravity Expeditions at Sea 1923–1938, *Neth. Geol. Comm., Delft,* **4.**

—— (1954): Indonesian Archipelago: A Geophysical Study, *Geol. Soc. Am. Bull.,* **65,** pp. 143–164.

Verbeek, A. A., and G. D. L. Schreiner (1967): Variations in ^{39}K:^{41}K Ratio and Movement of Potassium in a Granite-Amphibolite Contact Region, *Geochim. Cosmochim. Acta,* **31,** no. 11, pp. 2125–2134.

Verspyck, G. W. (1961): Zircons of Some Metamorphic and Intrusive Rocks from the Aston- and Hospitalet Massifs (Central Pyrenees), *Geol. Mijnbouw,* **40,** pp. 58–70.

Vincent, E. A., and R. Phillips (1954): Iron-Titanium Oxide Minerals in Layered Gabbros from the Skaergaard Intrusion, East Greenland, Part I: Chemistry and Ore-microscopy, *Geochim. Cosmochim. Acta,* **6,** pp. 1–26.

Vine, F. J. (1966): Spreading of the Ocean Floor: New Evidence, *Science,* **154,** p. 1405.

—— and D. H. Mathews (1963): Magnetic Anomalies over Oceanic Ridges, *Nature,* **199,** pp. 947–949.

—— and E. M. Moores (1969): Paleomagnetic Results for the Troodos Igneous Massif, Cyprus (abst.), *Am. Geophys. Union. Trans.*, **50**, p. 131.

—— and J. T. Wilson (1965): Magnetic Anomalies over a Young Oceanic Ridge off Vancouver Island, *Science*, **150**, pp. 485–489.

Virgo, D. (1968): Partition of Strontium between Coexisting K-Feldspar and Plagioclase in Some Metamorphic Rocks, *J. Geol.*, **76**, pp. 331–346.

Viswanathan, T. V., and M. V. N. Murthy (1968): Zircon Studies in Interpreting the Origin of Charnockites and Associated Rocks from Pallavaram, Madras and Puri District, Orissa, India, *Intern. Geol. Congr.*, **22d**, India, 1964, pt. 13, p. 97.

Vitaliano, C. J. (1971): Capping Volcanic Rocks: West Central Nevada, *Bull. Volcanol.*, **34**, pp. 617–635.

Vlesov, K. A., M. K. Kuz'menko, and E. M. Es'kova (1959): "Lovozerskiy shchelochnoy massiv (The Lovozero Alkali Massif)," *Izd. Akad. Nauk SSSR;* English trans., Oliver and Boyd, London, 1966.

Vogel, D. E. (1969): Catazonal Rock Complexes in the Polyorogenic Terrain at Cabo Ortegal (NW Spain), *in* "Age Relations in High-grade Metamorphic Terrains," H. Wynne-Edwards, editor, *Geol. Assoc. Can. Spec. Paper* **5**, pp. 83–88.

Vogel, T. A., B. L. Smith, and R. M. Goodspeed (1968): The Origin of Antiperthites from Some Charnockitic Rocks in the New Jersey Precambrian, *Am. Mineralogist*, **53**, pp. 1696–1708.

Vogt, J. H. L. (1916–1918): Die Sulfid-Silikatschmelzlosungen, *Norsk. Geol. Tidsskr.*, **4**, p. 151.

DeVore, G. W. (1955): The Role of Adsorption in the Fractionation and Distribution of Elements, *J. Geol.*, **63**, pp. 159–190.

—— (1956): Surface Chemistry as a Chemical Control on Mineral Association, *J. Geol.*, **64**, pp. 31–55.

—— (1959): Role of Minimum Interfacial Free Energy in Determining the Macroscopic Features of Mineral Assemblages, Part I: The Model, *J. Geol.*, **67**, pp. 211–227.

Vorobieva, O. A. (1960): Alkali Rocks of the U.S.S.R., *Intern. Geol. Congr.*, **21st**, Copenhagen, 1960, *Rept. Session, Norden*, pt. 13, pp. 7–18.

Vuagnat, M. (1946): Sur quelques diabases suisses. Contribution à l'étude du problème des apilites et des pillow lavas, *Schweiz. Mineral. Petrog. Mitt.*, **26**, pp. 116–228.

—— (1949): Sur les pillow lavas dalradiennes de la peninsule de Tayvallich (Argyllshire), *Schweiz. Mineral. Petrog. Mitt.*, **29**, pp. 524–536.

—— (1952): Le rôle des anciennes coulées volcaniques sous-marines dans les anciennes chaînes de montagnes, *Intern. Geol. Congr.*, **19ᵉ**, Algiers, 1952, *Compt. Rend.*, 17, pp. 53–59.

Wade, A., and R. T. Prider (1940): The Leucite-bearing Rocks of the West Kimberly Area, Western Australia, *Geol. Soc. Lond. Quart. J.*, **96**, pp. 39–98.

Wadsworth, W. J. (1961): The Ultrabasic Rocks of Southwest Rhum, *Phil. Trans. Roy. Soc. Lond.*, ser. B, **244**, pp. 21–64.

—— (1963): The Kapalagulu Layered Intrusion of Western Tanganyika, *in* "Symposium on Layered Intrusions," *Mineral. Soc. Am. Spec. Paper* **1**, pp. 108–115.

—— et al. (1966): Cryptic Layering in the Belhelvie Intrusion, Aberdeenshire, *Scot. J. Geol.*, **2**, pp. 54–66.

Wager, L. R. (1953): Layered Intrusions, *Medd. Dansk. Geol. Foren.*, bd. **12**, K, pp. 335–349.

—— (1959): Differing Powers of Crystal Nucleation as a Factor Producing Diversity in Layered Igneous Intrusions, *Geol. Mag.*, **96**, pp. 75–80.

—— (1960): The Major Element Variation of the Layered Series of the Skaergaard Intrusion and a Re-estimation of the Average Composition of the Hidden Layered Series and of the Successive Residual Magmas, *J. Petrol.* **1**, pp. 364–398.

—— (1963): The Mechanism of Adcumulus Growth in the Layered Series of the Skaergaard Intrusion, *in* "Symposium on Layered Intrusions," *Mineral. Soc. Am. Spec. Paper* **1**.

—— (1968): Rhythmic and Cryptic Layering in Mafic and Ultramafic Plutons, *in* "Basalts," vol. 2, H. H. Hess and A. Poldervaart, editors, Interscience, a division of Wiley, New York, pp. 573–622.

—— and G. M. Brown (1968): "Layered Igneous Rocks," Oliver and Boyd, London, 588 pp.

—— and W. A. Deer (1939): Geological Investigations in East Greenland, Part III: The Petrology of the Skaergaard Intrusion, Kangerdlugssuag, East Greenland, *Medd. Grønland*, **105**, no. 4, pp. 1–352. (Reissued in 1962.)

—— and R. L. Mitchell (1950): The Distribution of Cr, V, Ni, Co, and Cu during the Fractional Crystallization of a Basic Magma, *Intern. Geol. Congr.*, **18th**, London, 1948, pt. 2, pp. 14–150.

—— and —— (1951): The Distribution of Trace Elements during Strong Fractionation of Basic Magma—A Further Study of the Skaergaard Intrusion, East Greenland, *Geochim. Cosmochim. Acta*, **1**, pp. 129–208.

——, E. A. Vincent, and A. A. Smales (1957): Sulphides in the Skaergaard Intrusion, East Greenland, *Econ. Geol.*, **52**, pp. 855–903.

Wagner, G. A. (1968): Fission-track Dating of Apatites, *Earth Planet. Sci. Letters*, **4**, pp. 411–415.

Walker, F. (1940): Differentiation of the Palisade Diabase, New Jersey, *Geol. Soc. Am. Bull.*, **51**, pp. 1059–1106.

—— (1953): The Pegmatitic Differentiates of Basic Sheets, *Am. J. Sci.*, **251**, pp. 4–60.

—— (1957): Ophitic Texture and Basaltic Crystallization, *J. Geol.*, **65**, pp. 1–14.

—— (1961): The Causes of Variation in Dolerite Intrusions, *in* "Dolerite, a Symposium," S. W. Carey, editor, Convener, *Univ. Tasmania Geol. Dept.*, Symp. July, **1957**, pp. 1–25.

—— and A. Poldervaart (1949): Karoo Dolerites of the Union of South Africa, *Geol. Soc. Am. Bull.*, **60**, pp. 591–706.

Walker, G. P. L., and R. R. Skelhorn (1966): Some Associations of Acid and Basic Igneous Rocks, *Earth Sci. Rev.*, **2**, pp. 93–109.

Walker, G. W. (1970): Some Comparisons of Basalts of Southeast Oregon with Those of the Columbia River Group, *in* "Proceedings of the Second Columbia River Basalt Symposium," E. H. Gilmour and D. Stradling, editors, Eastern Wash. State Coll. Press, pp. 223–237.

Walker, K. R. (1970): The Palisades Sill, New Jersey: A Reinvestigation, *Geol. Soc. Am. Spec. Paper* **111,** 178 pp.

——, G. A. Joplin, J. F. Lovering, and R. Green (1960): Metamorphic and Metasomatic Convergence of Basic Igneous Rocks and Lime-Magnesia Sediments in the Precambrian of North-western Queensland, *Geol. Soc. Australia J.,* **6,** pp. 149–178.

Walter, L. S. (1963a): Equilibrium Studies on Bowen's Decarbonation Series, Part I: P-T Univariant Equilibria of the "Monticellite" and "Akermanite" Reactions, *Am. J. Sci.,* **261,** pp. 488–500.

—— (1963b): Experimental Studies on Bowen's Decarbonation Series, Part II: P-T Univariant Equilibria of the Reaction Forsterite + Calcite = Monticellite + Periclase + CO_2, *Am. J. Sci.,* **261,** pp. 763–770.

Wambecke, L. van (1966): Mineralogical and Geochemical Evolution of the Carbonatites of the Kaiserstuhl, Germany, *Intern. Mineral. Assoc.,* 1964 meeting, India.

Wanless, R. K., R. D. Stevens, and W. D. Loveridge (1969): Excess Radiogenic Argon in Biotites, *Earth Planet. Sci. Letters,* **7,** p. 167.

Wantanabe, T. (1943): Geology and Mineralization of the Surian District, Tyosen (Korea), *J. Fac. Sci. Hokkaido Univ.,* ser. 4, **6,** pp. 205–303.

Ward, R. F. (1959): Petrology and Metamorphism of the Wilmington Complex, Delaware, Pennsylvania, and Maryland, *Geol. Soc. Am. Bull.,* **70,** pp. 1425–1458.

Washington, H. S. (1900): Igneous Complex of Magnet Cove, Arkansas, *Geol. Soc. Am. Bull.,* **11,** pp. 389–416.

—— (1901): The Foyaite-Ijolite Series of Magnet Cove, *J. Geol.,* **9,** pp. 607–622, 645–670.

—— (1920): Italite, a New Leucite Rock, *Am. J. Sci.,* **200,** pp. 33–47.

Waters, A. C. (1955a): Volcanic Rocks and the Tectonic Cycle, *Geol. Soc. Am. Spec. Paper* **62,** pp. 703–722.

—— (1955b): Geomorphology of South-central Washington, Illustrated by the Yakima East Quadrangle, *Geol. Soc. Am. Bull.,* **66,** pp. 663–684.

—— (1960): Determining Direction of Flow in Basalts, *Am. J. Sci.,* **258-A,** pp. 350–366.

—— (1961): Stratigraphic and Lithologic Variations in the Columbia River Basalt, *Am. J. Sci.,* **259,** pp. 583–611.

—— (1962): Basalt Magma Types and Their Tectonic Associations: Pacific Northwest of the United States, *in* "The Crust of the Pacific Basin Symposium," *Pacific Sci. Congr.,* **10th,** *Am. Geophys. Union Monograph* **6,** pp. 158–170.

Watkinson, D. H. (1970): Experimental Studies Bearing on the Origin of the Alkalic Rocks-Carbonatite Complex and Niobium Mineralization at Oka, Quebec, *Can. Mineralogist,* **10,** pp. 350–361.

—— and P. J. Wyllie (1964): The Limestone Assimilation Hypothesis, *Nature,* **204,** no. 4963, pp. 1053–1054.

Watson, K. D. (1960): Eclogite Inclusions in Serpentine Pipes at Garnet Ridge, Northeastern Arizona (abs.), *Geol. Soc. Am. Bull.,* **71,** pp. 2082–2083.

—— (1967): Kimberlites of Eastern North America, *in* "Ultramafic and Related Rocks," P. J. Wyllie, editor, Wiley, New York, pp. 312–323.

—— and D. M. Morton (1969): Eclogite Inclusions in Kimberlite Pipes at Garnet Ridge, North-eastern Arizona, *Am. Mineralogist,* **54,** pp. 267–285.

Watson, T. L., and S. Taber (1913): Geology of the Titanium and Apatite Deposits of Virginia, *Virginia Geol. Surv. Bull.,* **3-A,** 308 pp.

Weaver, C. E. (1916): The Tertiary Formations of Western Washington, *Wash. Geol. Surv. Bull.,* **13,** 327 pp.

Webber, G. R. (1966): Relative Depth of Origin of Alkali and Tholeiitic Rocks, *Earth Planet. Sci. Letters,* **1,** pp. 183–184.

Weed, W. H., and L. V. Pirsson (1895): Igneous Rocks of Yogo Peak, Montana, *Am. J. Sci.,* **150,** pp. 467–479.

—— and —— (1896): Igneous Rocks of the Bearpaw Mountains, Montana, *Am. J. Sci.,* ser. 4, **1,** pp. 283–301, 351–362.

—— and —— (1901): Geology of the Shonkin Sag and Palisade Butte Laccoliths, *Am. J. Sci.,* **12,** pp. 1–17.

Weeks, W. P. (1956): A Thermochemical Study of Equilibrium Relations during Metamorphism of Siliceous Carbonate Rocks, *J. Geol.,* **64,** pp. 245–270.

Wegmann, C. E. (1935): Zur Deutung der Migmatite, *Geol. Rundsch.,* **26,** pp. 305–350.

—— (1938): Geological Investigation in Southern Greenland, Part 1, *Medd. Grønland,* **113,** pp. 98–121.

Weill, D. F. (1963): Hydrothermal Synthesis of Andalusite from Kyanite, *Am. Mineralogist,* **48,** pp. 944–947.

—— (1966): Stability Relations in the Al_2O_3-SiO_2 System Calculated from Solubilities in the Al_2O_3-SiO_2-Na_3AlF_6 System, *Geochim. Cosmochim. Acta,* **30,** pp. 223–237.

—— and W. S. Fyfe (1964): A Discussion of the Korzhinski and Thompson Treatment of Thermodynamic Equilibrium in Open Systems, *Geochim. Cosmochim. Acta,* **28,** pp. 565–576.

Wells, F. G., and A. C. Waters (1935): Basaltic Rocks in the Umpqua Formation, *Geol. Soc. Am. Bull.,* **46,** pp. 961–972.

Wentworth, C. K., and G. A. Macdonald (1953): Structures and Forms of Basaltic Rocks in Hawaii, *U. S. Geol. Surv. Bull.,* **994,** 98 pp.

Wescott, M. R. (1966): Loss of Argon from Biotite in a Thermal Metamorphism, *Nature,* **210,** no. 5031, pp. 83–84.

Wetherill, G. W. (1956a): An Interpretation of the Rhodesia and Witwatersrand Age Patterns, *Geochim. Cosmochim. Acta,* **9,** pp. 290–292.

—— (1956b): Discordant Uranium-Lead Ages, *Trans. Am. Geophys. Union,* **37,** pp. 320–326.

——, G. L. Davis, and C. Lee-Hu (1968): Rb-Sr Measurements on Whole Rocks and Separated Minerals from the Baltimore Gneiss, Maryland, *Geol. Soc. Am. Bull.,* **79,** pp. 757–762.

————, O. Kouvo, G. R. Tilton, and P. W. Gast (1962): Age Measurements on Rocks from the Finnish Precambrian, *J. Geol.,* **70,** pp. 74–88.

Weyl, R. (1967): Krustenbau und sialischer Magmatismus [with English, Russian, and Spanish abs.], *Geol. Rundschau,* **56,** no. 2, pp. 369–372.

Wheedon, D. S. (1961): Basic Igneous Rocks of the Southern Cuillin, Isle of Skye, *Trans. Geol. Soc. Glasgow,* **24,** pp. 190–212.

———— (1965): The Layered Ultrabasic Rocks of Sgurr Dubh, Isle of Skye, *Scot. J. Geol.,* **1,** pp. 42–68.

Wheeler, E. P. (1942): Anorthosite and Associated Rocks about Nain, Labrador, *J. Geol.,* **50,** pp. 611–642.

———— (1960): Anorthosite-Adamellite Complex of Nain, Labrador, *Geol. Soc. Am. Bull.,* **71,** pp. 1755–1762.

———— (1965): Fayalitic Olivine in Northern Newfoundland-Labrador, *Can. Mineralogist,* **8,** pp. 339–346.

Wheeler, E. P., II (1968): Minor Intrusives Associated with the Nain Anorthosite, *N. Y. State Mus. Sci. Serv. Mem.* **18,** pp. 189–206.

Wheeler, J. O. (1965): Big Bend Map-area, British Columbia, *Geol. Surv. Can. Paper* **64–32,** 37 pp.

Whetten, J. T. (1965): Carboniferous Glacial Rocks from the Werrie Basin, New South Wales, Australia, *Geol. Soc. Am. Bull.,* **76,** pp. 43–56.

White, A. J. R. (1959): Scapolite-bearing Marbles and Calc-Silicate Rocks from Tungkillo and Milendella, South Australia, *Geol. Mag.,* **96,** pp. 285–306.

———— (1966): Genesis of Migmatites from the Palmer Region of South Australia, *Chem. Geol.,* **1,** pp. 165–200.

White, D. A., D. H. Roeder, T. H. Nelson, and J. C. Crowell (1970): Subduction, *Geol. Soc. Am. Bull.,* **81,** pp. 3431–3432.

White, D. E., E. T. Anderson, and D. K. Grubbs (1963): Geothermal Brine Well, *Science,* **139,** pp. 919–922.

———— and G. A. Waring (1963): "Data of Geochemistry," 6th ed., Chap. K: Volcanic Emanations, *U. S. Geol. Surv. Prof. Paper* **440-K.**

White, R. W. (1966): Ultramafic Inclusions in Basaltic Rocks from Hawaii, *Contr. Mineral. Petrol.,* **12,** pp. 245–314.

White, W. S. (1960): The Keweenawan Lavas of Lake Superior and Example of Flood Basalts, *Am. J. Sci.,* Bradley vol., **258-A,** pp. 367–374.

Whittaker, E. J. W. (1967): Factors Affecting Element Ratios in the Crystallization of Minerals, *Geochim. Cosmochim. Acta,* **31,** pp. 2275–2288.

———— and R. Muntus (1970): Ionic Radii for Use in Geochemistry, *Geochim. Cosmochim. Acta,* **34,** pp. 945–956.

Whitten, E. H. T. (1966): "Structural Geology of Folded Rocks," Rand McNally, Chicago, 663 pp.

Wikstrom, A. (1970): Hydrothermal Experiments in the System Jadeite-Diopside, *Norsk Geol. Tidsskr.,* **50,** pp. 1–14.

Wilcox, R. E. (1954): Petrology of Paricutin Volcano, Mexico, *U.S. Geol. Surv. Bull.,* **965-C,** pp. 281–353.

——— and A. Poldervaart (1958): Metadolerite Dike Swarm in Bakersvill-Roan Mountain Area, North Carolina, *Geol. Soc. Am. Bull.,* **69,** pp. 1323–1368.

Wilkinson, J. F. G. (1956): Clinopyroxenes of Alkali Olivine Basalt Magma, *Am. Mineralogist,* **41,** pp. 724–743.

——— and J. T. Whetten (1964): Some Analcime-bearing Pyroclastic and Sedimentary Rocks from New South Wales, *J. Sediment. Petrol.,* **34,** pp. 543–553.

Willemse, J. (1959): The "Floor" of the Bushveld Igneous Complex and Its Relationships with Special Reference to the Eastern Transvaal, *Geol. Soc. S. Africa Proc.,* **72.**

——— (1969): The Geology of the Bushveld Igneous Complex, the Largest Repository of Magmatic Ore Deposits in the World, *in* "Magmatic Ore Deposits," Symp., H. D. B. Wilson, editor.

Williams, A. F. (1932): "The Genesis of the Diamond," 2 vols., Ernest Benn, London, 636 pp.

Williams, C. E. (1952): Carbonatite Structure, Tororo Hills, Eastern Uganda, *Geol. Mag.,* **89,** pp. 286–292.

Williams, C. R., and M. P. Billings (1938): Petrology and Structure of the Franconia Quadrangle, New Hampshire, *Geol. Soc. Am. Bull.,* **49,** pp. 1011–1044.

Williams, H. (1936): Pliocene Volcanoes of the Navajo-Hopi Country, *Geol. Soc. Am. Bull.,* **47,** pp. 111–172.

——— (1942): The Geology of Crater Lake National Park, Oregon, with a Reconnaissance of the Cascade Range Southward to Mount Shasta, *Carnegie Inst. Wash. Publ.* **540.**

———, F. J. Turner, and C. M. Gilbert (1954): "Petrography: An Introduction to the Study of Rocks in Thin Sections," Freeman, San Francisco, 406 pp.

Wilshire, H. G. (1959): Deuteric Alteration of Volcanic Rocks, *J. Proc. Roy. Soc. New South Wales,* **93,** pp. 105–120.

——— (1967): The Prospect Alkaline Diabase–Picrite Intrusion, New South Wales, Australia, *J. Petrol.,* **8,** pp. 97–163.

Wilson, A. E. (1959): The Charnockitic Rocks of Australia, *Geol. Rundsch.,* **47,** pp. 491–510.

——— (1960): Co-existing Pyroxenes: Some Causes of Variation and Anomalies in the Optically Derived Compositional Tie-lines, with Particular Reference to Charnockitic Rocks, *Geol. Mag.,* **97,** p. 1.

——— (1968): The Petrological Features and Structural Setting of Australian Granulites and Charnockites, *Intern. Geol. Congr.,* **22d,** India, 1964, pt. 13, pp. 21–44.

Wilson, G. (1961): The Tectonic Significance of Small Scale Structures, and Their Importance to the Geologist in the Field, *Ann. Soc. Geol. Belgique,* **LXXXIV,** pp. 423–548.

Wilson, I. F. (1943): Geology of the San Benito Quadrangle, California, *Calif. Div. Mines Geol. Bull.,* **39,** pp. 183–270.

Wilson, J. T. (1963a): Evidence from Islands on the Spreading of Ocean Floors, *Nature,* **197,** pp. 536–538.

—— (1963b): Hypothesis of Earth's Behavior, *Nature,* **198,** pp. 925–929.

—— (1965): Transform Faults, Oceanic Ridges and Magnetic Anomalies Southwest of Vancouver Island, *Science,* **150,** pp. 482–485.

—— (1966): Patterns of Growth of Ocean Basins and Continents, *in* "Continental Margins and Island Arcs, *International Upper Mantle Project Symposium*" W. H. Poole, editor, *Geol. Surv. Can. Paper* **66–15,** pp. 338–391.

—— (1970): Some Possible Effects If North America Has Overridden Part of the East Pacific Rise, *Geol. Soc. Am.,* Abst. with Prog., **2,** pp. 722–723.

Wilson, M. M., and C. I. Mathison (1968): The Eulogie Park Gabbro, a Layered Basic Intrusion from Eastern Queensland, *J. Geol. Soc. Australia,* **15,** pp. 139–158.

Wimmenauer, W. (1962): Zur petrogenese der eruptingesteine und karbonatite der Kaiser stuhls, *Neues Jahrb. Mineral. Mh.,* **1,** pp. 1–11.

—— (1966): The Eruptive Rocks and Carbonatites of the Kaiser Stuhls, Germany, *in* "Carbonatites," O. F. Tuttle and J. Gittins, editors, Interscience, a division of Wiley, New York, pp. 183–204.

Windley, B. F. (1967): On the Classification of the West Greenland Anorthosites, *Geol. Rundschau,* **56,** no. 3, pp. 1020–1026.

Winkler, H. G. F. (1957): Experimentelle Gesteinsmetamorphose–I. Hydrothermale Metamorphose Kárbonatfreier Tone, *Geochim. Cosmochim. Acta,* **13,** pp. 42–69.

—— (1964): Das T-P Feld der Diagenese und niedrigtemperierten Metamorphose aufgrund von Mineralreaktionen, *Beitr. Mineral. Petrog.,* **10,** pp. 70–93.

—— (1965): "Petrogenesis of Metamorphic Rocks," Springer-Verlag, New York, 220 pp.

—— (1967): "Petrogenesis of Metamorphic Rocks," 2d ed., Springer-Verlag, New York, 237 pp.

—— and H. von Platen (1968): Experimentelle Gesteinsmetamorphose–II. Bildung von anatektischer granitischen Schmelzen bei der Metamorphose von NaCl-fuhrenden kalkfreien Tone, *Geochim. Cosmochim. Acta,* **15,** pp. 91–112.

—— and —— (1960): Experimentelle Gesteinsmetamorphose–III. Anatektischen Ultrametamorphose Kalkhaltiger Tone, *Geochim. Cosmochim. Acta,* **18,** pp. 294–316.

—— and —— (1961a): Experimentelle Gesteinsmetamorphose–IV. Bildung anatektischer Schmelzen aus metamorphisierten Grauwacken, *Geochim. Cosmochim. Acta,* **24,** pp. 48–69.

—— and —— (1961b): Experimentelle Gesteinsmetamorphose–V. Experimentelle anatektischer Schmelzen und ihre petrogenetische Bedeutung, *Geochim. Cosmochim. Acta,* **24,** pp. 250–265.

Wise, W. (1959): Occurrence of Wairakite in Metamorphic Rocks of the Pacific Northwest, *Am. Mineralogist,* **44,** pp. 1099–1101.

—— (1969): Geology and Petrology of the Mt. Hood Area, a Study of High Cascade Volcanism, *Geol. Soc. Am. Bull.,* **80,** pp. 969–1006.

Wiseman, J. D. H. (1934): The Central and Southwest Highland Epidiorites, *Geol. Soc. Lond. Quart. J.,* **90,** pp. 354–417.

—— (1937): "Basalts from the Carlsberg Ridge, Indian Ocean," *Sci. Rept., John Murray Brit. Mus. Nat. Hist.*

Wolff, J. E. (1938): Igneous Rocks of the Crazy Mountains, Montana, *Geol. Soc. Am. Bull.,* **49,** pp. 1569–1628.

Wollenberg, H. A., and A. R. Smith (1970): Radiogenic Heat Production in Prebatholithic Rocks of the Central Sierra Nevada, *J. Geophys. Res.,* **75,** no. 2, pp. 431–438.

Wones, D. R. (1967): A Low Pressure Investigation of the Stability of Phlogopite, *Geochim. Cosmochim. Acta,* **31,** pp. 2248–2253.

—— and F. C. W. Dodge (1968): On the Stability of Phlogopite, *Geol. Soc. Am. Spec. Paper* **101.**

—— and H. P. Eugster (1965): Stability of Biotite: Experiment, Theory, and Application, *Am. Mineralogist,* **50,** pp. 1228–1272.

Wood, J. A., J. S. Dickey, Jr., U. B. Marvin, and B. N. Powell (1970): Lunar Anorthosites, *Science,* **167,** pp. 602–604.

Woodard, H. H. (1957): Diffusion of Chemical Elements in Some Naturally Occurring Silicate Inclusions, *J. Geol.,* **65,** pp. 61–84.

—— (1968): Contact Alteration in the North Wall of the Cape Neddick Gabbro, Maine, *J. Geol.,* **76,** pp. 191–204.

Woollard, G. P., and W. E. Strange (1962): Gravity Anomalies and the Crust of the Earth in the Pacific Basin, *Am. Geophys. Union Monograph* **6,** pp. 60–80.

Worst, B. G. (1958): The Differentiation and Structure of the Great Dike of Southern Rhodesia, *Geol. Soc. S. Africa Trans.,* **59,** pp. 283–358.

Woussen, G. (1970): La géologie du complexe igné du Mont Royal, *Can. Mineralogist,* **10,** pp. 432–451.

Wright, T. L. (1967): The Microcline-Orthoclase Transformation in the Contact Aureole of the Eldora Stock, Colorado, *Am. Mineralogist,* **52,** pp. 117–136.

—— (1970): Presentation and Interpretation of Chemical Data for Igneous Rocks, *Geol. Soc. Am.,* Abst. with Prog., **2,** pp. 727–729.

Wyllie, P. J. (1960): The System CaO-CO_2-H_2O and the Origin of Carbonatites, *J. Petrol.,* **1,** pp. 1–46.

—— (1961): Fusion of Torridonian Sandstone by a Diorite Sill in Soay (Hebrides), *J. Petrol.,* **2,** pp. 1–37.

—— (1965): Melting Relationships in the System CaO-MgO-CO_2-H_2O, with Petrological Applications, *J. Petrol.,* **6,** pp. 101–123.

—— (1966a): High Pressure Techniques, *in* "Methods and Techniques in Geophysics," Part 2, S. K. Runcorn, editor, Interscience, a division of Wiley, New York, pp. 33–79.

—— (1966b): Experimental Data Bearing on the Petrogenetic Links between Kimberlites and Carbonatites, *Intern. Mineral. Assoc. Papers,* **I.M.A. vol.,** 4th Genl. Meeting, New Delhi, India, pp. 67–82.

—— (1966c): Experimental Studies of Carbonatite Problems: The Origin and Differentiation of Carbonatite Magmas, *in* "Carbonatites," O. F. Tuttle and J. Gittins, editors, Interscience, a division of Wiley, New York.

——, editor (1967): "Ultramafic and Related Rocks," Wiley, New York, 464 pp.

——, K. G. Cox, and G. M. Biggar (1962): The Habit of Apatite in Synthetic Systems and Igneous Rocks, *J. Petrol.*, **3**, pp. 238–243.

—— and O. F. Tuttle (1959): Melting of Calcite in the Presence of Water, *Am. Mineralogist*, **44**, pp. 453–459.

—— and —— (1961a): Hydrothermal Melting of Shales, *Geol. Mag.*, **98**, pp. 56–66.

—— and —— (1961b): Experimental Investigation of Silicate Systems Containing Two Volatile Components, *Am. J. Sci.*, **259**, pp. 128–143.

—— and —— (1964): The Effects of SO_3, P_2O_5, HCl, and Li_2O, in Addition to H_2O, on the Melting Temperatures of Albite and Granite, *Am. J. Sci.*, **262**, pp. 930–939.

Wynne-Edwards, H. R. (1967a): Westport Map-area, Ontario, with Special Emphasis on the Precambrian Rocks, *Geol. Surv. Can. Mem.* **346**, 142 pp.

—— (1967b): The Frontenac Axis, *in* "Guidebook, Geology of Parts of Eastern Ontario and Western Quebec," *Geol. Assoc. Can. Mineral. Assoc. Can. Ann. Mtg.*, Kingston, Ont., pp. 73–86.

—— (1969): Tectonic Overprinting in the Grenville Province, Southwestern Quebec, *Geol. Assoc. Can. Spec. Paper* **5**, pp. 163–182.

—— and P. W. Hay (1963): Coexisting Cordierite and Garnet in Regionally Metamorphosed Rocks from the Westport Area, Ontario, *Can. Mineralogist*, **7**, pt. 3, pp. 453–478.

Yagi, K. (1964): Pillow Lavas of Keflavik, Iceland, and Their Genetic Significance, *J. Fac. Sci. Hokkaido Univ.*, ser. 4., **XII**, pp. 171–182.

—— (1967): Silicate Systems Related to Basaltic Rocks, *in* "Basalts: The Poldervaart Treatise on Rocks of the Basaltic Composition," A. Poldervaart and H. H. Hess, editors, Interscience, a division of Wiley, New York, pp. 359–400.

—— (1969): Petrology of the Alkalic Dolerites of the Nemuro Penninsula, Japan, *in* "Igneous and Metamorphic Geology," L. H. Larsen, editor, *Geol. Soc. Am. Mem.* **115**, pp. 103–147.

—— and H. Matsumoto (1966): Note on Leucite-bearing Rocks from Leucite Hills, Wyoming, U.S.A., *J. Fac. Sci. Hokkaido Univ.* ser. 4, **13**, no. 3, pp. 301–311.

Yashina, R. M. (1957): Shchelochnyye porody yugo-vostoke Tuvy (Alkalic Rocks in Southeastern Tuva), *Izv. AN SSSR*, ser. geol. no. **5**.

Yoder, H. S., Jr. (1950): The Jadeite Problem, *Am. J. Sci.*, **248**, pp. 225–248, 312–334.

—— (1952): The $MgO\text{-}Al_2O_3\text{-}SiO_2\text{-}H_2O$ System and the Related Metamorphic Facies, *Am. J. Sci.*, **250**, pp. 569–627.

—— (1954): The System Diopside-Anorthite-Water, *Carnegie Inst. Wash. Yearbook, Rept. Dir. Geophys. Lab.*, pp. 106–107.

—— (1955a): Almandine Garnet Stability Range (abs.), *Am. Mineralogist*, **40**, p. 342.

——— (1955b): Role of Water in Metamorphism, *Geol. Soc. Am. Spec. Paper* **62,** pp. 505–524.

——— (1964a): Genesis of Principal Basalt Magmas, *Carnegie Inst. Wash. Yearbook* **63,** pp. 97–101.

——— (1964b): Soda Melilite, *Carnegie Inst. Wash. Yearbook* **63,** pp. 86–89.

——— (1967): Spilites and Serpentinites, *Carnegie Inst. Wash. Yearbook* **65,** pp. 269–279.

——— (1968): Experimental Studies Bearing on the Origin of Anorthosite, *N. Y. State Mus. Sci. Serv. Mem.* **18,** pp. 13–22.

——— (1969a): Anorthite-Akermanite and Albite-Soda Melilite Reaction Relations, *Carnegie Inst. Wash. Yearbook* **67,** pp. 105–108.

——— (1969b): Calcalkaline Andesites: Experimental Data Bearing on the Origin of Their Assumed Characteristics, *in* "Proceedings of the Andesite Conference," A. R. McBirney, editor, *Oregon Dept. Geol. Mineral. Ind. Bull.,* **65,** pp. 77–89.

——— (1971): Contemporaneous Rhyolite and Basalt, *Carnegie Inst. Wash. Yearbook* **69,** pp. 141–145.

——— and H. P. Eugster (1954): Phlogopite Synthesis and Stability Range, *Geochim. Cosmochim. Acta,* **6,** pp. 157–185.

——— and ——— (1955): Synthetic and Natural Muscovites, *Geochim. Cosmochim. Acta,* **8,** pp. 225–280.

———, D. B. Stewart, and J. R. Smith (1957): Ternary Feldspars, *Carnegie Inst. Wash. Yearbook* **56,** pp. 206–214.

——— and C. E. Tilley (1962): Origin of Basalt Magmas: An Experimental Study of Natural and Synthetic Rock Systems, *J. Petrol.,* **3,** pp. 342–532.

Zaporoshtseva, A. S., T. N. Vishnevskaya, and G. D. Dubar (1961): Successive Change in Calcium Zeolites through Vertical Sections of Sedimentary Strata, *Dokl. Akad. Nauk SSSR,* **141,** pp. 1264–1266.

Zartman, R. E. (1965): Rubidium-Strontium Age of some Metamorphic Rocks from the Llano Uplift, Texas, *J. Petrol.,* **6,** pp. 28–36.

———, M. R. Brock, A. V. Heyl, and H. H. Thomas (1967): K-Ar and Rb-Sr Ages of some Alkalic Intrusive Rocks from Central and Eastern United States, *Am. J. Sci.,* **265,** pp. 848–870.

———, P. M. Hurley, H. W. Krueger, and B. J. Giletti (1970): A Permian Disturbance of K-Ar Radiometric Ages in New England: Its Occurrence and Cause, *Geol. Soc. Am. Bull.,* **81,** pp. 3359–3374.

Zen, E-an (1960): Metamorphism of Lower Paleozoic Rocks in the Vicinity of the Taconic Range in West-central Vermont, *Am. Mineralogist,* **45,** pp. 129–175.

——— (1961): Mineralogy and Petrology of the System Al_2O_3-$SiO_2$$H_2O$ in Some Pyrophyllite Deposits of North Carolina, *Am. Mineralogist,* **46,** pp. 52–66.

——— (1963): Components, Phases, and Criteria of Chemical Equilibrium in Rocks, *Am. J. Sci.,* **261,** p. 929.

——— (1969): The Stability Relations of the Polymorphs of Aluminum Silicate: A Survey and Some Comments, *Am. J. Sci.,* **267,** pp. 297–308.

———— and A. L. Albee (1964): Coexistent Muscovite and Paragonite in Pelitic Schists, *Am. Mineralogist,* **49,** pp. 904–925.

————, W. S. White, J. B. Hadley, and J. B. Thompson, Jr., editors (1968): "Studies of Appalacian Geology: Northern and Maritime," Interscience, a division of Wiley, New York, 475 pp.

Zhabin, A. G., and G. Y. Cherepivskaya (1965): Carbonatite Dikes as Related to Ultrabasic-Alkalic Extrusive Igneous Activity, *Dokl. Akad. Nauk SSSR,* **160,** pp. 135–138.

Zharikov, V. A. (1966): Coexisting Compositions of Scapolite and Plagioclase as a Function of Depth, *Dokl. Acad. Sci. SSSR,* **170,** pp. 212–214.

Zwart, H. J. (1960): Relations between Folding and Metamorphism in the Pyrenees and Their Chronological Succession, *Geol. Mijnbouw,* **39,** pp. 163–180.

———— (1962): On the Determination of Polymetamorphic Mineral Associations and Its Application to the Bosost Area (Central Pyrenees), *Geol. Rundschau,* **52,** no. 1, pp. 38–65.

———— (1963): Some Examples of the Relations between Deformation and Metamorphism from the Central Pyrenees, *Geol. Mijnbouw,* **42,** pp. 143–154.

———— (1967): The Duality of Orogenic Belts, *Geol. Mijnbouw,* **46,** pp. 283–309.

———— (1971): Metamorphism in Mobile Belts around the World, *in* "Metamorphism in the Canadian Cordillera," *Geol. Assoc. Can.,* Cordill. Sect. Prog. and Abst., p. 31.

Zyl, C. van (1959): An Outline of the Geology of the Kapalagula Complex, Kungwe Bay, Tanganyika, Territory, and Aspects of the Evolution of Layering in Basic Intrusives, *Geol. Soc. S. Africa Trans.,* **62,** pp. 1–31.

INDEX

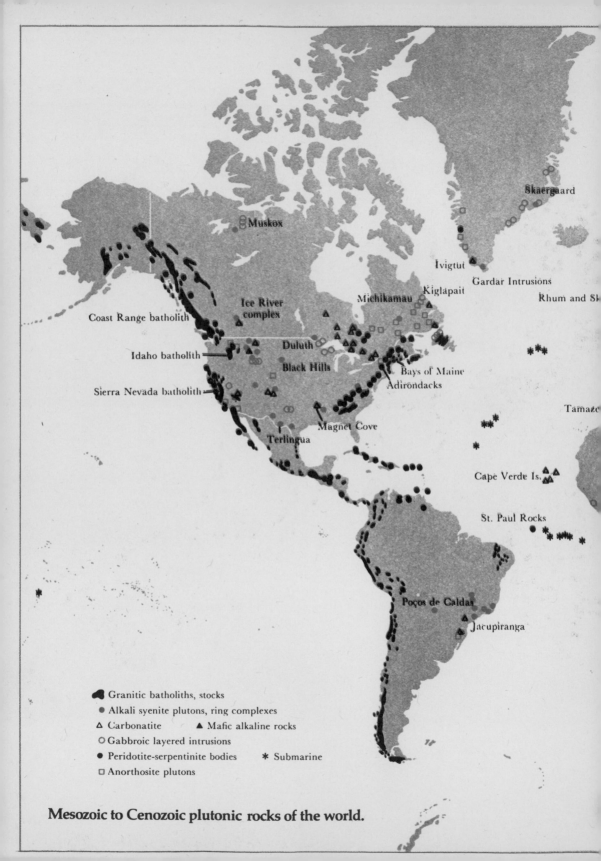

Mesozoic to Cenozoic plutonic rocks of the world.